Please return this book on or before the last date stamped below.

Fines will be charged on books returned after this date.

R568

Springer Series in Optical Sciences   Volume 23

Edited by Jay M. Enoch

Springer Series in Optical Sciences

Editorial Board: J. M. Enoch   D. L. MacAdam   A. L. Schawlow   T. Tamir

Vertebrate Photoreceptor Optics

Edited by J. M. Enoch and F. L. Tobey, Jr.

With Contributions by
H. E. Bedell G. D. Bernard B. Borwein J. M. Enoch
F. I. Harosi B. R. Horowitz J. A. C. Nicol
F. L. Tobey, Jr. W. Wijngaard R. Winston

With a Foreword by W.S. Stiles

With 164 Figures

Springer-Verlag Berlin Heidelberg New York 1981

Jay M. Enoch, Ph. D.

School of Optometry
University of California
Berkeley, CA 94720, USA

Frank L. Tobey, Jr. Ph. D.

Department of Physics, Williamson Hall
University of Florida
Gainesville, FL 32611, USA

ISBN 3-540-10515-8 Springer-Verlag Berlin Heidelberg New York
ISBN 0-387-10515-8 Springer-Verlag New York Heidelberg Berlin

Library of Congress Cataloging in Publication Data. Vertebrate photoreceptor optics. (Springer series
in optical sciences ; v. 23). Bibliography: p. Includes index. 1. Retina. 2. Photoreceptors. 3. Optics,
Physiological. I. Enoch, Jay M. II. Tobey, F. L. (Frank L.) III. Bedell, H. E. (Harold E.) IV. Series.
[DNLM: 1. Optics. 2. Photoreceptors. 3. Retina. 4. Vertebrates. 5. Vision. WW 103 V567]
QP479.V49 596'.01823 81-5091 AACR2

Offset printing: Beltz Offsetdruck, Hemsbach/Bergstr. Bookbinding: J. Schäffer oHG, Grünstadt.
2153/3130-543210

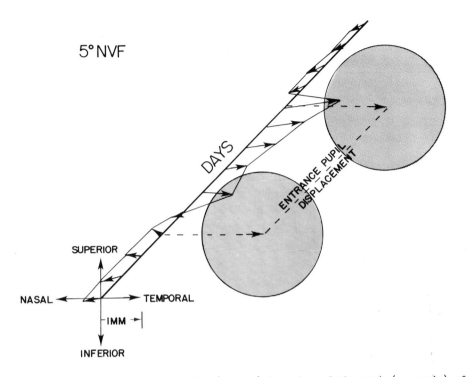

5° NVF

DAYS

ENTRANCE PUPIL DISPLACEMENT

SUPERIOR

NASAL ← → TEMPORAL

⊢ IMM ⊣

INFERIOR

This figure shows the day by day (z axis) location of the peak (x,y axis) of
the photopic directional sensitivity function relative to the center of the
entrance pupil of the eye of the senior editor. The retina is directionally
sensitive, i.e., it normally responds more efficiency to light entering near
the pupil center and is less sensitive to light entering the periphery
(see Chap.3).

In this experiment the pupil was dilated and accommodation greatly reduced
with cycloplegic/mydriatic instillation. Once baseline data was recorded, a
black panted iris contact lens was worn which had a displaced aperture (3 mm
diameter, center displaced temporally 2.5 mm). Note how the peak of the
Stiles-Crawford function shifted towards the displaced aperture. The period
of wear of this contact lens is characterized by the two half-toned schematics
of the aperture. The peak returned to its original location after removal of
the contact lens. This is presented as evidence of *phototropism* of that which
is directionally sensitive in the retina. By inference, this implies the
photoreceptors and their associated anatomical components (see Chap.4).
The test area was 5° in the nasal visual field (NVF) in the right eye.

Foreword

It is in the receptors of the vertebrate retina that the characteristic visual process — the transduction of radiational energy into physiological activity of a different kind — takes place. The way these receptors modify or redistribute the incident radiation and thereby control the light absorption by the visual pigments they contain, is the central theme of this book. As far back as 1843 Brucke put forward a well-reasoned model for the optics of a receptor, assuming simple ray optics, and it is already some forty-seven years since the dependence of receptor sensitivity on retinal angle of incidence was established experimentally as an important factor in human vision and as one by which the direction of alignment of receptors in the living eye might be determined. But it is to Professor J.M. Enoch, editor and author of several major contributions to this volume, that we owe the first experimental demonstration (in 1961) of the wave-mode propagation of light in vertebrate visual receptors, as well as the results of some thirty years devoted research concerned with all questions of receptor optics, particularly directional sensitivity and receptor alignment, both for normal vertebrate eyes and for pathologically modified eyes. His work on the latter has opened up a whole range of clinical possibilities.

Professor Enoch has selected a group of collaborating authors all of whom have made notable contributions to particular aspects of the subject, so that the book's coverage ranges from the difficult mathematical problems of radiation distribution in a number of models of single receptors and receptor arrays, to the observable effects on receptor alignment in pathological conditions before and after treatment, as well as like effects producable in normal eyes subjected to harmless modifications of everyday conditions of vision. One fascinating and quite new discovery of the latter kind that is dealt with is the change — in fact the virtual elimination — of the directional sensitivity phenomenon for cone receptors of a normal eye which has been kept in the dark continuously for several days by the wearing of a suitable eye patch. The effect is reversible, again taking several days after the removal of the eye patch.

The wide scope of the book means that any research worker or clinician in vision concerned with retinal properties will find some parts of prime interest, while the treatment of topics not his direct concern will help to give a balanced picture of the multivarious problems involved in receptor optics.

Surrey, 1981   *W.S. Stiles*

Contents

List of Contributors

Bedell, Harold E.
  School of Optometry, University of California,
  Berkely, CA 94720, USA

Bernard, Gary D.
  Department of Ophthalmology and Visual Science, Yale University,
  333 Cedar Street, New Haven, CT 06510, USA

Borwein, Bessie
  Department of Anatomy, University of Western Ontario,
  Ontario, Canada N6A 5C1

Enoch, Jay M.
  School of Optometry, University of California,
  Berkely, CA 94720, USA

Hárosi, Ferenc I.
  Laboratory of Sensory Physiology, Marine Biological Laboratory,
  Woods Hole, MA 02543, USA

Horowitz, Barry R.
  Perceptual Alternatives Laboratory, Belknap Campus, University
  of Louisville, Louisville, KY 40292, USA

Nicol, Jaldin A.C.
  Institute of Zoology, Academia Sinica, Nankang, Taipeh, Taiwan 115 R.O.C.

Tobey, Frank L.
  Department of Physics, Williamson Hall, University of Florida,
  Gainesville, FL 32611, USA

Wijngaard, Wopke
  Dahliastraat 9, 5271 HB St. Michielsgestel, The Netherlands

Winston, Roland
  Enrico Fermi Institute, University of Chicago,
  Chicago, IL 60637, USA

1. Introduction

J. M. Enoch

With 1 Figure

1.1 Vertebrate Photoreceptor Optics

It is only recently that we have come to appreciate that the retina is a complex optical system whose properties play an essential role in the visual
process. This book is intended to bring together our knowledge of these properties from several points of view. The individual vertebrate retinal receptors are optical waveguides and a retina is comprised of an enormous array
of these waveguides. For example, each human retina contains well over
100,000,000 receptors.

MILLER [1.1] recently gave a concise definition of receptor optics. He
termed it "the science which explores the optical and physiological consequences of the propagation of light with photoreceptor cells."

The history of studies of optics of vertebrate retinal receptors is relatively brief. In 1843, Brücke approached the question with considerable understanding. The following paragraph is quoted from HELMHOLTZ [1.2].

Brücke called attention to a peculiar advantage which must be afforded by
the bacillary layer of rods and cones in the reflection of light from the
retina. These elements are small cylinders, 0.030 mm long and 0.0018 mm
thick, made of some highly refracting substance. Packed close together
like palisades, they constitute the last layer of the retina. The axes of
those which cover the retina in the back of the eye are pointed towards
the pupil, so that all light that falls on these elements penetrates them
nearly parallel to their axes. Now when light, proceeding in a denser
medium, arrives at the boundary of a less dense medium at a very large
angle of incidence, it is totally reflected there. So we may infer that
light which has once been refracted into a retinal rod is mostly retained
there, and if it should fall on the curved cylindrical surface anywhere,
it is nearly all internally reflected. For example, suppose the index of
refraction of the substance of the rods is equal to that of oil (1.47),
and the index of refraction of the intervening substance is equal to that
of water (1.33); then any rays that make an angle with the surface less
than 25° will be totally reflected. But the rays that arrive through the
pupil must fall on the rod walls at angles that are never more than about
8°. If the light has at last reached the farther (outer) end of the rod,
and if here part of it is diffusely reflected from the choroid, nearly all

of this portion must be sent back again through the same rod. Such light as proceeds in a direction more inclined to the axis, may, of course, succeed in escaping from the rod; but not until it has been repeatedly reflected at the surface of the adjacent rod, will it contrive to get into the vitreous humor. On the other hand, the light that comes back nearly parallel to the axis of the rod suffers only a few total reflections at most, and hence will not have lost much when it emerges. Besides, this light will be directed towards the pupil and will issue through it. This function of the rods appears to be of importance, especially in the case of those animals that have a highly reflecting surface or tapetum instead of the layer of black pigment cells on the choroid. The first effect of it is that the light which was originally incident on the sensitive elements of the retina meets and stimulates them again on the rebound. And, secondly, on its return, it affects only the same elements of the retina or possibly in some measure those right next to them. Thus, the light that arrives at any one place is practically confined to a very minute region of the retina; a circumstance that must have an important bearing on the accuracy of vision. When the retinal image is sufficiently bright, diffusely scattered light of this sort may be noticeable in the field of view.

At about the same time, HANNOVER [1.3] had clearly observed waveguide modal patterns in the large "red" rod retinal receptors of the frog. Figure 1.1 is a reproduction of Hannover's drawing. For comparison, Fig.5.2 is a recent photograph of modal patterns in the frog [1.4]. Obviously, there is remarkable similarity. Not surprisingly for his time (first half of the nineteenth century), HANNOVER interpreted the modal pattern distributions as cellular fine structure.

There have been many other discussions of vertebrate retinal optical properties and from time-to-time optically based theories have been advanced to account for various visual phenomenan. Early examples are found in the work of IVES (1918)[1] and FORBES (1928)[2], who pustulated interesting (but unsupported) color vision theories based on receptor optical properties. Many modern examples can be cited. The receptor fiber optics element as a discrete response "unit" has also been invoked in discussions of retinal summation and visual resolution, e.g., HECHT [1.7] and PIRENNE [1.8]. In this context it should be pointed out that growing evidence for interreceptor synapses may limit future arguments based on unit or discrete response concepts.

---

1 IVES considered the possbility of different wavelengths being focused at different depths in the receptor outer segment (due to chromatic aberration). He, in turn, used this notion to account for apparent differences in visual response-time rates for the different wavelength mechanisms (See [Ref.1.5, note added in proof]).

2 FORBES [1.6] postulated an interaction between incident waves within the receptor and backreflected waves as giving different position dependent maxima of response within the receptor outer segment as a function of wavelength. He implicated these maxima in color theory.

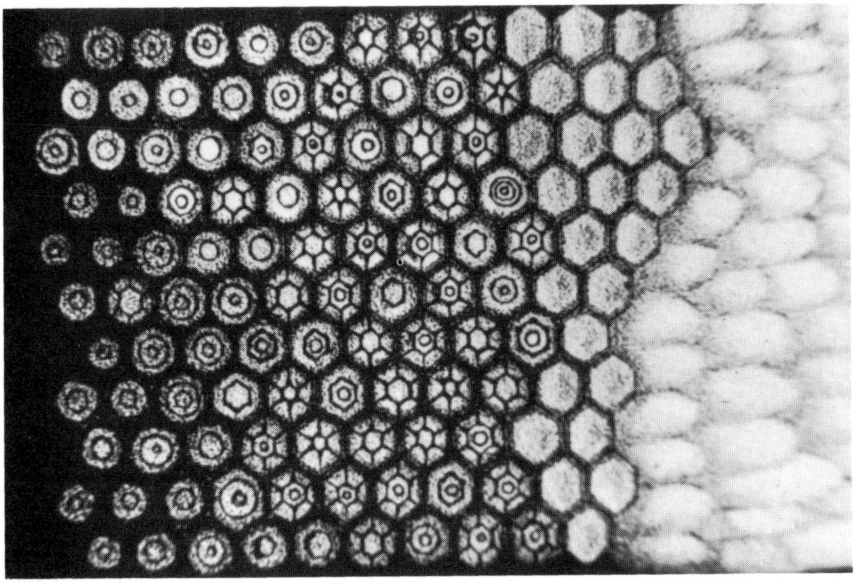

Fig.1.1. Drawing of frog outer segment viewed end-on. Compare Fig.5.2
(from Ref. [1.3])

Many added optical properties present in vertebrate retinas are well documented. Examples are retinal polarization effects (Haidinger's brushes,
e.g., see [1.9,10], and Boehm's brushes, e.g., see [1.11]), back-reflections
from the tapetum in many species (e.g., the green eye shine we have all observed in cats' eyes), photomechanical (axial) movements of receptors and of
neighboring (sheathing) pigment cells associated with different light levels
(very marked in numerous fish and amphibians), and the presence of colored
oil droplets in photoreceptors of birds and turtles. Birefringence and dichroism of receptor outer segments were well described in a classic paper by
SCHMIDT [1.12].

1.2 Stiles-Crawford Effect

Modern interest in the optics of vertebrate retinal receptors can be traced
directly to a single paper, that of STILES and CRAWFORD [1.13]. They were
associated with the National Physical Laboratory (Teddington) and were interested in a broad range of photometric and colorimetric problems. As part of
their research program they chose to build a device for measuring pupil size.

Assuming that each part of the pupillary aperture contributed equally to the retinal image, they argued that if a uniform beam filled the entrance pupil of the eye, retinal illuminance would be proportional to the area of the pupil. They added a second light beam that passed through a tiny area centered in the pupil (hence, uninfluenced by pupil size) and performed a brightness (photometric) match between an image derived from the entire pupil area and the image of the small measuring beam passing through the pupil center. They found that as pupil size increased, significant departures occurred from the expected photometric relationship. This discrepancy was the effective starting point for modern studies of receptor optics.

STILES and CRAWFORD found that visual sensitivity was greatest for light entering near the center of the eye pupil and that response fell off from this peak roughly symmetrically across the pupil. Since varying the position of entry of a beam of light in the entrance pupil of the eye varies the angle of incidence of light striking the retina, their finding implied that the visual system is sensitive to the direction of incidence of light falling on the retina. This discovery of directional sensitivity has become known as the Stiles-Crawford effect of the first kind. Later, it was found that there were also associated changes in apparent color (hue and saturation). The color effects became known collectively as the Stiles-Crawford effect of the second kind.

1.3 Some Definitions and Distinctions

When STILES and CRAWFORD first described the directional sensitivity effect, they suspected that the effect they described, based on ocular transmissivity data, was largely retinal in origin. Proof followed when CRAWFORD [1.14] showed that there was much less directional sensitivity under scotopic viewing conditions than under photopic conditions. This difference could not be due to the lens elements of the eye; therefore, it was deduced that the directional effect must be based primarily at the retina.[3] (Footnote see next page).

On the basis of Crawford's study and some others, the erroneous notion that rods are not directionally sensitive has entered the literature. Brücke and Helmholtz had already recognized that each receptor must have a limiting aperture and derives advantage by pointing at the pupil (this point will be considered below). When considering such questions, please distinguish between the directional sensitivity of a single receptor and that of an array or population of receptors having some distribution of alignments. Further, please distinguish between response to the pertinent visual stimulus — that is, light having origin in the observer's environment and entering the eye through his pupil and directed to the retinal image (signal), and response to stray light (noise) present in the eye. Realize that stimuli impinging on the receptor are not limited to the pupillary aperture, but that the receptor lies in an ocular sphere which, without its dark pigment, would be an effective integrating sphere. Think of the white, leathery sclera-cornea as a ping-pong ball pierced by an aperture. Anyone who has had a slit lamp examination in an eye doctor's office quickly learns that light does pass through the sclera — a problem exacerbated in the partial and complete albino. To describe the response of the receptor to only the convergent (approximately $\pm 10°$ maximum) cone of energy passing through the eye pupil is to overlook a substantial integrated distribution of energy which may strike the photoreceptor from any direction. Single rod and cone directionality and integrated directionality differ quantitatively, not qualitatively. This question will be treated in appropriate detail below.

CRAIK [1.15] dissected a hole in the rear of a cat's eye and measured radiant flux reaching the retinal plane having passed through different parts of the eye pupil. He found that there was only a small directional factor which could not account for the Stiles-Crawford effect measured in humans. Since that time, WEALE [1.16] and more morecently MELLERIO [1.17] have measured transmission of the human eye for different points of pupil entry at different wavelengths and have shown that the longer path through the center of the blue-absorbing yellow pigment in the eye lens tends to *reduce* directional sensitivity at shorter wavelengths. Eye lens fluorescence also can influence the result in the short-wavelength end of the spectrum.

Electrophysiological recording techniques and fundus reflectometric methods also have been used to localize the directional effect at the retina. Further,

---

3 Scotopic vision operates at low light levels, for example, vision by star- or moonlight, and is served by the so-called rod receptors. In the human, scotopic vision is achromatic, there is poor visual acuity and in any part of the field there are larger areas of neural interaction than in photo-topic vision. Visual acuity at low light levels is best several degrees from the observer's point of fixation. Rod receptors are not found in the central retinal area (the center of fixation and highest resolution), called the fovea (pit) centralis. Photopic vision operates at roughly day-time light levels, and is served by cone receptors. It is responsible for color vision and fine resolution. Mesopic vision refers to transitional states between photopic and scotopic vision. There are many recorded instances of rod-cone neural interactions.

in my own laboratory we have observed anomalous directional sensitivity functions associated with known retinal pathology and have noted remarkable recoveries in a limited number of cases. In addition, we have demonstrated that retinal stress associated with accommodation can transiently translate the peak of the Stiles-Crawford function [1.18]. RICHARDS [1.19] has provided some evidence for a comparable result due to eye movements. Recently it has been noted that the Stiles-Crawford function is altered significantly by monocular occlusion [1.20]. In short, this seems to be a rather active milieu!

Since the definition, localization, and elaboration of the Stiles-Crawford effect there have been two major advances in this field. First, it was recognized that the receptor is actually a waveguide, and secondly, it was found that the receptors align with a point approximating the center of the exit pupil of the eye.

Brücke and others had suggested that the presence of fiber-optics-like light-guiding effects. TORALDO DI FRANCIA [1.21] first suggested that the vertebrate receptor might serve as a dielectric waveguide, i.e., a device for guiding or conducting electromagnetic radiation in single or multiple modal patterns. JEAN and O'BRIEN [1.22] obtained estimates of cone directionality using microwave radiation and scaled-up styrofoam plastic models of cone receptor ellipsoids. With the development of fiber-optics technology and the demonstration of modal patterns in glass fibers by SNITZER and OSTER-BERG [1.23] and KAPANY [1.24] and their co-workers it became evident that it would be possible to evaluate waveguide effects in receptors within the visible spectrum. I duplicated O'Brien's cone model (and built a rod model as well) while I was a graduate student with Fry and spent many months during my postdoctoral fellowship in England with Stiles studying Snitzer's glass fibers. After my return from England, it was a matter of only weeks before comparable observations of waveguiding in retinal receptors were made on a number of different mammalian preparations [1.25,26], including human. Over the years we have attempted to further elaborate upon these observations.

1.4 Major New Developments

In 1968, LATIES published the results of an experiment which required a frank reevaluation of our notions of receptor optics [1.27]. Although thousands upon thousands of histological sections of eye preparations had been cut in eye pathology laboratories over the years, LATIES was the first

to notice that receptors were not uniformly oriented normal to the retinal
surface; rather, he noted that they were aligned with an anterior point in
the eye. Together with LIEBMAN, LATIES formulated a simple model which he and
I later developed in detail [1.28,29]. Anterior receptor pointing has now
been observed in a broad array of species. In this laboratory we have been
able to demonstrate the presence of the same phenomenon in humans psychophysically, and have established that the point of convergence of the axes
of the receptors is near the center of the plane of the exit pupil of the eye.

WEBB [1.30] has also confirmed the anterior pointing of frog receptors
using X-ray crystallography techniques. There are many fascinating aspects
to the problem of receptor alignment — in particular, these properties help
us understand the role of the receptor as a fiber-optics element or waveguide.

There are several other characteristics which have only been partially
elaborated on to date which, taken individually or in combination, are of
great importance to receptor optics and visual function. Thus, the discrete
division of the retinal receptors into unitary response areas, the light-collecting ability of many photoreceptors, the cascading of aligned visual pigments in the outer segment to more effectively trap quanta in the image
plane, the waveguide properties of the single fiber or array of fibers, the
back-reflectance properties of tapeta and/or the absorbing properties of
screening pigment, the wavelength-biasing properties of oil droplets and possibly of receptor waveguides, the polarization detection properties of some
retinas, the alignment and maintenance of alignment of receptors, etc., all
play a role in defining the optics of the retina. They constitute the subject
matter of this book. The whole takes on particular importance when we consider the close association of these collective properties with the primary
transduction and coding stage of the visual system. This means that visual
response is of necessity influenced by these complex optical properties.
Thus, whether one talks of probability of quantum absorption as a function
of angle of incidence, or of alterations in properties caused by anomalies
and/or pathology, response in individual and groups of cells has to be affected. We are only at the beginning of these studies. More and more it is
becoming evident that we must consider the eye lens/iris aperture complex
and the retina/receptor/choroid optical system as an integrated optical device for the optimization of presentation of the pertinent physical stimulus
for transduction into neural excitation.

In this brief introduction, one cannot hope to do justice to the many
individuals and developments which have contributed to the analysis of

optical properties of receptors — hopefully this oversight will be partially corrected in the body of the manuscript.

Acknowledgement. This report was supported in part by a National Eye Institute Grant number EY 01418, NIH, Bethesda, Maryland

References

1.1 W.H. Miller: Receptor-optic waveguides effects. Invest. Ophthalmol. *13*, 556-559 (1974)

1.2 H. von Helmholtz: *Treatise on Physiological Optics 2* (Dover, New York 1962) p.229

1.3 A. Hannover: Vid. Sel. Naturv. og Math. Sk. *X* (1843) [see Ref.1.4]

1.4 J.M. Enoch, F.L. Tobey: A special microscope-microspectrophotometer: optical design and application to the determination of waveguide properties of frog rods. J. Opt. Soc. Am. *63*, 1345-1356 (1973)

1.5 H.E. Ives: The resolution of mixed colors by differential visual diffusivity. Philos. Mag. *35*, Series 6, 413-421 (1918)

1.6 W.T.M. Forbes: An interference theory of color vision. Am. J. Psychol. *40*, 1-25 (1928)

1.7 S. Hecht: The relation between visual acuity and illumination. J. Gen. Physiol. *11*, 255-281 (1928)

1.8 M.H. Pirenne: *Vision and the Eye* (Chapman and Hall, London 1948)

1.9 H. de Vries, A. Spoor, J. Jielof: Properties of the eye with respect to polarized light. Physica *19*, 419-432 (1953)

1.10 E.J. Naylor, A. Stanworth: Retinal pigment and the Haidinger effect. J. Physiol. *124*, 543-552 (1954)

1.11 J.J. Vos, M.A. Bouman: Contribution of the retina to entoptic scatter. J. Opt. Soc. Am. *54*, 95-100 (1964)

1.12 W.J. Schmidt: Polarisationsoptische Analyse eines Eiweiss-Lipoid-Systems erläutert am Aussenglied der Sehzellen. Kolloid-Z. *85*, 137-148 (1938)

1.13 W.S. Stiles, B.H. Crawford: The luminous efficiency of rays entering the eye pupil at different points. Proc. R. Soc. Londong B*112*, 428-450 (1933)

1.14 B.H. Crawford: The luminous efficiency of light entering the eye pupil at different points and its relation to brightness and threshold measurement. Proc. R. Soc. Londong B*124*, 81-96 (1937)

1.15 K.J.W. Craik: Transmission of light by the eye media. J. Physiol. London *98*, 179-184 (1940)

1.16 R.A. Weale: Notes on the photometric significance of the human crystalline lens. Vision Res. *1*, 183-191 (1961)

1.17 J. Mellerio: Light absorption and scatter in the human lens. Vision Res. *11*, 129-141 (1971)

1.18 J.M. Enoch: Marked accommodation, retinal stretch, monocular space perception, and retinal receptor orientation. Am. J. Optom. Physiol. Opt. *52*, 375-392 (1975)

1.19 W. Richards: Saccadic suppression. J. Opt. Soc. Am. *59*, 617 (1969)

1.20 J.M. Enoch, D.G. Birch, E. Birch: Monocular light exclusion for a period of days reduces directional sensitivity of the human retina. Science *206*, 705-707 (1979)

1.21 G. Toraldo di Francia: Retinal cones as dielectric antennas. J. Opt. Soc. Am. *39*, 324 (1949)

1.22 J. Jean, B. O'Brien: Microwave test of a theory of the Stiles-Crawford effect. J. Opt. Soc. Am. *39*, 1057 (1949)

1.23 E. Snitzer, H. Osterberg: Observed dielectric waveguide modes in the visible spectrum. J. Opt. Soc. Am. *51*, 499-505 (1961)

1.24 N.S. Kapany: *Fiber Optics* (Academic, New York 1962)

1.25 J.M. Enoch: Visualization of waveguide modes in retinal receptors. Am. J. Ophthalmol. *51*, 1107/235-1118/246 (1961)

1.26 J.M. Enoch: Nature of transmission of energy in the retinal receptors. J. Opt. Soc. Am. *51*, 1122-1126 (1961)

1.27 A.M. Laties: Histological techniques for the study of photoreceptor orientation. Tissue Cell *1*, 63 (1968)

1.28 A. Laties, P. Liebman, C. Campbell: Photoreceptor orientation in the primate eye. Nature *218*, 172 (1968)

1.29 J.M. Enoch, A.M. Laties: An analysis of retinal receptor orientation II. Predictions for psychophysical tests. Invest. Ophthalmol. *10*, 959 (1971)

1.30 N. Webb: X-ray diffraction from outer segments of visual cells in intact eyes of the frog. Nature *235*, 44 (1972)

2. The Retinal Receptor: A Description

B. Borwein

With 18 Figures

Although the vertebrate rods and cones vary in size and shape both with
species and with location within one retina, there is, nonetheless, a very
uniform basic pattern of organization of vertebrate photoreceptor cells.
This chapter describes mainly mammalian and primate rods and cones, and when
those of other animals are described, they are clearly specified.

2.1 Background

2.1.1 Gross Structure of the Retina

The eye concentrates light of the visible spectrum onto the photoreceptors
(the rods and cones) of the retina. The retina is the innermost of the three
coats of the eyeball (the other two being the choroid and sclera) and it is
a soft tissue less than 0.5 mm in thickness, transparent in the living state
except for the blood vessels that pass through it (Fig.2.1). It is firmly
attached to the underlying choroidal coat only at the retinal periphery (the
ora serrata) and the optic disk (the optic nerve head).

The retina is usually described as containing 10 layers. Starting at the
scleral end these are 1) the pigmented epithelium, consisting of a single
layer of cells; 2) the photoreceptor outer and inner segments (the dendritic
processes); 3) the outer limiting membrane (OLM), a junctional complex, not
a true membrane; 4) the outer nuclear layer, composed of the nuclei of the
photoreceptors; 5) the outer plexiform layer, where the photoreceptor axons
synapse with processes of the bipolar and horizontal cells of the 6) inner
nuclear layer; 7) the inner plexiform layer, where bipolar and amacrine cells
synapse with the dendrites of the 8) ganglion cell layer, whose axons form
9) the nerve fiber layer; 10) the inner limiting membrane, which delimits the
vitreal surface of the retina. There are, thus, 4 cell layers in the retina

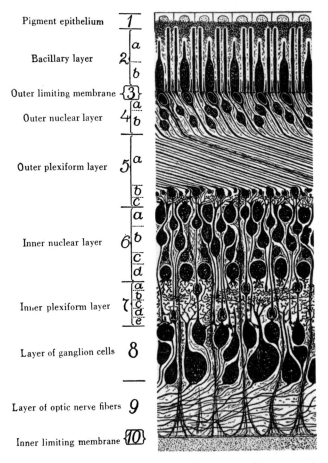

Pigment epithelium — *1*

Bacillary layer — *2* ⎰ a
⎱ b

Outer limiting membrane — ⎰*3*⎱

Outer nuclear layer — *4* ⎰ a
⎱ b

Outer plexiform layer — *5* ⎰ a
⎱ b/c

Inner nuclear layer — *6* ⎰ a
⎪ b
⎪ c
⎱ d

Inner plexiform layer — *7* ⎰ a/b/c/d/e

Layer of ganglion cells — *8*

Layer of optic nerve fibers — *9*

Inner limiting membrane — ⎰*10*⎱

Fig.2.1. Layers of the adult human retina (parafovea) shown in schematic form based on Golgi reproductions. Light traverses the retina from below. Slightly modified from [2.1]. Reproduced from [2.2] by permission

(the pigment epithelium, the rods and cones, the bipolar cell layer and the ganglion cell layer) and 5 types of neurons (the photorecptors, the horizontal cells, the bipolar cells, the amacrines and the ganglion cells) and these interact (mainly) in 2 synaptic layers, the outer and the inner plexiform layers. A supporting framework made up largely of the neuroglial Müller cells fills the spaces between the retinal cells.

The retina is thinnest at its periphery, the *ora serrata*, and in its central part. The central part of the primate retina, the *macula lutea*, with its yellow pigment, contains the fovea and the foveola. In the human eye the fovea is a shallow pit, about 4 mm temporal to the optic disk and 0.8 mm above the horizontal meridian. It contains a very high preponderance of cones which are thinner, more elongated and less tapered than elsewhere in the retina. It is the area of highest visual acuity and spectral sensitivity [2.3].

2.1.2 Embryology of the Retina

The retina derives embryologically from the neural ectoderm of the developing forebrain, the diencephalon, which evaginates to reach the surface ectoderm of the embryo and there induces lens formation in the ectoderm. The retina always remains connected to the brain and its cavities by the developing optic stalk, which becomes the adult optic nerve, invested in the same coverings (meninges) as is the brain. The retina is thus an outgrowth of the brain near the surface of the body.

2.2 The Photoreceptor Cell

2.2.1 History: The Duplicity Theory

MÜLLER [2.4-6] provided an early reliable description of the vertebrate photoreceptor and established that they receive the light stimulus, thus being the first element in the visual processing. Rods and cones as two distinct types of photoreceptors were first described by van Leeuwenhoek in 1684 (quoted in [2.7]). SCHULTZE recognized that the rods and cones were neurons and constituted the terminal organs of the optic nerve [2.8]. He studied both nocturnal and diurnal vertebrate species and formulated his Duplicity Theory [2.8-11]. He separated the scotopic rods, which predominate in the retinas of nocturnal species, from the photopic cones, which predominate in diurnal species, on the basis of outer segment shape (elongate cylinders in rods and short cones in the cones), nuclear position, and size and shape of the receptor terminals. Ultrastructural studies have, in general, emphasized and reinforced the Duplicity Theory, as the great majority of visual cells fall into these two classes remarkably well [2.12-16]. However, anomalies and exceptions have been reported, and these are reviewed and discussed by PEDLER [2.17].

There appear to be specific adaptations in the photoreceptors, and also in their arrangement and concentration, in relation to habit and habitat. The ratios of rods to cones vary. Vertebrate retinas are duplex, containing both rods, operating maximally at low light intensities, and cones, operating maximally at high light intensities and in color vision. The cones are relatively useless at night. Some retinas had been reported to contain rods only, e.g., the rat, but a few cones have now been shown to be present [2.18], and in the nocturnal owl monkey 5% of the photoreceptors are cones [2.19]. Previously reported pure-cone retinas, e.g., ground squirrel [2.20, 21] are now known to contain sparse rods (4-40% of photoreceptors [2.22,23]) which is also true of diurnal squirrels [2.24,25]. HUGHES doubts if any pure-cone or pure-rod retinas exist [2.18]. Even 1 cone to 1,000 rods contributes to diurnal activity. Both rod and cone receptors have been found in every retina where they have been diligently looked for [2.13,26], and even in the most primitive vertebrates such as lampreys (Cyclostomata) there are differentiated retinal receptors that are rodlike and conelike [2.27].

In the cone the inner segment and outer segment taken together constitute a cell with an overall tapered structure, [2.28-31]. The rod receptor is more nearly cylindrical.

2.2.2 The Generalized Photoreceptor Cell

The photoreceptor cell is an elongated structure which is divisible into several reasonably well-demarcated regions (Figs.2.2,3).

KRAUSE first used the terms outer segment and inner segment to describe the distal parts of the cell that lie scleral to the outer limiting membrane [2.32]. They may be considered homologous with the dendrites of a neuron. The inner conducting fibers, homologous with unmyelinated nerve axons, are thin and tortuous in rods, and thicker in cones. They lie in the outer plexiform layer of the retina. They are variable in length and connect the nuclei to the synaptic terminals (Fig.2.4). The outer cone fiber connects the inner segment to the nucleus and may be very short in cones.

The inner and outer segments are connected by a cilium, the "intermediate link" of POLYAK, which forms a narrowed zone (Figs.2.2,5) [2.3]. The cilium arises from a centriole in the distal inner segment and it is usually associated with a second centriole [2.34]. It is interesting that in several of the major sensory systems the detector cell contains a cilium or "hair". It is important to emphasize that the apparent space between the inner and outer receptor segments, shown for illustrative reasons, does not exist in the normal, healthy, living state.

The nuclei, each with a small amount of surrounding cytoplasm, lie in the outer nuclei layer of the retina. Generally the more numerous, rounder, and

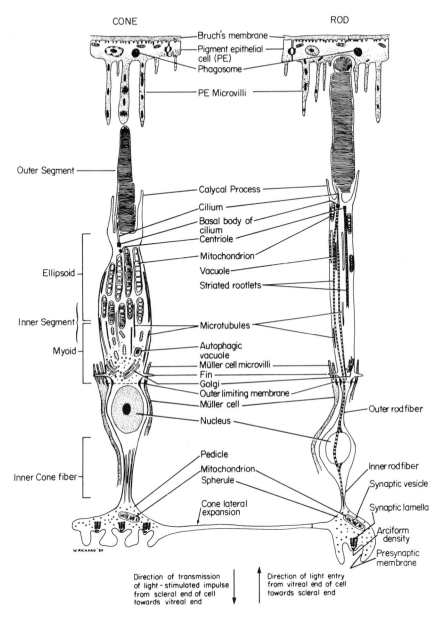

CONE   ROD

Bruch's membrane
Pigment epithelial cell (PE)
Phagosome
PE Microvilli

Outer Segment

Calycal Process
Cilium
Basal body of cilium
Centriole
Mitochondrion
Vacuole
Striated rootlets

Ellipsoid

Inner Segment

Myoid

Microtubules

Autophagic vacuole
Müller cell microvilli
Fin
Golgi
Outer limiting membrane
Müller cell

Outer rod fiber

Nucleus

Inner Cone fiber

Pedicle
Mitochondrion
Spherule

Inner rod fiber

Synaptic vesicle

Cone lateral expansion

Synaptic lamella

Arciform density

Presynaptic membrane

W.RKKARD '80

Direction of transmission of light-stimulated impulse from scleral end of cell towards vitreal end

Direction of light entry from vitreal end of cell towards scleral end

Fig.2.2. Diagram of a generalized mammalian rod and cone showing the specialized parts of the elongated cell and the basic pattern of organization

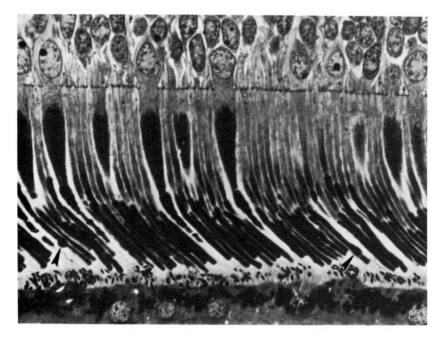

Fig.2.3. Light micrograph of thick section of human peripheral retina prepared by Dr. John Marshall, Institute of Ophthalmology, London. The outer limiting membrane appears as a dark line. The larger and lighter staining cone nuclei are nearest the outer limiting membrane. The inner segments are vertically aligned. The outer segments are also aligned but at an angle of roughly 45° to the inner segments. The cone outer segments end some distance from the pigment epithelial cell bodies. The rod outer segments reach to the pigment epithelium. The cone inner segment stains very darkly. Note that the rod and cone outer segments often display irregular knobs (arrows) but the overall shape of the rod remains cylindrical

darker-staining rod nuclei lie vitreal in several layers; the sparser, ovoid, larger, lighter-staining cone nuclei lie in one or two rows close to the OLM (Fig.2.3) and may even project a little from it, in a scleral direction [2.3]. In the primate retina, only within the *macula lutea* do cone nuclei become more abundant than rod nuclei. In amphibian retinas the rod nuclei abut on the outer limiting membrane [2.35].

The inner and outer segments and the polymucosaccharides that surround them, together with the microvillous processes of the pigment epithelium, fill the region between the external limiting membrane and the pigment epithelial cells (Fig.2.3). This region was the embryonic ventricular space and remains postnatally as a potential space. Normally there is a firm association of receptors and pigment epithelium, but this association breaks down in retinal detachment.

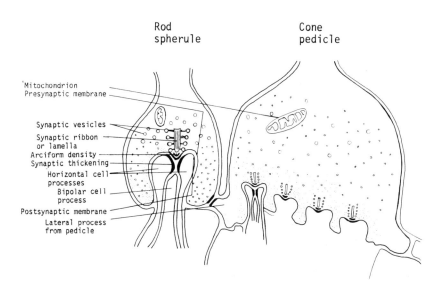

Rod
spherule

Cone
pedicle

Mitochondrion
Presynaptic membrane

Synaptic vesicles
Synaptic ribbon
or lamella
Arciform density
Synaptic thickening
Horizontal cell
processes
Bipolar cell
process
Postsynaptic membrane
Lateral process
from pedicle

Fig.2.4. Diagram of the synaptic terminals of the photoreceptors illustrating the details of the synaptic ribbon and arciform density and the differences between the rod spherule and the cone pedicle. Note that the synaptic vesicles are attached by filamentous material to the synaptic ribbon. (Adapted from [2.33])

The enlarged synaptic terminals, the spherule in rods and the pedicle in cones, lie in the outer plexiform layer, where synapses occur with bipolar and horizontal cells (Fig.2.4). Vitreal to the outer limiting membrane the photoreceptor cells are, for the most part, physically separated by Müller cell extensions, but photoreceptor cells also make some direct contacts.

The highly specialized and oriented photoreceptor cells develop from ciliated ependymal cells of the embryonic forebrain. They contain the visual pigment in membranous disks within their most distal portions, known as the outer segment. Protein (opsin) and retinal (a vitamin A derivative) are assembled into visual pigment, and light falling on the outer segment triggers an effect which is passed along the length of the photoreceptor to its most proximal part, the synaptic terminal (Figs.2.2,4), across which the neural stimulus is chemically transmitted to a relay of other cells involved in the visual pathway to the visual cortex.

The inner segment contains the synthetic apparatus of the cell and is distal to the cell body and nucleus; the inner fiber (equivalent to an axon) connects the cell body to the synaptic terminal. The light enters the retina at its vitreal interface and passes through the photoreceptor from the vit-

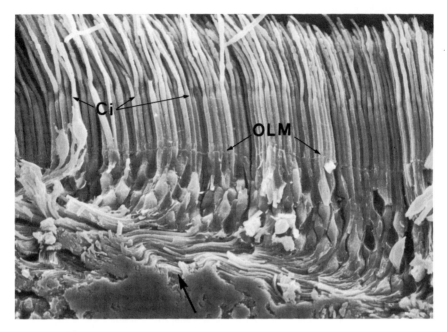

Fig.2.5. Scanning electron micrograph of *M. mulatta* foveal cones. The outer segments bear knobs and show some foldings and convolutions. There is a constriction of the photoreceptor at the cilium (Ci). The outer limiting membrane (OLM) appears as a thin line. The inner cone fibers of Henle (arrow) turn sharply at an angle to the photoreceptors and their nuclei

real end (synaptic terminal) to its scleral end (the tip of the outer segment) which abuts the pigment epithelium. The shape, size, contents, and packing arrangements of the photoreceptors, as well as their refractive indices, influence how light passes through them.

The photoreceptor cell has a high metabolic demand. It is nourished mainly by the choriocapillaris in the choroid and also by the retinal vessels, and is sustained by its intimate association with the pigment epithelium. It has some self-repairing ability, displays considerable turnover, especially of its outer segment disks, and has all the homeostatic devices that simultaneously maintain equilibrium and permit responses to environmental needs.

In the following description distal, scleral, and external are synonymous, and so are proximal, vitreal, and internal.

2.2.3 Multiple Cones

Paired or double cones occur widely in all groups below the placental mammals, including the mammalian Marsupials [2.36] (Fig.2.6). These cones may be a) twin cones with equal-sized components, each similar to a single cone, a teleost fish monopoly, and b) unequal double cones, where the 2 components differ, one being the principal cone, and the other the accessory cone [2.35]. The two components are closely apposed at their inner segments and their abutting surfaces have underlying subsurface cisterns along their contiguous surfaces [2.37,38]. Both fish that feed sluggishly and those that feed actively were found to have multiple cones [2.39]. Double cones are common; triplet cones, arranged linearly or in triangles, are less common; and quadruplet cones are known, but rare [2.40]. The outer segment disks in the two members of the twin double cone are of different dimensions in the four-eyed fish, *Anableps* [2.37]. SJÖSTRAND and ELFVIN found that the disks differed in the unequal double-cone outer segments of the toad [2.41]. The two members stain differently [2.8,35,37]. Double cones do not necessarily connect to the same ganglion cell [2.17].

The photoreceptors may be bundled into groups of 20 rods and 20 cones in the Goldeye and Mooneye (fish) [2.42]. In some deep-sea fish there may be a complex, many-layered receptoral organization as well as grouped photoreceptors [2.43]. There are unique fused units in the anchovy retina where complex bifid cones have conjoint outer segments and form triplet units of 3 cones, composed of one outer segment of 1 long cone and one outer segment from each of 2 adjoining bifid cones. In all other double cones the outer segments lie free from each other; only the inner segments are apposed or joined [2.44]. These authors suggested that the anchovy triplets may be channel analyzers for linearly polarized light, as the fish usually swim in shallow, somewhat turbid waters.

The double cones develop last in the fundus [2.45] and their development may be temperature dependent.

2.2.4 Mosaic Patterns

A mosaic formed from a regular arrangement of double cones and single cones is common in teleost fish retinas [2.35,40,46]. These mosaics are best seen in cross sections at the inner segment level and may vary in different regions of the retina. The pattern appears to form at the ora, in rows directed towards the center of the retina. There may be parallel rows, or alternating

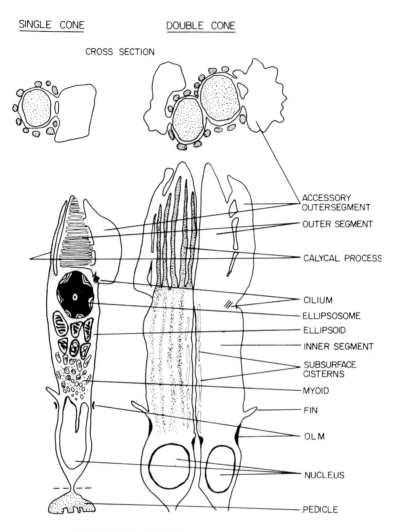

SINGLE CONE   DOUBLE CONE

CROSS SECTION

ACCESSORY OUTERSEGMENT
OUTER SEGMENT
CALYCAL PROCESS
CILIUM
ELLIPSOSOME
ELLIPSOID
INNER SEGMENT
SUBSURFACE CISTERNS
MYOID
FIN
OLM
NUCLEUS
PEDICLE

LONGITUDINAL SECTION

Fig.2.6. Diagram of the single and double cone of a teleost retina, based on *Anableps anableps L.* [2.37] illustrating, *inter alia*, the subsurface cisterns, the accessory outer segments, and the ellipsosome

rows of single cones and double cones at the ora, the anterior boundary of the retina in the eye, and square units at the center; or there may be square units throughout. The pattern is constant within one species but the degree of its perfection is influenced by environment and habit. The patterns are best developed in fish which rely on vision for food capture and are absent, or nearly so, in bottom dwellers [2.40].

Mosaic patterns are also found in reptiles [2.47], amphibians [2.48], and birds [2.35], and the patterns extend to the organization of other cell layers in the retina. In Gecko, as in fish, the mosaic patterns are best developed in the retinal center where there are very regular parallel rows, whereas at the periphery there are only suggestions of pattern [2.49].

It has been suggested that there may be a degree of integration of information within the retina at the photoreceptor cell level [2.49]; that the ordered mosaic patterns may be implicated in the detection of polarized light [2.50]; that the cone patterns are involved in the detection of the direction of moving objects, with a row mosaic adapted to register movements in two directions and square mosaics adapted to register movements in all directions [2.45].

2.2.5 Retinomotor (Photomechanical) Movements of Photoreceptors

Photomechanical movements of rods and cones in submammalian species have been known for more than one hundred years, and their widespread occurrence is well documented [2.7,35,51]. They have been mostly extensively studied in teleost fish [2.40,52] which do not generally have pupillary control of light entrance to the eye. These movements are absent or poorly developed in deep sea fish [2.35].

The photomechanical movements appear to provide protection for the rods in conditions where the cones are exposed to bright light [2.53] as the photo-receptors undergo positional changes in response to light and dark. In the dark the retinal epithelium pigment moves toward the pigment epithelial cell base and in light towards the OLM, so that in the light the rods are well surrounded by pigment, and vice versa in the dark (Fig.2.7). In the light, rod myoids elongate and cone myoids shorten, while in the dark rod myoids contract and the cone myoids elongate. In the light, then, the cones lie closer to the outer limiting membrane and away from the shielding effect of the pigment epithelial cell processes, while the pigment granules migrate within the pigment epithelial processes. Actinlike filaments have been found in association with migrating pigment granules in frog retinal pigment epithelium [2.55]. The light-dark adaptation in fishes is nearly complete in 30 minutes at room temperature, the cells altering in length at the rate of 1-2 μm/min [2.52].

Double cones generally move as one unit in the photomechanical migrations but in some teleost fish double cones do not alter in length, showing very little or no retinomotor response to light [2.56]. In some fish an accessory

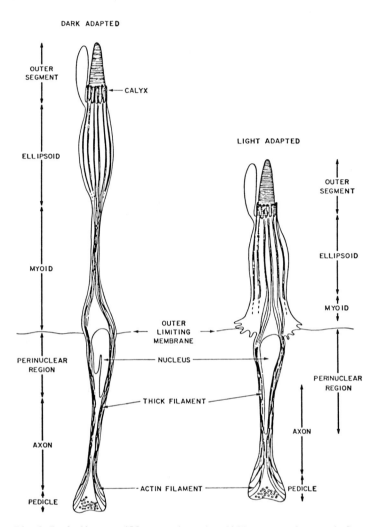

Fig.2.7. A diagram illustrating the differences in morphology of the light-
and dark-adapted teleost cone, the distribution of thick and thin filaments,
and the fate of the fins. (By courtesy of Dr. B. BURNSIDE and the J. Cell
Biol., see [2.54])

outer segment without membrane disks is found alongside the regular outer
segment in both single and double cones (Sect.2.3.3). ALI suggested that the
accessory outer segment may act as a spring, pulling the cone sclerad in the
dark and allowing it to take a vitreal position near the OLM in the light
[2.52].

Rods develop later than cones, and their development is associated with
the onset of the retinomotor response. In some flatfish the cones move very
much less and more irregularly than do the rods [2.57] which, in general,
seem to show the most pronounced movements. There are no photomechanical
movements in the larval eye, and the close connection between the development
of rods and retinomotor responses suggests that rods may in some way initiate
or control this process [2.45,58].

AREY found that optic nerve section blocked these movements in one
fish, but not completely in others, and not at all in a few [2.59]. This observation, the established presence of interreceptor contacts, and the presence of melanin within the retina [2.60] add plausibility to an hypothesis
of intercellular local control [2.61]. EASTER and MACY [2.61] feel that
their data indicate that retinomotor movements result from signals elicited
by light falling on a particular cell or its neighbors. Temperature affects
the myoid and pigment epithelium expansions and contractions [2.59].

2.2.6 Receptor Alignment and Orientation

The receptors are not parallel to each other clear across the retina but are
aligned with their long axes so arranged that they are directed towards the
center of the pupillary aperture, in line with the incoming light rays [2.62-
64] which increases their capacity for light capture [2.65]. In the posterior
pole of the eye the foveal cones are perpendicular to the retinal ventricular
surface and point directly towards the pupil center.

Towards the periphery of the retina the inner and outer segments tilt
away from the perpendicular at progressively larger angles [2.66]. This tilt
of peripheral photoreceptors has been seen in a large range of vertebrates,
and the degree of inclination varies with retinal position [2.67]. The outer
segment itself is apparently oriented towards the pupil, which orientation
enhances light capture (see Chap.4), and the human rod outer segments in particular "show a marked angular deviation between the ciliary region and the
pigment epithelim" [2.68] (Fig.2.3). In LATIES's studies of receptor orientation (e.g., [2.62,67]) (Chap.4), alignment has its origin at the external
limiting membrane, while in Fig.2.2 it is suggested that alignment originates
at the cilium. What role do microtobules and microfilaments play in this
orientation, and can this orientation alter with lighting conditions? There
are changes in the photoreceptor position and possibly orientation in res-
pose to light entry. Changes in this anterior pointing occur with prolonged

darkness which alters the photoreceptor orientation [2.69]. Changes also occur in photomechanical movements in many nonmammalian species so there is both a capacity for elongation and for changes in orientation direction in response to external lighting.

For proper alignment of the photoreceptors their normal apposition to the pigment epithelium must be maintained. In retinal detachment there is a loss of receptor alignment which may be restored with reattachment of the pigment epithelium. Even with detachment some semblance of receptor alignment may remain [2.70]. HOPE pointed out the difficulty of sectioning the retina so as to see an accurate representation [2.71]. This problem arises from the fact that the "receptors tilt progressively relative to the tangent to the retinal curve at progressively more peripheral loci." For there to be no distortion of the apparent length of the receptors or of the apparent thickness of the inner layers the cross section "must be simultaneously parallel to the long axis of the receptor outer limbs and perpendicular to the inner layers."

Perhaps fins, which are absent in foveolar cones [2.30], the abundant microtubules of the inner segment, and the calycal processes may play a part in enabling the mammalian receptor to effect its orientation and alignment by contracting parts of the cell. FITZGERALD et al. considered that "receptor alignment is most probably an active process" [2.70].

2.2.7 Refractive Indexes

There is not much information available on the refractive indexes of the primate photoreceptors and their surrounding extracellular space. SIDMAN for fresh monkey retina estimated the average refractive index of the rod myoid (1.36) and rod outer segment (ROS) (1.4) to be higher than that of the mucopolysaccharide (MPS) in the extracellular space between the photoreceptors (scleral to the outer limiting membrane) [2.72]. It lies between that of saline (1.334) and serum (1.347). ENOCH et al. measured a value of 1.40 for frog ROS [2.73]. ENOCH and TOBEY measured an index *difference* between rat ROS and the interstitial material of 0.06 [2.74]. Subtracting this value from Sidman's value of the index for rat ROS yields 1.348 for the index of the interstitial material (see Chap.5). Table 2.1 summarizes the available information.

Table 2.1. Refractive indexes of photoreceptors

| | RODS | | | CONES | | | Foveal cone COS | Extracellular MPS |
| | RIS | | ROS | CIS | | COS | | |
	myoid	ellipsoid		myoid	ellipsoid			
SIDMAN primate [2.72] frog	<1.3638	1.36	1.4	<1.3638	1.3978	1.3883	1.419	between 1.334 + 1.347
salamander		1.3978	1.4056	1.3638	1.3939	<1.3865		
chicken					1.3902	<1.3958		
pigeon			1.4076					
rat			1.4076					
M. rhesus			1.4076	1.3605		1.3985		1.34
average for many species	1.36	1.40	1.41	1.36	1.39	1.385		
ENOCH et al. frog [2.73]			1.40					
ENOCH, TOBEY rat [2.74]								1.347

RIS - rod inner segment   CIS - cone inner segment   MPS - mucopolysaccharide
ROS - rod outer segment   COS - cone outer segment

2.3 The Outer Segment

The outer segment lies closest to and is intimately associated with the pigment epithelium. The rods reach to the pigment epithelial cell itself, as do the foveal cones, but the tapered ends of the shorter nonfoveal cones are vitreal to the pigment epithelium cells and long pigment epithelial cell processes extend to them (Fig.2.3) [2.75,76].

Only the central part of the outer segment is without a cytoplasmic wrapping. Its distal part is surrounded by processes of the pigment epithelium and its proximal part by calycal processes extending from the inner segment (Fig.2.8). The pigment epithelial cell processes extend to 40% of the rod outer segment length, but only just reach to the extrafoveal cone outer segment [2.77].

Some of the cone outer segment tips may be encircled by a sheath of concentrically arranged pigment epithelial processes containing pigment granules and little else [2.30,78]. The portions of the photoreceptor cells scleral to the OLM are surrounded by mucopolysaccharide (MPS), the inner segment being the probable site of much of its synthesis [2.79]. The pigment epithelium may also contribute some of this material [2.80,81]. With retinal detachment MPS disappears from the subretinal space and returns with reattachment [2.82]. The primate outer segment has the highest concentration of solids recorded in biological material (rods 40-43%, cones 29-34%) [2.72] (see Chap.5).

2.3.1 Disk Structure

The outer segment is filled with double-membrane disks, or flattened sacs, piled one upon the other like a stack of pennies (Fig.2.2). These disks were first reported by SCHULTZE [2.10] but their existence was disputed until the advent of electron microscopy [2.34,83,84]. They are derived from the plasma membrane at the base of the outer segment [2.86]. The regularity of the arrangement of the disks is very constant in almost all species of vertebrates. They are very few exceptions [2.44,86].

The highly ordered regularity of the disk arrangement and structure may be disrupted, more easily in cones than in rods, by a variety of traumas such as fixation artifact [2.87], visible light of too long duration or too great an intensity [2.87,88], retinal detachment [2.82], vitamin A deficiency [2.89,90], thermal trauma as a result of non-Q-switched laser irradiation [2.91], low-energy-level laser damage [2.92], aging [2.68], and osmotic shock [2.93].

FOVEOLAR CONES

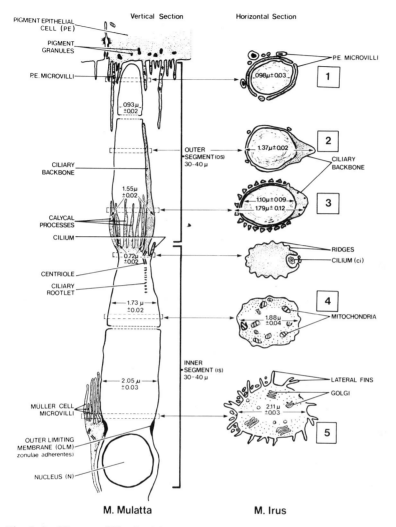

Fig.2.8. Diagram illustrating the morphology and dimensions of the foveolar cone of Macaca spp [2.30]

The guinea pig rod has 700 double-membrane disks, the perch rod 1400 [2.94]. The rod outer segment in rhesus monkeys is longest in the parafovea and decreases gradually in length towards the periphery. The parafoveal rhesus rod has 1100 disks, the perifoveal rod has 920 disks, and the peripheral rod 790 disks. An average of 80-90 disks are assembled daily in each rod [2.77] (Sect.2.3.2).

Most of the normal rhodopsin molecules are oriented so that the dipole movement of the rhodopsin chromophore lies transverse to the receptor axis enhancing light absorption [2.95]. When the normal horizontal orientation of outer segment disks has been disturbed, one might expect the outer segment to become less sensitive to light.

The light-sensitive visual pigment, rhodopsin, is an integral part of the lipid bilayer of the disk membrane. Exposure to light triggers a molecular alteration in the rhodopsin molecule that initiates visual excitation [2.96]. Rhodopsin has also been found in frog rod outer segment plasma membrane, as well as in the newest disk [2.97], indicating a very rapid association of opsin and retinal in the outer segment.

The disk edge differs from the rest of the disk saccule in that it stains more heavily and it resists deformation [2.94]. In hypotonic solutions where the disks swell, the edge remains unchanged [2.98,99]. With treatment with osmium tetroxide and TRIS [tris(hydroxymethyl)aminomethane], the frog rod disks disappear but the saccule edge remains intact and in normal relationship to the plasma membrane; the cone infoldings are not affected [2.100]. In a freeze-fracture study, SJÖSTRAND and KREMAN found that there is a particulate structure at the edge of the disk that looks different from the disk membrane and they suggest that calcium-binding protein may be located here close to the plasma membrane [2.101]. Calcium has been implicated in the transmitter process, being released when light acts on rhodopsin.

a) Incisures of the Disks

Cone outer segments have smooth circular margins [2.102,103], whereas rod outer segments have scalloped margins or one or more incisures, which are aligned for considerable distances [2.13] (Figs.2.9,10). A single incisure is found in rodents (guinea pig [2.84], mouse [2.104], rat [2.89]) and in fish (deep seafish [2.105,106], perch [2.84], *Anableps* [2.37]. Shallow scalloping of the rod disk margin is reported in man [2.107,108], rhesus monkey [2.30,109], owl monkey [2.110], squirrels [2.28], and pigeon [2.86]. There are many deep incisions in the rods of amphibia (mudpuppy [2.111], Rana pipiens [2.112], newt [2.113a]) and catfish [2.12], but only two to three shallow incisures in Labrid fish [2.40] and parrot fish [2.113b] (see Chap.5 for a discussion of the effects of these incisure zones on wave guiding in aging frog receptor preparations).

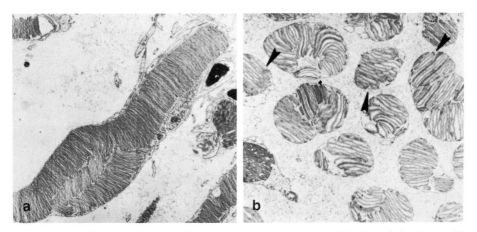

Fig.2.9a,b. Electronmicrographs of human retina prepared by Dr. John Marshall of the Institute of Ophthalmology, London, showing (a) the rod outer segment containing a stack of membrane disks. This particular rod has undergone a convolution of the disks. These convolutions appear as knobs in Fig.2.3. (b) Rod outer segments in tangential sections displaying the rod incisures (see arrows) and the convolutions

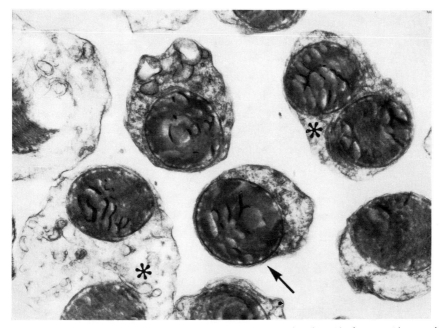

Fig.2.10. Cone outer segments of rhesus monkey showing their smooth margins (arrow). There are also present two cone disk stacks within one cone plasma membrane (*) indicating that cones, like rods, undergo convolutions with age

The plasma membrane may be slightly indented over the incisures but it does not follow into them. At the base of the outer segment the degree of scalloping in the rod outer segment disks is minimal and nonsymmetrical. The alignment of the incisure also becomes more regular away from the base [2.114]. The incisures increase the surface area of the disks, and BROWN et al. suggested that they are related to the need for metabolite exchange, as the rod disks, but not the cone disks, are separated from the extracellular space [2.111]. Why so many forms of incisure exist, and what their function is, is unknown.

STEINBERG and WOOD report a single disk cleft in cat and ground-squirrel cones, the only report of a cone incisure [2.115].

b) *Origin of the Disks*

The outer segment is a ciliary derivative which arises in development from a pair of centrioles in the distal inner segment of the growing photoreceptor. The disks form one at a time, throughout life, in the region of the apex of the cilium in the basal (vitreal) outer segment [2.48,116] (Figs. 2.2,11). For this orderly formation light is not required.

One of the most consistent differences found between rods and cones is that between their outer segment disks. These form the same way in both by repeated infoldings of the plasma membrane of the outer segment according to COHEN and NILSSON [2.86,48], or by evagination of the membrane of the connecting cilium outwards along the lowest edge of the disk stack towards the opposite periphery according to ANDERSON et al. [2.78]. The rod disk soon becomes independent and free-floating by pinching off from the plasma membrane [2.16,48,94,114,117,118].

c) *Disks in Relation to the Extracellular Space*

Many (or most?) cone disks retain the continuity between disk and plasma membrane and, hence, are open to the extracellular space. Only about 10 disks at the base of the rod retain continuity with the plasma membrane at the site of formation, whereas, in the rhesus monkey, about one-third of all the cone disks are open widely to extracellular space [2.109].

Extracellular markers such as lanthanum do not cross the plasma membrane. COHEN found that lanthanum did not infiltrate the rods, in which the sealed intradisk space derives from the extracellular space [2.119]. It did fill the spaces within the cone disks, even those that looked as if they were in-

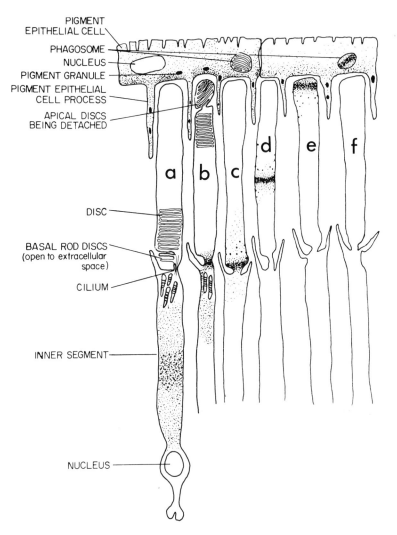

Fig.2.11. I. Diagram of rods illustrating: (a) the formation of basal disks; (b) the detachment of apical disks; (c) the appearance of the shed disks in the pigment epithelium as a phagosome.

II. The path taken by protein (opsin) following an injection of radio-active amino acid. (a) In the myoid; (b) through the cilium; (c) incorporated into disks at the base of the outer segment; (d) the displacement of the labeled disks sclerally in the outer segment; (e) the labeled disks appear at the apex of the outer segment; (f) the labeled disks appear in a phagosome in the pigment epithelium.

Some scattered diffuse labeling of the disks and the plasma membrane is seen from stage (c) to (f)

dependent, discrete disks, indicating that their intradisk spaces were in continuity with the extracellular space. Since open rod saccules are very rarely seen except at the very base of the outer segment [2.120], the disks must be very rapidly isolated after formation [2.24,121]. Procion yellow, a fluorescent dye binding to molecules that cannot cross the plasma membrane, selectively stains the cones, but not the rods [2.122]. Rod saccules show osmotic expansion and shrinkage, indicating again that they are sealed off from the extracellular space [2.99].

d) The Intradisk Space

The disk membrane consists of a lipid bilayer and one monolayer of photopigment molecules dispersed in a small volume of aqueous medium.

Each disk is a double unit membrane enclosing a less dense (intradisk) space. At their edges the membranes are further apart then centrally (Fig. 2.2). The sizes of the intra- and interdisk spaces may well vary with the osmotic strength of their fluid media [2.65], with fixation methods [2.101] as well as with species [2.65,102,103,110,113b]. There are conflicting reports about the relative sizes of the intradisk spaces in rods and cones. MARSHALL found that the cone intradisk spaces were wider than those of the rod in primates [2.65]; KROLL and MACHEMER found the rod intradisk spaces to be wider [2.82]. SJÖSTRAND and KREMAN [2.101], using freeze fracturing, infiltration, and embedding in hydroxypropylmethacrylate at $-30^{\circ}C$, could not see any intradisk space and they questioned whether an appreciable intradisk space really exists; if it exists, it must be only about 10-15 Å. CLARK and BRANTON were also unable to find an intradisk space in freeze-fractured material without any previous cross-linking with glutaraldehyde [2.123].

2.3.2 Turnover in Outer Segments

BAIRATI and ORZALESI [2.124] and ISHIKAWA and YAMADA [2.125] first suggested that the outer segment disks are constantly shed and renewed (Fig.2.11). By autoradiography it has since been established that rod disks are renewed one at a time throughout life, displaced towards the outer segment apex, and shed intermittently, in groups, and in a balanced way at the apex. They are then phagocytosed by the pigment epithelium and degraded within phagosomes [2.77, 118,126-130].

Proteins involved in the visual pigments of the outer segment are manufactured in the myoid of the inner segment [2.131]. They migrate to the inner segment apex and through the cilium [2.132] into the outer segment, where they are incorporated into disk membranes at the base of the rod outer segments within hours after the injection of tritiated amino acids, intraperitoneally, in the dorsal lymph sac or in the saphenous vein [2.77,118,130,133]. The newly formed, labeled disk moves towards the outer segment apex. A weak, diffuse labeling occurs simultaneously with the intensely labeled, newly formed disk. In cones, the labeled protein is incorporated into the preexisting disks in a diffuse pattern throughout the length of the outer segment indicating that only parts of the material of the disks are replaced, whereas in rods entire new disks are formed at the outer segment base while some subunits are also replaced [2.77,118,134,135]. However, cones are known to make new disks after retinal reattachment surgery [2.82], during posthibernation arousal [2.136], and in recovery from light-induced damage [2.137], but no distinct bands such as is seen in rods ever appear, not even in embryonic development [2.138-140].

It appears that there is a basically different organization of the outer segments in rods and cones, but that light stimulates the rate of membrane addition in both rods and cones; the outer segments of both photoreceptors develop most rapidly in constant light and slowest in constant dark [2.141]. The distinctive pattern of protein incorporation, which differs in rod and cone outer segments, has now been seen in a wide variety of vertebrates [2.24,77,116,126,132,133,139,140,142].

It was thought that only rods shed disk packets from their apices [2.16, 126], and that this constituted a major distinction between rods and cones and accounted for their different outer segment shapes [2.127,134]. The apparent absence of phagosomes in the pigment epithelium of cone-rich retinas and over primate foveas was taken as evidence that cones do not shed disks. However, phagosomes have since been reported in the foveolar region of primates [2.76], and it has been established that in cyclic light rods shed most of their disk packets soon after the onset of the light period, i.e., in the daytime when most experiments are performed [2.143-145] and cones shed their disk packets soon after the beginning of the dark period [2.146]. Shedding tends to be synchronous. Cone disk shedding has been reported in man [2.76,147], in Rhesus [2.78], in cats [2.148], in squirrels [2.24], in chicken [2.149], in lizards [2.150], and goldfish [2.146]. The process of shedding by the photoreceptor and phagocytosis in the pigment epithelium is very similar in cones [2.76,134,147] and rods [2.24,134,151]. Light appears to initiate synchronous shedding in frogs [2.144] but there is a circadian rhythm of shedding in the rat [2.143].

Normally the outer segment remains of constant length [2.128]. A single rod disk takes 10 days in rodents to migrate the length of the outer segment, 7 weeks in frogs, and 9-13 days in monkeys [2.116]. The rod outer segments of rhesus monkey are longer in the parafovea and shortest in the periphery so the renewal time varies with retinal region, from nearly 13 days on average in the parafovea, about 11 1/2 days in the perifovea to 8-9 days in the periphery [2.77].

In Rhesus monkeys each pigment epithelial cell engulfs 2,000-4,000 disks daily, and 39-45 rods and 2-3 cones are associated with each pigment epi-thelial cell in most of the retina, and 24-25 rods and 4-5 cones in the para-fovea [2.77]. In frogs 25-36 rod disks are formed per rod per day at 22.5°C [2.130].

In the shedding process a group of apical disks curl sclerally and the outer segment plasma membrane invaginates gradually below the curled disks and separates them from the rest of the outer segment [2.24,78,126,134,151]. Sometimes disks that are not curled up are also separated off [2.78,151]. It appears that the addition of sugar residues to the outer segment tip may be the signal that triggers the start of the membrane invagination that produces an isolated disk packet [2.152] and makes the oldest disks available for phagocytosis by the pigment epithelium.

Rods and cones can develop in total darkness but the rate of development differs under different lighting conditions [2.153]. The normal maintenance of balance in disk assembly and disk shedding depends on a daily cyclic alternation of light and dark. The rate of disk assembly and shedding may be affected by the availability of polyunsaturated fatty acids [2.154], a rise in environmental temperature [2.155], and lighting [2.141,143,145,156,157].

Disk formation had been thought to occur as a continuous process [2.127], but BESHARSE et al. have shown that in *Xenopus* and *Rana pipiens* the rate of disk formation is enhanced in constant light compared to the dark; that re-newal, like shedding, occurs in bursts in response to light, discontinuously; and that different lighting conditions elicit different rates of disk assembly [2.156,157]. In general, similar factors affect the rate of disk assembly and shedding in rods and cones [2.78,146,149,150].

The pineal gland is considered to be a major center controlling circadian rhythms [2.158]. However, it does not exert a major effect on the daily rhythm of shedding and phagocytosis, since pinealectomized rats kept in con-stant darkness over a 24-hour period showed a normal circadian shedding pat-tern, as do animals that have had the superior cervical ganglion removed

[2.159,160]. The shedding response occurs even if both optic nerves are severed and after pinealectomy and hypophysectomy [2.161].

There appears to be a mechanism *intrinsic* in the eye that initiates the daily rhythm of disk shedding which is synchronized with eye opening and the start of visual function [2.162,163]. Enzymes for melatonin synthesis have been found in the pineal gland and in the rat retina [2.164,165], and melatonin and its precursors have been localized by immunohistochemistry in the rat in the outer nuclear layer; in the optic nerves, chiasma and tracts; and in the Harderian glands [2.60]. In utero fetal guinea pigs, near-term, shed disks which are phagocytosed and degraded in the pigment epithelium [2.166]. Is the stimulus for shedding transmitted from the mother or is shedding and phagocytosis an intrinsic and continuous process which is only stimulated and altered by external factors?

2.3.3 The Accessory Outer Segment

This structure was named by ENGSTROM who first described it by electron microscopy in Labrid fish cones [2.40]. It originates at the lateral edge of the cone outer segment, close to the base of the cilium, as if formed jointly from it and the outer segment (Fig.2.6). It extends alongside the outer segment for most of its length, and it tapers. The cytoplasm is homogeneous looking and devoid of any obvious organelles. The axis of the outer segment is generally aligned with that of the inner segment of the cone cell, while the axis of the accessory outer segment is aligned with that of the ciliary shaft [2.167]. The ciliary microtubules end in the accessory outer segment as they do in the ciliary backbone of the regular outer segment. In longitudinal sections the accessory outer segment may be confused with an enlarged ciliary backbone which is not separated by membranes from the disk zone. ENGSTROM [2.40] said it arose from a second centriole, but STELL [2.168], BORWEIN and HOLLENBERG [2.37] and FINERAN and NICOL [2.113b] found only one cilium associated with both the outer segment disks and the accessory outer segment. In the long single cone of Parrot fish both the cilium and the accessory outer segment were absent [2.113b].

FINERAN and NICOL referred to the accessory outer segment as a lateral sac and this name has the merit of greater brevity [2.113b]. They describe it as a lateral expansion of the cell membrane of the outer segment, connected to it by a narrow tapering cytoplasmic strip running longitudinally almost the full length of the outer segment. Several authors reported that the accessory outer segment is connected to the outer segment for most of its length [2.44,169], but BORWEIN and HOLLENBERG [2.37] in *Anableps* cones saw at all levels in cross sections a number of short cytoplasmic bridges, forming a weblike connection between the outer segment and the accessory outer segment (lateral sac), as was seen in Labrid fish, where the connecting links are either narrow or quite broad [2.40]. The accessory outer segment of long single cones are found deeply inserted into the pigment epithelium. (In some literature, the shorter of a pair of double cones is called the accessory cone. This is different from the use of the term "accessory" here).

Accessory outer segments have been reported in cones of European freshwater fish [2.170], goldfish [2.168], the S. American four-eyed fish [2.37], perch [2.45], catfish [2.167], pike [2.169] and parrot fish [2.113b]. Its embryonic development has not been studied and its function is unknown (see [2.52]). It may be that they play a role in light capture [2.171] as the photoreceptor has a light-capture area larger than its geometric cross section.

2.3.4 Aging and Convolutions in Outer Segments

With aging there is an increased incidence of nodular excrescences in outer segment rods [2.68]. Similar excrescences are also seen in cones [BORWEIN, unpublished] (Figs.2.5,12). Localized convolutions are found starting in man in the fourth decade, perimacularly, on the distal one-third of the outer segment. By the seventh decade 20% of all human rods were found to show distortions such as nodules, convolutions, and groups of disorganized disks. In cross sections the localized convolutions appear as fused associations of independent outer segments. In longitudinal sections one can see that the disk stack corrugates in its long axis and curves so that ascending and descending limbs become apposed. The outer segment increases both in length (by 6-10 μm) and disk content (20%-40%), but the distance between the cilium and the pigment epithelium remains constant [2.68]. Has the phagocytosis rate in this situation decreased while disc production has continued unabated? How does this accumulation of disks affect the size of the catchment area for incident light? POLYAK assumed that the "bulbous swellings and other deformities" he often saw on the outer segments were evidence of fragility [Ref. 2.3, pp.212-221].

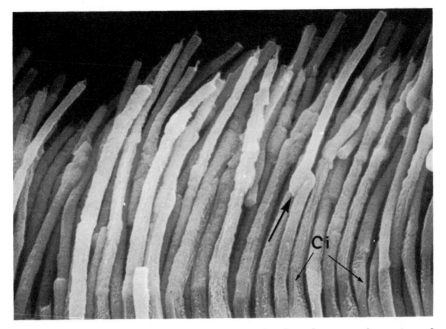

Fig.2.12. Scanning electron micrograph of the foveal cones of a mature female rhesus monkey showing the ciliary connective (Ci) and the tapering outer segments which bear excresences or knobs and also convolutions (arrow) in the middle region of the outer segment, and not at their most proximal or distal ends

2.4 The Connecting Cilium

The existence of a cilium was reported by Henle in 1861 [2.172]. SJÖSTRAND, using electron microscopy, saw that the outer and inner segments were connected by an eccentrically positioned stalk containing 9 fibril bundles [2.83, 173a] (Figs.2.13-15). DE ROBERTIS described and named "the connecting cilium" [2.34]. During embryogenesis the cilium arises from one centriole of a diplosome in the distal inner segment, and all the ciliary tubules end in a centriole which is associated with dense satellite bodies. The second centriole is often seen to give rise to a striated rootlet (Sect.2.5.6).

There is a marked uniformity of ciliary structure throughout the animal kingdom [2.174], consisting of 9 pairs of microtubules arranged circumferentially and 2 microtubules placed centrally in motile cilia [9(2) + 2], e.g., flagella; or with the 2 central microtubules widely reported to be absent in nonmotile cilia, including those of photoreceptors [9(2) + 0] [2.34].

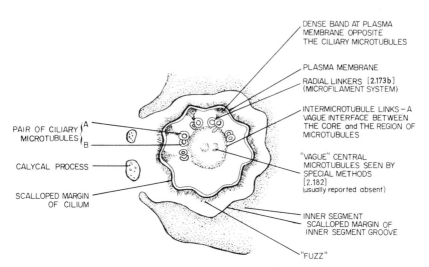

DENSE BAND AT PLASMA
MEMBRANE OPPOSITE
THE CILIARY MICROTUBULES

PLASMA MEMBRANE

RADIAL LINKERS [2.173b]
(MICROFILAMENT SYSTEM)

INTERMICROTUBULE LINKS – A
VAGUE INTERFACE BETWEEN
THE CORE and THE REGION OF
MICROTUBULES

"VAGUE" CENTRAL
MICROTUBULES SEEN BY
SPECIAL METHODS
[2.182]
(usually reported absent)

INNER SEGMENT
SCALLOPED MARGIN OF
INNER SEGMENT GROOVE

"FUZZ"

PAIR OF CILIARY {A
MICROTUBULES }B

CALYCAL PROCESS

SCALLOPED MARGIN
OF CILIUM

Fig.2.13. Diagram of the cone cilium lying within the ciliary groove of the
inner segment. Note that the two ill-defined central microtubules are thought
to form a helix. 5 of the 9 pairs of ciliary microtubules are illustrated

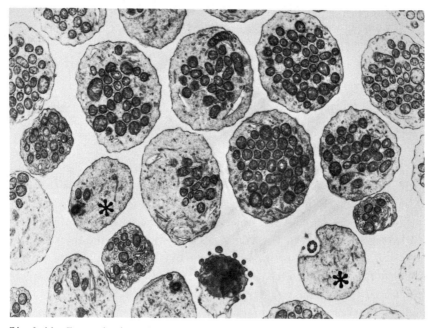

Fig.2.14. Transmission electron micrograph through the apical part of the cone
inner segment showing the mitochondria that fill the ellipsoid. The mitochondria
are sparse or absent at the very apex of the inner segment where the cilium (*)
arises. One outer segment is present surrounded by calycal processes; it bears
a large ciliary backbone in which 9 singlet microtubules of the cilium can be
seen, 5 of them close to the disks, the other 4 at the plasma membrane

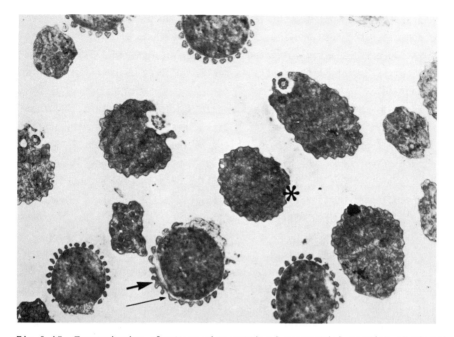

Fig.2.15. Transmission electron micrograph of tangential section of photoreceptors of *Macaca irus*. The cilium lies in the inner segment ciliary groove and only one calycal process is apparent immediately opposite it. The inner segment margin is ridged and grooved (*). Calycal processes arise from these ridges and are shown still attached to each other (thin arrow). One calycal process is seen bifurcating (broad arrow)

The connecting cilium of rods and cones is enclosed by a ciliary membrane which is continuous with, but thicker than the plasma membrane of the inner and outer segments. The ciliary membrane is covered by "fuzz" [2.175], an amorphous material which also outlines the microvillous processes of the pigment epithelium and the most basal rod disks. The fuzz is interpreted tentatively by BUNT as a sleeve of sugar-containing material as it can be labeled with ^{3}H-fucose, and this sleeve may play a role in ionic regulation in this zone [2.176].

The cilium is the only connection between the site where light is absorbed (the outer segment) and the rest of the cell, and it forms an obvious zone of constriction (Fig.2.5). However, direct continuity between the inner and outer segments has been seen without any ciliary connective in the long single cone of parrotfish [2.113b], gecko cones [2.177], monkey rods [2.178a], and in human rods [2.179]. The absence of a cilium has been interpreted as an abnormality [2.179]. Where the cilium was absent, new disks were seen forming at the apex of the inner segment [2.178a]. In human peripheral rods, where inner and outer segments abutted, the outer segment was short and wide and the disks were sometimes disoriented in various patterns [2.179]. The important optical point is that there is really no space between the inner and outer segment.

In general, the rod cilium appears longer and more complete [2.17] than the cone cilium [2.37,44,98,109]. In humans the cone cilium is very short and the cone inner segment lies very close to the outer segment [2.102]. In rodent and primate rods the cilium is about 1 µm long from inner to outer segment [2.13,91,108,175]. In frogs the ciliary region is very short and the inner and outer segments abut [2.13]. FAWCETT [2.174] gave the diameter of cilia as 0.2-0.25 µm; MATSUSAKA [2.180] found that the rod cilium in albino rat is 0.3 µm, as it is in mammals [2.65,181].

The cilium may appear isolated and unsurrounded over most of its length [2.84,99,173], but it often lies on one side of the inner segment in a recess or groove which partially surrounds the cilium, and in cross section its outline is seen to be scalloped (Figs.2.13,15) [2.15]. In *Macaca fascicularis* and *M. rhesus* cones an independent cilium was never seen. As observed in sequential tangential sections, it is always surrounded in its groove by inner segment, and the gap between the inner and outer segments must, therefore, be very narrow indeed [2.30].

In certain fish the cone cilium is associated with an "elaborately evaginated structure" [2.168], the accessory outer segment, a connecting channel between inner and outer segments (Fig.2.6).

The ciliary process penetrates eccentrically and quite deeply into the outer segment and forms a stalk or backbone structure. This ciliary backbone was seen extending to the apex of the cone outer segment, but only halfway up the rod outer segment [2.111]. In Macaque monkeys the ciliary backbone was reported of variable length [2.107] but it does not extend to the apex of the outer segment [2.30]. The 9 microtubule doublets splay out rapidly to become 9 singlets in the base of the ciliary backbone, losing their precise arrangement, and end somewhere in the vitreal half of the cone outer segment [2.37,107] or in the accessory outer segment. In general these microtubules extend further in the cone outer segment than in the rod outer segment. The microtubules in primate rods are seen in the crypts between the disk lobules formed by the incisures [2.30,108]. In the cat, single microtubules face individual clefts very close to the outer segment tip in both rods and cones [2.115].

There is a well-defined inner core to the cilium, separated from the radially arranged microtubule doublets by a thin membrane [2.107,175] or an interface [2.13]. The inner core is described as being less dense than the outer zone containing the microtubule doublets [2.107], or as empty looking [2.13], but MATSUSAKA said it is definitely not empty [2.180].

No protein synthesis occurs in the cilium itself. It acts as a pipeline. Labelled amino acids incorporated into proteins in the myoid pass to the base of the cilium, where they are briefly detained, and then pass through the cilium to the base of the outer segment where they are incorporated into disks [2.132].

The microtubular doublets made up of subfibers A and B, which appear electron lucent, as in motile cilia, do not always attach as a complete figure-of-eight, one side of subfiber B being sometimes separated. Each subfiber is an electron-lucent circle, of which only the surfaces stain with ruthenium red [2.180,182]. Rarely, however, central single microtubules, as seen by DE ROBERTIS [2.34], and electron-dense central structures, as found by MATSU-SAKA [2.180,182], stain with ruthenium red throughout. Ruthenium red with osmium tetroxide oxidizes polysaccharides to make insoluble products [2.183]. These structures were interpreted as two central singlets forming a double helix with lengthening pitch towards the outer segment [2.180]. Central microtubules of motile cilia are known to disappear after dialysis, and MATSU-SAKA suggested that conventional histological methods fail to preserve these structures which may be important in the functions of the ciliary connective [2.180,182].

The cilium may be involved in 1) transmitting the light-triggered response, 2) the transfer of metabolites forming a passage way from the inner segment to the outer segment [2.131,132], 3) the growth and maintenance of the outer segment, 4) embryogenesis, where it appears to have an inductionlike role [2.184], and 5) regeneration, which is possible only as long as the inner segment and cilium are intact [2.82].

2.5 The Inner Segment

2.5.1 General

The inner segment is generally somewhat barrel shaped. It is constricted at the OLM and is narrow at its distal end, the "intermediate link" of POLYAK [2.3]. At this region a cilium connects the inner and outer segment. The proximal part of the inner segment, vitreal to the outer limiting membrane, is called the myoid, and the scleral part is the ellipsoid. These two regions intergrade into each other.

The inner segment has often been illustrated as a smooth cylinder but in many species it is reported to have a surface that is longitudinally ridged and grooved [2.8,12,40,44,109,177,178b,185]. ENGSTROM wrote that the "plasma strains" (calycal processes) "do not really begin at the outer tip of the ellipsoid but run like mouldings along the surface of the entire ellipsoid" [2.40]. BORWEIN and HOLLENBERG [2.37] found this ridging to be more marked in the cones and it is especially marked in primate cones [2.30] (Figs.2.8, 15). The ridges become very prominent nearest the outer segment where they separate from each other and from the inner segment to become the calycal processes around the proximal outer segment [2.8,12,30,109] (Figs.2.15,16). The ridges often contain longitudinally oriented microtubules, and there is a peripheral lining of microtubules or microfilaments in both the myoid and the ellipsoid [2.37,186] which continues into the calycal processes [2.37] (Fig.2.17).

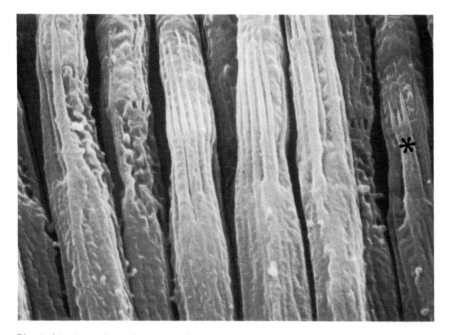

Fig.2.16. Scanning electron micrograph of monkey cones showing the tapering calycal processes originating from the inner segment ridges. They are thicker and fewer over the ciliary zone and then they taper, often bifurcating or, more rarely, splitting into three (*), over the outer segment itself

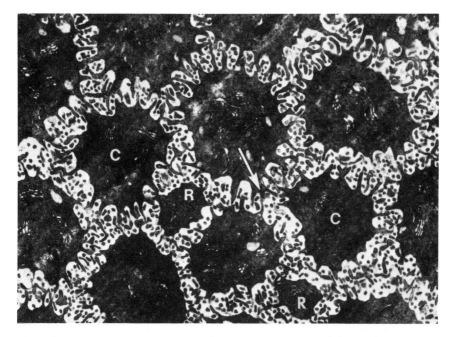

Fig.2.17. Transmission electron micrograph of tangential section through the photoreceptors of *Macaca irus* immediately scleral to the OLM, within the fovea. The Golgi apparatus is prominent in both rods (R) and cones (C). The fins are large, sometimes bifurcated, and they approach those of neighboring cells (arrows). Müller cell microvilli fill the spaces between the photoreceptors and their fins

In the inner segment in some amphibians, reptiles, and birds there is a glycogen-rich organelle, the paraboloid, found vitreal to the ellipsoid [2.35,187]. It is composed of smooth-surfaced membranes of endoplasmic reticulum, which form first, and an abundance of glycogen granules, which accumulate later in development. The organelle may affect optical function [2.86,188-190]. It retains its shape and integrity when expelled from the living cell [2.35]. PEDLER, however, found that in the nocturnal gecko the paraboloid merged with the surrounding cytoplasm, but that it was definitively demarcated in diurnal geckos [2.172]. The glycogen of the paraboloid does not seem to be depleted in frogs even after long starvation [2.191,192] or by light and dark stimulation [2.188]. Abundant glycogen has been found in the myoid of man and monkeys. It is depleted within twenty minutes if the choroidal circulation is cut off [2.193].

RNA synthesized in the nucleus [2.194] travels to the myoid, which is rich in ribosomes and the site of protein synthesis [2.131]. Also present are many microtubules, a few poorly differentiated mitochondria, and most of the cell's complement of autophagic vacuoles, organelles in which the discarded constituents of the cell's own cytoplasm are enclosed by membranes and digested by enzymes [2.195].

The myoid contracts and elongates in photomechanical movements and can shorten by up to 90% of its length within two minutes to one hour in fish, amphibians and birds, depending on species [2.35,52,59].

2.5.2 The Ellipsoid

a) Ellipsoid Mitochondria

The ellipsoid area of the inner segment is filled with mitochondria. ENOCH has shown that there are staining differences in the ellipsoid mitochondria which depend on the activity or inactivity of a given photoreceptor, exposed to light or not [2.196].

In mammals the ellipsoid mitochondria are typically long, slender, in rows approximately parallel to the axis of the inner segment, and along the full length of the ellipsoid [2.108,136]. In primate cones there appear to be few or no mitochondria in the inner segment apex [2.30,120], and KROLL and MACHEMER and DE ROBERTIS illustrated this condition [2.82,34]. Primates have fewer mitochondria than lower species, and primate foveal cones have even fewer mitochondria [2.197] than the extrafoveal cones. However, larger cones contain more mitochondria.

In human cones and in rod cells the mitochondria are all approximately the same size [2.125]. In many nonmammalian cones there is a linear or a radial gradient of mitochondrial size. In forms with a linear vitrealscleral gradient, the smallest mitochondria with the fewest cristae are at the vitreal end of the ellipsoid (guppy [2.198], lamprey, snakes, gekko, carp [2.125], frog [2.199], fish [2.40], chick [2.186], newt [2.113a], parrot-fish [2.113b], gecko [2.178], and Anableps [2.37].

Independent of or alongside a linear gradient there may be a radial gradient The smallest mitochondria with the fewest cristae are peripheral, and the largest are central (gecko [2.178b,189], pike [2.200], newt [2.113a]. In some fish the mitochondria undergo changes, becoming progressively denser as they enlarge in a vitreal-scleral gradient until the cristae are barely recognizable or entirely obscured, and a large "oil-drop-like" structure, derived from mitochondria, forms in the inner segment apex (guppy [2.198], Anableps [2.37], and guppy [2.201]. ISHIKAWA and YAMADA described many structurally modified atypical mitochondrial configurations [2.125]. In the lamprey, mitochondria in the rod inner segment enlarge in a vitreal-scleral gradient; in the snake a very large, central, granule-filled mitochondrion forms surrounded by normal mitochondria; and in gecko there are a number of very large mitochondria-like bodies filled with lipoidal material. ENOCH (personal communication) has made the observation that in species with active longitudinal alteration in length with light and darkness, mitochondria are round and can tumble, as compared to human cigar-shaped mitochondria with their long axes roughly parallel to the cell axis.

It has been suggested that the ellipsoid mitochondrial mass acts as a focusing device, helping to condense light onto the outer segments [2.94,172]. Clearly there is a higher concentration of solids in this zone than in the myoid, but a lesser concentration of solids (in most species) than in the outer segment. Thus, at a minimum there is an index step or gradient [2.72].

b) Oil Droplets

Colored or colorless oil droplets are permanent inclusions of the adult visual cell in some nonmammalian species. They are found in the most scleral part of the inner segment, mainly in cones, in amphibians, reptiles, and birds [2.35]. They are membrane bound, show no discernible internal structure, and contain unsaturated lipids [2.202], mainly phospholipids [2.187]. Carotenoid pigments have been identified in chick oil droplets [2.203].

Adult colors of oil droplets are present at birth [2.204]. MEYER et al. found that adult Japanese quail fed for 11 months with a carotenoid-free, vitamin-supplemented diet retained their colored oil droplets but their offspring had colorless oil droplets in cells that were normal in all other respects [2.205,206]. The carotenoids of the oil droplet appear to be derived from the egg and do not contribute to the visual pigments. They show no turnover [2.129] except possibly for triglycerides [2.207] and are not affected by the nutritional state of the animal [2.208].

Although their function is uncertain [2.172], HAILMAN has suggested that the main function of Anuran oil droplets is chemical storage [2.202]. They have very high indexes of refraction. Light must pass through the oil droplet to reach the outer segment so that cone sensitivity depends on both the visual pigments in the outer segment and the transmission of light through the inner segment. Colored oil droplets act as filters, altering the wavelength of light reaching the outer segment. They remove the violet and ultraviolet light [2.202] for which the eye has considerable chromatic aberration, thus reducing glare [2.209] and increasing contrast and acuity [2.7,35,210].

Eighty percent of the incoming light is removed by the oil droplets in turkey, turtles, and pigeon, irrespective of color [2.211].

c) Ellipsosomes

Mitochondria are often associated with the oil droplet. Large structures, at first called oil droplets, were seen in cone inner segments of the teleost fish, *Lebistes* [2.198] and *Anableps* [2.37], and have since been seen in other fish [2.212,213]. ISHIKAWA and YAMADA had reservations about using the term "oil droplets" for structures that have rudimentary cristae and bordering double membranes [2.125]. Those structures in fish which superficially resemble oil droplets of birds, reptile, and amphibian cones have been shown to stain differently and to have a different ultrastructure [2.214], reminiscent of an unusually large mitochondrion with a double-layered bounding membrane and cristaelike structures, with fibrous electron-dense material present [2.37,198,201]. They are all probably what MACNICHOL et al. have called ellipsosomes [2.214] (Fig.2.6). They are found in the oil droplet position in the cones of certain fish and do not contain carotenoid pigments. They are dense, spherical bodies containing a higher concentration than do mitochondria of a heme pigment (resembling pure cytochrome C spectroscopically). They resemble mitochondria ultrastructurally and also in that they contain the succinic dehydrogenase system. They may, like oil droplets, serve as intracellular filters as well as serving optical and also metabolic functions. DUNN suggested that the oil droplet of geckos, as in fish, may also be the end product of transmuted mitochondria [2.15]. Two types of oil droplets have been reported by BERGER [2.198] and KUNZ and REGAN [2.215], and are reflections of altered metabolic states [2.213], those of light-adapted retinas being denser than those in the dark [2.214].

2.5.3 Subsurface Cisterns

Alongside and closely parallel to the contiguous plasma membranes of the inner segments of equal double cones of teleosts are subsurface cisterns [2.37,38,40,45,113b,168,200] (Fig.2.6). These membranous sheets are also found in rod spherules opposite invaginated processes and alongside membranes where cones contact rod myoids near the outer limiting membrane [2.168], where rod knoblike evaginations penetrate cone inner segments, and where rods abut directly on cones [2.37]. They have been looked for but not seen in squirrel, *Xenopus*, and cat retinas, but they were seen in cones of the developing human retina [2.216]. They seem to originate from rough endoplasmic reticulum and may bear ribosomes. Their function is unknown but it has been suggested that they play a role in coordination of impulses, transference of water and ion metabolites, in cell-to-cell contact adhesion, electrical insulation, and intercellular communication [2.38,40,216,217]. Perhaps the subsurface cisterns provide a low index of refraction separation of receptors important for wave guiding [Enoch, personal communication].

The cell opposite the subsurface cistens may be neuronal or glial, but they themselves occur only within neurons and are considered unique to neuronal cells. There is evidence that they modify the cell membrane and, therefore, could have a role in neuronal function. Subsurface cisterns have also been described in the amacrine, bipolar, and ganglion cells of the vertebrate retina [2.216].

2.5.4 Fins

Cytoplasmic prolongations (or fins or pleats) are frequently seen extending from the cell surface, in tangential sections, at the vitreal end of the inner segment immediately scleral to the outer limiting membrane. They interdigitate in a gearlike way with those of neighboring cells. Müller cell microvilli are seen between these fins, and some fins show simple branching (Fig.2.17). They become shallower in a vitreal-scleral gradient as they merge into the inner segment ridges. In the turtle they are vesicle laden [2.218]. They are present in double cones except for the region where the cone components are apposed [2.37].

The dimensions of the fins are given as "a few microns" in length in toads [2.219]. YAMADA and ISHIKAWA [2.220] said that in the lamprey "the height of each pleat increases towards the base of the inner segment (up to 1-1.5 µm) and the thickness is about 50-70 nm" with the Müller cells between them about 60 nm in diameter. DUNN reported widths of 27-154 nm and lengths of 0.2-1.7 µm [2.15]. The numbers of fins described are given in Table 2.2.

Table 2.2

Animal	Type of Receptor	No. of fins per cell	Author (References)
Toad	red rods	up to 36	FAIN et al. [2.219]
Gecko	(diurnals)	30-35	PEDLER, TANSLEY [2.177]
Frog	red rods	36	RAVIOLA, GILULA [2.221]
Gecko	(nocturnals)	44-55	PEDLER, TILLY [2.178b]
Macaca irus	cones	18-25	BORWEIN et al. [2.30]

The fins vary in number depending on whether they are counted nearest to the outer limiting membrane or more sclerally, on the size of the cell, and, it seems, on the state of the cell (see below). They are much better developed in *Macaca irus* cones than in rods, and are not seen in the foveal center [2.30]. RAVIOLA and GILULA [2.221] found in toad that 30 of 36 fins of each rod cell made contact with fins from neighboring red rod cells, involving small gap junctions; larger gap junctions were seen between green rods and cones. Some fins make contact by simple apposition [2.219,221]. CARASSO, who first described them, reported that the fins were 200-300 nm apart from each other [2.222].

The functions of these fins are unknown. Obviously, they increase the surface area. Various functions have been suggested, such as active transport and exchange of metabolites and water [2.16,177,218] and assistance in holding retinal cells in position [2.15,188,189]. It was thought that they were present in avascular retinas only [2.177] or certain classes of vertebrates [2.15], but they have now been reported in all classes of vertebrates: lampreys [2.220,223], birds [2.85,224], frogs and toads (toad, on all receptors) [2.219,225,226], lizards [2.177], turtles [2.188,218], gecko, where they were first reported [2.177,189,222], snake [2.298], grey squirrels [2.177], monkeys *M. irus* and *M. rhesus* [2.30], fish [2.37,54,227,228].

BURNSIDE studied the actin- and myosinlike filaments in the inner segments of fish photoreceptors and postulated that a sliding mechanism, dependent on the microtubules, effects cone contraction and elongation [2.54,227, 228]. When colchicine was injected into the vitreous, disrupting the microtubules of the myoid, the normal myoid elongation of the cone in the dark was prevented, and no fins were then seen on the dark-adapted cone [2.229].

The fins may be the source of cytoplasm and plasma membrane available for rapid elongation of the inner segment in retinomotor movements in response to lighting conditions [2.54] (Fig.2.7). If this proves to be the case it may be that photoreceptors show some photomechanical movement in all classes of vertebrates; it may also account for the absence of fins in the light-adapted foveolar cones [2.30].

2.5.5 Microtubules and Microfilaments

Longitudinal microtubules and filaments are abundant, running the full length of the inner segment [2.168,230,231] and grouped in bundles around the periphery of the cell (chick [2.186]), parallel to the long axis. The only microtubules found in the outer segment are associated with the cilium and lie in the ciliary backbone. The microtubules of the inner segment are not known to connect directly to the striated rootlet or to the terminal bars of the OLM.

The inner fiber (connecting nucleus to synaptic terminal) and the outer fiber (connecting myoid to nucleus) are of variable length and contain abundant microtubules and filaments parallel to the long axis of the cell [2.231]. The inner and outer fibers are known to alter in length in photomechanical movements in teleosts [2.15]. Microtubules have also been seen in the spherule [2.231], approaching the synaptic ribbon [2.232].

In teleosts, photoreceptors can contract in those cell parts that lie between the pigment epithelium and the OLM at 2-3 μm/minute, proportional to the intensity of light [2.52]. PIETZSCH-ROHRSCHNEIDER [2.56] found actin-like 70 Å diameter parallel filaments in the myoid of teleost cones that she believed could be involved in the contractility associated with retinomotor responses.

BURNSIDE [2.54,228] found both thin actin- and thick myosinlike filaments in the inner segments of teleost cones, in the zone (the myoids) where contraction occurs in response to light adaptation. No contraction occurred in the light when these filaments were disrupted by the application of cytochalasin B. The injection of colchicine into the vitreous disrupted the microtubules in the myoid, and no elongation of the cone myoid occurs in dark adaptation [2.229].

DRENCKHAHN and GROSCHEL-STEWART found, by indirect immunofluorescence microscopy using specific antibodies, thin actin filaments with a diameter of 50-80 Å, similar to skeletal muscle actin, and also myosin in virtually all ocular nucleated nonmuscle cells [2.233]. The presence of actin and myosin provides an apparatus for a cytoplasmic contractile system and the authors suggested that "certain motile activities might also occur in the photoreceptor inner segments" of mammals. They saw thin filaments 1) disoriented and dispersed throughout the rabbit rod inner segment, 2) forming parallel bundles as a long rootlet from the basal body of the connecting cilium to the outer limiting membrane, and 3) extending into the calycal processes. They likened the saccules around the rootlet to the tubular system associated with muscle fibrils. MATSUSAKA found ATPase activity in the ciliary rootlets of human rods [2.299].

Actinlike filaments binding myosin are found in some other ocular cells [2.234]. In teleost retinal cones [2.227] there are both thin actin- and thick myosinlike filaments, with overlapping sets of actin filaments, resembling the two actin halves of a muscle sarcomere [2.54]. She postulated a sliding hypothesis for cone contraction, with thick myosinlike filaments producing interdigitations of two sets of oppositely directed actin filaments.

The pigment epithelial cell processes (microvilli) are known to be able to undergo length changes [2.82], and it has long been known that it is easier to effect detachment of the retina from the pigment epithelium if it is dark adapted [2.235]. Actin filaments have been located in primate and amphibian pigment epithelial processes [2.55,234].

The abundant microtubules, the calycal processes, fins, striated rootlets, and the inner segment ridges may all be involved in one way or another with movement of the photoreceptor cells either in contraction and expansion in relation to lighting conditions, or in alignment of the cells towards the pupillary center, or both.

2.5.6 The Striated Rootlet

SJÖSTRAND first reported a cross-striated "fibril" in the inner segment of the guinea pig rod extending to the outer limiting membrane [2.173a]. This cross-striated ciliary rootlet originates near or from the pair of centrioles in the apex of the inner segment, and it has been described in a wide variety of animals (Fig.2.2).

There have been many brief descriptions of the ciliary rootlet but the most detailed account is given by SPIRA and MILLMAN [2.236] in guinea pigs. They described it as one continuous organelle constructed from thin filaments which aggregate to form a single cross-striated fibril running the full length of the "nonreceptor part of the cell." It originates from densities around the basal body of the connecting cilium diplosome and passes down the ellipsoid in close contact with the mitochondria. Along the Golgi zone it narrows to a ribbon shape. It separates into two to three discrete strands which curve around the nucleus and reunite in a single bundle on the vitreal side of the nucleus. It then travels down the cell axon to end deep in the synaptic terminal among the synaptic vesicles. The fibrils are bordered by two separate membrane-bound vacuoles which have attached ribosomes in the myoid region. COHEN saw these and referred to them as a "rootlet vacuole system" [2.104,109,120].

Earlier workers had seen the striated rootlet in the rod inner segment only [2.104,120,185,299,237a]. MURRAY et al. saw them well developed in rods and less so in cones, all the way to the nucleus, and closely associated with mitochondria and large cisternae of endoplasmic reticulum [2.19]. The major and minor bands of the cross striations showed a distinct polarity, with tubular cristae of the mitochondria related to the major bands. PEDLER noted that they are more rudimentary in cones and much better formed in rods and associated this with the high sensitivity of the rod [2.17]. SCHUSCHEREBA and ZWICK [2.237b] in monkey rods reported branched bundles of striated root-let filaments emerging from the basal body, running the length of the inner segment to the outer limiting membrane, and beyond, and merging with neuro-tubules to end in the synaptic terminals. The proximity of the organelle to the mitochondria and to glycogen suggested to the authors that the striated rootlet organelle may be involved in active optical alignment of the photo-receptor. BORWEIN (unpublished) has seen a number of apparently separate rootlets running longitudinally and parallel in the inner segments of rhesus monkey rods. SJÖSTRAND found that the striated filament could be pulled out of the inner segment when the outer segment was separated mechanically, thus showing a very intimate attachment to the outer segment [2.114].

It has been suggested that the striated rootlet forms an anchor for the cilium. The cilium and its rootlet may be the site of transmission of the light-triggered impulse from the outer segment to the synapse [2.104,173a] or a conductor route for other intracellular stimuli. It may be that the stri-ated rootlet is implicated in the movement of photoreceptor cells which ex-pand and contrast in response to lighting conditions [2.107,231,299]. BURN-SIDE also ascribed to it a role in photomechanical movements [2.227]. It may also be involved in the alignment of the rods and cones. The ability to con-tract and elongate some of its parts may be a property of the mammalian photoreceptor.

2.5.7 The Calycal Processes

The calycal processes were so named by COHEN [2.12]. They are thin, cyto-plasmic extensions from the distal end of the inner segment and they form a palisade around the proximal outer segment (Fig.2.2). These processes con-tain longitudinally oriented microfilaments or microtubules [2.45,112,136, 186] continuous with those of the inner segment [2.37,233]. Calycal processes had been seen earlier by light microscopists. SCHULZE [2.8] reported that if the outer segment fell off a well-preserved photoreceptor (in perosmic acid), there remained an array of fine fibrils continuous with the "striae" of the inner segment, encircling the base of the outer segment and extending over the outer segment for variable distances. Their presence was widely reported after BROWN et al. described them in some detail, thus conferring importance on them [2.111]. They described the "dendrites" of the mudpuppy as a circlet of 27-30 fibriller microvillous processes extending more than half-way up the outer segment, shorter in the rods than the cones. One "dendrite" was seen opposite each fissure in the rod saccule. Calycal processes seem to be present in all visual cells investigated with the exception of albino rat rods [2.89,111,181] and mouse rods [2.104]. On anchovy rods they are few and irregularly distributed [2.44]. They are better developed in cones than in rods (primate [2.30], fish [2.56]). They were seen to be forming embryologically at the time that the disks were forming [2.48].

The calycal processes are usually seen single and separate from each other, but they have been seen forked in Macaca [2.30] (Figs.2.15,16). In the deep sea fish *Platytroctes* [2.106] the calycal processes form a continuous low wall with an irregular scleral border, arising vitreal to the apex of the inner segment. However, another deep sea fish, *Poromitra*, has short, discrete calycal processes [2.105]. In the light-adapted condition in anchovy [2.44] and in Macaca [2.30] the calycal processes are seen to be continuous with the low ridges of the ellipsoid. In macaque monkeys the calycal processes surround the vitreal part of the outer segment, and the pigment epithelial cell processes surround the scleral part. Only a small central area has no sheathing [2.30,107,109] and is enclosed only by the mucopolysaccaride that fills the ventricular space [2.238].

The calycal processes are thickest at their origin at the ellipsoid apex [2.30,136]and taper [2.30,37,113b]. They appear first on the side of the eccentrically placed cilium, and these are also the shortest calycal processes [2.30,108]. They do not lie alongside the ciliary backbone. The calycal processes are usually fairly regularly arranged and spaced around the outer segment, but in anchovy the calycal processes form two separate groups, one on the nasal and the other on the temporal face of the cone [2.44].

In the ground squirrel, where over 90% of the visual cells are cones, there are 5-25 calycal processes, but none of them is around the ciliary backbone. During hibernation the outer segments shorten and the calycal processes thicken markedly at their origin at the apex of the inner segment and they contain mainly clear cytoplasm, a few vesicles, and a few ribosomes. The thickened calycal processes form a collar around the outer segment at the level of the connecting cilium. This collar compresses the disks to one side. Mitochondria, which reduce in number during hibernation, do not move into the enlarged calycal process collar. On recovery from hibernation the calycal processes become thin and normal again in seven days, as the outer segment lengthens to reach its normal length in nine days [2.136].

Calycal processes have been said to have a supportive mechanical role, to prevent eccentric rotation, to align the inner and outer segments, to be a channel for uptake of metabolites from the ventricular space for transfer to the inner segment, to assist in conduction of the light-triggered stimulus and the exchange of material between the inner segments and outer segments, to be involved in photomechanical movements, and to help separate the receptor light guides. The retraction of the calycal processes in hibernation to form collars, the shorter calycal processes seen in the rods, and their reported absence in albino rat and mice rods may be clues to their functions, at present unknown.

2.6 The Outer Limiting Membrane (OLM)

The outer limiting membrane (OLM) was so named because it shows distinctively as a clear line by light microscopy and was thought to be a fenestrated membrane (Figs.2.3,5,8). It was later described by AREY as a densely staining array of terminal bars between the membranes of the photoreceptor cells and the Müller cells [2.239]. Generally, Müller cells separate the photoreceptors, but not always. That the OLM is a series of junctional complexes has been confirmed by electron microscopy [2.117,240,241] and identified as "zonulae adhaerentes" [2.242]. It forms a barrier separating the ventricular space and the retina proper from the mucopolysaccharides that are normally found in this subretinal area, and from exudates and haemorrhages that occur in abnormal conditions of pathology and/or trauma [2.243]. There are no "zonulae occludens" between Müller cells and photoreceptor inner segments [2.244]. The zonulae adhaerentes OLM allows current to pass and does not restrict the passage of ions and small size particles as completely as do the tight junctions of the zonulae occludens pigment epithelium [2.13,245]. Tight junctions occur here only where two Müller cells abut, but are not found between receptors and Müller cells [2.245]. Ferritin, horseradish peroxidase and toluidine blue injected into the brain ventricles enter between the Müller cells and the photoreceptors to the neural tissue [2.29]. TONUS and DICKSON have described the junctional complexes of the OLM in newts [2.246]. They found that the most scleral components of the interglial junctions are large gap junctions which permit the movement of ions and penetration of lanthanum, and the vitreal components are zonulae adhaerens (terminal bars). The neuroglial junctions are zonulae adhaerens, but with intercellular clefts that may be extremely narrowed.

Gap junctions between Müller cells at the OLM have been reported in *Necturus* [2.247], frog, carp, and teleosts [2.248]. Zonulae adhaerens between photoreceptors and Müller cells have been reported in many species [2.245, 247,248].

2.7 The Outer and Inner Fibers

Extrafoveally the cone outer fiber is very wide and short, as the cone cell body lies at the OLM, very close to the myoid. The rod outer fiber is of greater and very variable length as the rod nuclei lie vitreal to the cone nuclei (Figs.2.2,5). Foveally, the outer cone fibers are also of variable lengths because the cone nuclei are multilayered, and not one-layered as they are extrafoveally. HOGAN et al. defined the outer fiber as the part of the cell between the OLM and the cell body [2.108]. It contains some mitochondria, smooth-surface endoplasmic reticulum vesicles, some free ribosomes, and many microtubules. The inner fiber is of variable length and runs from the nucleus to the synaptic terminal. In the foveal center they are long and all run horizontally, parallel to the retinal surface (i.e., meridionally) (Fig.2.5) towards their synaptic terminals which are displaced to the periphery of the fovea. These oblique fibers in the macular area together with their 'enveloping' Müller cells constitute the fiber layer of Henle of the outer plexiform layer. Extrafoveally the outer fibers run perpendicular to the retinal surface [2.1].

2.8 The Synaptic Terminals

Synapses of rods and cones were first described by SJÖSTRAND [2.94,173,240], DE ROBERTIS [2.249], DE ROBERTIS and FRANCHI [2.250], and LADMAN [2.251]; subsequent studies [2.252-259] have shown that their basic morphology is remarkably uniform in all classes of vertebrates (Fig.2.4), even in Cyclostomes with degenerate eyes [2.27,260]. The synaptic terminal is the photoreceptor site where information is exchanged or transmitted, and its structure is so consistent that PEDLER, who studied 71 vertebrate species, regarded it as the most reliable indicator of whether the photoreceptor is conelike or rodlike [2.172].

Synapse formation in human retina develops early and is well under way
at 12 weeks (83 mm) of gestation [2.261], whereas the outer segments develop
at 23 weeks of gestation and the disks are not well developed until 32 weeks
[2.262].

2.8.1 The Rod and Cone Synaptic Terminals Differ

The synaptic terminals of the rods and cones differ. The rod spherule is
considerably smaller than the cone pedicle and, with conventional methods
of electron microscopy staining, it appears denser [2.37,113a]. The rod
spherule has a pyramidal shape with a flat base; the cone pedicle is roughly
circular.

In the pedicle each invaginated synaptic unit [2.253] or triad [2.33,252]
consists of one to three central cone-bipolar dendritic processes and, on
either side of it, more deeply inserted horizontal cell processes [2.263],
probably from two different cells [2.256]. These processes are assumed to
synapse at a region near the base of the synaptic lamella and its arciform
density [2.251], which resembles a three-pronged fork in shape and which
anchors the ribbon to the presynaptic membrane [2.258] (Fig.2.4).

More superficial contacts from other bipolar cells make conventional
(surface) synapses on cones [2.256,257] and on rods, and all retinas have
both superficial and deeply invaginated contacts [2.264]. Typically each
cone pedicle contains many triads [2.257], 25 reported in humans [2.33] and
12 in goldfish [2.253] and monkey [2.103], whereas the rod spherule has only
one or, rarely, a few [2.104] with invaginated dendrites from only one bipolar cell. The rod synaptic ribbon (see below) tends to be larger than that
of the cone. COHEN [2.12] suggested that the invagination in the receptor
base may act in part to shield the terminating processes from the activity
of neighboring receptor terminals.

The cone pedicles, which tend to get larger towards the periphery of the
retina, are generally arranged in one row, vitreal to the multilayered array
of rod spherules [2.108], except in the fovea where the cone pedicles are
too numerous for only one row, the foveola having pedicles only.

Five types of processes contact the presynaptic membranes; they are
1) processes from the same or 2) adjacent receptors, 3) processes from horizontal cells, 4) Müller cells, and 5) dendrites of bipolar cells. Müller
cells envelope the synaptic terminals and may appear as whorls of thin processes around them [2.37]. The direction of transmission is from synaptic

pedicle to bipolar cell. In the case of the other contacting processes or cells the direction of transmission is not certain.

2.8.2 Synaptic Ribbons and Synaptic Vesicles

The synaptic terminal is always associated with synaptic vesicles close to the presynaptic membrane and with increased membrane densities on both the pre- and postsynaptic membranes [2.244]. It may contain a few mitochondria [2.33,104,109,251], but these are not a constant feature. The distinctive presynaptic organelle of the photoreceptor synaptic terminal is the synaptic ribbon, a pentalaminate [2.265] or trilaminate [2.266] structure composed of protein with no measurable carbohydrate [2.267]. The ribbons lie in synaptic grooves [2.258] associated with the deeply inserted bipolar cell dendrites and horizontal cell processes which invaginate in a very regular and orderly way. The synaptic ribbon is surrounded by a halo of synaptic vesicles parallel to it but a few hundred angstrom units distant from it [2.12,13,253]; in the intervening space there is fibrillar material present which seems to attach to both the vesicles and the ribbon [2.108,258].

It has been suggested that synaptic ribbons store transmitter substances from synaptic vesicles [2.266,268] or act as conveyor belts or orienting structures guiding the synaptic vesicles in progression to the synaptic zone for transmitter release [2.258,267]. The ribbons, closely associated with synaptic function, appear in embryological development at the time that the b wave appears [2.269,270].

There is a continuous release of transmitter substance in the dark, which ceases in the light. In the dark, transmitter substance is depleted and there is an associated significant reduction in synaptic ribbons in fish cones [2.266], in the receptors of albino rats [2.271], and in the cones of hibernating ground squirrels [2.136]. The synaptic ribbons are largest at the time that they are most numerous, i.e., in the light period [2.271].

There is evidence that the synaptic ribbons are dynamic, polymorphic structures capable of rapid changes in number and morphology, depending on lighting conditions, with an average life span of eight hours [2.266,272]. For example, WAGNER found that in the pedicles of a male cichlid fish the synaptic ribbons are very sparse only half an hour before the onset of the light period, but numerous within the light period [2.266]. He found, however, that irrespective of lighting conditions, there were always two synaptic ribbons in each rod spherule, so that only the cone ribbons are affected by the lighting conditions. There is also a suggestion that an endogenous circadian rhythm of synaptic ribbon function exists, accentuated by environmental lighting [2.271,272].

SPADARO et al. described a model of synaptic ribbon formation from a ballooning membrane at the invagination, the membranes of which meet and fuse to form a pentalaminate structure [2.271]. The stalk of the invagination becomes the arciform density. SCHAEFFER and RAVIOLA proposed that, in transmitter release, the synaptic vesicle membrane is reversibly incorporated into the plasma membrane, and is later returned through coated vesicles and vacuoles. There may also be a constant flow of new vesicles, originating from the cell body, which move along the microtubule-filled inner fiber to the synaptic terminal. They found that, in the dark and in cold, dyads and triads are deeply invaginated into pedicles but in the light they barely indent the basal surface [2.273].

Microtubules (150-250 Å) are closely related to the synaptic ribbons and are found in the terminals. MISSOTTEN saw convoluted microtubules in the human cone pedicle and assumed that they contain transmitter substance [2.33]. It has also been mooted that they have a role in the formation of vesicles [2.258].

2.9 Interreceptor Contacts

Interreceptor contacts were first reported by SJÖSTRAND [2.240]. There is a basic general pattern of interreceptor contacts in vertebrates (man [2.33 , goldfish [2.253], mudpuppy [2.254], carp [2.255], primates [2.256,274]). Photoreceptors are connected to each other in various ways from the inner segment level (with or without membrane densities) to the synaptic terminal (without the involvement of synaptic vesicles). Photoreceptors are not connected at their outer segments (for an exception, see Sect.2.2.3) nor at their nuclei. Scleral from the OLM the photoreceptors tend to be separated from each other by 1) the glial cells of Müller, but not invariably, 2) the lateral fins, 3) the calycal process, and 4) the microvilli of the pigment epithelium.

Contacts are made in various ways. 1) In teleost fish paired double cones are associated by the close approximation of their mutually apposed inner segments with their underlying subsurface cisterns. The principal and accessory cones of the unequal double cones make contact from their distal inner segments up to their synaptic terminals [2.86,226,275]. 2) There are rod indentations into cone inner segments, with wide separation of membranes

[2.26,37]. 3) Membrane thickenings and densities at inner segment contacts
have been seen by PEDLER and TANSLEY [2.177], MORRIS and SHOREY [2.186],
BOROVYAGIN [2.200], and UGA et al. [2.231]. In the toad there are gap junctions between inner segments [2.219,257]. 4) Most fins make contact with
those from other cells by membrane apposition [2.219]. 5) At the level of
the OLM, photoreceptors may contact each other without the Müller cell
processes intervening [2.276]. These contacts are zonulae adhaerentes [2.151].
There are no electronic coupling (gap) junctions at the outer limiting membrane between photoreceptors and Müller cells [2.277]. 6) Horizontally
oriented lateral expansions from the synaptic terminals make various interreceptor contacts [2.33,86,200,219,257,278]. Cones have long basal processes
which superficially contact adjacent cone cells [2.65] and make shallow de-
pressions on the surface of their synaptic terminals [2.108]. However, many
spherules do not contact any pedicles, and spherules never contact other
spherules [2.231]. 7) Synaptic terminals may abut directly without lateral
expansions [2.12,21,33,86,104,109,186,200,226,231,243].

In general, mammalian cones make many more connections than do rod spherules, and the processes from cone to cone are larger than those from cone to
rod junctions [2.257]. The rods receive many contacts but do not give rise
to lateral processes [2.29]. In the human fovea pedicles are linked to all
adjacent pedicles and in the periphery contact nearby spherules.

The junctions involved take various forms. 1) Some are gap or electronic
junctions, forming a pathway from cell to cell for ions and small molecules
[2.219,221,278]. 2) Some are simple membrane appositions where there are no
intervening glia [2.226]. 3) Desmosomes are sometimes associated with membrane appositions [2.33,240,252]. They provide mechanical anchoring and may
thus help in maintaining in register sets of gap junctions [2.221]. 4) Tight
junctions, of integrading forms from very tight to somewhat leaky, function
as diffusion barriers and prevent lanthanum, for example, from penetrating
the junction.

The retina is an epithelial derivative and may have the same sort of
functional associations as epithelia elsewhere. The interreceptor contacts
may be strengthening and attachment devices, function in assisting the integration of information within the retina between photoreceptors, synapti-
cally and electronically, and play a role in diffusion processes. The widespread occurrence of a large range of interreceptor contacts between photoreceptors suggests that they play an important role in the transfer of information.

2.10 The Foveal Photoreceptors

2.10.1 The Foveal Rods

The fovea contains a great preponderance of cones, but rods are also present (Fig.2.18).

The rodless central area is smaller than the foveal excavation. There are rods on the foveal slope [2.3] and immediately outside the "central bouquet of cones" [2.279], which is the only rod-free area. The measurements given for the rod-free area of the human fovea vary from 50 µm [2.197] to 400 µm [2.3] and 600-800 µm [2.280]. ØSTERBERG said that the first rod is 100-130 µm from the foveal center, and that the rods first appear in small bunches [2.281]. However, POLYAK [2.3] and BORWEIN et al. [2.30] found that, within the fovea, the rods first appear singly and widely scattered; as they gradually increase in frequency they form small chains or groups, until the rods are more numerous than the cones on the foveal slope (the clivus) (Fig.2.18).

About 400 µm from the foveal center each cone is encircled by a row of rods (Fig.18). This region is defined as the parafovea and here the rods easily outnumber the cones [2.77,108]. In the far periphery three to four circles of rods surround each cone [2.77]. At 5-6 mm from the foveal center there is a circular zone where rods reach their maximum concentration.

2.10.2 The Foveal Cones

In the very center of the foveal pit is the foveola, containing the central bouquet of cones with 2,000 [2.3] or 2,500 cones [2.279,280] and no rods. Here are the longest cones, very closely crowded together. Their outer segments, longer than their own inner segments (Fig.2.5), are twice as long as those outside the fovea, They provide a very long path length for the incoming light which increases their sensitivity [2.172]. Unlike cones elsewhere, the scleral ends of the foveal cone outer segments are embedded in the pigment epithelium.

Vitread from this central bouquet of densely packed, thin, elongated cones the inner retinal layers are absent. The long Henle fibers connect the nuclei to their pedicles crowded in a ring around the foveal margin, where the innermost pedicles are on the slope itself [2.280]. These fibers, 2 µm in diameter at the foveal center [2.197], make up the outer fiber layer of Henle and run almost horizontally to the retinal surface (Fig.2.5). The photoreceptor nuclei are nearer the vitreous here than anywhere else in the retina [2.3]. The

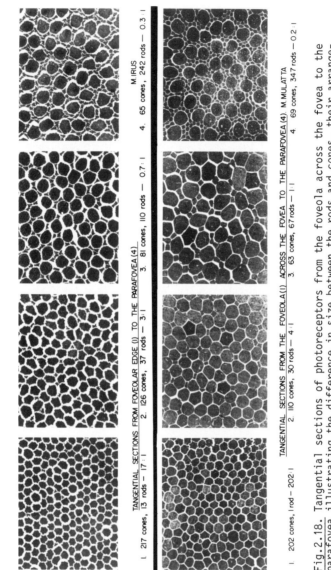

TANGENTIAL SECTIONS FROM FOVEOLAR EDGE (1) TO THE PARAFOVEA (4)

1. 217 cones, 13 rods — 17:1   2. 126 cones, 37 rods — 3:1   3. 81 cones, 110 rods — 0.7:1   4. 65 cones, 242 rods — 0.3:1   M.IRUS

TANGENTIAL SECTIONS FROM THE FOVEOLA(1) ACROSS THE FOVEA TO THE PARAFOVEA (4) M.MULATTA

1. 202 cones, 1 rod — 202:1   2. 110 cones, 30 rods — 4:1   3. 63 cones, 67 rods — 1:1   4. 69 cones, 347 rods — 0.2:1

Fig.2.18. Tangential sections of photoreceptors from the foveola across the fovea to the parafovea illustrating the difference in size between the rods and cones, their arrangements, and numbers. Even within the fovea, towards its margin, rods outnumber cones

innermost pedicles are 100 μm from the foveal center and are found irregularly until they form a continuous layer 800-100 μm from the center, where
the first rod spherules appear [2.280].

The nuclei of the foveal cones are smaller than those of the extrafovea
and are banked in up to 12 layers [2.280], as are the rod nuclei in the extrafoveal regions. Pyknotic nuclei increase in number with age [2.68]. In the
foveolar center itself the cone nuclei form two to three layers [2.280].
Extrafoveally, the cone nuclei form one row directly beneath, or at, the
outer limiting membrane [2.3,30].

The foveolar cones are distinguished by their great length, their small
diameters, and the very small taper of their outer segments. They are closely
packed, with glial cells of Müller between them. They are said to lack the
usual well-defined accumulation of mitochondria in the inner segment ellipsoid [2.197]. They have the highest refractive index of any cones [2.72].
Their inner segments are shorter and only a little thicker than their outer
segments, whereas outside this area the cone inner segments widen towards the
periphery to become considerably thicker than their own outer segments, to
give the cell an overall conical shape. Their morphology alters across the
retina, being longest and thinnest in the fovea and shortest and fattest at
the *ora serrata* [2.3,8,282].

In most of the literature, cone sizes are given without reference to retinal position. The shape, size, packing pattern, and density of the photoreceptors depend on retinal location [2.1,3,30,31,281,283]. Cone density
falls rapidly with distance from the foveal center [2.30,284] as does acuity
[2.281].

There is a rapid change of cone inner segment diameter from the foveola,
across the fovea, and to its periphery [2.30,285] (Table 2.3). As the diameter increases, the cone length decreases and the distance from one cone
center to the neighboring cone center increases. The gradual increase in
cone diameter towards the periphery along all radii is consistent but irregular [2.3,30].

The cone cells are arranged in rows "cutting across one another in three
definite ways at 60° approximately" [2.3].

Cone outer segments become more markedly conical and tapered towards the
periphery. In the literature, primate foveal cone outer segments are repeatedly said to be rodlike and cylindrical, with the explicit or implicit
assumption that they lack any taper. ENOCH reported that central foveal cone
ellipsoids are tapered [2.286]. A slight taper of foveal cone outer segments
has sometimes been reported [2.7,8,280]. According to COHEN it is hard to

Table 2.3.

M. IRUS - Foveal cone dimensions [2.30]

Ratio $\frac{Cones}{Rods}$	Average cone-to-cone-center distances [µm] D	No. of cones [Sq • mm] × 10^3	CIS (Ellipsoid) diameters [µm] d
∞ a	2.2	209	2.1 ± 0.09
∞ a	2.5	159	
∞ a	2.8	123	2.0 ± 0.04
56	2.4	180	1.9 ± 0.05
40	3.3	90	2.1 ± 0.05
21	3.5	81	2.5 ± 0.04
20	3.7	71	2.5 ± 0.04
18	3.5	81	2.4 ± 0.03
15	3.5	81	2.6 ± 0.04
11	3.8	68	2.7 ± 0.04
11	3.7	73	2.5 ± 0.04
11	3.6	77	2.4 ± 0.05
11	3.7	73	3.0 ± 0.10
10	4.0	64	2.8 ± 0.04
8	3.8	68	2.8 ± 0.08
7	3.2	98	2.8 ± 0.08
6	4.3	55	3.0 ± 0.06
3	4.4	51	3.2 ± 0.07
3	4.6	48	3.0 ± 0.05
2	4.6	46	3.3 ± 0.10
2	5.0	40	3.8 ± 0.06
1.5	5,0	40	3.7 ± 0.06
1.4	4.7	45	3.5 ± 0.13
1.4	5.1	39	3.5 ± 0.08
1.3	5.0	40	3.7 ± 0.17
1.2	5.3	36	4.0 ± 0.17
1	5.4	34	4.1 ± 0.15
0.8	5.3	35	4.0 ± 0.10
0.8	5.5	34	4.1 ± 0.19
0.7	5.8	30	4.6 ± 0.13
0.7	6.2	26	5.0 ± 0.18
0.7	5.8	30	4.6 ± 0.07
0.6	6.1	27	4.6 ± 0.09
0.3	6.3	25	4.6 ± 0.09

a ∞ - no rods

rule out a slight tapering of the foveal cone outer segment, and he described it as "minimally tapered or cylindrical" [2.29]. In longitudinal sections of foveal cone outer segments it is very difficult to detect a slight taper as it is practically impossible to section a very long and thin structure through its perfectly central vertical axis. In a study of monkey retinas in which the central foveal cones were sectioned transversely in sequence, from their scleral ends to the outer limiting membrane, it was shown that the inner and

Table 2.4. Foveolar cone dimensions — monkey [2.30]

	M. mulatta		M. irus	
	No. of cones measured	Average diameter [µm]	No. of cones measured	Average diameter [µm]
(1) Inner segment (at Golgi)	480	2.05 ± 0.03	190	2.11 ± 0.03
(2) Inner segment (ellipsoid)	3417	1.73 ± 0.02	68	1.88 ± 0.04
(3) Outer segment (disks only)	264	0.72 ± 0.02	15	1.10 ± 0.09
(4) Outer segment (calycal processes included)	260	1.55 ± 0.02	15	1.79 ± 0.12
(5) Outer segment (ciliary backbone)			289	1.37 ± 0.02
(6) Outer segment (among pigment epithelial processes)	422	0.93 ± 0.02	93	0.98 ± 0.03

outer segments are independently tapered structures, and the entire cell forms a clearly tapered structure (Table 2.4)[2.30].

In the rod-free foveal area each cone cell probably connects to two to three bipolar cells. Here the ratio of bipolars to foveal cones is 2.5:1. There are three types of bipolars, flat midget, invaginating midget, and diffuse cone bipolars; all three of these contact with horizontal cells one foveal cone [2.280,287,288]. In the rhesus fovea, one horizontal cell contacts 6-9 cones, and the ratio increases in a graduated way to 30-40 cones at the periphery [2.288].

The fovea may be abnormal in albinos. Albinos are known to have abnormal visual pathways [2.289].

2.10.3 Embryology and Development of the Fovea

The full adult-form development of the rods and cones in rodents and other lower mammals occurs sometime after birth.

In the Rhesus monkey, rods and cones are present at birth, but are not fully developed until two months after birth [2.290,291]. The outer segment is short at birth, but, while the rod disks appear to be fully formed, the disks in the cones are often vesiculated at their lateral margins and do not reach all the way to the plasma membrane. After birth, the cones of the fovea are the last retinal structure to achieve adult form, density, and orientation [2.291]. The last part of the retina to be developed embryologically is the fovea. At birth a fovea is present and there is considerable visual acuity. It is unpigmented and has a small and shallow excavation with some of the retinal layers present; in the matura fovea the inner retinal layers are absent. At birth the inner nuclear layer and ganglion cell layer are present until the third month although thinned compared to the rest of the retina. Then foveal cone density, better orientation of the outer segments, adult visual acuity, and the adult form, depth, and pigmentation of the fovea develop [2.292,293].

Before and at birth, in *Macaca irus*, the pure cone area is larger than that found 3 months after birth. Starting 50 days before birth the cones are displaced towards the fovea, to concentrate there four fold. After birth the cones elongate *in all regions* as the inner segment narrows to half what it was was at birth. It is known that in monkeys the first few months are the critical period for the development of the effects of visual deprivation [2.293]. In monkeys reared in the dark for 6 months there is no sign that postnatal foveal or photoreceptor development is disturbed or reduced [2.294] nor does lid closure have any such effect [2.295]. It seems, then, that the known visual deprivation syndromes are located in the visual centers and not in the developing photoreceptors.

2.11 The Size, Shape and Density of Rods

Where rods and cones appear together in primates, the cone inner segment is much bulkier than the rod inner segment; the difference in diameter between the cone and rod outer segments is, however, much smaller [2.30,77]. Rods outnumber cones in all regions of the peripheral retina [2.77] and even in the peripheral fovea [2.30]. Only within the central fovea is there a rod-free area surrounded by an area with a preponderance of cones (Fig.2.18). The concentration of cones drops sharply at the parafovea and then declines slowly across the rest of the retina. Within the fovea the rods appear singly at first, then in small chains; eventually they form in the parafovea a circle of a single row of rods around each cone (Fig.2.18). In the periphery the cones are separated from each other and encircled by two to four rows of rods [2.3,30,77].

Rods have been widely assumed to be of the same dimensions everywhere in the retina, but ØSTERBERG [2.281], DUKE-ELDER [2.46], and YOUNG [2.77] draw attention to their change in diameter. They are smaller close to the fovea and widest near the retinal periphery. The rod outer segments are longest at two locations — just beyond the parafovea and near the periphery. Between these two reginons the rods are slightly shorter and wider [2.8]. Based on Laties's preparations ENOCH notes (Chap.4) that rods in the mid-to-far periphery show inner segment enlargement and taper in several species (rabbit, squirrel monkey, etc.). This point should be further evaluated.

The greatest concentration of rods is found immediately beyond the parafovea. YOUNG found in Rhesus monkey that the rod outer segment has 1100 disks in the parafovea, 920 disks in the perifovea, and 790 disks in the periphery [2.77]. Rods appear nearer the temporal side of the foveal center than the nasal side [2.281,284]. The variation of the size of rods is much less than in the cones. Though there is a small change of dimension with retinal position, rods are more nearly the same overall shape across the retina than are the cones. In any one *small* area the cones are similar but not identical [2.3], but the changes in shape and size with retinal position are much more dramatic in cones than in rods.

The boundary between the rod inner segment and outer segment is more or less in the same plane throughout the retina; in the cones this boundary line is more vitreal than in the rods, the inner segment being about 6 μm shorter than the rod inner segment in the posterior part of the primate eye [2.8]. Where rods and cones occur together, the cone outer segment is always shorter than the rod outer segment and often ends at a level where the rod outer segment begins [2.31]. Only in the fovea are the cone outer segments as long as rod outer segments.

2.12 Pigment Epithelial Cells

The pigment epithelium is intimately associated with the metabolism, support, and protection of the photoreceptors [2.296], and with phagocytosis and turnover of outer segment material.

The microvilli of the pigment epithelium serve to separate the receptors. The dark pigment absorbs stray light, and in some species there are back-reflecting tapetal elements present (see Chap.10). Based on analysis of changes in pathology, FITZGERALD et al. claimed that the pigment epithelium plays some role in maintaining alignment [2.70]. It also participates in the regeneration

of photolabile visual pigments, and has a role in the production of the mucopolysaccharides that fill the ventricular space of the retina. On illumination
the free radical content of melanin (in the pigment granules) changes, with
suggests that it may have photoactivity [2.297].

During light adaptation in lower vertebrates, melanin granules of the pigment epithelium migrate into the microvilli between the outer segments of
the photoreceptors. In dark adaptation the melanin granules return to a basal
position. "This distribution could increase visual activity in bright light
by preventing reflection, and heighten sensitivity in low-intensity illumination by permitting reflections" 2.297 . The melanin granules of the pigment
epithelium may act as a photopigment involved in protecting the retina from
sudden bright light by communication, through a system of tight junctions,
directly with the iris muscles 2.297 .

In general, pigment epithelial cells are broader and shorter towards the
periphery. In Rhesus monkey they are three-fold larger in the periphery, while
the area occupied by each pigment epithelial cell is smallest in the fovea.
In the parafovea there are 24-25 rods and 4-5 cones intimately associated with
each pigment epithelial cell, and in the peripheral retina there are 39-45
rods and 2-3 cones to each pigment epithelial cell. Each pigment epithelial
cell daily destroys 2000 disks in the parafovea, 3500 in the perifovea, and
4000 in the periphery [2.77]. In the developing retina the pigment epithelial
pigmentation, which is of crucial significance to retinal function, is the
last element to mature [2.292].

2.13 Comparison of the Morphology of Rods and Cones – A Summary

When carefully examined, no vertebrate retina has yet been found that does
not have both rods and cones, however sparse one of these photoreceptor types
may be.

The rods have long cylindrical outer segments and their diameters, while
not constant across retina, change little compared to the diameter changes
with retinal position in cone outer segments. Both are shorter and wider in
the periphery compared to the central retina. The rod ellipsoids in the midperiphery are wider and taper slightly. The cone outer segment is short and
conical in peripheral retina in most vertebrates. It is elongated and almost
cylindrical in the primate fovea. Overall, the inner segment and outer segment together form a clearly conical structure, and the foveal cone outer
segment taken alone has a small and gradual, but distinct, taper. Where rods

and cones are present together, the cone outer segment is always shorter than the rod outer segment, often ending where the rod outer segment starts.

The rod outer segment disks are variously incised, while the cone outer segment disk has a smooth circumference. Except for a very few disks at the base of the outer segment, rod disks are independent of each other and free floating. Many or most of the cone disks show continuities with the plasma membrane, and thus with each other, and are open to extracellular space. Generally, cone disks are more easily disrupted.

The rod outer segment incorporates labeled aminoacids in two ways, most of them as one distinct band at its vitreal end, but some is diffusely incorporated throughout. In the cone outer segment all the label is incorporated diffusely. Thus, rod disks are replaced by formation of entire disks, one at a time, and cone disks are renewed by replacement of molecules in situ. Cone disks are shed shortly after the onset of the dark period. Rod disks are shed in a burst shortly after the onset of the light period in many species.

Double cones, equal or unequal, are found in all vertebrate groups below that of placental mammals including marsupials. Paired equal double cones are a teleost monopoly; accessory outer segments may also be found, associated with both single and double cones.

In lower vertebrates photomechanical movements of the rods and cones are well established. The cone myoids elongate when dark adapted and shorten when light adapted, while the rod myoids lengthen in the light and shorten in the dark.

The connecting cilium is more complete and longer in the rods than in the cones.

In the cones calycal processes tend to be more numerous, more massive, and better developed, extending over a larger proportion of the length of the outer segment than in the rods.

Compared to rod inner segments, the inner segments of nonfoveal cones are larger in diameter and more barrel shaped, carry a more distinct surface ridging, and contain many more mitochondria in their ellipsoids.

The cross-sectional diameter of the nonfoveal cone inner segment is larger than that of its outer segment, whereas in the rods the two diameters are approximately equal (see Chap.4) and the cell is therefore nearly cylindrical. Cytoplasmic fins extending from the distal inner segment in a narrow zone very near to the OLM are generally better developed in the cones, than in the rods. In many animals below placental mammals the cone inner segment is likely to bear an oil droplet at its distal end.

In general, the cone nuclei lie in one layer at the level of the OLM, and are larger, more elongated, sometimes even lobed, and paler staining, and have more dispersed "tigroid" chromatin and a longer inner fiber than that of the rod. The rod nuclei are spherical, tend to be banked in several layers, and are surrounded by very little cytoplasm. The frog cone nuclei are exceptional in lying closer to the outer plexiform layer than do the rod nuclei.

The synaptic terminals are different sizes and shapes. The larger cone pedicle is usually conical with a wide base and looks paler than the rod spherule, which is spherical or ovoid. The pedicle contains many more invaginated units than the spherule.

Cones vary in size and shape with position in the retina in a graduated way, being most numerous, elongate, and slender in the fovea and shorter and squat near the *ora serrata*. Rods are thinner near the central retina, but their change of dimension with retinal position is small by comparison with the cones.

References

2.1   S. Polyak: *The Retina* (Univ. Chicago Press, Chicago 1941)
2.2   W. Bloom, D.W. Fawcett: *A Textbook of Histology*. 8th ed. (Saunders, Philadelphia 1962)
2.3   S. Polyak: *The Vertebrate Visual System* (Univ. Chicago Press, Chicago 1957)
2.4   H. Müller: Zur Histologie der Netzhaut. Z. Wiss. Zool. *3*, 234-237 (1851)
2.5   H. Müller: Über einige Verhältnisse der Netzhaut bei Menschen und Tieren. Verhand. Phys. Med. Ges. Würzburg *4*,96-100 (1853)
2.6   H. Müller: Anatomisch-physiologische Untersuchungen über die Retina des Menschen und der Wirbeltiere. Z. Wiss. Zool. *8*, 1 (1857)
2.7   W.S. Duke-Elder: *The Eye in Evolution*, System of ophthalmology, Vol.I (Kimpton, London 1958)
2.8   M. Schultze: "The Retina", in *Manual of Human and Comparative Histology*, Vol.III, ed. by S. Stricker (New Syndenham Soc., London 1873) pp.218-298
2.9   M. Schultze: Zur Anatomie und Physiologie der Retina. Arch. Mikrosk. Anat. *2*, 175-286 (1866)
2.10 M. Schultze: Über Stäbchen und Zapfen der Retina. Arch. Mikrosk. Anat. *3*, 215-247 (1867)
2.11 M. Schultze: Neue Beiträge zur Anatomie und Physiologie der Retina des Menschen. Arch. Mikrosk. Anat. *7*, 244-259 (1872)
2.12 A.I. Cohen: Vertebrate retinal cells and their organization. Biol. Rev. Cambridge Philos. Soc. *38*, 427-459 (1963)
2.13 A.I. Cohen: "Rods and Cones and the Problem of Visual Excitation", in *The Retina, Morphology, Function and Clinical Characteristics*, ed. by B.R. Straatsma, M.O. Hall, R.A. Allen, F. Crescitelli (Univ. California Press, Berkeley 1969) pp.31-62
2.14 H.J.A. Dartnall, K. Tansley: Physiology of vision: retinal structure and visual pigments. Annu. Rev. Physiol. *25*, 433-458 (1963)

2.15 R.F. Dunn: "The Ultrastructure of the Vertebrate Retina", in *The Ultrastructure of Sensory Organs*, ed. by I. Friedmann (Elsevier, New York 1973) pp.155-222

2.16 R.W. Young: The Organization of Vertebrate Photoreceptor cells", in *The Retina, Morphology, Function and Clinical Characteristics*, ed. by B.R. Straatsma, M.O. Hall, R.A. Allen, F. Crescitelli (Univ. California Press, Berkeley 1969) pp.177-210

2.17 C. Pedler: "Duplicity Theory and Microstructure of the Retina. Rods and Cones — a New Approach", in *Color Vision*, ed. by A.V.S. de Reuck, J. Knight (Little Brown, Boston 1965) pp.52-88

2.18 A. Hughes: "The Topography of Vision in Mammals of Contrasting Life Styles: Comparative Optics and Retinal Organization", in *The Visual System in Vertebrates*, Handbook of Sensory Physiology, Vol.7/5, ed. by F. Crescitelli (Springer, Berlin, Heidelberg, New York 1977) pp.613-756

2.19 R.G. Murray, A.E. Jones, A. Murray: Fine structure of photoreceptors in the owl monkey. Anat. Rec. *175*, 673-696 (1973)

2.20 J.E. Dowling: "Structure and Function in the All-Cone Retina of the Ground Squirrel", Symposium on the Physiological Basis for Form Discrimination (Brown University, Providence, RI 1964) pp.17-23

2.21 M.J. Hollenberg, M.H. Bernstein: Fine structure of the photoreceptor cells of the ground squirrel (*Citellus tridecemlineatus tridecemlineatus*). Am J. Anat. *118*, 359-374 (1966)

2.22 T. Samorajski, J.M. Ordy, J.R. Keefe: Structural organization of the retina in the tree shrew (*Tupaia glis*). J. Cell Biol. *28*, 489-504 (1966)

2.23 G.H. Jacobs, S.K. Fisher, D.H. Anderson, M.S. Silverman: Scotopic and photopic vision in the California ground squirrel: physiological and anatomical evidence. J. Comp. Neurol. *165*, 209-228 (1976)

2.24 D.H. Anderson, S.K. Fisher: The photoreceptors of diurnal squirrels: Outer segment structure, disk shedding and protein renewal. J. Ultrastruct. Res. *55*, 119-141 (1976)

2.25 R.W. West, J.E. Dowling: Anatomical evidence for cone and rodlike receptors in the grey squirrel, ground squirrel, and prairie dog retinas. J. Comp. Neurol. *159*, 439-460 (1975)

2.26 W.K. Stell: "The Morphological Organization of the Vertebrate Retina", in *Physiology of Photoreceptor Organs*, Handbook of Sensory Physiology, Vol.7/2, ed. by M.G.F. Fuortes (Springer, Berlin, Heidelberg, New York 1972) pp.111-213

2.27 K. Holmberg: The Cyclostome Retina", in *The Visual System in Vertebrates*, Handbook of Sensory Physiology, Vol.7/5, ed. by F. Crescitelli (Springer, Berlin, Heidelberg, New York 1977) pp.47-66

2.28 A.I. Cohen: Some observations on the fine structure of the retinal receptors of the American gray squirrel. Invest. Ophthalmol. *3*, 198-216 (1964)

2.29 A.I. Cohen: "Rods and Cones", in *Physiology of Photoreceptor Organs*, Handbook of Sensory Physiology, Vol.7/2, ed. by M.G.F. Fuortes (Springer, Berlin, Heidelberg, New York 1972) pp.63-110

2.30 B. Borwein, D. Borwein, J. Medeiros, J. McGowan: The ultrastructure of monkey foveal photoreceptors, with special reference to the structure, shape, size and spacing of the foveal cones. Am. J. Anat. *159*, 125-146 (1980)

2.31 W.H. Miller, A.W. Snyder: Optical function of human peripheral cones. Vision Res. *13*, 2185-2194 (1973)

2.32 W. Krause: Über den Bau der Retina-Stäbchen beim Menschen. Z. Rat. Med., 3. Reihe *11*, 175 (1861)

2.33 L. Missotten: *The Ultrastructure of the Human Retina* (Arscia, Brussels 1965)

2.34 E. De Robertis: Electron microscope observations on the submicroscopic organization of the retinal rods. J. Biophys. Biochem. Cytol. *2*, 319-330 (1956)

2.35 G.L. Walls: *The Vertebrate Eye and Its Adaptive Radiation* (Cranbrook, Bloomfield Hills 1942)

2.36 C.R. Braekevelt: Fine structure of the retinal pigment epithelium and photoreceptor cells of an Australian marsupial *Setonix brachyurus*. Can. J. Zool. *51*, 1093-1111 (1972)

2.37 B. Borwein, M.J. Hollenberg: The photoreceptors of the "four-eyed" fish, *Anableps anableps L.* J. Morph. *140*, 405-442 (1973)

2.38 E.R. Berger: Subsurface membranes in paired cone photoreceptor inner segments of adult and neonatal *Lebistes* retinae. J. Ultrastruct. Res. *17*, 220-232 (1967)

2.39 B.A. Collins, E.F. MacNichol: Triple cones found in retinas of 3 fish species. Experientia *35*, 106-108 (1979)

2.40 K. Engstrom: Structure, organization and ultrastructure of the visual cells in the teleost family, *Labridae*. Acta. Zool. *44*, 1-41 (1963)

2.41 F.S. Sjöstrand, L.G. Elfvin: "Some Observations on the Structure of the Retinal Receptors of the Toad Eye as Revealed by the Electron Microscope", in *Electron Microscopy*. Proceedings, Stockholm Conf. 1956, ed. by F.S. Sjöstrand, J. Rhodin (Academic, New York 1957) pp.194-196

2.42 H.J. Wagner, M.A. Ali: Retinal organization in goldeye and mooneye (*Teleostei Hiodontidae*). Rev. Can. Biol. *37*, 65-83 (1978)

2.43 N.A. Locket: Deep-sea fish retinas. Br. Med. Bull. *26*, 107-111 (1970)

2.44 B.A. Fineran, J.A.C. Nicol: Studies on the photoreceptors of *Anchoa mitchelli* and *A. hepsetus* (*Engraulidae*) with particular reference to the cones. Philos. Trans. R. Soc. London B*283*, 25-60 (1978)

2.45 I.-B. Ahlbert: Ontogeny of double cones in the retina of perch fry (*Perca fluviatilis*, Teleostei). Acta Zool. Stockholm *54*, 241-254 (1973)

2.46 W.S. Duke-Elder, K.C. Wybar: *The Anatomy of the Visual System*, System of Ophthalmology, Vol.II (Kimpton, London 1961)

2.47 G. Underwood: Reptilian retinas. Nature London *167*, 183-185 (1951)

2.48 S.E.G. Nilsson: Receptor cell outer segment development and ultrastructure of the disk membranes in the retina of the tadpole (*Rana pipiens*). J. Ultrastruct. Res. *11*, 581-620 (1964)

2.49 R.F. Dunn: Studies in the retina of the gecko *Coleonyx variegatus* II. The rectilinear visual cell mosaic. J. Ultrastruct. Res. *16*, 672-678 (1966)

2.50 T.H. Waterman, R.B. Forward: Field evidence for polarized light sensitivity in the fish *Zenarchopterus*. Nature London *228*, 85-87 (1970)

2.51 L.B. Arey: The occurrence and the significance of photomechanical changes in the vertebrate retina — an historical survey. J. Comp. Neurol. *25*, 535-554 (1915)

2.52 M.A. Ali: Les réponses rétinomotrices: caractères et mécanismes. Vision Res. *11*, 1225-1288 (1971)

2.53 C.P. O'Connell: The structure of the eye of *Sardinops caerulea*, *Engraulis mordax*, and four other pelagic marine teleosts. J. Morphol. *113*, 287-323 (1963)

2.54 B. Burnside: Thin (actin) and thick (myosinlike) filaments in cone contraction in the teleost retina. J. Cell Biol. *78*, 227-246 (1978)

2.55 R.L. Murray, M.W. Dubin: The occurrence of actin-like filaments in association with migrating pigment granules in frog retinal pigment epithelium. J. Cell. Biol. *64*, 705-710 (1975)

2.56 I. Pietzsch-Rohrschneider: Scanning electron microscopy of photoreceptor cells in the light and dark adapted retina of *Haplochromis burtoni* (Cichlidae, Teleostei). Cell Tissue Res. *175*, 123-130 (1976)

2.57 J.A.C. Nicol: Retinomotor changes in flat fishes. J. Fish. Res. Board Can. *22*, 513-520 (1965)

2.58 J.H.S. Blaxter, M.P. Jones: The development of the retina and retino-motor responses in the herring. J. Mar. Biol. Assoc. U.K. *47*, 677-697 (1967)

2.59 L.B. Arey: The movements in the visual cells and retinal pigment of the lower vertebrates. J. Comp. Neurol. *26*, 121-202 (1916)

2.60 G.A. Bubenik, G.M. Brown, L.G. Grota: Differential localization of N-acetylated indole alkylamines in CNS and the Harderian gland using immunohistology. Brain Res. *118*, 417-427 (1976)

2.61 S.S. Easter, A. Macy: Local control of retinomotor activity in the fish retina. Vision Res. *18*, 937-942 (1978)

2.62 A.M. Laties: Histological techniques for study of photoreceptor orientation. Tissue Cell *1*, 63-81 (1969)

2.63 J.M. Enoch: Retinal receptor orientation and the role of fiber optics in vision. The Glenn Fry Lecture. Am. J. Optom. Arch. Am. Acad. Optom. *49*, 455-471 (1972)

2.64 J.M. Enoch, G.M. Hope: An analysis of retinal receptor orientation: III Results of initial psychophysical tests. Invest. Ophthalmol. *11*, 765-782 (1972); Directional sensitivity of the foveal and parafoveal retina. Invest. Ophthalmol. *12*, 497-503 (1973)

2.65 J. Marshall: The Retinal Receptors and the Pigment Epithelium", in *Scientific Foundations of Ophthalmology*, ed. by E.S. Perkins, D.W. Hill (Heinemann Med. Books, London 1977) pp.8-17

2.66 A.M. Laties, P.A. Liebman, C.E.M. Campbell: Photoreceptor orientation in the primate eye. Nature London *218*, 172-173 (1968)

2.67 A.M. Laties, J.M. Enoch: An analysis of retinal receptor orientation. I. Angular relationship of neighboring photoreceptors. Invest.Ophthalmol. *10*, 69-77 (1971)

2.68 J. Marshall, J. Grindle, P.L. Ansell, B. Borwein: Convolution in human rods: an aging process. Br. J. Ophthalmol. *63*, 181-187 (1979)

2.69 J.M. Enoch, D.G. Birch: Evidence for alteration in photoreceptor orientation. Ophthalmologica (in press)

2.70 C.R. Fitzgerald, J.M. Enoch, D.G. Birch, M.D. Benedetto, L.A. Temme, W.W. Dawson: Anomalous pigment epithelial photoreceptor relationships and receptor orientation. Invest. Ophthalmol. *19*, in press (1980)

2.71 G.M. Hope: Graded differential photoreceptor orientation: ramifications for ultramicrotomy of retina. Stain Technol. *54*, 205-211 (1979)

2.72 R.L. Sidman: The structure and concentration of solids in photoreceptor cells studied by refractometry and interference microscopy. J. Biophys. Biochem. Cytol. *3*, 15-30 (1957)

2.73 J.M. Enoch, J. Scandrett, F.L. Tobey: A study of the effects of bleaching on the width and index of refraction of frog rod outer segments. Vision Res. *13*, 171-183 (1973)

2.74 J.M. Enoch, F.L. Tobey, Jr.: Use of the waveguide parameter V to determine the difference in the index of refraction between the rat rod outer segment and the interstitial matrix. J. Opt. Soc. Am. *68*, 1130-1134 (1978)

2.75 G.L. Walls: Human rods and cones. Arch. Ophthalmol. *12*, 914-930 (1934)

2.76 M.J. Hogan, I. Wood, R. Steinberg: Cones of human retina: phagocytosis by pigment epithelium of human retinal cones. Nature London *252*, 305-307 (1974)

2.77 R.W. Young: The renewal of rod and cone outer segments in the rhesus monkey. J. Cell Biol. *49*, 303-318 (1971)

2.78 D.H. Anderson, S.K. Fisher, R.H. Steinberg: Mammalian cones: disk shedding, phagocytosis and renewal. Invest. Ophthalmol. Visual Sci. *17*, 117-133 (1978)

2.79 M.O. Hall, J. Heller: "Mucopolysaccharides of the Retina", in *The Retina, Morphology, Function and Clinical Characteristics*, ed. by B.R.

Straatsma, M.O. Hall, R.A. Allen, F. Crescitelli (Univ. California Press, Berkeley 1969) pp.211-224

2.80 E.R. Berman: The biosynthesis of mucopolysaccharides and glycoproteins in pigment epithelial cells of bovine retina. Biochem. Biophys. Acta *83*, 371-373 (1964)

2.81 E.R. Berman: Isolation of neutral sugar containing mucopolysaccharides from cattle retina. Biochem. Biophys. Acta *101*, 358-360 (1965)

2.82 A.J. Kroll, R. Machemer: Experimental retinal detachment and re-attachment in the Rhesus monkey. Electron microscopic comparison of rods and cones. Am. J. Ophthalmol. *68*,58-77 (1969)

2.83 F.S. Sjöstrand: An electron microscope study of the retinal rods of the guinea pig eye. J. Cell. Comp. Physiol. *33*, 383-403 (1949)

2.84 F.S. Sjöstrand: The ultrastructure of the outer segments of rods and cones of the eye as revealed by the electron microscope. J. Cell. Comp. Physiol. *42*, 15-44 (1953)

2.85 F.S. Sjöstrand: Electron microscopy of the retina. Anat. Rec. *136*, 278 (Abstract) (1960)

2.86 A.I. Cohen: The fine structure of the visual receptors of the pigeon. Exp. Eye Res. *2*, 88-97 (1963)

2.87 T. Kuwabara, R.A. Gorn: Retinal damage by visible light: an electron microscopic study. Arch. Ophthalmol. *79*, 69-78 (1968)

2.88 A. Grignola, N. Orzalesi, R. Castellazzo, P. Vittone: Retinal damage by visible light in albino rats: An electron microscope study. Ophthalmologica *157*, 43-59 (1969)

2.89 J.E. Dowling, J.R. Gibbons: "The Effect of Vitamin A Deficiency on the Fine Structure of the Retina", in *The Structure of the Eye*, ed. by G.K. Smelser (Academic, New York 1961) pp.85-99

2.90 K.C. Hayes: Retinal degeneration in monkeys induced by deficiences of vitamin E or A. Invest. Ophthalmol. *13*, 499-510 (1974)

2.91 J. Marshall: Thermal and mechanical mechanisms in laser damage to the retina. Invest. Ophthalmol. *9*, 97-115 (1970)

2.92 D.O. Adams, E.S. Beatrice, R.B. Bedell: Retina: Ultrastructural alterations produced by extremely low levels of coherent radiation. Science *177*, 58-60 (1972)

2.93 G. Falk, P. Fatt: Changes in structure of the disks of retinal rods in hypotonic solutions. J. Cell Sci. *13*, 787-799 (1973)

2.94 F.S. Sjöstrand: The ultrastructure of the retinal receptors of the vertebrate eye. Ergeb. Biol. *21*,128-160 (1959)

2.95 G. Wald, P.K. Brown, I.R. Gibbons: The problem of visual excitation. J. Opt. Soc. Am. *53*, 20-35 (1963)

2.96 E.A. Dratz, G.P. Miljanich, P.P. Nemes, J.E. Gaw, S. Schwartz: The structure of rhodopsin and its deposition in the rod outer segment membrane. Photochem. Photobiol. *29*, 661-670 (1979)

2.97 S. Basinger, D. Bok, M. Hall: Rhodopsin in the rod outer segment plasma membrane. J. Cell Biol. *69*, 29-42 (1976)

2.98 E. De Robertis, A. Lasansky: Submicroscopic organization of retinal cones of the rabbit. J. Biophys. Biochem. Cytol. *4*, 743-746 (1958)

2.99 E. De Robertis, A. Lasansky: "Ultrastructure and Chemical Organization of Photoreceptors", in *The Structure of the Eye*, ed. by G.K. Smelser (Academic, New York 1961) pp.29-49

2.100 G. Falk, P. Fatt: Distinctive properties of the lamellar and disk-edge structures of the rod outer segment. J. Ultrastruct. Res. *28*, 41-60 (1969)

2.101 F.S. Sjöstrand, M. Kreman: Molecular structure of outer segment disks in photoreceptor cell. J. Ultrastruct. Res. *65*, 195-226 (1978)

2.102 L. Missotten: L'ultrastructure des cones de la rétine humaine. Bull. Soc. Belge Ophthalmol. *132*, 472-502 (1963)

2.103   J.E. Dowling: Foveal receptors of the monkey retina: fine structure. Science *147*, 57-59 (1965)

2.104   A.I. Cohen: The ultrastructure of the rods of the mouse retina. Am. J. Anat. *107*, 23-48 (1960)

2.105   N.A. Locket: The retina of *Poromitra nigrofulvus* (Garman). Exp. Eye Res. *8*, 265-275 (1969)

2.106   N.A. Locket: Retinal structure in *Platytroctes apus*, a deep-sea fish with a pure rod fovea. J. Mar. Biol. Assoc. U.K. *51*, 79-91 (1971)

2.107   A.I. Cohen: New details of the ultrastructure of the outer segments and ciliary connectives of the rods of human and macaque retinas. Anat. Rec. *152*, 63-80 (1965)

2.108   M.J. Hogan, J.A. Alvarado, J.E. Weddell: *Histology of the Human Eye.* (Saunders, Philadelphia 1971)

2.109   A.I. Cohen: The fine structure of the extrafoveal receptors of the rhesus monkey. Exp. Eye Res. *1*, 128-136 (1961)

2.110   A.J. Kroll, R. Machemer: Experimental retinal detachment in the owl monkey. III. Electron microscopy of retina and pigment epithelium. Am. J. Ophthalmol. *66*, 410-427 (1968)

2.111   P.K. Brown, I.R. Gibbons, G. Wald: The visual cells and visual pigment of the mudpuppy *Necturus*. J. Cell Biol. *19*, 79-106 (1963)

2.112   S.E.G. Nilsson: The ultrastructure of the receptor outer segments in the retina of the leopard frog (*Rana pipiens*). J. Ultrastruct. Res. *12*, 207-231 (1965)

2.113a   D.H. Dickson, M.J. Hollenberg: The fine structure of the pigment epithelium and the photoreceptor cells of the newt, *Triturus viridescens dorsalis*. J. Morphol. *135*, 389-432 (1971)

2.113b   B.A. Fineran, J.A.C. Nicol: Studies on the eyes of New Zealand parrot-fishes (*Labridae*). Proc. R. Soc. London B*186*, 217-247 (1974)

2.114   F.S. Sjöstrand: Fine structure of cytoplasm: the organization of membranous layers. Rev. Mod. Phys. *31*, 301-318 (1959)

2.115   R.H. Steinberg, I. Wood: Clefts and microtubules of photoreceptor outer segments in the retina of the domestic cat. J. Ultrastruct. Res. *51*, 397-403 (1975)

2.116   R.W. Young: Visual cells and the concept of renewal. Invest. Ophthalmol. *15*, 700-725 (1976)

2.117   L. Missotten: Étude des bâtonnets de la rétine humaine au microscope électronique. Ophthalmologica *140*, 200-214 (1960)

2.118   R.W. Young: A difference between rods and cones in the renewal of outer segment protein. Invest. Ophthalmol. *8*, 222-231 (1969)

2.119   A.I. Cohen: New evidence supporting the linkage to extracellular space of outer segment saccules of frog cones but not rods. J. Cell Biol. *37*, 424-444 (1968)

2.120   A.I. Cohen: "Some Preliminary Microscopic Observations on the Outer Receptor Segments of the Retina of the *Macaca rhesus*", in *The Structure of the Eye*, ed. by G.K. Smelser (Academic, New York 1961) pp.151-158

2.121   A.I. Cohen: Further studies on the question of the patency of saccules in outer segments of vertebrate photoreceptors. Vision Res. *10*, 445-453 (1970)

2.122   A.M. Laties, P.A. Liebman: Cones of living amphibian eye: Selective staining. Science *168*, 1475-1477 (1970)

2.123   A.W. Clark, D. Branton: Fracture faces in frozen outer segments from the guinea pig retina. Z. Zellforsch. Mikrosk. Anat. *91*, 586-603 (1968)

2.124   A. Bairati, Jr., N. Orzalesi: The ultrastructure of the pigment epithelium and of the photoreceptor-pigment epithelium junction in the human retina. J. Ultrastruct. Res. *9*, 484-496 (1963)

2.125   T. Ishikawa, E. Yamada: Atypical mitochondria in the ellipsoid of the photoreceptor cells of vertebrate retinas. Invest. Ophthalmol. *8*, 302-316 (1969)

2.126 R.W. Young: Shedding of disks from rod outer segments in the rhesus monkey. J. Ultrastruct. Res. *34*, 190-203 (1971)

2.127 R.W. Young: Biogenesis and renewal of visual cell outer segment membranes. Exp. Eye Res. *18*, 215-223 (1974)

2.128 R.W. Young, D. Bok: Participation of the retinal pigment epithelium in the rod outer segment renewal process. J. Cell Biol. *42*, 392-403 (1969)

2.129 R.W. Young, D. Bok: Autoradiographic studies on the metabolism of the retinal pigment epithelium. Invest. Ophthalmol. *9*, 524-536 (1970)

2.130 R.W. Young, B. Droz: The renewal of protein in retinal rods and cones. J. Cell Biol. *39*, 169-184 (1968)

2.131 B. Droz: Dynamic condition of proteins in the visual cells of rats and mice as shown by autoradiography with labelled aminoacids. Anat. Rec. *145*, 157-167 (1963)

2.132 R.W. Young: Passage of newly formed protein through the connecting cilium of retinal rods in the frog. J. Ultrastruct. Res. *23*, 462-473 (1968)

2.133 R.W. Young: The renewal of photoreceptor cell outer segments. J. Cell Biol. *33*, 61-72 (1967)

2.134 R.W. Young: An hypothesis to account for a basic distinction between rods and cones. Vision Res. *11*, 1-5 (1971)

2.135 D. Bok, R.W. Young: The renewal of diffusely distributed protein in the outer segments of rods and cones. Vision Res. *12*, 161-168 (1972)

2.136 C.E. Remé, R.W. Young: The effects of hibernation on cone visual cells in the ground squirrel. Invest. Ophthalmol. Vis. Sci. *16*, 815-840 (1977)

2.137 M.O.M. Tso: Photic maculopathy in the rhesus monkey. Invest. Ophthalmol. *12*, 17-34 (1973)

2.138 M. Ditto: A difference between developing rods and cones in the formation of outer segment membranes. Vision Res. *15*, 535-536 (1975)

2.139 J.C. Besharse, J.G. Hollyfield: Removal of normal and degenerating photoreceptor outer segments in the Ozark cave salamander. J. Exp. Zool. *198*, 287-352 (1976)

2.140 M.S. Kinney, S.K. Fisher: The photoreceptors and pigment epithelium of the larval *Xenopus* retina: morphogenesis and outer segment renewal. Proc. R. Soc. London B*201*, 149-167 (1978)

2.141 J.G. Hollyfield, M.E. Rayborn: Photoreceptor outer segment development: light and dark regulate the rate of membrane addition and loss. Invest. Ophthalmol. Vis. Sci. *18*, 117-132 (1979)

2.142 N. Buyukmihci, G.D. Aguirre: Rod disk turnover in the dog. Invest. Ophthalmol. *15*, 579-584 (1976)

2.143 M.M. Lavail: Rod outer segment disk shedding in rat retina: relationship to cyclic lighting. Science *194*, 1071-1074 (1976)

2.144 S. Basinger, R. Hoffman, M. Matthes: Photoreceptor shedding is initiated by light in the frog retina. Science *194*, 1074-1076 (1976)

2.145 J.G. Hollyfield, J.C. Besharse, M.E. Rayborn: The effect of light on the quantity of phagosomes in the pigment epithelium. Exp. Eye Res. *23*, 623-635 (1976)

2.146 W.T. O'Day, R.W. Young: Rhythmic daily shedding of outer segment membranes by visual cells in the goldfish. J. Cell. Biol. *76*, 593-604 (1978)

2.147 R.H. Steinberg, I. Wood, M.J. Hogan: Pigment epithelial ensheathment and phagocytosis of extrafoveal cones in human retina. Phil. Trans. R. Soc. London B*277*,459-474 (1977)

2.148 R.H. Steinberg, I. Wood: Pigment epithelial cell ensheathment of cone outer segments in the retina of the domestic cat. Proc. R. Soc. London B*187*, 461-478 (1974)

2.149 R.W. Young: The daily rhythm of shedding and degradation of rod and cone outer segment membranes in the chick retina. Invest. Ophthalmol. Vis. Sci. *17*, 105-116 (1978)

2.150 R.W. Young: The daily rhythm of shedding and degradation of cone outer segment membranes in the lizard retina. J. Ultrastruct. Res. *61*. 172-185 (1977)

2.151 M. Spitznas, M.J. Hogan: Outer segments of photoreceptors and the retinal pigment epithelium. Interrelationship in the human eye. Arch. Ophthalmol. *84*, 810-819 (1970)

2.152 P.J. O'Brien: Rhodopsin as a glycoprotein: a possible role for the oligosaccharide in phagocytosis. Exp. Eye Res. *23*, 127-137 (1976)

2.153 R.M. Eakin: Differentiation of rods and cones in total darkness. J. Cell. Biol. *25*, 162-165 (1965)

2.154 R.E. Anderson, R.M. Benolken, P.A. Dudley, D.J. Landis, T.G. Wheeler: Polyunsaturated fatty acids of photoreceptor membranes. Exp. Eye Res. *18*, 205-213 (1974)

2.155 J.G. Hollyfield, J.C. Besharse, M.E. Rayborn: Turnover of rod photoreceptor outer segments. I. Membrane addition and loss in relationship to temperature. J. Cell Biol. *75*, 490-506 (1977)

2.156 J.C. Besharse, J.G. Hollyfield, M.E. Rayborn: Turnover of rod photoreceptor outer segments. II. Membrane addition and loss in relationship to light. J. Cell Biol. *75*, 507-527 (1977)

2.157 J.C. Besharse, J.G. Hollyfield, M.E. Rayborn: Photoreceptor outer segments: accelerated membrane renewal in rods after exposure to light. Science *196*, 536-538 (1977)

2.158 J. Axelrod: The pineal gland: a neurochemical transducer. Science *184*, 1341-1348 (1974)

2.159 M. Tamai, P. Teirstein, A. Goldman, P. O'Brien, G. Chader: The pineal gland does not control rod outer segment shedding and phagocytosis in the rat retina and pigment epithelium. Invest. Ophthalmol. Vis. Sci. *17*, 558-562 (1978)

2.160 M. Tamai, G.J. Chader: The early appearance of disk shedding in the rat retina. Invest. Ophthalmol. Vis. Sci. *18*, 913-917 (1979)

2.161 J.R. Currie, J.G. Hollyfield, M.E. Rayborn: Rod outer segments elongate in constant light: darkness is required for normal shedding. Vision Res. *18*, 995-1003 (1978)

2.162 P.S. Teirstein, P.J. O'Brien, A.I. Goldman: Nonsystemic regulation of rat rod outer segment disk shedding. Invest. Ophthalmol. Vis. Sci. *17* (Arvo Suppl.), 134 (1978)

2.163 J.G. Hollyfield, S.F. Basinger: Photoreceptor shedding can be initiated within the eye. Nature London *274*, 794-796 (1978)

2.164 W.B. Quay: Retinal and pineal hydroxyindole-o-methyl transferase activity in vertebrates. Life Sci. *4*, 983-991 (1965)

2.165 D.P. Cardinale, J.M. Rosner: Retinal localization of the hydroxyindole-o-methyl transferase (HIOMT) in the rat. Endocrinology *89*, 301-303 (1971)

2.166 A.W. Spira, P.T. Huang: Phagocytosis of photoreceptor outer segments during retinal development in utero. Am. J. Anat. *152*, 523-528 (1978)

2.167 H.J. Arnott, A.C.G. Best, S. Ito, J.A.C. Nicol: Studies on the eyes of catfishes with special reference to the tapetum lucidum. Proc. R. Soc. London B*186*, 13-36 (1974)

2.168 W. Stell: Some ultrastructural characteristics of goldfish retinal cones. Am Zool. *5*, 716-717 Abstr. 435 (1965)

2.169 C.R. Braekevelt: Photoreceptor fine structure in the northern pike. (*Esox lucius*). J. Fish. Res. Board Can. *32*, 1711-1721 (1975)

2.170 H.H. Dathe: Vergleichende Untersuchungen an der Retina mitteleuropäischer Süßwasserfische. Z. Mikrosk. Anat. Forsch. *80*, 269-319 (1969)

2.171 A.W. Snyder, M. Hamer: The light capture area of a photoreceptor. Vision Res. *12*, 1749-1753 (1972)

2.172 C. Pedler: Rods and cones — a new approach. Int. Rev. Gen. Exp. Zool. *4*, 219-274 (1969)

2.173a F.S. Sjöstrand: The ultrastructure of the inner segment of the retinal rods of the guinea pig eye as revealed by electron microscopy. J. Cell Comp. Physiol. *42*, 45-70 (1953)

2.173b R. Röhlich: The sensory cilium of retinal rods is analogous to the transitional zone of motile cilia. Cell Tissue Res. *161*, 421-430 (1975)

2.174 D.W. Fawcett: *An Atlas of Fine Structure. The Cell, Its Organelles and Inclusions* (Saunders, Philadelphia 1966)

2.175 J.M. Richardson: Cytoplasmic and ciliary connections between the inner and outer segments of mammalian visual receptors. Vision Res. *9*, 727-731 (1969)

2.176 A.H. Bunt: Fine structure and radioautography of rabbit photoreceptor cells. Invest. Ophthalmol. Visual Sci. *17*, 90-104 (1978)

2.177 C. Pedler, K. Tansley: Fine structure of the cones of a diurnal gecko (*Phelsuma inunguis*). Exp. Eye Res. *2*, 39-47 (1963)

2.178a C. Pedler, R. Tilly: Ultrastructural variations in the photoreceptors of the Macaque. Exp. Eye Res. *4*, 370-373 (1965)

2.178b C. Pedler, R. Tilly: The nature of the gecko visual cell. A light and electron microscopic study. Vision Res. *4*, 499-510 (1964)

2.179 K. Tokuyasu, E. Yamada: The fine structure of the retina. V. Abnormal retinal rods and their morphogenesis. J. Biophys. Biochem. Cytol. *7*, 187-190 (1960)

2.180 T. Matsusaka: "Fine Structure of the Connecting Cilium in the Rat Eye", in *The Structure of the Eye*, ed. by E. Yamada, S. Mishima (Univ. Tokyo, Tokyo 1975) pp.261-271

2.181 H.A. Hansson: Scanning electron microscopy of the rat retina. Z. Zellforsch. Mikrosk. Anat. *107*, 23-44 (1970)

2.182 T. Matsusaka: Cytoplasmic fibrils of the connecting cilium. J. Ultrastruct. Res. *54*, 318-324 (1976)

2.183 J.H. Luft: Ruthenium red and violet I. Chemistry, purification, methods of use for electron microscopy and mechanism of action. Anat. Rec. *171*, 347-368 (1971);
J.H. Luft: Ruthenium red and violet.II. Fine structural localization in animal tissues. Anat. Rec. *171*, 369-415 (1971); and
B. Szubinska, J.H. Luft: Ruthenium red and violet. III. Fine structure of the plasma membrane and extraneous coats in amoebae (A. proteus and Chaos chaos). Anat. Rec. *171*, 417-441 (1971)

2.184 K. Tokuyasu, E. Yamada: The fine structure of the retina studied with the electron microscope. IV. Morphogenesis of outer segments of retinal rods. J. Biophys. Biochem. Cytol. *6*, 225-230 (1959)

2.185 J.N.H. Anh: Ultrastructure des recepteurs visuels chez les vertébrès. Arch. Ophthal. *29*, 795-822 (1969)

2.186 V.B. Morris, D.C. Shorey: An electron microsope study of the types of receptors in the chick retina. J. Comp. Neurol. *129*, 313-349 (1967)

2.187 R.L. Sidman, G.B. Wislocki: Histochemical observations on rods and cones in retinas of vertebrates. J. Histochem. Cytochem. *2*, 413-433 (1954)

2.188 E. Yamada: Observations on the fine structure of photoreceptor elements in the vertebrate eye. J. Electron Microsc. *9*, 1-14 (1960)

2.189 R.F. Dunn: Studies on the retina of the gecko *Coleonyx variegatus*. I. The visual cell classification. J. Ultrastruct. Res. *16*, 651-671 (1966)

2.190 T. Amemiya, S. Ueno: Electron microscopic and cytochemical study on development of the paraboloid of accessary cone of chick retina. Cell Mol. Biol. *22*, 313-321 (1977)

2.191 K. Majima: Studien über die Struktur der Sehzellen und der Pigment-Epithelzellen der Froschnetzhaut. Albrecht von Graefes Arch. Ophthal. *115*, 286-304 (1925)

2.192 A. Gourévitch: La localisation histologique du glycogène dans la rétine des poissons. C.R. Soc. Biol. *148*, 213-215, 345-346 (1954)

2.193 K. Mizuno, Y. Takei: "Retinal glycogen", in *Structure of the Eye*, ed. by E. Yamada, S. Mishima (Univ. Tokyo, Tokyo 1975) pp.341-350

2.194 D. Bok: The distribution and renewal of RNA in retinal rods. Invest. Ophthalmol. *9*, 516-523 (1970)

2.195 C.E. Remé: Autophagy in visual cells and pigment epithelium. Invest. Ophthalmol. Vis. Sci. *16*, 807-814 (1977)

2.196 J.M. Enoch: The use of tetrazolium to distinguish between retinal receptors exposed and not exposed to light. Invest. Ophthalmol. *2*, 16-23 (1963)

2.197 E. Yamada: Some structural features of the fovea centralis in the human retina. Arch. Ophthalmol. *82*, 151-159 (1969)

2.198 E.R. Berger: On the mitochondrial origin of oil drops in the retinal double cone inner segments. J. Ultrastruct. Res. *14*, 143-157 (1966)

2.199 E. Yamada: The fine structure of retina studied with electron microscopy. I. The fine structure of frog retina. Kurume Med. J. *4*, 127-147 (1957)

2.200 V.L. Borovyagin: Submicroscopic morphology and structural connection of the receptor and horizontal cells of the retina of a number of lower vertebrates. Biofizika *11*, 90-111 (1966)

2.201 Y.W. Kunz, C. Wise: Ultrastructure of the oil droplet in the retinal twin cone of *Lebistes reticulatus* (Peters). Preliminary results. Rev. Suisse Zool. *80*, 694-698 (1974)

2.202 J.P. Hailman: Oil droplets in the eyes of adult anuran amphibians: a comparative survey. J. Morphol. *148*, 453-468 (1976)

2.203 G. Wald, H. Zussman: Carotenoids of the chicken retina. J. Biol. Chem. *122*, 449-460 (1937/1938)

2.204 T.G. Cooper, D.B. Meyer: Ontogeny of retinal oil droplets in the chick embryo. Exp. Eye Res. *7*, 434-442 (1968)

2.205 D.B. Meyer: The effect of dietary carotenoid deprivation on avian retinal oil droplets. Ophthalmic. Res. *2*, 104-109 (1971)

2.206 D.B. Meyer, S.R. Stuckey, R.A. Hudson: Oil droplet carotenoids of avian cones. Comp. Biochem. Physiol. B*40*, 61-70 (1971)

2.207 C. Bibb, R.W. Young: Renewal of fatty acids in the membranes of visual cell outer segments. J. Cell Biol. *61*, 327-343 (1974)

2.208 E.L. Craig, J.A. Eglitis, D.G. McConnell: Observations on the oil droplets of the principal cone cells of the frog retina. Exp. Eye Res. *2*, 268-271 (1963)

2.209 A.M. Granda, K.W. Haden: Retinal oil globule counts and distributions in two species of turtles: *Pseudemys scripta* elegans (Wied) and *Chelonia mydas mydas* (Linnaeus). Vision Res. *10*, 79-84 (1970)

2.210 C. Pedler, M. Boyle: Multiple oil droplets in the photoreceptors of the pigeon. Vision Res. *9*, 525-528 (1969)

2.211 G.K. Strother: Absorption spectra of retinal oil globules in turkey, turtle and pigeon. Exp. Cell Res. *29*, 349-355 (1963)

2.212 M. Anctil, M.A. Ali: Cone droplets of mitochondrial origin in the retina of *Fundulus heteroclitis* (Pisces: *Cyprinodontidae*) Zoomorphologie *84*, 103-111 (1976)

2.213 Y.W. Kunz, C. Wise: Structural differences of cone oil droplets in the light- and dark-adapted retina of *Poecilia reticulata* P. Experientia *34*, 246-249 (1978)

2.214 E.F. MacNichol, Y.W. Kunz, J.S. Levine, F.I. Harosi, B.A. Collins: Ellipsosomes: organelles containing a cytochrome-like pigment in the retinal cones of certain fishes. Science *200*, 549-551 (1978)

2.215 Y.W. Kunz, C. Regan: Histochemical investigations into the lipid nature of the oil droplet in the retinal twin cones of *Lebistes reticulatus* (Peters). Rev. Suisse Zool. *80*, 699-703 (1974)

2.216 S.K. Fisher, K. Goldman: Subsurface cisterns in the vertebrate retina. Cell Tissue Res. *164*, 473-480 (1975)

2.217 K.A. Siegesmund: The fine structure of subsurface cisterns. Anat. Rec. *162*, 187-196 (1968)

2.218 W.G. Bush, P.D. Langer: Ultrastructural survey of the retina of *Pseudemys scripta elegans*. Brain Behav. Evol. *5*, 114-123 (1972)

2.219 G.L. Fain, G.H. Gold, J.E. Dowling: Receptor coupling in the toad retina. Cold Spring Harbor Symp. Quant. Biol. *40*, 547-561 (1975)

2.220 E. Yamada, T. Ishikawa: The so-called "synaptic ribbon" in the inner segment of the lamprey retina. Arch. Histol. Jpn. *28*, 411-417 (1967)

2.221 E. Raviola, N.B. Gilula: Gap junctions between photoreceptor cells in the vertebrate retina. Proc. Nat. Acad. Sci. U.S.A. *70*, 1677-1681 (1973)

2.222 N. Carasso, N.: Mise en évidence de prolongements cytoplasmiques inframicroscopiques au niveau du segment interne des cellules visuelles du Gecko (Reptile). C.R. Acad. Sci. *242*, 2988-2991 (1956)

2.223 P. Öhman: Fine Structure of photoreceptors and associated neurons in the retina of the *Lampetra fluviatilis* (Cyclostomi), Vision Res. *16*, 659-662 (1976)

2.224 G. Yasuzumi, O. Tezuka, T. Ikeda: The submicroscopic structure of the inner segments of the rods and cones in the retina of *Uroloncha striata var. domestica* flower. J. Ultrastruct. Res. *1*, 295-306 (1958)

2.225 N. Carasso: Etude au microscope electronique des synapses des cellules visuelles chez le tetard *d'Alytes obstetricans*. C.R. Acad. Sci. D*245*, 216-219 (1957)

2.226 S.E.G. Nilsson: Interreceptor contacts in the retina of the frog (*Rana pipiens*). J. Ultrastruct. Res. *11*, 147-165 (1964)

2.227 B. Burnside: Microtubules and actin filaments in retinal cone elongation and contraction. J. Cell Biol. *67*, 50a (1975)

2.228 B. Burnside: Microtubules and actin filaments in teleost visual cone elongation and contraction. J. Supramol. Struct. *5*, 257-276 (1977)

2.229 R.H. Warren, B. Burnside: Microtubules in cone myoid elongation in the teleost retina. J. Cell Biol. *78*, 247-259 (1978)

2.230 T. Kuwabara: "Microtubules in the Retina", in *The Structure of the Eye*. II. Symposium, ed. by J.W. Rohen (Schattauer, Stuttgart 1965) pp.69-84

2.231 S. Uga, F. Nakao, M. Mimura, H. Ikui: Some new findings on the fine structure of the human photoreceptor cells. J. Electron Microsc. *19*, 71-84 (1970)

2.232 E.G. Gray: Microtubules in synapses of the retina. J. Neurocytol. *5*, 361-370 (1976)

2.233 D. Drenckhahn, U. Gröschel-Stewart: Localization of myosin and actin in ocular nonmuscle cells. Cell Tissue Res. *181*, 493-503 (1977)

2.234 B. Burnside, A.M. Laties: Actin filaments in apical projections of the primate pigmented epithelial cell. Invest. Ophthalmol. *15*, 570-579 (1976)

2.235 W. Kühne: "Chemische Vorgänge in der Netzhaut", in *Handbuch der Physiologie*, Vol.3, ed. by L. Hermann (Vogel, Leipzig 1879) pp.235-337

2.236 A.W. Spira, G.E. Milman: The structure and distribution of the cross-striated fibril and associated membranes in guinea pig photoreceptors. Am. J. Anat. *155*, 319-338 (1979)

2.237a N. Orzalesi, A. Bairati: Filamentous structures in the inner segment of human retinal rods. J. Cell Biol. *20*, 509-514 (1964)

2.237b S. Schuschereba, H. Zwick: " The Striated Rootlet System of Primate Rods - a Candidate for Active Photoreceptor Alignment", Recent Advances in Vision, Topical Meeting, Optical Society 1980

2.238 L.E. Zimmerman: Applications of histochemical methods for the demonstration of acid mucopolysaccharides to ophthalmic pathology. Trans. Am. Acad. Ophthalmol. Otolaryngol. *62*, 697-703 (1958)

2.239 L.B. Arey: "Retina, Choroid, Sclera", in *Special Cytology*, Vol.3, ed. by E.B. Cowdry (Hoeber, New York 1932) pp.1213-1304

2.240 F.S. Sjöstrand: Ultrastructure of retinal rod synapses of the guinea pig eye as revealed by three-dimensional reconstructions from serial sections. J. Ultrastruct. Res. *2*, 122-170 (1958)

2.241 B.S. Fine: Limiting membranes of the sensory retina and pigment epithelium. Arch. Ophthalmol. *66*, 847-860 (1961)

2.242 A.I. Cohen: Some electron microscopic observations on interreceptor contacts in the human and macaque retinae. J. Anat. *99*, 595-610 (1965)

2.243 B.S. Fine, L.E. Zimmerman: Observations on the rod and cone layer of the human retina. Invest. Ophthalmol. *2*, 446-459 (1963)

2.244 J.E. Dowling: Organization of vertebrate retinas. Invest. Ophthalmol. *9*, 655-680 (1970)

2.245 A. Lasansky: Functional implications of structural findings in retinal glial cells. Prog. Brain Res. *15*, 48-72 (1965)

2.246 J.G. Tonus, D.H. Dickson: Neuroglial relationships at the external limiting membrane of the newt retina. Exp. Eye Res. *28*, 93-110 (1979)

2.247 R.F. Miller, J.E. Dowling: Intracellular responses of the Müller (glial) cells of mudpuppy retina: their relation to b-wave of the electroretinogram. J. Neurophysiol. *33*, 323-341 (1970)

2.248 S. Uga, G.K. Smelser: Comparative study of the fine structure of retinal Müller cells in various vertebrates. Invest. Ophthalmol. *12*, 434-448 (1973)

2.249 E. De Robertis: Submicroscopic morphology and function of the synapse. Exp. Cell Res. Suppl. *5*, 347-369 (1958)

2.250 E. De Robertis, C.M. Franchi: Electron microscope observations on synaptic vesicles in synapses of the retinal rods and cones. J. Biophys. Biochem. Cytol. *2*, 307-318 (1956)

2.251 A.J. Ladman: The fine structure of the rod-bipolar cell synapse in the retina of the albino rat. J. Biophys. Biochem. Cytol. *4*, 459-466 (1958)

2.252 J.E. Dowling, B.B. Boycott: Organization of the primate retina: electron microscopy. Proc. R. Soc. London B*166*, 80-111 (1966)

2.253 W.K. Stell: The structure and relationship of horizontal cells and photoreceptor bipolar synaptic complexes in goldfish retina. Am. J. Anat. *121*, 401-424 (1967)

2.254 J.E. Dowling, F.S. Werblin: Organization of retina of the mudpuppy *Necturus maculosus*. I. Synaptic Structure. J. Neurophysiol. *32*, 315-338 (1969)

2.255 P. Witkovsky, J.E. Dowling: Synaptic relationships in the plexiform layers of carp retina. Z. Zellforsch. Mikrosk. Anat. *100*, 60-82 (1969)

2.256 H. Kolb: Organization of the outer plexiform layer of the primate retina. Philos. Trans. R. Soc. London B*258*, 261-283 (1970)

2.257 H. Kolb: The organization of the outer plexiform layer in the retina of the cat. Electron microscopic observations. J. Neurocytol. *6*, 131-153 (1977)

2.258 E.G. Gray, H.L. Pease: On understanding the organization of the retinal receptor synapses. Brain Res. *35*, 1-15 (1971)

2.259 A. Lasansky: "Synaptic Organization of Retinal Photoreceptors", in *Vertebrate Photoreception*, ed. by H.B. Barlow, P. Fatt. Symp. 1976, R. Soc. London (Academic, New York 1977) pp.275-290

2.260 K. Holmberg: The hagfish retina: Electron microscopic study comparing receptor and epithelial cells in the Pacific hagfish, *Polistotrema stouti*, with those in the Atlantic hagfish, *Myxine glutinosa*. Z. Zellforsch. Mikrosk. Anat. *121*, 249-269 (1971)

2.261 A.W. Spira, M.J. Hollenberg: Human retinal development: ultrastructure of the inner retinal layers. Dev. Biol. *31*, 1-21 (1973)

2.262 E. Yamada, T. Ishikawa: "Some Observations on the Submicroscopic Morphogenesis in the Human Retina", in *The Structure of the Eye*, ed. by J.W. Rohen. 2nd Symp. (Schattauer, Stuttgart 1965) pp.5-16

2.263 W.K. Stell: Correlation of retinal cytoarchitecture and ultrastructure in Golgi preparations. Anat. Rec. *153*, 389-398 (1965)

2.264 F.S. Werblin, J.E. Dowling: Organization of the Retina of the mudpuppy *Necturus maculosus*. II. Intracellular recording. J. Neurophysiol. *32*, 339-355 (1969)

2.265 G. Lanzavecchia: Ultrastruttura del coni e dei bastoncelli nella retina di *Xenopus laevis*. Arch. Ital. Anat. Embriol. *65*, 417-435 (1960)

2.266 H.J. Wagner: Darkness induced reduction of the number of synaptic ribbons in fish retina. Nature London New Biol. *246*, 53-55 (1973)

2.267 A.H. Bunt: Enzymatic digestion of synaptic ribbons in amphibian retinal photoreceptors. Brain Res. *25*, 571-577 (1971)

2.268 M.P. Osborne, R.A. Thornhill: The effect of monoamine-depleting drugs upon the synaptic bars in the inner ear of the bullfrog (*Rana catesbeiana*) Z. Zellforsch. Mikrosk. Anat. *127*, 347-355 (1972)

2.269 S.E.G. Nilsson, F. Crescitelli: A correlation of ultrastructure and function in the developing retina of the frog tadpole. J. Ultrastruct. Res. *30*, 87-102 (1970)

2.270 H.J. Wagner, M.A. Ali: Cone synaptic ribbons and retinomotor changes in the brook trout, *Salvelinus fontinalis* (Salmonidae, Teleostei), under various experimental conditions. Can. J. Zool. *55*, 1684-1691 (1977)

2.271 A. Spadaro, I. De Simone, D. Puzzolo: Ultrastructural data and chronobiological patterns of the synaptic ribbons in the outer plexiform layer in the retina of albino rats. Acta Anat. *102*, 365-373 (1978)

2.272 L. Vollrath: Synaptic ribbons of a mammalian pineal gland - circadian changes. Z. Zellforsch. Mikrosk. Anat. *145*, 171-183 (1973)

2.273 S.R. Schaeffer, E. Raviola: Membrane recycling in the cone cell endings of the turtle retina. J. Cell Biol. *79*, 802-825 (1978)

2.274 B.B. Boycott, J.E. Dowling: Organisation of primate retina. Phil. Trans. R. Soc. London B*255*,109-184 (1969)

2.275 T. Matsusaka: Lamellar bodies in the synaptic cytoplasm of the accessory cone from the chick retina as revealed by electron microscopy. J. Ultrastruct. Res. *18*, 55-70 (1967)

2.276 E. Raviola: Intercellular junctions in the outer plexiform layer of the retina. Invest. Ophthalmol. *15*, 881-895 (1976)

2.277 E. Reale, L. Luciano, M. Spitznas: Communicating junctions of the human sensory retina. Albrecht von Graefes Arch. Klin. Exp. Ophthalmol. *208*, 77-92 (1978)

2.278 P. Witkovsky, M. Shakib, H. Ripps: Interreceptoral junctions in the teleost retina. Invest. Ophthalmol. *13*, 996-1009 (1974)

2.279 A. Rochon-Duvigneaud: *Les Yeux et la Vision des Vertébrès* (Masson, Paris 1943)

2.280 L. Missotten: Estimation of the ratio of cones to neurons in the fovea of the human retina. Invest. Ophthalmol. *13*, 1045-1049 (1974)

2.281 G. Østerberg: Topography of the layer of rods and cones in the human retina. Acta Ophthalmol. Suppl. *VI*, 1-102 (1935)

2.282 R. Cajal: *The Structure of the Retina* (Thomas, Springfield, Ill. (1955)

2.283 R. Von Greeff: "Die Mikroskopische Anatomie des Sehnerven und der Netzhaut", in *Handbuch der Gesamten Augenheilkunde*, ed. by A. Graefe, Th. Saemisch, C. Hess (Springer, Berlin 1926)

2.284 E.T. Rolls, A. Cowey: Topography of the retina and striate cortex and its relationship to visual acuity in rhesus monkeys and squirrel monkeys. Exp. Brain Res. *10*, 298-310 (1970)

2.285 B. O'Brien: Vision and resolution in the central retina. J. Opt. Soc. Am. *41*, 882-894 (1951)

2.286 J.M. Enoch: Optical properties of the retinal receptors. J. Opt. Soc. Am. *53*, 71-85 (1963)

2.287 J.E. Dowling: "Synaptic Arrangements in the Vertebrate Retina: The Photoreceptor Synapses", in *Synaptic Transmission and Neuronal Interaction*, ed. by M.V.L. Bennett (Raven, New York 1974) pp.87-104

2.288 B.B. Boycott, H. Kolb: The horizontal cells of the Rhesus monkey retina. J. Comp. Neurol. *148*, 115-140 (1973)

2.289 A.B. Fulton, D.M. Albert, J.L. Craft: Human albinism. Light and electron microscopy study. Arch. Ophthalmol. *96*, 305-310 (1978)

2.290 I. Mann: *Development of the Human Eye*, 3rd ed. (Grune and Stratton, New York 1964)

2.291 J.M. Ordy, T. Samorajski, R.L. Collins, A.R. Nagy: Postnatal development of vision in a subhuman primate (*Macaca mulatta*). Arch. Ophthalmol. *73*, 674-686 (1965)

2.292 T. Samorajski, J.R. Keefe, J.M. Ordy: Morphogenesis of photoreceptor and retinal ultrastructure in a subhuman primate. Vision Res. *5*, 639-648 (1965)

2.293 A. Hendrickson, C. Kupfer: The histogenesis of the fovea in the macaque monkey. Invest. Ophthalmol. *15*, 746-756 (1976)

2.294 A. Hendrickson, R. Boothe: Morphology of the retina and dorsal lateral geniculate nucleus in dark-reared monkeys. (*Macaca nemestrina*) Vision Res. *16*, 517-521 (1976)

2.295 G.K. Von Noorden: Histological studies of the visual system in monkeys with experimental amblyopia. Invest. Ophthalmol. *12*, 727-738 (1973)

2.296 M.H. Bernstein: "Functional Architecture of the Retinal Epithelium", in *The Structure of the Eye*, ed. by G.K. Smelser (Academic, New York 1961) pp.139-150

2.297 F.H. Moyer: "Development, Structure and Function of the Retinal Pigmented Epithelium", in *The Retina, Morphology, Function and Clinical Characteristics*, ed. by B.R. Straatsma, M.O. Hall, R.A. Allen, F. Crescitelli (Univ. California Press, Berkeley 1969) pp.1-30

2.298 E. Yamada, T. Ishikawa, T. Hatae: "Some Observations on the Retinal Fine Structure of the Snake *Elaphe clinocophora*", in *Electron Microscopy*, Vol.2, ed. by R. Uyeda (Maruzen, Tokyo 1966) pp.495-496

2.299 T. Matsusaka: ATPase activity in the ciliary rootlet of human retinal rods. J. Cell Biol. *33*, 203-208 (1967)

3. The Stiles-Crawford Effects

J. M. Enoch and H. E. Bedell

With 15 Figures

In their pioneering experiments, described in Chap.1, STILES and CRAWFORD
[3.1] determined that light entering the periphery of the dilated pupil was
a less effective stimulus for vision than a physically equal light which
entered near the pupil center. Under conditions of foveal viewing, a beam
which entered the eye near the edge of the pupil required a luminance of
from 5 to 10 times that of a beam entering near the center, in order to be
judged equally bright.[1]

STILES and CRAWFORD defined the parameter η as the ratio of the luminance
of the standard (at the pupil center) and the displaced beams at the photometric match point. When η was plotted against the entry position of the
displaced beam in the pupil, the resulting curves were essentially symmetrical about a maximum value at or near the pupil center and decreased monotonically to the pupil margins (Fig.3.1a).

STILES [3.4] empirically fit the psychophysical photopic retinal directional sensitivity function, or Stiles-Crawford (S-C) function, with the
equation

$$\log_{10}\eta = \log_{10}\eta_{max} - \rho r^2 \quad \text{(Fig.3.1b)} \quad .$$

The parameter η is as defined above and η_{max} is the value of η at the function maximum. The distance within the entrance pupil from the function peak
in mm is represented as r. The shape parameter, ρ, provides an index of the
directionality, or the spread, represented by the function.

---

1 Throughout this discussion, luminance will be used to refer to the physical
  properties of the light stimulus, as corrected for the spectral response
  characteristics of a standard observer, whereas brightness will be reserved
  for the subjective impression occasioned by the stimulus. Thus, the Stiles-
  Crawford effect represents a situation in which luminance and brightness,
  as defined here, are dissociated. That is, a source of constant luminance
  is judged to change in brightness depending upon the region of the pupil
  through which light from the source enters the eye. Some authors have attempted to incorporate a correction for the Stiles-Crawford effect into the
  definition of retinal illuminance (e.g., [3.2,3]).

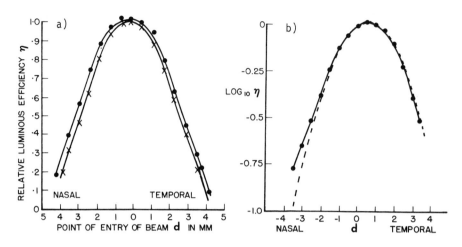

Fig.3.1. (a) The relative luminous efficiency (η) of a foveally viewed target as a function of the horizontal pupillary entry position of the beam. The data are for W.S. Stiles's left eye using a flicker photometric technique. The two curves were taken six months apart [3.1]. (b) Log relative luminous efficiency (log η) of a foveally viewed target as a function of the horizontal pupil entry position of the beam. Data are for W.S. Stiles's left eye using a photometric matching technique. The solid line is drawn through the data points; the dashed line is the best-fitting parabolic equation as described in the text [3.4]

The underlying basis of the S-C function is apparently the differential sensitivity of the retinal photoreceptors and associated structures to light impinging upon them at different angles. Based on a ray trace through the Gullstrand schematic eye, O'Brien [3.5] calculated that a 1 mm displacement in the entrance pupil corresponds to a change in the angle of incidence at the fovea of about 2.5°. This relation remains a fair approximation out to at least the midperipheral retina [3.6].

Since light which enters the receptors parallel to their long axes is presumably the most effective for producing visual excitation, as discussed in Chaps.2, 5, and 6, the position of the peak of the psychophysical S-C function is apparently an indicator of the orientational properties of groups of retinal photoreceptors. However, it is not clear to what extent the directionality of the S-C function, as expressed for example by the parameter ρ, is indicative of the directionality of *individual* receptors. It is clear that the psychophysically determined S-C function does not sample the directionality of single receptors. It is also apparent that the psychophysical S-C function must be modified from the true retinal directionality effect by the optical elements of the eye interposed between the stimulus and the retinal photoreceptors. The single most important of these elements is the lens. These questions will be taken up below.

3.1 Measurement of the Psychophysical Stiles-Crawford Function

3.1.1 General Considerations

STILES and CRAWFORD employed classic photometric techniques, i.e., direct bipartite field matching and flicker photometry, to measure the S-C function. CRAWFORD later showed that the function could be measured without a change of its form using sensitive increment *threshold* techniques [3.7]. The insensitivity of the S-C function to measurement by threshold versus suprathreshold techniques suggests a primarily physical and optical rather than a neural basis. Unlike the S-C function, the majority of visual functions, which presumably reflect neural processing or interaction contributions, typically show a change in form when measured by threshold as compared with suprathreshold techniques.

In this section, the apparatus and procedures employed to measure the S-C function in this laboratory are presented in some detail. The purpose of this presentation is to illustrate, by reference to this instrument, the types of controls which are required for accurate determinations of the S-C function.

The measurement of the S-C function requires 1) accurate control of stimulus energy and accurate specification and control of the pupil entry position of that energy, 2) minimization or elimination of changes in stimulus appearance for different stimulus entry positions in the pupil *not* due to the retinal directionality effect, and 3) the use of a sensitive psychophysical procedure. The incorporation of these requirements within the apparatus and the procedures used in this laboratory will be considered in turn.

In general, measurement of the S-C function requires two beams to be imaged in the plane of the entrance pupil of the eye, i.e., presented in Maxwellian view [3.8] (Fig.3.2). One beam is held in a fixed position in the pupil, usually the center, while the second is subject to displacement within the pupil.

The use of the Maxwellian view insures that neither beam is limited by the pupil and hence is unaffected by changes in pupil size. The Maxwellian view lens also incorporates the Badel optometer principle [3.9]. That is, a target or aperture moved toward or away from the Maxwellian view lens on the object side varies in optical distance from the observer whose pupil is

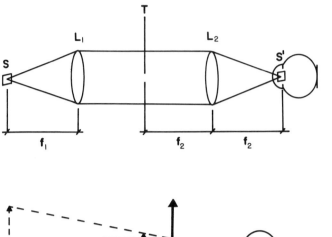

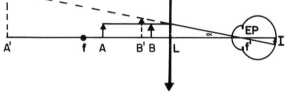

Fig.3.2.

Fig.3.3.

Fig.3.2. The Maxwellian view. The source S is placed at the anterior focal plane of lens L1 and the target T is transilluminated by a parallel beam. For the image of the source (S') to fall in the plane of the eye's entrance pupil, the latter should be in the posterior focal plane of lens L2 [3.8]

Fig.3.3. The Badal principle. The observer's entrance pupil (EP) is located in the posterior focal plane of lens L. A target located either at A or at B, within the anterior focal plane of the lens, appears as if at A' or B', respectively. Equal displacements of the target result in equal dioptric changes in apparent target distance. When the target is placed at the anterior focal point, f, it appears at optical infinity. Note that the target always subtends same visual angle (α) and hence provides the same retinal image size (I)

at the focal point of the lens (Fig.3.3). This is accomplished without changing target image size or luminance. The Badal principle is used to advantage in order to provide a well-imaged target to the observer, i.e., to correct spherical refractive error. The range of a Badal optometer is determined by the focal length of the lens used and by how closely the target can be made to approach the back surface of the lens.

A recurring question with regard to the S-C function has been whether the brightness of a source filling the pupil in non-Maxwellian view could be predicted from the integration of the S-C function over the total pupillary area; that is, whether the S-C functions determined for Maxwellian view sources are additive across the pupil [3.1,10-13]. In his dissertation, ENOCH compared direct matching and flicker photometric techniques for S-C effect curve determinations [3.12,14]. He found that blurring introduced by the peripheral pupil path of the displaced beam caused a brightness decrement for beams passing eccentrically in the pupil *in addition to* that produced by the S-C effect itself. Enoch's study demonstrated that some, and perhaps most, of the departures from additivity found by previous authors resulted from the blurring of off-axis images. He found much closer approaches to perfect additivity when blur was corrected. For this reason, it is of the utmost importance to insure that the observer views a well-imaged target for all pupil entry positions of the stimulus. The use of narrow-band spectral stimuli, rather than white light, minimizes the effect of the chromatic aberrations of the eye.

3.1.2 Principles of Instrumental Design

The instrument itself (Fig.3.4) incorporates several important features.

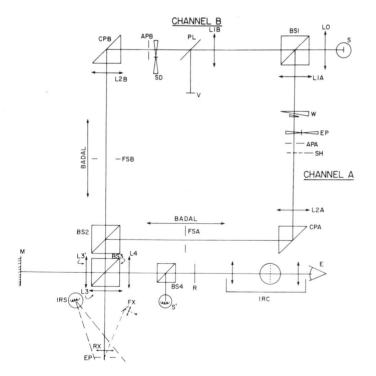

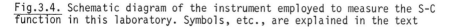

Fig.3.4. Schematic diagram of the instrument employed to measure the S-C function in this laboratory. Symbols, etc., are explained in the text

1) The observer effectively compares light from two channels, both of which are derived from the same light source. Fluctuations in the output of the source are thereby compensated, much as electrical noise is rejected by differential amplification techniques.

2) The output of the light source is continuously monitored.

3) The observer's pupil is illuminated with infrared (IR) energy which is reflected back through the optical system and onto on IR image converter tube. This permits the position of the observer's entrance pupil to be continuously monitored. An infrared sensitive television system might equally well be used.

4) A first surface mirror conjugate with the observer's pupil reflects energy from the stimulus beams backward through the optical system, allowing the experimenter to monitor stimulus position within the plane of the entrance pupil. Images of the observer's entrance pupil and of the stimulus beams are continuously viewed by the experimenter using the image converter.

5) The observer's position can be adjusted in the x, y, and z directions in order to make the initial alignment to the instrument. Continuous adjustment of the observer's pupil position maintains centration during measurements. In appropriate situations the corneal reflex can be used as a centering reference without altering the apparatus except for a small focal adjustment.

6) Adjustable position fixation lamps permit measurement of the S-C function at any visual field location horizontally and to ca. 45° in the inferior field.

7) The field stops within the instrument, which define the targets seen by the observer, may be moved toward and away from the Maxwellian view lens, thus acting as Badal optometers (discussed above). These field stops may also be moved horizontally and vertically with respect to the optical axis, thereby shifting the position of their retinal images. This permits compensation to be made for shifts in the target-image position due to spherical and off-axis aberrations of the eye, encountered with peripheral pupil entry positions of the stimulus beams.

8) An interferometric visual acuity device, which permits estimation of the minimum angle of resolution for a grating target formed by two-beam interference, is folded into the S-C instrument. Interferometric visual acuity can therefore be assessed under direct visual control of the position of the observer's pupil and of the interfering sources.

A more complete description of the instrument follows (Fig.3.4).

3.1.3 Stiles-Crawford Function Instrument

A cube beam splitter (BS1) divides the collimated beam (lens L0) from a tungsten ribbon filament source (S) into test (A) and background (B) beam channels. Within both channels the ribbon filament is imaged by lenses L1A and L1B at four-times lateral magnification onto approximately 0.30-mm diameter round apertures (APA, APB), mounted on mechanical stages. These apertures, which can be moved by remote control through the homogenous ($\pm$ 0.05 log units) filament image, serve as secondary sources. A portion of the beam in channel B is diverted by a thin glass pellicle (PL) and falls on a photovoltaic cell (V), the output of which is monitored. The apertures are collimated by lenses L2A and L2B and channels A and B are rejoined at a second beam splitting cube (BS2). The test and surround field dimensions are determined by the adjustable field stops FSA and FSB. After passing through beam splitter BS3, lens L3 forms unit lateral magnification images of apertures APA and APB in the observer's entrance pupil (EP). Movement of either aperture by means of its mechanical stage mounting causes an equal and opposite movement of its image in the entrance pupil.

At cube beam splitter BS3, a portion of both test and surround field beams are deflected and imaged by lens L3' onto a first surface mirror (M) conjugate with the observer's entrance pupil. Reflected images pass backward through lens L3' and join infrared radiation reflected from the observer's eye, provided by tungsten infrared sources (IRS), at cube BS3. Lens L4 forms an image of the test and background field beams and of the observer's pupil on a reticle (R), marked in concentric circles. The reticle is retroilluminated by infrared source S' reflected in cube beam splitter BS4. An interferometric acuity channel (not shown) is folded into the optical system at BS4. The reticle and images of the entrance pupil and both the test and the background field beams as they enter the pupil are viewed by the experimenter (E) in an infrared image converter system (IRC, RCA #6914A).

Provision for filtering of both test and surround field beams is made in the collimated portions prior to lenses L1A and L1B. Light restricted to a narrow spectral range is preferred in order to minimize the effects of the chromatic aberration of the eye. Typically, light from the red or red-orange region of the spectrum is used to favor photopic (cone) vision (e.g., Kodak Wratten Filter No. 23A). Moreover, long-wavelength light is less absorbed than short-wavelength light by the yellow eye lens pigments. The test field beam passes through a Kodak neutral wedge and balance filter (W) and is interrupted twice per second by an episcotister (chopper, EP). An infrared pass band is needed also for the image converter.

Field stops FSA and FSB are adjustable toward and away from lens L3, thereby functioning as Badal optometer systems with a range of approximately 2.0 diopters. Supplemental lens corrections, based on retinoscopic examination of the retinal test location, can be placed (at RX) close to the observer's eye. Optical compensation for the corrective lens is made by refocussing the entrance pupil in the viewing system. When such corrections are used, the distance from the lens to the eye (vertex distance) is carefully measured and the correction is centered on the optical axis of the system. A correction for the power of the spectacle must be applied to the displacements of the background beam in the entrance pupil. For data taken at other than fixation, the observer's gaze is directed to a dim red collimated fixation source (FX).

The observer is held in position by means of a dental impression and forehead rest in order to minimize head and eye movement. Both are attached to a frame which is adjustable in the x, y, and z directions. The experimenter positions the observer using these controls, while observing the image of his entrance pupil upon the reticle in the IR image converter. It is crucial that the pupil position be monitored continuously and adjusted during experiments to maintain proper alignment with respect to the test and background field beams.

Photopic S-C functions are typically determined by an increment threshold procedure, the *test beam being fixed* at the pupillary center and the background beam displaced in successive steps across the pupillary aperture [3.15]. The observer sees a test field limited by aperture FSA (typically 0.50° of visual angle or less) superimposed upon the center of a larger (4.5-5°) background field, defined by FSB. This technique has the advantage that the larger background field, rather than the smaller test field, is subjected to the off-axis optical aberrations of the eye. Thresholds for small fields are highly dependent upon blur whereas those for larger fields are less so [3.16-19]. Additionally, since the background field beam is displaced across the pupil, its image undergoes whatever shifts of position on the retina may accrue from spherical and other aberrations, instead of the test field's image. This insures that thresholds are always determined for a test field at the same retinal locus with respect to a stationary fixation target. Since both FSA and FSB are mounted on mechanical stages, aperture FSB can be shifted to compensate for the change in position of its retinal image, as the result of occular aberrations, when the background beam is displaced from the pupillary center. The test and background fields are thereby maintained in concentric alignment for all pupil entry positions of the background field beam.

The increment threshold technique just described measures the S-C function indirectly. What is determined is the change of the threshold for the incremental field, produced by the change of brightness of the background field due to the S-C effect. Fortunately, in normal observers the Weber relation is valid, e.g., the ratio of the incremental field luminance to the background field luminance at threshold is a constant ($\Delta L/L$ = constant) over a considerable range of background luminances. Within this range a logarithmic change of the background field brightness due to the S-C effect results in an equal change in the log luminance of the increment field at threshold. Thus, on the Weber region of the increment threshold curve (Fig. 3.5), measurement of changes in increment threshold is equivalent to measurement of changes in the background field brightness. It can be seen, however, that for regions of the increment threshold curve other than the Weber portion, increment threshold changes are not equal to changes in the brightness of the background field. Hence, outside the Weber region increment thresholds as a function of the pupil entry position of the background beam do not adequately represent the S-C effect. In practice, an increment threshold function is measured, with the entry position of the background beam held fixed, in order to determine a background field luminance for which a valid S-C function may be obtained [3.19,20].

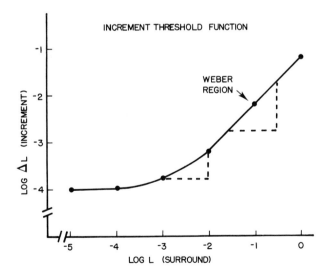

Fig.3.5. A hypothetical increment threshold curve for a single-response mechanism. Note that on the linear (Weber) portion of the curve the change in log L equals the change in log ΔL. This equality breaks down in the lower left-hand portion of the curve. Here changes in log L are not longer precisely reflected as changes in log ΔL

3.1.4 Other Approaches to Measurement

As noted above, a variety of nonincrement threshold procedures, including both threshold and suprathreshold techniques, have been applied to the determination of S-C functions. The advantages of the increment threshold procedure discussed above are 1) the control of test field blur and 2) its applicability at both foveal and extrafoveal test locations and during both photopic and scotopic adaptation.

BLANK et al. described a rapid method for determining the location of the peak of the S-C function [3.21,22]. Two beams are made to enter the pupil at a fixed separation (2 mm in practice); light from each beam forms one half of a bipartite photometric field. The pupillary entry positions of the two beams, which move in unison, are adjusted to give a photometric match. (A threshold judgment might be substituted for the photometric matching task.) If symmetry of the S-C function about its peak is assumed or has previously been determined, then at the match or null point the entry positions of the two beams in the pupil straddle the position of the S-C peak in the tested pupillary meridian. In principle, this peak-finding technique might be modified to generate an estimate of S-C function spread or of entire S-C functions.

STILES, WESTHEIMER, and HEATH and WALRAVEN have described procedures by which the S-C effect can be viewed entoptically [3.23,24a,24b]. Presumed local variations in receptor orientation can also be seen entoptically (see below).

Objective techniques have been applied to the measurement of the S-C function in humans, giving results consistent with the psychophysical data. Thus, changes in the amplitude of the electroretinogram [3.25,26], the visual evoked response [3.27], and the consensual pupillary response [3.28,29] can be measured as the pupil entry position of a Maxwellian view stimulus is varied. Fundus reflectometry techniques permit the measurement of retinal photopigment before and after exposure to pigment-bleaching lights. The relative photopigment-bleaching efficacy of monochromatic lights which enter the pupil at different locations has been found comparable to the relative efficacy of these lights in producing brightness [3.30,31]. The latter results further support the localization of the S-C effect to the retinal photoreceptors and to events prior to and including the phototransduction process. GOLDMANN performed an experment, similar in principle, which anticipated fundus reflectometry [3.32] . He noted ophthalmoscopically that retinal images formed by central and peripheral pupillary beams were unequally bright when the observer, whose retina he was observing, had photometrically

matched the two fields. When the examiner matched the brightness of the two
retinal images seen through the ophthalmoscope, the observer judged the image
formed by the central pupillary beam to the brighter.

3.2 Stiles-Crawford-Like Effects in Animals

A number of important questions concerning retinal directional sensitivity
and receptor orientation might be addressed with an animal model in which
the effects of possible invasive manipulations upon receptor alignment, etc.,
could be assessed. In such an animal model, histological and physiological
data from the same retina could be compared. For these reasons the use of
objective techniques for the measurement of S-C-like effects in infrahuman
species is of great interest. Unfortunately, there have been relatively few
attempts to identify S-C-like effects in animal models. The existing data
are in the form of retinal electrophysiological responses to different angles
of incidence of a test illumination.

DONNER and RUSHTON recorded ganglion cell responses to a flashed incremental field presented against a steady adapting background from a frog
eyecup preparation [3.33]. The ganglion cell threshold was found as the angle
of incidence of the incremental beam at the retina was varied. The authors
were careful to direct the test field image onto a portion of the ganglion
cell receptive field remote from the recording electrode, in order to avoid
possible shadowing artifacts. The test illumination necessary to produce a
criterion ganglion cell response changed by 0.4-0.8 log units over a 15°
range of angles of incidence when the preparation was photopically adapted.
Scotopically, changes in the angle of test beam incidence had no *measurable*
effect upon the ganglion cell threshold. Similar experiments were performed
by Reynauld and Laviolette, in which directional sensitivity was found under
photopic conditions for goldfish retina, as assessed by ganglion cell responses (see [3.34]).

BAYLOR and FETTIPLACE, using the turtle eyecup, recorded slow potentials
intracellularly from photoreceptors while changing the angle of the incident
light [3.35]. The authors reported clear sensitivity changes in "red" and
"green" cones as a function of the angle of incidence and possibly smaller
effects in "blue" cones and rods, which were encountered less frequently.
PAULTER had earlier found evidence for directional sensitivity of turtle receptors in an isolated retinal preparation [3.36]. He noted changes in the
magnitude of the S potential when the angle of incident light was varied.

S-C-like effects in this preparation are modified by the presence of high-refractive index, absorbing oil droplets between cone inner and outer segments. These produce differential absorption of incident light depending upon its angle of incidence, due to changes in path length through the oil droplets. However, the oil droplet absorption might be expected to reduce rather than enhance measured Stiles-Crawford-like effects in turtle cones. BAYLOR and FETTIPLACE reported that the magnitude of the directional sensitivity effect, for a given cell, depends upon the wavelength of the incident light, directional sensitivity becoming less or "inverting" (i.e., a greater sensitivity to presumed off-axis light) for wavelengths away from the cell's sensitivity maximum. These curious wavelength effects may be largely or wholly attributable to the directionality of the oil droplets, that is, their lesser absorption of obliquely incident light.

BAYLOR and FETTIPLACE concluded that the directional sensitivity of individual turtle receptors is fairly broad, corresponding well with the angular subtense of the turtle eye pupil at the retina. This conclusion is in contrast to optically determined directionalities of single photoreceptors in other species (see Chap.4). ENOCH and co-workers have determined the angular *radiation* patterns of single goldfish, frog, and rat photoreceptors when retroilluminated in isolated retina preparations (see [3.34]). On the basis of Helmholtz's theorem of optical reciprocity, the equivalence of forward and backward passage through an optical system, ENOCH has argued that the angular radiation pattern specifies the angular acceptance pattern, or the optical directional sensitivity. Directional sensitivities of single receptors, derived in this way, are considerably narrower than either the angular subtense of the pupil or of the optical directional sensitivities of large *groups* of receptors. Differences between rods and cones are quantitative rather than qualitative. ENOCH [3.37,38] has also made similar observations when light is passed through receptors in the physiologic direction, i.e., from inner retina to outer retina. That is, varying the angle of incidence of a light beam at the retina by only a few degrees greatly alters the transmissivity of individual rods and cones as observed in numerous mammalian species, including humans. It should be stressed that these optical measurements and observations were intentionally made in species without receptor oil droplets.

Whereas electrophysiological determinations of S-C-like effects in animals are potentially a very powerful tool for understanding basic recetor directional sensitivity processes, the difficulties involved in such studies are also tremendous. In an eyecup or isolated retina preparation, specification of receptor alignment relative to the eye's pupil is difficult. Of the electrophysiological studies cited in this section, only the report by BAYLOR and FETTIPLACE provided histological evidence that the overall receptor alignment in their preparation was directed toward the eye pupil. However, the effects of the electrode itself upon the stimulating light are uncertain. This is an especially difficult problem when intrareceptor recordings are attempted since the mere presence of the recording electrode in proximity to the photoreceptor must perturb light propagation in and along the receptor. Obviously, electrode penetration can influence cellular orientation. It is not clear what kind of controls are possible to avoid potential artifacts from such sources. The optical properties of the receptors studied should be monitored during such experiments (see Chap.5). It is also clear that whatever procedure is employed to measure S-C-like effects in animal preparations, the experimental and instrumental controls that were outlined above for psychophysical testing must be observed.

3.3 Experimental Data

3.3.1 The Effect of Luminance

CRAWFORD measured S-C functions by finding the incremental threshold for a small test field against a large background field for several luminances of the background field [3.7]. Some of Crawford's measurements were made with the background field extinguished, i.e., as the absolute threshold for the test beam as a function of its pupillary entry position. (In Crawford's study the test beam was displaced in the pupil and the background beam held at the pupil center.) For test fields confined to the foveal region, absolute and increment threshold determinations of the S-C function gave similar results. Crawford's study indicated that the foveal S-C function was largely independent of the background luminance.

Both STILES [3.39] and CRAWFORD [3.7] showed a similar independence of background luminance for the extrafoveal S-C effect provided the level was relatively high (greater than ca. 1.2 log photopic trolands at 5° in the visual field). At low background intensities or at absolute threshold, extrafoveal S-C functions were reported to be much more nearly flat [3.7,15,39], indicating that in scotopic vision, brightness is much less dependent upon entry position of the beam in the pupil. This marked difference between photopically and scotopically measured S-C functions has misled some authors to the oversimplification that there is *no* scotopic S-C function.

Using a sensitive psychophysical signal detection procedure, VAN LOO and ENOCH definitively demonstrated the existence of a scotopic S-C function, on the order of 0.2-0.3 log units [3.40]. This scotopic S-C effect had previously been hinted at in both STILES's and CRAWFORD's [3.39,7] data as well as in subjects whose S-C functions were decentered in the pupil [3.15,41]. Additionally, styrofoam models of rod receptors irradiated with microwave radiation evidenced directional sensitivity only quantitatively different from that of cone models [3.42]. Individual rods and groups of rod receptors have been found to be directionally sensitive in in vitro optical studies (see Chap.4). When Van Loo and Enoch's data were corrected for the differential path lengths of central and peripheral beams through absorbing lens pigment (see below), the magnitude of the scotopic S-C effect was found to be quite similar at all wavelengths tested (Fig.3.6). Moreover, both photopic and scoptic S-C functions taken at the same locus (6° nasal visual field) had common axes of symmetry with reference to the pupil. Results of BEDELL and ENOCH for one observer at 35° in the temporal visual field are also indicative of a common alignment tendency of photopic and scotopic receptors ([3.6], see also [3.15,41]). These findings are highly significant in terms of the retinal receptor orientation properties which the S-C function apparently reflects.

In the mesopic range, the S-C function makes a gradual transition between the smaller scotopic effect and the larger photopic effect. This so-called rising-sun phenomenon is illustrated in Fig.3.7 from CRAWFORD [3.7].

At very high photopic levels, the magnitude of the S-C effect at the fovea increases, i.e., there is a greater difference in brightness between beams passing through the center and periphery of the pupil [3.4,43]. WALRAVEN has confirmed this effect for midspectral monochromatic lights and proposed that it may be due to cone pigment bleaching [3.44]. He measured a recovery period for the increased directionality of the S-C effect following light adaptation, finding it to have a time constant of ca. 30 s, consistent with that of cone photochemical regeneration following bleaching. BRINDLEY noted that the Stiles-Crawford hue change (S-C II; see below) effectively disappeared for yellow light when the eye was first exposed to a bright adapting stimulus [3.45]. Earlier, WRIGHT [3.46] had determined that the foveal luminosity curve for a Maxwellian view beam entering the pupil 3 mm from its center was similar to the luminosity curve for a *higher* luminance target entering at the pupil center. Taken together, these results suggest that eccentric pupil entry and light adaptation have comparable influences on (at least midspectral) luminosity. BRINDLEY [3.45], following

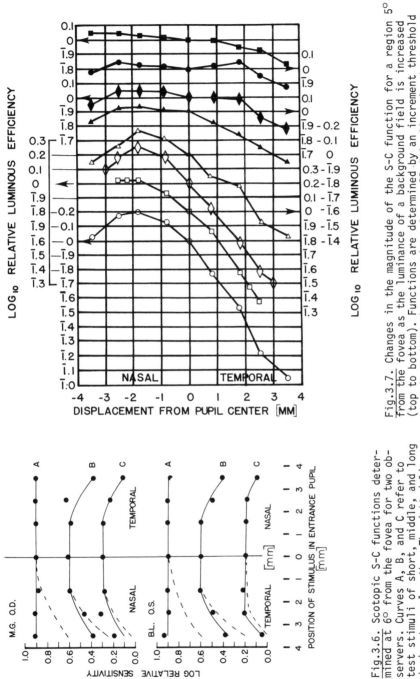

Fig.3.7. Changes in the magnitude of the S-C function for a region 5° from the fovea as the luminance of a background field is increased (top to bottom). Functions are determined by an increment threshold procedure. Note the suggestion of a small S-C effect in the topmost curve, for which increment thresholds were against a background of zero luminance, i.e., at absolute scotopic threshold. The bottommost curves are determined against photopic background fields [3.7]

Fig.3.6. Scotopic S-C functions determined at 6° from the fovea for two observers. Curves A, B, and C refer to test stimuli of short, middle, and long dominant wavelengths. The dotted lines replot the function incorporating MELLERIO's [3.102] correction for lens pigment absorption [3.40]

a proposal made by STILES [3.4], suggested that this similarity might be due to reduced effective receptor photopigment concentration in both instances, to a partial bleaching from light adaptation, and to a diminished effective absorbing path length within the receptor in the case of oblique incidence.

3.3.2 Directionality at Different Retinal Loci

WESTHEIMER [3.47] argued that if the S-C effect reflected retinal photoreceptor optical properties, and if these were physical optical rather than geometrical optical in character, due to the small size of receptor apertures [3.48], then foveal and extrafoveal photopic S-C functions might be expected to differ, reflecting anatomical differences between foveal and extrafoveal cones. In fact, he found that the foveal S-C effect was of lesser magnitude (smaller ρ value) than that measured 3.75° parafoveally. Similar results can also be seen in the data of VOS and HUIGEN taken at 0° and 4°, although these authors failed to comment on this finding [3.49]. STILES's [3.39] results also indicate a larger S-C effect parafoveally (5°) than foveally for *long*-wavelength data. ENOCH and HOPE measured S-C functions for orange test and surround fields at 0°, 2°, 3.75°, and 10° and found that the change in magnitude of the S-C effect occurred between 0° and 2° with no further increase thereafter out to 10° [3.20]. Recently, BEDELL and ENOCH [3.6,50] replicated this finding of increased directionality parafovelly but found the S-C effect at 35° to be similar in magnitude to that of the fovea. Taken together, the results of these studies indicate that directionality first increases upon leaving the fovea, and then decreases once again in the near to midperiphery (Fig.3.8).

It is probably reasonable to suggest that psychophysically determined S-C functions represent 1) the directionality of individual receptors and their related structures, 2) some distributive orientation factor a) between receptors and b) between groups of receptors, and 3) possible neural integrative processes by which the outputs of groups of receptors are combined. Additionally, one may posit perceptual, criterion, and/or judgmental factors, inherent in the nature of psychophysical measurements, which might derive from possible differences in the appearance of stimuli entering the pupil at different locations. Since directionality at 35° in the peripheral visual field has been found to be similar to that measured at the fovea, despite marked differences in cone morphology at these locations, it is likely that the contribution of some or all of these factors, as well as their relative importance, changes with position on the retina and perhaps with other observer or stimulus variables as well.

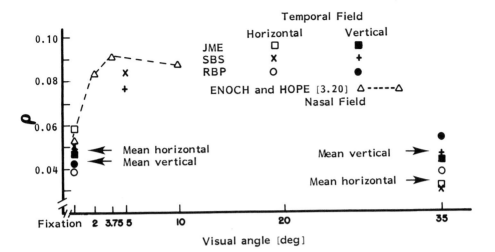

Fig.3.8. S-C function directionality (ρ) is plotted as a function of test-ing location in the visual field. The results of two studies are presented. Mean directionality data for 3 observers between 0° and 10° nasal field are shown as triangles connected by the dashed line (from [3.20]). Individual data for 3 observers between 0° and 35° temporal field are from [3.6]

Studies aimed at identifying the differential contributions of the above factors at different retinal locations, and perhaps within different indi-viduals or species, are required. Optical studies of single receptors and of receptor groups in situ offer an approach toward the estimation of the receptor alignment distribution. In addition to differences in the alignment of neighboring receptors, there are apparently definable groups of receptors which share a more or less common alignment, slightly different from that of other nearby groups. ENOCH (e.g., [3.51]) has noted alignment subgroups in excised retina preparations. Such groups may also be apparent entopti-cally. This was first recognized by O'Brien [3.52], who constructed an aperture designed to compensate the photopic S-C effect and saw residual bright and dark patches in what was expected to be a uniform field. Brighter and darker regions shifted with respect to one another as the entry position of the compensated beam in the pupil was altered. The effect rapidly faded when the beam was not moved, presumably due to image stabilizaion. The same phenomenon has been reported by ENOCH and by HEATH and WALRAVEN for short flashes [3.24b,51]. ENOCH has observed 10-15 such subareas within a centrally fixed 5° field. ENOCH et al. reported that a similar, if somewhat more dif-ficult to observe, effect also occurs in scotopic vision, apparently reflect-ing local variations in orientation of groups of rods [3.53].

Further psychophysical evidence for a degree of variability in receptor alignment is the transient S-C effect. MAKOUS reported that if two fields, which had been equated for the S-C effect and which entered the eye at opposite sides of the pupil, were interchanged, this exchange was marked by a sudden increase in brightness, followed by a slow decline to a lower, steady-state value [3.54]. When these two fields were alternately employed as the background upon which a small incremental test field appeared, a sudden rise in the increment field threshold, on the order of 0.7 log units, was found to occur when the background beams were exchanged. Thereafter, the increment threshold returned to an asymptotic value with an apparently exponential course and a half time of 10-25 s, depending upon the surround field luminance. Curiously, the recovery was slower for dimmer background fields. The increment threshold results seemed to mirror the subjective perception in all respects. (MAKOUS has recently reported that a transient reduction in visual resolution capability occurs under similar experimental conditions. He reported the magnitude of this reduction to be about a factor of two [3.55]).

Confirmation of this so-called transient S-C effect appeared in the work of HEATH, BAILEY, and SANSBURY et al. [3.56-59]. HEATH noted that fields (compensated for the S-C effect and entering the pupil at disparate points) demonstrated flicker when interchanged at a moderate rate. BAILEY pursued this finding in his dissertation and was able to define directional sensitivity functions much narrower than the standard psychophysical S-C function utilizing the criterion of critical flicker frequency (Fig.3.9) [3.57,58]. SANSBURY et al. replicated Makous's original experiment using monochromatic, rather than white, test and surround fields and confirmed his report in all particulars. These authors also demonstrated, as had MAKOUS, that both the magnitude and time course of the transient S-C effect wer independent of the pupil entry position of the incremental test beam. Finally, they showed that in the steady-state condition, the threshold raising properties of surround fields entering at opposite sides of the pupil were linearly additive. Both MAKOUS and SANSBURY et al. argued on the basis of their experiments that the transient S-C effect indicated the existence of receptors or channels with directional sensitivities narrower than the aperture of the dilated pupil. However, they concluded that the effect was not the result of a preferential light adaptation of different groups of receptors by background beams incident at the retina at different angles. COLBE and RUSHTON measured the fraction of cone pigment bleached in the fovea densitometrically, using a measuring beam which was varied in its pupil entry position, and also con-

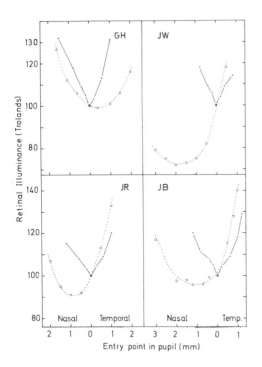

Fig.3.9. A target alternately illuminated by two beams entering the pupil at different locations and matched for brightness appears to flicker at slow-alternation rates. The narrower curves (———) were determined by finding the alternation rate at each of several beam separations at which this flicker just disappeared. The dashed functions are conventional S-C functions for the same observers. The solid curves thus represent a retinal directionality effect with narrower tuning than the conventional S-C function [3.58]

cluded that bleaching did not occur differentially in groups of cones with dissimilar orientations [3.31]. While MACLEOD also failed to find psychophysical evidence of a directionally selective light adaptation at the fovea, he did elicit such an effect at 6° in the parafovea [3.60]. In contrast to the foveal results, he found that the entire S-C function shifted slightly depending upon the entry position of a background beam in the pupil for peripheral pupil entry positions. Differences between the curves were on the order of 0.3 log units.

There is as yet no adequate explanation for the transient S-C effect. MAKOUS [3.54] noted that the transient S-C effect was similar in time course, magnitude, and the effects of varying luminance to a phenomenon described by BAKER [3.61] in conjunction with rapid light adaptation. It would seem that, contrary to Makous's argument that the two phenomena are unrelated, sufficient similarity exists to actively pursue a connection. In particular, both effects may reflect processes initiated by a transient signal from (groups of) photoreceptors, but which occur at a more proximal site.

3.3.3 The Stiles-Crawford Function in Retinal Pathology

a) Normative Data

Histological investigations as well as photoreceptor X-ray diffraction pattern analyses have shown in a variety of species that photoreceptors do not normally point toward the center of the eye, but rather toward some anterior location, presumably within the pupil ([3.35,62-67b], also see Chap.4).

Inasmuch as the S-C function represents the orienting properties of the photoreceptors [3.68], one would expect that peaks of S-C functions measured at different retinal loci would mirror these histological findings, i.e., S-C function peaks should maintain an approximately constant relation to the pupil center. ENOCH and HOPE carefully measured S-C functions from 5° in the temporal visual field, approximately the posterior pole, to 20° in the nasal visual field [3.19]. Their findings were that both horizontal and vertical traverses of the pupil yielded S-C function peaks clustered about a point, slightly different for each of their three observers, but for each near the center of the entrance pupil of the eye. They confirmed these results for three more observers at loci between 0° and 10° in another study [3.20]. BEDELL and ENOCH extended this analysis to 35° in the temporal peripheral field and found evidence that receptor alignment is maintained within the pupil [3.6,50].

The stability of the S-C functions of normal observers over time has also been addressed. In 1939, STILES reported the results of six years of S-C effect determinations on his own eye. During that period the peak of his foveal S-C function had shifted from 0.2 mm nasal of pupil center in 1933 to 0.6 mm temporal in 1937 and to 0.9 mm temporal in 1939, giving a total shift in the horizontal meridian of just over 1 mm. Very little or no change occurred in the position of the S-C function peak for a vertical pupillary traverse. Crawford's eye showed essentially no change in the position of the horizontal peak from 1933 to 1937 [3.4]. SAFIR et al. failed to find any horizontal shift in the S-C function peaks of two observers over a two-year interval [3.69]. BEDELL and ENOCH [3.6] measured both vertical and horizontal foveal S-C functions for two observers and compared these with earlier data. In one case, no change in peak location could be detected after a lapse of five years; in the second, a 0.4 mm horizontal shift was detected over a 17 year interval. Thus, not only do receptors at different retinal locations tend to align to the pupil center, but, with the exception of Stiles's eye, normal S-C function peak positions within the pupil are remarkably stable over time. (See end of Chap.4 for a further discussion).

b) Results in Pathology

Clearly the precise receptor alignment which is seen in careful histology and is presumably reflected in the normal S-C function, may be altered or disrupted by the action of pathological or aphysiological challenges to the retina or pigmented epithelium. Thus, FANKHAUSER et al. documented anomalies of the S-C function in patients with retinal detachment, angiomatosis of the retina (proliferation of retinal blood vessels) and retinoschisis (separation of retinal layers) 3.7o . In cases in which photocoagulation therapy was performed, marked alterations were observed in postoperative S-C functions (Fig.3.10). Scarring subsequent to photocoagulation presumably introduces marked tractional forces upon the retina (Fig.3.11). Abnormal S-C functions have been noted in other types of retinal pathology as well [3.71-78]. However, not all retinal diseases or even all those affecting the photoreceptor and pigment epithelial layers disturb retinal receptor alignment. For example, an essentially normal S-C function has been measured in a case of Best's disease [3.79], in which a large egg-yolk-like lesion apparently lifts the retina above the pigment epithelium. In this case ρ may be somewhat reduced. Additionally, fairly normal S-C functions, which have been found to be stable over time, have been obtained in cases of observed changes in the pigment epithelium [3.80].

What is more instructive from cases of pathology is that following disturbance, a realignment of photoreceptors, as reflected in the recovery of the S-C function, may occur in tandem with the resolution of the pathological condition. The first indications that recovery was possible were seen in a case of retinal degeneration in which photocoagulation was performed [3.70], and more clearly in cases of serous retinal detachment [3.73] and of subretinal fluid subsequent to trauma 3.75 (Fig.3.12). In the latter case, there was evidence that receptor reorientation occurred within the macular region at the same time that the S-C function of a paramacular location, apparently just outside the region of the primary lesion, showed deteriorative changes. Subsequently, this paramacular test location also showed the recovery of a normal S-C function. Thus, in this case the alteration of alignment and the receptor realignment occurred with different schedules at two retinal loci just a few degrees apart. Marked improvement of severely disturbed S-C functions has also recently been seen in cases of senile macular degeneration [3.76], placoid pigment epitheliopathy [3.77], and central serous choroidopathy [3.78].

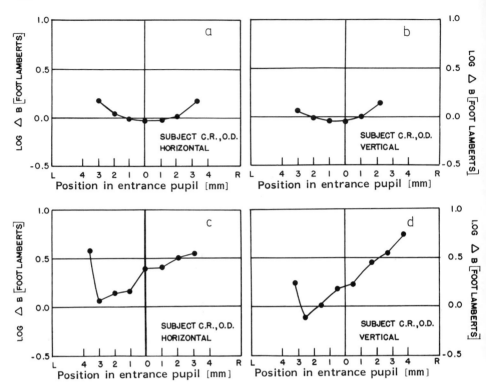

Fig.3.10a-d. Horizontal (on left) and vertical (on right) S-C functions determined in a subject with retinoschisis. The top functions were measured prior to treatment and are flatter than normal. The curves on the bottom were measured after photocoagulation was performed in this eye and indicate a further disturbance of receptor orientation [3.70]. Because threshold is plotted, these functions are inverted relative to other figures in this chapter

DUNNEWOLD reported a patient with an iris coloboma, resulting in a displaced pupil, in which the S-C function peaked near the center of the displaced, rather than physiologic pupil [3.72]. BONDS and MACLEOD documented a similar case, in which a displaced pupil resulted from insult to the eye at an early age [3.81]. Both photopic and scotopic S-C function curves were symmetrical about a point near the displaced pupil center. It is possible that both of these instances also represent a "recovery" of receptor orientation to abnormal situations of retinal illumination (see also Chap.4).

It is significant that in eyes in which anomalous S-C functions have been demonstrated within regions of observable retinal pathology, other visual functions, including visual acuity, have also shown adverse changes. In cases in which the S-C function has been found to recover, visual acuity has also shown improvement.

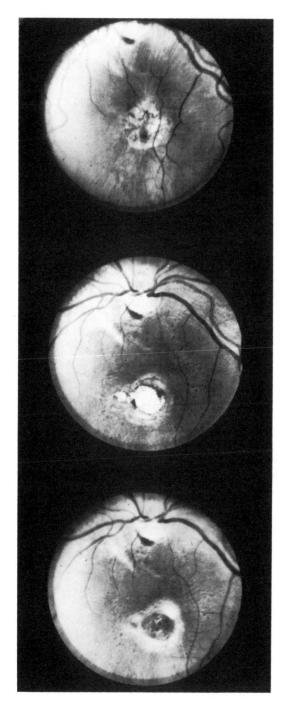

Fig.3.11. Photographs of the fundus of the eye immediately after photocoagulation of a macular hole (top left), one week later (top right) and three weeks later (bottom). Note that the retinal blood vessels undergo marked shifts in position, presumably due to tractional forces on the retina. This is a different subject than the one whose S-C functions are shown in Fig.3.10 [3.70]

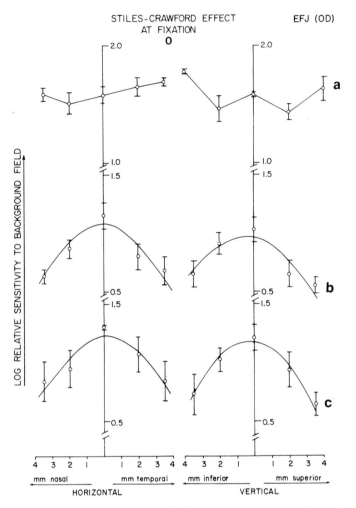

STILES-CRAWFORD EFFECT AT FIXATION   EFJ (OD)

Fig.3.12a-c. Recovery of the S-C function measured at the fixation point in a case of subretinal serous fluid secondary to trauma. Horizontal (left) and vertical (right) traverses of the pupil are shown for three testing dates (top to bottom). Note the severely anomalous S-C functions determined on 5 May (a) and the remarkable recovery in these functions as determined on 26 May (b). No further change was seen from 26 May to 28 July (c). Changes in this subject's S-C functions paralleled the resolution of the retinal pathological [3.75]

OHZU et al. [3.82,83] observed that the modulation transfer function, indicative of the optical resolution capability, of isolated retinas, is markedly inferior in areas in which receptors are poorly oriented as compared with regions of well-oriented receptors. These observations have been made in rod-dominant rat retinas and in cone-dominant squirrel monkey, and human foveas. Unfortunately, the extent of malorientation of receptors in poorly oriented areas can not readily be specified quantitatively.

CAMPBELL reported psychophysical evidence for decreased resolution of grating targets entering the pupil at peripheral locations and presumably imaged obliquely onto foveal photoreceptors [3.84]. Resolution was poorer for targets oriented perpendicularly to the pupillary test meridian than for targets oriented parallel to it. Later work indicated that much of the effect could be attributed to aberrations suffered along peripheral optical paths in the eye [3.85-87]. However, a small effect was found to persist even with interferometrically formed targets [3.85], see also [3.86]. ENOCH and GLISMANN [3.88,87] noted that the optical resolution capability of groups of well-oriented receptors in rat and monkey isolated retinas was reduced for obliquely incident light. Both meridional and nonmeridional effects seemed to contribute to the measured resolution decrement. The change in observed resolution for an 8° shift in angle of incidence was on the order of 10-20%, much smaller than the resolution losses reported by OHZU et al. [3.82] in areas of poor receptor orientation.

The relationship between visual acuity and photoreceptor orientation is of special interest in functional amblyopia. Functional amblyopia is characterized by a diminished visual acuity, usually in one eye only, which is not resolved by optimal refractive correction. Gross indications of pathology are either absent or are of insufficient magnitude to account for the acuity loss. Conditions of abnormal visual experience during early life, due to 1) ocular opacities or scattering elements which interfere with image formation, 2) a significantly different refractive status in the two eyes, or 3) misalignment of the visual axes of the two eyes, can be documented or inferred in many cases.

ENOCH detected anomalies of the S-C function measured at the point of fixation in *some* amblyopic eyes and suggested that photoreceptor malorientation, which could be inferred from such measurements, might play a part in the decreased visual acuity of these amblyopic eyes [3.22,51,89,90]. In a recent study, BEDELL utilized the S-C function to assess photoreceptor orientation at a range of retinal locations in both eyes of several functional amblyopic observers [3.50]. This population was not derived from clinical sources. Test locations spanned 30° of visual angle in the horizontal meridian and included the foveal region. In each of the amblyopic eyes of this study, normal-appearing S-C functions were measured at all test locations. In each case, S-C function peaks were found to cluster within a subregion of the dilated pupil, indicative of receptor orientation toward the pupil at all test sites (see above and Chap.4). Thus, in these cases, the amblyopic eye acuity deficits are apparently determined at sites in the visual system proximal to the photoreceptors.

One of the observers of Bedell's study is of interest from another point of view. Whereas S-C function peaks measured in this observer's amblyopic eye are indicative of photoreceptor alignment toward the exit pupil of that eye, S-C function peaks of the other, *nonamblyopic* eye indicate photoreceptor orientation toward the center of the globe. The data obtained from this unique observer are more fully presented in Chap.4.

The reduction in resolution which could be expected to result from photoreceptor orientation anomalies is modest in terms of that found in profound amblyopias. Despite this, and although receptor malorientation is not common to all amblyopias [3.50,51,90,91], the demonstration of cases of amblyopia in which there is evidence of anomalous receptor alignment is significant in that a component of the acuity loss in such cases may be attributable to a localized and quantifiable pathophysiological situation. However, it is unclear whether the receptor orientation abnormalities, sometimes detected in amblyopes, represent a sign of microscopic pathology in the outer retina, a failure of the normal receptor alignment mechanism, or the sequelae of transient retinal pathology during infancy, which has otherwise resolved [3.92]. To confound the situation even further, S-C functions which are significantly decentered with respect to the pupil have occasionally been reported for presumably normal eyes [3.15,24a,50,93]. At least some eyes with anomalous S-C functions at the locus of fixation maintain excellent (better than 20/20) visual acuity. Since normal observers typically show visual acuities substantially better than 20/20 [3.94], a slight acuity reduction resulting from a modest amount of receptor misalignment might easily escape notice on standard acuity tests. Based upon Enoch and Glismann's results, significant resolution losses would not be expected for a modest simple tilt of the receptors, i.e., the maintenance of a good alignment with neighboring receptors, but an overall orientation tendency toward a noncentral region of the entrance pupil. The issue is confounded by the lack of a clear relationship between the measured S-C function and detailed alignment properties of small groups of receptors. The histological evidence presented above, the results in retinal pathology, and microwave simulation studies [3.95] all suggest that a *general malorientation* of the receptors, a loss of alignment between neighboring elements, would more significantly disturb resolution capacity. As yet there has been no psychophysical study directly relating visual acuity and receptor orientation at the same *small* retinal area. We hope to pursue this question.

3.3.4 The Effect of Wavelength

STILES reported that the magnitude of the foveal S-C effect changed systematically with the wavelength of the test stimulus [3.4]. He found the smallest S-C effect (smallest ρ value) in the green region and larger effects at both ends of an equal luminance spectrum. In later work, STILES [3.39] confirmed this general wavelength dependence; however, he measured a smaller S-C effect at all wavelengths using different measurement techniques (see Fig.3.13). Both sets of measurements were on Stiles's own eye.

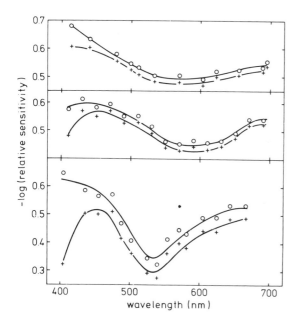

Fig.3.13. The variation of the magnitude of the foveal S-C effect with wavelength (bottom curve of each panel) and corrected for absorption by the human eye lens (top curve of each panel). From top to bottom, data are taken from [3.39,4,96]. Lens correction is due to [3.103]

Foveal S-C data as a function of wavelength have also been published by ENOCH and STILES [3.96] for two subjects, as raw data only, by SAFIR and HAYMS [3.97] for two observers, and by WIJNGAARD and VAN KRUYSBERGEN for one observer [3.93]. In all cases, a larger S-C effect is observerd for long and short wavelength than for midspectral monochromatic targets.

STILES also measured the wavelength dependence of the photopic S-C effect for parafoveal targets [3.4]. Here, the S-C effect was largest at the red end of the spectrum, falling to a rather constant (nonscotopic) value in the green and blue regions. This experiment has apparently never been repeated. It is curious that the wavelength dependence of the S-C effect for the foveal and parafoveal photopic retina should differ. Although the parafoveal curves are not at all scotopic in appearance, a possible rod contribution at the more eccentric pupil entry positions needs to be considered. It can be said that the wavelength dependence of the photopic S-C effect outside the fovea is an open question. On the other hand, the magnitude of the scotopic S-C effect seems to be less dependent on wavelength, after correction has been made for absorption by the ocular media [3.15,40].

Returning to the foveal data, the function of S-C magnitude versus wavelength has been suggested to represent a composite of the somewhat different directional sensitivities of three cone types posited for trichromatic vision [3.4,39,96,98,99]. STILES approached this problem by determining increment threshold functions for a variety of test field, background field wavelength combinations, and either central or peripheral pupil entry of the test beam [3.39]. For some combinations of test and background wavelengths, the difference between increment threshold curves for central and peripheral pupil entry of the test beam changed more or less abruptly as a function of the background luminance, indicating a change in the directional sensitivity of the detecting mechanism. STILES interpreted these data in terms of three cone mechanisms, the "blue" cones being more directionally sensitive than either the "red" or "green" cones, and the "green" cones slightly more so than the "red" cones for wavelengths under 630 mm (also see [3.100]).

ENOCH and STILES calculated the directional sensitivities of three cone types based on color matching functions for central and peripheral pupil entry of test fields [3.96]. The results indicate that all three receptor types have directional sensitivities which change as a function of wavelength; the "blue" receptors were found to be more directionally sensitive than eithe the "red" or "green" receptors.

3.3.5 Corrections for the Ocular Media

While the S-C effect would seem to be primarily retinal in origin, the "retinal function" is modified somewhat by differential absorption of beams passing through the center and periphery of the pupil within the eye medium. The bulk of the absorption within the eye is attributable to the lens pigment which most effectively absorbs light at the short-wavelength end of the spectrum. Macular pigment absorption is relatively unimportant since the path length differences involved are only about 2%.

WEALE and later MELLERIO corrected the magnitude of the S-C effect as a function of wavelength for lens absorption, based upon a homogeneous pigment concentration throughout the lens [3.101,102]. Since the path length through the lens is longer for beams passing through the center than through the periphery, the correction results in a modest increase in the magnitude of the S-C effect at middle and long wavelengths; at short wavelengths the increase in the (presumed retinal) S-C effect is quite substantial. WEALE also noted that after lens path corrections, Stiles's 1939 data inciated a small S-C effect in scotopic vision; this effect was later measured by VAN LOO and ENOCH [3.40].

VOS and VAN OS reexamined the WEALE and MELLERIO corrections for short wavelengths, arguing that they failed to account for the slight decentration of the S-C function peak in the pupil generally found in normal observers [3.103]. They noted that the effect of lens pigment at short wavelengths is to displace the psychophysically determined S-C function peak and distort the function's shape, rather than to diminish its magnitude by as much as the WEALE and MELLERIO corrections suggest. Vos and Van Os's own correction, incorporating the position of the S-C function peak with respect to the lens, is shown in Fig.3.13. It would seem that the validity of the various lens corrections or, conversely, a closer approximation to the retinal effect, could be experimentally examined in an aphakic observer, corrected with a corneal contact lens, as suggested by BAILEY [3.104].

3.4 The Stiles-Crawford Effect of the Second Kind

When the S-C effect is measured for monochromatic lights, using a direct photometric matching technique, a comparison field which enters the pupil displaced from the center may undergo not only a change in brightness but in perceived hue as well. This hue shift has come to be known as the Stiles-Crawford effect of the second kind (S-C II).

The first detailed description of the S-C II was by STILES [3.4]. He presented observers with a bipartite field which was to be matched for both hue and brightness. Standard and comparison half fields derived from two monochromators, the exit slit of the first imaged at the center of the observer's pupil. The comparison field beam, from the second monochromator, was imaged successively at half mm intervals across the width of the dilated pupil. The observer controlled both the luminance and wavelength of the comparison beam, thereby permitting him to match the standard half field, which was set at one of 11 wavelengths spanning the visisble spectrum.

In measuring the S-C II, a brightness match between comparison and standard half fields, in order to compensate for the S-C effect (of the first kind) is required in order to avoid contamination of the S-C II by the Bezold-Brücke hue effect. The latter is the change in apparent hue which occurs with changes in luminance. In general, as luminance is decreased, hues shift toward a greener or redder appearance (see, e.g., [3.105]).

The S-C II functions reported by STILES (change in apparent wavelength versus pupil entry position of the comparison beam) are complex in shape (Fig.3.14). Despite a general resemblance to S-C (of the first kind) curves, several of the functions seem not to be symmmetrical about the peak of the S-C function. Since all lights were essentially monochromatic, lens or other preretinal absorptions cannot be responsible for these asymmetries. With a few exceptions in midspectrum, most of the hue shifts are toward longer apparent wavelengths.

WALRAVEN and BOUMAN presented S-C II hue shift data for comparison beams entering the pupil 3.5 mm from the S-C function peak [3.98]. WIJN-GAARD and VAN KRUYSBERGEN also presented S-C II data for a single observer [3.93]. Neither study gives any particulars as to measurement techniques; however, the results agree qualitatively with both STILES and later work by ENOCH and STILES [3.4,96].

STILES noted that for most wavelengths, comparison and standard fields differed not only in (brightness and) hue, but in saturation as well. In the blue-green region of the spectrum, a beam passzng through the periphery of the pupil appeared more saturated than one entering at the center. ENOCH and STILES thus addressed the problem of specifying a *complete* color match (hue, brightness, and saturation) between fields entering at the center and periphery of the pupil. Standard and comparison monochromatic beams of the same wavelength were brought to match by the addition of small amounts of fixed primary lights to one or both fields, i.e., by trichromatic color-imetry. Standard fields were adjusted to an equal luminance spectrum. Enoch and Stiles's results, specifying the complete S-C II color change, are given in terms of Wright's u, v, w color coordinate system. From these data the apparent shift of the spectrum locus for the principle observer for a beam 3.5 mm from the S-C function peak could be specified. Corroborative data were obtained for a second observer as well. The calculated shifts of dominant

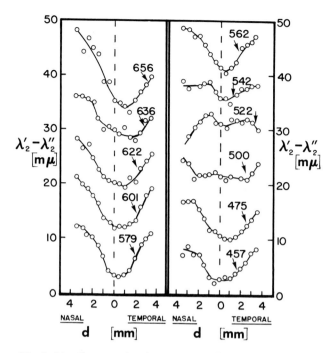

Fig.3.14. Changes in the apparent hue of foveally viewed monochromatic stimuli are plotted as a function of their pupillary entry positions (S-C II). An increase in the value of the ordinate indicates an apparent shift toward a longer wavelength hue. Several of these functions are complex in shape; some are apparently asymmetrical [3.4]

wavelengths coincide well in a qualitative way with S-C II hue shifts reported for other observers (Fig.3.15).

BRINDLEY investigated the effects of intense adapting lights on color matching [3.45]. In one of his experiments, he matched a monochromatic yellow with a mixture of red and green lights, all beams entering at the center of the pupil. When the pupil entry position of all stimuli was shifted to 3 mm from center, the match no longer held. This phenomenon was, of course, much more extensively studied in Enoch and Stiles's later work. However, BRINDLEY further observed that after adaptation to an intense (10^4 trolands) yellow field, a S-C II hue shift was no longer seen for a monochromatic yellow light when its entry position was shifted in the pupil. The observer could still discriminate actual wavelength changes of the stimulus after the intense light adaptation, however. Brindley's results indicate that, at least for a single wavelength, the S-C II hue shift could largely be nullified by intense light adaptation.

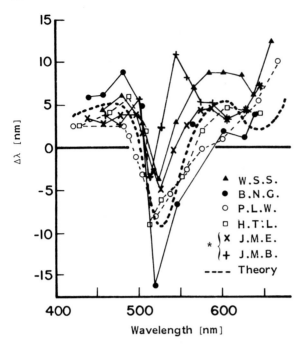

Fig.3.15. Changes in apparent hue for foveally viewed monochromatic stimuli at a pupillary entry position 3.5 mm from the peak of the S-C function (of the first kind). An increase in the value of the ordinate indicates an apparent shift toward a longer wavelength hue. Each curve is the data of a single subject; three separate studies are represented [3.4,96,98] [3.128]

FULD et al. reported that an S-C II effect does persist at high bleaching levels [3.106a,106b]. The high-intensity S-C II apparently differs both qualitatively and quantitatively from that reported for lesser retinal illuminances.

WALRAVEN and LEEBECK measured the S-C II hue shift in single deuteranomalous and protanomalous observers [3.107]. The hue shifts of both observers were larger than those obtained in color-normals. This result presumably indicates that the color anomalous' hue discrimination functions were not as acute as those of color-normals [3.46]. The color anomalous observers reported increased saturation as well as hue changes for some wavelengths entering peripherally in the pupil. The most curious aspect of the data is that for wavelengths greater than 500 nm the hue shifts were pronouncedly toward *shorter* wavelengths for both anomalous observers, in contrast to the shift toward the red described by color-normals above ca. 560 nm.

Very early on, STILES entertained the idea that S-C II hue changes resulted from different directional sensitivities of the fundamental cone mechanisms [3.4]. Although this must certainly play some role in the S-C II, Stiles's calculations indicated that another factor or factors must also contribute to the color changes.

One possibility is contained in the so-called self-screening hypothesis. Briefly, this is based on the observation that pigment absorption characteristics change with the effective density of the pigment in its substrate medium, or alternatively, with the path length through a solution of given pigment concentration [3.98]. STILES, and later BRINDLEY, WALRAVEN and BOUMAN, and ENOCH and STILES noted that light entering the pupil away from the center would strike well-oriented receptors at an oblique angle and hence, rather than pass down the whole absorbing length of the outer segment might "leak" out after traversing only some fraction of this distance [3.4, 45,96,98]. Hence, for nonsmall pigment densities, the spectral absorption of such receptors would depend upon the angle at which incident light struck the receptors. The self-screening hypothesis was tested by STARR, who measured foveal S-C functions (of the first kind) in protanopic and deuteranopic observers for a range of wavelengths and luminance levels [3.108]. In agreement with the self-screening prediction, the increase in directionality at high luminance levels was greatest for the wavelength at which each dichromat's single midspectral cone pigment was presumably the most sensitive.

ENOCH reported that the outer segments of bleached mammalian photoreceptors illuminated with white light appeared as a multicolored mosaic of modal patterns [3.37,42,109]. Changes of the angle of the incident light not only altered the modal patterns but also the distribution of hues observed. In general, increased obliquity of incident light resulted in an increased transmission of long and short wavelengths and lesser transmission of middle wavelengths in bleached receptors. Enoch's observations indicate that receptor waveguiding properties must be taken into account not only in the S-C brightness phenomenon but in S-C II color effects as well.

3.5 Theoretical Considerations

The first important theoretical issue concerning the S-C effect was whether it is primarily retinal in origin or an effect of preretinal absorption. STILES and CRAWFORD's calculations indicated that estimated preretinal absorption (and reflection) differences between beams passing through the center and periphery of the pupil were small compared to the magnitude of the psychophysically measured effect [3.1]. Experiments by CRAIK [3.110] and GOLDMANN [3.32] further indicated that the phenomenon was probably retinal in origin. WEALE's analysis of the effect of lens absorption indicated that this, in fact, *reduced* the magnitude of the psychophysical S-C effect [3.101]. Differences between normal photopic and scotopic S-C functions and changes seen over time in the S-C functions of observers with retinal pathology also indicate a retinal basis.

WRIGHT and NELSON proposed, virtually in passing, that the directional sensitivity of the retina might derive from the light trapping properties of retinal receptors [3.111]. That is, light incident upon receptors parallel to their long axis, or at very shallow angles from the parallel, might be expected to suffer total internal reflection and hence remain within the receptor for its entire length. On the other hand, light incident at more oblique anlges would no longer be contained by internal reflection, but rather be refracted out of the receptor into the intercellular space. Rays leaving the receptor would, of course, have a lower probability of encountering, and therefore exciting, visual pigment, rendering these rays visually less efficacious. This same concept had, in fact, been proposed by Brücke in the mid nineteenth century (see Chap.1).

O'BRIEN further developed the concept of total internal reflection within receptors as a possible basis for the S-C effect, stressing the importance of the ellipsoid, coupling the inner and outer segments, and its taper angle [3.5,112]. Within his theoretical framework, light losses from receptors occurred pirmarily within or near the ellipsoid region (also see [3.113]). O'Brien's model brought attention to the possibility that photoreceptors might concentrate light from the larger-diameter inner segments into the narrower outer segments. In fact, light collection has been directly observed in retinal photoreceptors [3.37,114]. Psychophysical evidence, which suggests that human receptors concentrate light, has been provided by BRINDLEY and RUSHTON [3.115,116].

TORALDO DI FRANCIA [3.48] pointed out that the dimensions of retinal receptors are on the order of the wavelengths of visible light; hence, receptor optical characteristics could not be adequately specified by the geometrical optical approaches such as WRIGHT and NELSON, and O'BRIEN had proposed. TORALDO DI FRANCIA drew an analogy between photoreceptors and dielectric antennae, suggesting not only the application of physical wave optics but also the usefulness of the optical reciprocity theorem, i.e., that angular radiation and acceptance patterns of receptors are identical.

Further analyses of the S-C effect have generally begun with the presumption that receptors act as optical waveguides (see Chap.6). Waveguide properties are ascribed on the basis of the dimensions of receptors, which in cross section are on the order of wavelengths of light, and on the higher refractive indices measured for receptors than for surrounding extracellular material [3.117-119].

Energy passes along waveguide structures in characteristic interference or modal patterns, which depend upon the receptor geometry (morphology), size, and refractive index relative to the surrounding material as well as upon the wavelength and angle of incidence of the exciting light [3.37,66, 109,120]. Such modal patterns have been observed in the receptors of many vertebrate species [3.34,37,120-122]. Moreover, observed modal patterns may be altered by a change of wavelength or of the angle of incidence of the exciting light [3.37,109,119,120]. The modal patterns seen are those which would be predicted on the basis of photoceptor dimensions and indices of refraction (Chaps.5 and 6).

Waveguides evidence directionality in the sense that energy incident at some angles is accepted preferentially over that at others. As noted above, optical studies of vertebrate receptors indicate a high degree of directionality, both in rods and in cones. Less directionality is found for groups of receptors than for single elements [3.123]. Waveguides also tend to concentrate light, in that their effective light capturing area is slightly larger than their geometrical cross-sectional dimension [3.124]. Psychophysically determined retinal directionality is thus seen to be in good qualitative agreement with the optical waveguiding properties of retinal receptors based upon their physical structure. SNIDER and PASK [3.125,126] and WIJNGAARD and co-workers [3.93,127] have proposed models of the S-C function based on physical wave optical treatments. WIJNGAARD has also proposed a model of the S-C II, based upon similar principles. In such analyses, receptor dimensions, refractive indices of the receptors, of their subregions, and of the extracellular interstitial matrix and associated structures are important parameters. These models require specification of the physical parameters noted above with high accuracy as well as simplifying assumptions as to receptor geometry and homogeneity. Unfortunately, the physical measurements made to date of photoreceptors and their natural retinal milieu fail to reach the accuracy demanded, essentially because of limitations inherent in available measurement techniques (but see [3.117-119] and Chap.5). This makes an evaluation of such models difficult, since all of them can be made to reasonably fit the psychophysical data by suitable adjustment of parameters.

An important issue for all models of the S-C effect is the extent to which the psychophysical function represents the directionality of single receptors, rather than an averaged directional tendency of a population of small acceptance angle receptors with interreceptor scatter in orientation [3.7,97,99]. Present evidence seems to favor the latter view, since 1) optical specification of acceptance angles of individual receptors of several species indicates small (2-4°) half angles [3.123]; 2) apparent local variations in receptor pointing can be demonstrated entoptically [3.24b,5-153]; 3) the transient S-C effect and directional light adaptation demand at least a modest nonuniform directional sensitivity at some level in the visual system (although the nonuniformity might lie *within* individual receptors, see [3.59,128,129]; and 4) BAILEY [3.57,58] has measured narrow directionality S-C functions using a flicker technique. On the other hand, fundus reflectometry indicates no measurable receptor scatter at the fovea in terms of bleaching [3.31],and the single electrophysiological study of the directional sensitivity of single retinal receptors (which is subject to the limitations discussed above) finds broad acceptance angles which fill the pupil [3.35].

LATIES and ENOCH have demonstrated that retinal histology can be of little help in assessing the small angular differences in neighbouring receptor orientation which may be involved [3.64]. Further intrareceptor electrophysiology and optical studies of single receptors and groups of receptors are important. Psychophysically, an understanding of the transient S-C effect might aid in the resolution of this question. A note of caution: it is not unreasonable to assume that, in different species, at different retinal locations within the same species, or at different adaptation levels, contributions of the acceptance angles of individual receptors, of interreceptor orientational scatter, and of neural summing across groups of receptors to the S-C effect might differ in relative importance.

3.6 Significance of the Stiles-Crawford Effect

The S-C effect or S-C-like effects have been demonstraqed in frog [3.33], goldfish [3.34], turtle [3.35,36] and humans. Optical studies of isolated retinas of goldfish, frog, rat, and humans have indicated directional sensitivity of individual receptors or of small groups of receptors in these species (e.g., [3.123]). Directionality has been shown for both rods and cones. It thus seems that directional sensitivity may be ubiquitous in vertebrate photoreceptors.

LATIES and co-workers found that photoreceptors at all positions of the retina tended to align to a common region at the anterior part of the eye, presumably near the center of the exit pupil. Such anterior pointing of photoreceptors has been seen in fish, amphibian, reptile, bird, mammal, and primate retinas [3.35,62-66]. The peak of the S-C function of normal human observers has been shown to remain near the center of the exit pupil of the eye for test locations up to 20° in the nasal visual field and 35° in the temporal visual field [3.6,19,50]. The psychophysically determined S-C function thus apparently reflects the retinal photoreceptor orientation and complements the histological findings. Additionally, in some species with reflecting tapeta, the tapetal reflecting surfaces have been demonstrated to be oriented approximately perpendicular to a hypothetical ray passing through the center of the exit pupil [3.65,130,131]. These questions are treated more fully in other chapters of this volume.

An obvious consequence of the S-C effect is that the effective stimulus for visual response cannot be simply specified in terms of retinal illuminance. The distinction between distal and proximal stimuli (e.g., [3.132]) becomes rather complicated. LEGRAND [3.3] has proposed the effective troland as a unit of retinal illuminance, in which a compensation for the S-C effect has been incorporated (see also [3.2,133]). Correction for the S-C effect, integrated across the pupil assumes additivity as well as a standard observer with the S-C function centered in the pupil. Markedly different S-C effect corrections are necessary for photopic and scotopic vision. The change from scotopic to photopic S-C function directionality occurs over a considerable range of mesopic illuminance, ca. 1.5 to 2 log units [3.7,39]. It is not immediately apparent how S-C corrections might be incorporated into photometry in a standardized, and yet meaningful, way.

VOS [3.134,135] has pointed out that a S-C effect peak which is decentered in the pupil tends to shift the *effective* pupil center, i.e., the location of the center of gravity of all the light entering the pupil when individual rays have been weighted for their luminous efficiencies, from the geometric pupil center toward the location of the S-C function peak. In fact, perfect centration of the S-C function in the pupil is not often found in normal observers. VOS has argued that individual differences in the magnitude and the direction of chromostereoscopic (apparent distance of colored lights) effects may be explained when the disparity between the effective and geometrical

pupillary centers is considered. He has presented S-C data for several observers which tend, at least qualitatively, to support these conclusions.

Since visual resolution is more dependent upon the presence of optical
aberrations than upon the small changes in target brightness which might occur on either side of the S-C function peak, maximal resolution is found for
light entering at the pupil center [3.84-87]. Thus, to a limited extent, the
effective pupil centers for brightness and for resolution may differ in
position. In one case of a displaced pupil, the best visual acuity was
achieved for target beam entry 2.5 mm from the S-C peak [3.81]. Clearly,
this disparity may be of importance in the use of subjective alignment to
artificial pupils or to a Maxwellian view optical system.

It is seen that the orientation of photoreceptors and associated structures with respect to the source of relevant visual stimuli is apparently
a common characteristic of vertebrate, and also of a large number of invertebrate visual organs (see Chap.11, also [3.136]). On the basis of their
physical properties, the photoreceptors of these diverse species apparently
all exhibit a directional sensitivity to incident light as well. The S-C
effect is a sensitive psychophysical function which, according to the current evidence as presented in this volume, reflects the underlying directionality and orientation of retinal photoreceptors.

Teleologically speaking, the prevalence of photoreceptor orientation and
directionality across numerous species indicates a highly significant role
for these specializations in the visual process. At this time, neither the
mechanisms which establish and maintain directionality and orientation nor
their complete role in vision are well worked out. Because the orientation
and directionality of photoreceptors must intimately bear upon the primary
phases of the neural visual response, these factors and their significance
must be understood in order to meaningfully evaluate their influence upon
subsequent visual processing, both in normal and in pathological conditions.

References

3.1   W.S. Stiles, B.H. Crawford: The luminous efficiency of rays entering
        the eye pupil at different points. Proc. R. Soc. London B*112*, 428-450
        (1933)
3.2   G. Wyszecki, W.S. Stiles: *Color Science* (Wiley, New York 1967)
        pp.561-564
3.3   Y. Le Grand: *Colour and Vision* (Chapman and Hall, London 1968)
        pp.103-108

3.4   W.S. Stiles: The luminous efficiency of monochromatic rays entering the eye pupil at different points and a new colour effect. Proc. R. Soc. London B*123*, 90-118 (1937)

3.5   B. O'Brien: A theory of the Stiles and Crawford effect. J. Opt. Soc. Am. *36*, 506-509 (1946)

3.6   H.E. Bedell, J.M. Enoch: A study of the Stiles-Crawford (S-C) function at 35° in the temporal field and the stability of the foveal S-C function peak over time. J. Opt. Soc. Am. *69*, 435-442 (1979)

3.7   B.H. Crawford: The luminous efficiency of light entering the eye pupil at different points and its relation to brightness threshold measurements. Proc. R. Soc. London B*124*, 81-96 (1937)

3.8   G. Westheimer: The Maxwellian view. Vision Res. *6*, 669-682 (1966)

3.9   K.N. Ogle: *Optics* (Thomas, Springfield, IL (1968)

3.10 L. Ronchi: On the influence of mydriatics and miotics on visual function. Atti Fond. Giorgio Ronchi *10*, 285-308 (1955)

3.11 A.M. Ercoles, L. Ronchi, G. Toraldo di Francia: The relation between pupil efficiencies for small and extended pupils of entry. Opt. Acta *3*, 84-89 (1956)

3.12 J.M. Enoch: Summated response of the retina to light entering different parts of the pupil. J. Opt. Soc. *48*, 392-405 (1958)

3.13 B. Drum: Additivity of the Stiles-Crawford effect for a Fraunhofer image. Vision Res. *15*, 291-298 (1975)

3.14 J.M. Enoch: "Summated response of the retina to light entering different parts of the pupil; Doctoral Dissertation, Ohio State University (1956)

3.15 F. Flamant, W.S. Stiles: The directional and spectral sensitivities of the retinal rods to adapting fields of different wavelengths. J. Physiol. *107*, 187-202 (1948)

3.16 J.C. Armington: Pupil entry and the human electroretinogram. J. Opt. Soc. Am. *57*, 838-839 (1967)

3.17 K.N. Ogle: Foveal contrast thresholds with blurring of the retinal image and increasing size of test stimulus. J. Opt. Soc. Am. *51*, 862-869 (1961)

3.18 K.N. Ogle: Peripheral contrast thresholds and blurring of the retinal image for a point light source. J. Opt. Soc. Am. *51*, 1265-1268 (1961)

3.19 J.M. Enoch, G.M. Hope: An analysis of retinal receptor orientation III. Results of initial psychophysical tests. Invest. Ophthalmol. *11*, 765-782 (1972)

3.20 J.M. Enoch, G.M. Hope: Directional sensitivity of the foveal and parafoveal retina. Invest. Ophthalmol. *12*, 497-503 (1973)

3.21 K. Blank, R.R. Provine, J.M. Enoch: Shift in the peak of the photopic Siles-Crawford function with marked accomodation. Vision Res. *15*, 499-507 (1975)

3.22 J.M. Enoch: Receptor amblyopia. Am. J. Ophthalmol. *48*, 262-273 (1959)

3.23 W.S. Stiles: Personal communication to Jay Enoch (1959)

3.24a G. Westheimer: Entopic visualization of Stiles-Crawford effect. Arch. Ophthalmol. *79*, 584-588 (1968)

3.24b G.G. Heath, P.L. Walraven: Receptor orientations in the central retina. J. Opt. Soc. Am. *60*, 733-734 (1970) (Abstract)

3.25 J.C. Armington: Pupil entry and the human electroretinogram. J. Opt. Soc. Am. *57*, 838-839 (1967)

3.26 C.E. Sternheim, L.A. Riggs: Utilization of the Stiles-Crawford effect in the investigation of the origin of the electrical responses of the human eye. Vision Res. *8*, 25-33 (1968)

3.27 R.G. Devoe, H. Ripps, H.G. Vaughan: Cortical responses to stimulation of the human fovea. Vision Res. *8*, 135-147 (1968)

3.28 K.H. Spring, W.S. Stiles: Variation of pupil size with change in the angle at which the light stimulus strikes the retina. Br. J. Ophthalmol. *32*, 340-346 (1948)

3.29 M. Alpern, D.J. Benson: Directional sensitivity of the pupillomotor photoreceptors. Am. J. Optom. Arch. Am. Acad. Optom. *30*, 569-580 (1953)

3.30 H. Ripps, R.A. Weale: Photo-labile changes and the directional sensitivity of the human fovea. J. Physiol. *173*, 57-64 (1964)

3.31 J.R. Coble, W.A.H. Rushton: Stiles-Crawford effect and the bleaching of cone pigments. J. Physiol. *217*, 231-242 (1971)

3.32 H. Goldmann: Stiles-Crawford Effekt. Ophthalmologica *103*, 225-229 (1942)

3.33 K.O. Donner, W.A.H. Rushton: Rod-cone interaction in the frog's retina analysed by the Stiles-Crawford effect and by dark adaptation. J. Physiol. *149*, 303-317 (1959)

3.34 F.L. Tobey, J.M. Enoch, J.N. Scandrett: Experimentally determined optical properties of goldfish cones and rods. Invest. Ophthalmol. *14*, 7-23 (1975)
      J.J. Vos: Some new aspects of color stereoscopy. J. Opt. Soc. Am. *50*, 785-790 (1960)

3.35 J.J. Vos: The color stereoscopic effect. Vision Res. *6*, 105-107 (1966)
      D.A. Baylor, R. Fettiplace: Ligth path and photon capture in turtle photoreceptors. J. Physiol. *248*, 433-464 (1975)

3.36 E.L. Paulter: Directional sensitivity of isolated turtle retinas. J. Opt. Soc. Am. *57*, 1267-1269 (1967)

3.37 J.M. Enoch: Nature of the transmission of energy in the retinal receptors. J. Opt. Soc. Am. *51*, 1122-1126 (1961)

3.38 J.M. Enoch: "The retina as a Fiber Optics Bundle", in *Fiber Optics, Principles and Applications*, ed. by N.S. Kapany (Academic, New York 1967) pp.372-396

3.39 W.S. Stiles: The directional sensitivity of the retina and the spectral sensitivities of the rods and cones. Proc. R. Soc. London B*127*, 64-105 (1939)

3.40 J.A. Van Loo, J.M. Enoch: The scotopic Stiles-Crawford effect. Vision Res. *15*, 1005-1009 (1975)

3.41 N.W. Daw, J.M. Enoch: Contrast sensitivity, Westheimer function and Stiles-Crawford effect in a blue cone monochromat. Vision Res. *13*, 1669-1680 (1973)

3.42 J.M. Enoch: Optical properties of the retinal receptors. J. Opt. Soc. Am. *53*, 71-85 (1963)

3.43 N.D. Miller: The changes in the Stiles-Crawford effect with high luminance adapting fields. Am. J. Optom. Arch. Am. Acad. Optom. *41*, 599-608 (1964)

3.44 P.L. Walraven: Recovery from the increase of the Stiles-Crawford effect after bleaching. Nature *210*, 311-312 (1966)

3.45 G.S. Brindley: The effects on colour vision of adaptation to very bright lights. J. Physiol. *122*, 332-350 (1953)

3.46 W.D. Wright: *Researches on Normal and Defective Colour Vision* (Kimpton, London 1946) pp.88-95

3.47 G. Westheimer: Dependence of the magnitude of the Stiles-Crawford effect on retinal location. J. Physiol. *192*, 309-315 (1967)

3.48 G. Toraldo di Francia: Retina Cones as dielectric antennas. J. Opt. Soc. Am. *39*, 324 (1949)

3.49 J.J. Vos, A. Huigen: A clinical Stiles-Crawford apparatus: Am. J. Optom. Arch. Am. Acad. Optom. *39*, 68-76 (1962)

3.50   H.E. Bedell: "Retinal receptor orientation in amblyopic and non-amblyopic eyes assessed at several retinal locations using the psychophysical Stiles-Crawford function"; Doctoral Dissertation, University of Florida (1978)

3.51   J.M. Enoch: The current status of receptor amblyopia. Doc. Ophthalmol. *23*, 130-148 (1967)

3.52   B. O'Brien: Local variations of the Stiles and Crawford effect. J. Opt. Soc. Am. *40*, 796 (1950) (Abstract)

3.53   J.M. Enoch, H.E. Bedell, E.C. Campos: Local variations in rod receptor orientation. Visual Res. *18*, 123-124 (1978)

3.54   W.L. Makous: A transient Stiles-Crawford effect. Vision Res. *8*, 1271-1284 (1968)

3.55   W. Makous: Some functional properties of visual receptors and their optical implications. J. Opt. Soc. Am. *67*, 1362 (1977) (Abstract)

3.56   G.G. Heath: Directional sensitivity of retinal receptors. J. Opt. Soc. Am. *60*, 736 (1970) (Abstract)

3.57   J.E. Bailey: "Directional sensitivity of retinal receptors"; Doctoral Dissertation, Indiana University (1974)

3.58   J.E. Bailey, G.G. Heath: Flicker effects on receptor directional sensitivity. Am. J. Optom. Physiol. Opt. *55*, 807-812 (1978)

3.59   R. Sansbury, J. Zacks, J. Nachmias: The Stiles-Crawford effect: two models evaluated. Vision Res. *14*, 803-812 (1974)

3.60   D.I.A. MacLeod: Directionally selective light adaptation: a visual consequence of receptor disarray? Vision Res. *14*, 369-378 (1974)

3.61   H.D. Baker: The course of foveal light adaptation measured by the threshold intensity increment. J. Opt. Soc. Am. *39*, 172-179 (1949)

3.62   A.M. Laties, P.A. Leibman, C.E.M. Campbell: Photoreceptor orientation in the primate eye. Nature *218*, 172-173 (1968)

3.63   A.M. Laties: Histological techniques for study of photoreceptor orientation. Tissue Cell *1*, 63-81 (1969)

3.64   A.M. Laties, J.M. Enoch: An analysis of retinal receptor orientation I. Angular relationship of neighboring photoreceptors. Invest. Ophthalmol. *10*, 69-77 (1971)

3.65   J.M. Enoch: Retinal receptor orientation and the role of fiber optics in vision. Am. J. Optom. Arch. Am. Acad. Optom. *49*, 455-470 (1972)

3.66   J.M. Enoch, B.R. Horowitz: "The Vertebrate Retinal Receptor as a Waveguide", in *Proceedings of the Polytechnic Institute of New York Microwave Research Institute Symposium* (Polytechnic Press, New York 1974) Vol.23, pp.133-159

3.67a   N.G. Webb: X-ray diffraction from outer segments of visual cells in intact eyes of the frog. Nature *235*, 44-46 (1972)

3.67b   N.G. Webb: Orientation of retinal rod photoreceptor membranes in the intact eye using X-ray diffraction. Vision Res. *17*, 625-631 (1977)

3.68   J.M. Enoch, A.M. Laties: An analysis of retinal receptor orientation II. Predictions for psychophysical tests. Invest. Ophthalmol. *10*, 959-970 (1971)

3.69   A. Safir, L. Hyams, J. Philpot: Movement of the Stiles-Crawford effect. Invest. Ophthalmol. *9*, 820-825 (1970)

3.70   F. Fankhauser, J.M. Enoch, P. Cibis: Receptor orientation in retinal pathology. Am. J. Ophthalmol. *52*, 760-783 (1961)

3.71   F. Fankhauser, J.M. Enoch: The effects of blur upon perimetric thresholds. Arch. Ophthalmol. *86*, 240-251 (1962)

3.72   C.J.W. Dunnewold: *On the Campbell and Stiles-Crawford effects and their clinical importance* (The Institute for Perception, Soesterberg RVO-TNO 1964)

3.73   J.M. Enoch, J.A. Van Loo, E. Okun: Realignment of photoreceptors disturbed in orientation secondary to retinal detechment. Invest. Ophthalmol. *12*, 849-853 (1973)

3.74   J. Pokorny, V.C. Smith, K.R. Diddie, F. Newell, F. Mausolf: "Color Matching and Stiles-Crawford Effect in Early Senile Macular Degeneration-Pilot Data", presented at Association for Research in Vision and Ophthalmology, Spring (1977)

3.75   E.C. Campos, H.E. Bedell, J.M. Enoch, C.R. Fitzgerald: Retinal receptive field-like properties and Stiles-Crawford effect followed in a patient with traumatic choroidal rupture. Doc. Ophthalmol. *45*, 381-395 (1978)

3.76   C.R. Fitzgerald, J.M. Enoch, E.C. Campos, H.E. Bedell: Comparison of visual function studies in two cases of senile macular degeneration. Klin. Monatsbl. Augenheilk. (1979)

3.77   V.C. Smith, J. Pokorny, J.T. Ernest, S.J. Starr: Visual function in acute posterior multifocal placoid pigment epitheliopathy. Am. J. Ophthalmol. *85*, 192-199 (1978)

3.78   V.C. Smith, J. Pokorny, K.R. Diddie: Color matching and Stiles-Crawford effect in central serous choroidopathy. Mod. Probl. Ophthalmol. *19*, 284-295 (1978)

3.79   W.E. Benson, A.E. Kolker, J.M. Enoch, J.A. Van Loo, Y. Honda: Best's vitelliform macular dystrophy. Am. J. Ophthalmol. *79*, 59-66 (1975)

3.80   J.M. Enoch: Quantitative layer-by-layer perimetry. Invest. Ophthalmol. Visual Sci. *17*, 208-257 (1978)

3.81   A.B. Bonds, D.I.A. MacLeod: A displaced Stiles-Crawford effect associated with an eccentric pupil. Invest. Ophthalmol. Visual Sci. *17*, 754-761 (1978)

3.82   H. Ohzu, J.M. Enoch, J.C. O'Hair: Optical modulation by the isolated retina and retinal receptors. Vision Res. *12*, 231-244 (1972)

3.83   H. Ohzu, J.M. Enoch: Optical modulation by the isolated human fovea. Vision Res. *12*, 245-251 (1972)

3.84   F.W. Campbell: A retinal acuity direction effect. J. Physiol. *143*, 25P-26P (1958) (Abstract)

3.85   F.W. Campbell, A.H. Gregory: The spatial resolving power of the human retina with oblique incidance. J. Opt. Soc. Am. *50*, 831 (1960)

3.86   D.G. Green: Visual resoltuion when light enters the eye through different parts of the pupil. J. Physiol. *190*, 583-593 (1967)

3.87   J.M. Enoch: "Retinal Directional Resolution", in *Visual Science*, ed. by J. Pierce, J. Levine (Indiana Univ. Press, Bloomington, IN 1971)

3.88   J.M. Enoch, L.E. Glismann: Physical and optical changes in excised retinal tissue. Invest. Ophthalmol. *5*, 208-221 (1966)

3.89   J.M. Enoch: Amblyopia and the Stiles-Crawford effect. Am. J. Optom. Arch. Am. Acad. Optom. *34*, 298-309 (1957)

3.90   J.M. Enoch: Further studies on the relationship between amblyopia and the Stiles-Crawford effect. Am. J. Optom. Arch. Am. Acad. Optom. *36*, 111-128 (1959)

3.91   R.L. Marshall, M.C. Flom: Amblyopia, eccentric fixation and the Stiles-Crawford effect. Am. J. Optom. Arch. Am. Acad. Optom. *47*, 81-90 (1970)

3.92   H.M. Burian, G.K. Von Noorden: *Binocular Vision and Ocular Motility* (Mosby, St. Louis 1974) pp.221-222

3.93   W. Wijngaard, J. Van Kruysbergen: "The Function of the Nonguided Light in Some Explanations of the Stiles-Crawford Effects", in *Photoreceptor Optics*, ed. by A.W. Snyder, R. Menzel (Springer, Berlin, Heidelberg, New York 1975) pp.175-183

3.94   F.W. Weymouth: "Visual Acuity of Children", in *Vision of Children*, ed. by M.J. Hirsch, R.E. Wick (Chilton, Philadelphia 1963) pp.119-143

3.95   J.M. Enoch: Optical interaction effects in models of the visual receptors. Arch. Ophthalmol. *63*, 548-558 (1960)

3.96   J.M. Enoch, W.S. Stiles: The colour change of monochromatic light with retinal angle of incidence. Opt. Acta *8*, 329-358 (1961)

3.97   A. Safir, L. Hyams: Distribution of cone orientations as an ex-
       planation of the Stiles-Crawford effect. J. Opt. Soc. Am. *59*,
       757-765 (1969)
3.98   P.L. Walraven, M.A. Bouman: Relation between directional sensitivity
       and spectral response curves in human cone vision. J. Opt. Soc. Am.
       50, 780-784 (1960)
3.99   A. Safir, L. Hyams, J. Philpot: The retinal directional effect:
       a model based on the Gaussian distribution of cone orientations.
       Vision Res. *11*, 819-831 (1971)
3.100 G.A. Geri, G.L. Kandel, C.R. Genter, H.E. Breed: Directional sensi-
       tivities of the human π-mechanisms. J. Opt. Soc. Am. *68*, 1399 (1978)
       (Abstract)
3.101 R.A. Weale: Notes on the photometric significance of the human crystal-
       line lens. Vision Res. *1*, 183-191 (1961)
3.102 J. Mellerio: Light absorption and scatter in the human lens. Vision
       Res. *11*, 129-141 (1971)
3.103 J.J. Vos, F.L. Van Os: The effect of lens density on the Stiles-
       Crawford effect. Vision Res. *15*, 749-751 (1975)
3.104 J.E. Bailey: The orientation properties of retinal rods. Optom. W.
       63(19), 32-33 (1972)
3.105 R.M. Boynton, J. Gordon: Bezold-Brücke hue shift measured by color-
       naming technique. J. Opt. Soc. Am. *55*, 78-86 (1965)
3.106a K. Fuld, B.R. Wooton, L. Katz: The Stiles-Crawford hue shift follow-
       ing photo pigment depletion. Nature *279*, 152-154 (1979)
3.106b B.R. Wooten, K. Fuld, L. Katz, M. Moore: The Stiles-Crawford II ef-
       fect at high bleaching levels. J. Opt. Soc. Am. *67*, 1362 (1977)
       (Abstract)
3.107 P.L. Walraven, H.J. Leebeck: Chromatic Stiles-Crawford effect of
       anomalous trichromats. J. Opt. Soc. Am. *52*, 836-837 (1962)
3.108 S.J. Starr: "Effect of Luminance and Wavelength on the Stiles-Craw-
       ford Effect in Dichromats"; Doctoral Dissertation, University of
       Chicago (1977)
3.109 J.M. Enoch: Visualization of waveguide modes in retinal receptors.
       Am. J. Ophthalmol. *51*, 1107-1118 (1961)
3.110 K.J.W. Craik: Transmission of light by the eye media. J. Physiol. *98*,
       179-184 (1940)
3.111 W.D. Wright, J.H. Nelson: The relation between the apparent intensity
       of a beam of light and the angle at which the beam strikes the retina.
       Proc. Phys. Soc. London *48*, 401-405 (1936)
3.112 B. O'Brien: Vision and resolution in the central retina. J. Opt. Soc.
       Am. *41*, 882-894 (1951)
3.113 W.H. Miller, A.W. Snyder: Optical function of human peripheral cones.
       Vision Res. *13*, 2185-2194 (1973)
3.114 K. Tansley, B.K. Johnson: The cones of the grass snake's eye. Nature
       178, 1285-1286 (1956)
3.115 G.S. Brindley, W.A.H. Rushton: The colour of monochromatic light when
       passed into the human retina from behind. J. Physiol. *147*, 204-208
       (1959)
3.116 G.S. Brindley: The deformation phosphene and the funnelling of light
       into rods and cones. J. Physiol. *147*, 24P-25P (1959) (Abstract)
3.117 R.L. Sidman: The structure and concentration of solids in photorecep-
       tor cells studied by refractometry and interference microscopy. J.
       Biophys. Biochem. Cytol. *3*, 15-30 (1957)
3.118 R. Barer: Refractometry and interferometry of living cells. J. Opt.
       Soc. Am. *47*, 545-556 (1957)
3.119 J.M. Enoch, F.L. Tobey: Difference in the index of refraction between
       the rat rod outer segment and the interstitial matrix. J. Opt. Soc.
       Am. *68*, 1130-1134 (1978)

3.120 J.M. Enoch: Waveguide modes in retinal receptors. Science *133*,
       1353-1354 (1961)
3.121 E.F. MacNichol: How to apply system analysis properly to biological
       systems. IEEE Int. Conv. Rec. *15*, 212-228 (1967)
3.122 J.M. Enoch, F.L. Tobey: Special microscope microspectrophotometer;
       optical design and application to the detemination of waveguide
       properties of frog rods. J. Opt. Soc. Am. *63*, 1345-1356 (1973)
3.123 J.M. Enoch: "Vertebrate Rod Receptors Are Directionally Sensitive",
       in *Photoreceptor Optics*, ed. by A.W. Snyder, R. Menzel (Springer,
       Berlin, Heidelberg, New York 1975) pp.17-37
3.124 A.W. Snyder, M. Hamer: The light capture area of a photoreceptor.
       Vision Res. *12*, 1749-1753 (1972)
3.125 A.W. Snyder, C. Pask: The Stiles-Crawford effect — explanation and
       consequences. Vision Res. *13*, 1115-1137 (1973)
3.126 C. Pask, A.W. Snyder: "Theory of the Stiles-Crawford Effect of the
       Second Kind", in *Photoreceptor Optics*, ed. by A.W. Snyder, R. Menzel
       (Springer, Berlin, Heidelberg, New York 1975) pp.145-158
3.127 W. Wijngaard, M.A. Bouman, F. Budding: The Stiles-Crawford colour
       change. Vision Res. *14*, 951-957 (1974)
3.128 B.H. Crawford: "The Stiles-Crawford Effects and Their Significance
       in Vision", in *Visual Psychophysics*, ed. 2y D. Jameson, L.M. Hurvich,
       Handbook of Sensory Physiology, Vol.7/4 (Springer, Berlin, Heidelberg,
       New York 1972) pp.407-483
3.129 P.E. King-Smith: The Stiles-Crawford effect and wave-guide modes: an
       explanation of MacLeod's paradox in terms of local adaptation within
       outer segments. Vision Res. *14*, 593-595 (1974)
3.130 E.J. Denton, J.A.C. Nicol: The choroidal tapeta of some cartilaginous
       fishes. J. Mar. Biol. Assoc. U.K. *44*, 219-258 (1964)
3.131 J.A.C. Nicol: The tapetum lucidum of the sturgeon. Contrib. Mar.
       Sci. *14*, 5-18 (1969)
3.132 L.A. Riggs: "Light as a Stimulus for Vision", in *Vision and Visual
       Perception*, ed. by C.H. Graham (Wiley, New York 1965) pp.1-38
3.133 P. Moon, D.E. Spencer: On the Stiles-Crawford effect. J. Opt. Soc.
       Am. *34*, 319-329 (1944)
3.134 A.W. Snyder, R. Menzel (eds.): *Photoreceptor Optics* (Springer, Berlin,
       Heidelberg, New York 1975)

4. Retinal Receptor Orientation and Photoreceptor Optics

J. M. Enoch

With 20 Figures

Let us consider the roles which the optical properties of retinal receptors play in the visual process. One can be overwhelmed when one examines the array of anatomical forms found in invertebrate and vertebrate species. In addition to differing receptor shapes and mosaics, in many species one must contend with the presence of tapetal (back-reflecting) mechanisms and screening dark pigments, photomechanical movements, double receptors, varying wavelength sensitivity and resolution characteristics, as well as differing forms of pupillary apertures. In deep-sea fish, complexes of amazingly different receptors may be observed in a single retina [4.1]. In addition, if one considers invertebrate eyes, there are numerous variants of superposition and apposition eyes which have evolved in those species.

When faced with this magnitude of complex forms, it is necessary to concentrate upon and investigate common properties rather than myriad specialized characteristics [4.2]. That which is common in all vertebrates and many invertebrates is the presence of some form of a fiber optics element associated with the retinal transducer. This presence implies the existence of a limiting aperture and therefore of directional transmissivity and sensitivity and the selection of certain parts of the incident wave front for transmittance to the transducer. The presence of the retinal fiber optics bundle allows the effective dissection of the retinal image into elements in a retinal plane and the guidance of these discrete signals to and along an extended length of absorber. The latter concept is of consequence since pigments in retinal receptors are highly oriented, and this orientation favors absorption of light passing along the axis of the fiber. Furthermore, in certain deep-sea fish, receptors are literally cascaded one behind the other into ranks of several layers. Quanta accepted for transmission are virtually assured of being trapped by some one cell in the sequential chain, although there is growing evidence that only the inner element is functional. The presence of a back-reflecting tapetum enhances the probability of entrapment of quanta in a local retinal area in applicable species. Given the presence of a limiting receptor waveguide aperture, that aperture also serves to eliminate the more aberrated rays as well as the stray light contained in the system. Thus rays striking the receptor obliquely (in excess of the effective cell aperture), regardless of cause, have a lower probability of stimulating the detector.

4.1 Pupillary Aperture as Source of Pertinent Visual Stimulus

In the vertebrate eye the retina is contained within the white, translucent, diffusing, integrating-sphere-like sclera. Light may impinge on the receptor from virtually any direction. Contrary to common belief, the world of the receptor is not limited to the pupillary boundary. Rather, the pupillary aperture is the source of the pertinent visual stimulus [4.2]. To prevent veiling glare from destroying image quality, special features have had to be evolved. These include the screening dark pigment in the choroid and pigment epithelium, and the fiber optic properties of the receptor. In certain species, photomechanical movements of parts of the pigment epithelium and receptors enhance or limit receptor exposure to light. Coupled with photomechanical movement in receptors, there are shape (morphological) changes which influence light transfer or guiding. For example, not only does the goldfish rod bury itself in dense pigment at high light levels, but the myoid portion of the rod inner segment thins to an almost nonexistent strand. The possibility of acceptance and guidance of energy to the now dark-pigment-shielded transducer portion of the cell is thus very limited [4.3].

From such considerations, one may hypothesize that the primary purpose of the receptor fiber optics element is to be a selective mechanism that accepts only that signal (information) pertinent to the organism. Of course, in addition to this primary function it is necessary to consider added specialized characteristics such as light collection, presence of colored oil drops, and so on. Inherent in this hypothesis is a key concept. Given that the receptor has a limiting aperture, exhibits directionality, and favors absorption of light passing down its axis, this integrated unit will be most effective if it is somehow aligned with the appropriate source of the pertinent luminous stimulus, the pupillary aperture.

4.1.1 Gibson's Figure

Some years ago, J.J. GIBSON sought to convey how space perception was organized in both invertebrate and vertebrate eyes [4.4]. He argued that each
position on the receptive surface of the eye (Fig.4.1) was associated with
a direction in real (or physical) space. He penned in an orientation of the
vertebrate receptors and assumed they were oriented as drawn.[1] Today we know
that he may have been right! This key feature of retinal receptor orientation had been overlooked for too long.

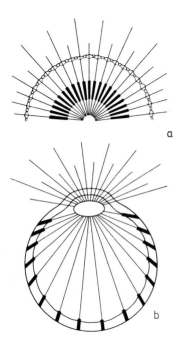

a

b

Fig.4.1. (a) The acceptance of light rays by
a convex compound eye of the type commonly
found in invertebrates. The receptors embrace
a converging sheaf of "pencils" of light and
register the differences in different directions. (b) The acceptance of light rays by a
concave vertebrate-type eye. Here also the
eye accepts a converging sheaf of (chief)
light rays, but the receptors register differences in different directions by an inverted image formed by the lens system of the
eye. (Courtesy of J.J. Gibson and Houghton
Mifflin Co., Boston)

Let us examine some of the underlying concepts considered in Fig.4.1. Located in physical space around any animal is an array of primary and secondary light sources which serve as stimuli for the visual system. The animal
needs to detect and locate those stimuli falling within its field of view
and to relate those sensed objects to its own position in space.

For simplicity, assume there is a fixed relationship between sense of
direction and retinal locus. As is implied in Fig.4.1, the animal is able

---

1 Personal communication from J.J. Gibson, 1972

to differentiate between different directions in an orderly manner, and it can interpret where it is in space in relation to other objects. This ability implies that the pertinent signal exciting the retina at a given point is identifiable and is retained for central processing. Furthermore, directionality-sensitive rods and cones function most efficiently if they are axially aligned with the aperture of the eye. Thus it becomes necessary to relate the aperture of the receptor as a fiber optics element relative to the center of the iris aperture. It is important to think of the retinal receptor waveguide-pupillary aperture system (and, where applicable, tapetal systems) as a functional unit designed for optimization of light capture for quanta having an origin in physical space and for rejection of straylight noise [4.2]. As was suggested in the Gibson figure, a fiber optics system has evolved twice, that is, in separate fashion for the rather different invertebrate and vertebrate eye forms.

An impressive point is how much evolutionary development takes place for modest gains. The back-reflecting tapetum, if optimally tuned, only doubles the probability of detection or increases signal up to a maximum of 0.3 l.u. These response systems function over several log units (about 10.0 l.u. in the human). The amazing thing is how many forms of tapetum have evolved in several species. Clearly it has had to be meaningful in survival.

If one assumes that 1) there exists a relationship between retinal locus of stimulation and sense of direction (spatial organization), 2) there exists a clear need to optimize the pertinent visual stimulus and distinguish it from stray light and aberrated rays, and 3) there exist directionally transmissive and hence selective detecting elements, it is amazing that others did not consider the orientational relationship which seemed so obvious to Gibson. We missed it because those of us close to the field were not aware that a relationship existed between receptor alignment and the pupil of the eye. A large number of eyes are sectioned daily for histological analysis. Part of the problem is that when histological sections are cut, the less rigid imbedded tissue block contacts a more rigid knife for cutting, and alterations and variations in receptor alignment that were encountered were regarded simply as cutting artifacts. It was assumed in all modern texts that all receptors were perpendicular to the pigment epithelial layer of the eye and hence pointed toward the center of the retinal sphere.

4.2 Retinal Receptor Orientation Studied by Means of Histological Techniques

Recently, LATIES examined the question of retinal receptor orientation by using histological techniques [4.5]. Using extremely fine techniques and great care in the preservation of tissue, he showed that receptors in a number of different vertebrate species did not point toward the center of the eye. He embedded his freshly obtained eye specimens in plastic and used metallurgical tools and polishing techniques rather than cutting in order to isolate a plane through the specimen. Hence, he minimally disturbed the alignment in his specimens and thus made possible the study of the alignment of several portions of the same retina. Figure 4.2 contains sections taken from different areas of a squirrel monkey retina at different distances from the posterior pole of the eye. From a set of such records, Laties formed a photomontage and simply drew lines along the axes of the receptors and noted that these tended to intercept each other at an anterior point in the eye [4.5]. This tendency implies that the receptors vary in orientation in relation to the pigment epithelium in a graded orderly manner across the retina. Comparable findings have been made in a series of mammalian, amphibian, reptilian, fish, and bird retinas [4.2,5-8]. Anterior receptor pointing has been observed in all warm- and cold-blooded vertebrates sampled to date. The sample included daylight-active species as well as nocturnal species, i.e., the property was common to rod-dominant, mixed-dominant, and cone-dominant species. Receptors from near the posterior pole of the nocturnal, rod-dominant, reptilian *Gekko gekko* are seen in Fig.4.3a. In Fig. 4.3b a more peripheral retinal locus is shown [4.6]. Comparable photomicrographs are shown for the rod-dominant, mammalian albino rabbit in Fig.4.4. On the basis of these histological studies, LATIES et al. evolved a simple mathematical model for describing the anterior pointing receptor characteristics [4.8]. Although these studies suggested that the eye pupil or eye lens might be the anterior pointing locus, this analysis cannot be made with great accuracy because of the limitiations of histology and the possibility of the induction of some disruption of the receptors even when the greatest of care is taken during the handling of specimens.

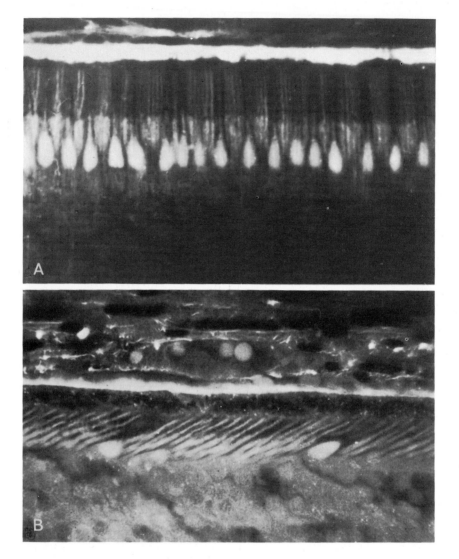

Fig.4.2a,b. Fluorescence micrographs of monkey retinas. (a) Receptors near the posterior pole of the eye. Cones can be readily distinguished from rods and are more plentiful in this part of the retina than in the periphery. In this retinal area the receptors are nearly perpendicular to the pigment epithelial layer. (b) In the periphery the receptors are aligned with an anterior ocular point, most probably the pupillary aperture. Note the relative dimensions of inner and outer segments of rods. Light would pass from below. (Courtesy of A. Laties, University of Pennsylvania, Philadelphia, and the Polytechnic Institute of New York, NY)

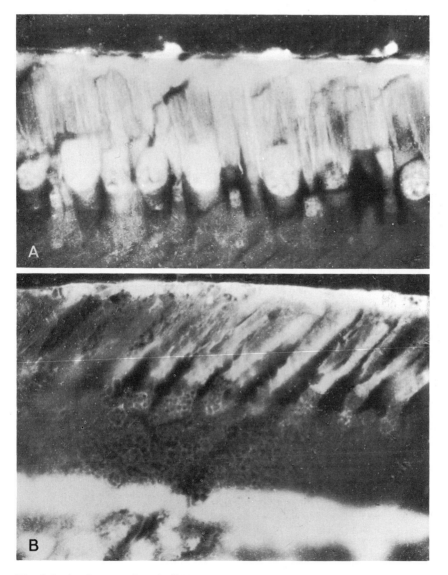

Fig.4.3a,b. An example of differential orientation of photoreceptors at two points along the same meridian of a Gekko *gekko* retina as shown by fluorescence microscopy. (a) Posterior pole. (b) Peripheral area. Light would strike the receptors from below. (Courtesy of A. Laties, University of Pennsylvania, Philadelphia, and C.V. Mosby Co., St. Louis, MO)

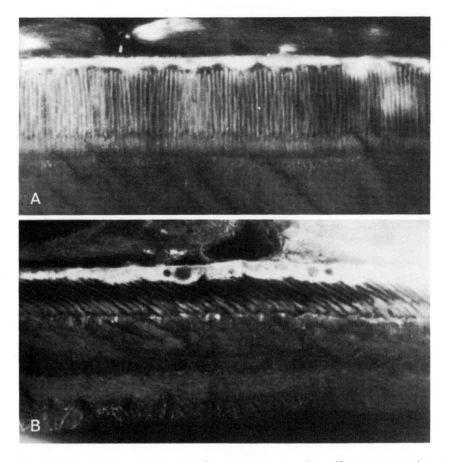

Fig.4.4a,b. Albino rabbit retinal receptors viewed by fluorescence microscopy.
(a) Receptors lying near the posterior pole of the eye. Here, the receptors
are virtually perpendicular to the highly fluorescent pigment epithelial
layer. Although cones are present, virtually all receptors appear as rods as
observed by light microscopy. (b) Fluorescence micrograph taken in the
reasonably far periphery of the same albino rabbit retina. In the peripheral
retina the receptors are no longer perpendicular to the pigment epithelial
layer. Note again that the inner segments of these mainly rod cells seem to
have a larger diameter than the outer segments. This point needs to be further
verified histologically. Light would normally pass from below upward. (Courtesy of A. Laties, University of Pennsylvania, Philadelphia, and the Polytechnic Institute of New York, NY)

4.3 Retinal Receptor Orientation Studied by Means of Psychophysical Techniques

LATIES and ENOCH [4.6,9] joined efforts, seeking to define a second experimental method which would verify anterior pointing and more accurately locate the point of convergence of receptor axes across the retina. Specifi-
cally, they asked whether further refinement of histological technique would
be fruitful, and whether a psychophysical study of the human photopic Stiles-
Crawford function [4.10] would allow independent verification of histological
studies. Qualitative evidence (histological) for anterior pointing in humans
was reported in [4.6]. The directional characteristics of the individual receptor cannot yet be properly related to the psychophysically or electro-
physiologically determined Stiles-Crawford function. However, the study of
the directional sensitivity of the retina (Stiles-Crawford function of the
first type) can be a powerful tool if it is assumed that this relationship
is indicative of the presence of an aligned retinal mechanism and that the
peak of the Stiles-Crawford function represents the central orientational
tendency of elements contained within the sampling area examined [4.9]. Given
this assumption, it is possible to investigate where the peak of the distribution is pointed or aligned at different points across the retina. If the
receptors represent a key component of the aligned mechanism, and if they
tend to point toward the center of the pupillary aperture of the eye, then
the peak of the Stiles Crawford function measured at different positions
across the retina should not change from a point approximating pupil center
regardless of where the Stiles-Crawford function is tested in the normal retina. On the other hand, if the alignment mechanism has as its orientational
locus the center of the retinal sphere (located near the center of rotation
of the eye), the peak of the directional sensitivity function would become
progressively displaced from the center of the pupil as the retinal test
area is moved away from the posterior pole of the eye. Such a model is implied if receptors are oriented perpendicular to the pigment epithelial layer
at all points across the retina. As a third alternative, let us assume that
all receptors in the eye are parallel to each other and to a reference receptor located at the posterior pole of the eye and that that reference receptor is aligned with the chief ray (ray passing through the center of the
aperture stop). Then the peak of the Stiles-Crawford function should move
out of the pupillary aperture within a certain displacement of retinal test
area from the posterior pole of the eye. However, in the latter case, the peak
would be displaced in a direction exactly opposite that which would be en-

countered for a center of the retinal sphere pointing hypothesis. Simply looking at Laties's figures eliminates any notion of random alignment.

4.3.1 Experimental Model

The three hypotheses, anterior pointing (center of the exit pupil pointing), center of the retinal sphere pointing, and all-parallel receptors can be quantitatively analyzed. These arguments have been sequentially developed [4.6,8,9,11-13]. The model used is shown in Fig.4.5. That part comprising the anterior point D, the retinal test locus P_n, and the center of the retinal sphere 0 is identical with that defined by Laties and associates. The Gullstrand human schematic eye has been grafted onto that model, and point D is assumed to be located at the center of the exit pupil of the eye lens system. The constants of the Gullstrand eye made emmetropic were employed (i.e., the retinal image was located at the secondary focal point of the Gullstrand eye lens system) [4.14]. For simplicity, the optic axis was as-

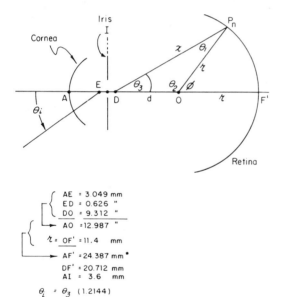

AE = 3.049 mm
ED = 0.626 "
DO = 9.312 "
AO = 12.987 "
$\mathcal{L} =$ OF' = 11.4 mm
AF' = 24.387 mm *
DF' = 20.712 mm
AI = 3.6 mm
θ_i = θ_3 (1.2144)

*Gullstrand Schematic Eye Made Emmetropic

Fig.4.5. Constants and angles used in the analysis are shown on this schematic drawing. A ray is incident at the center of the entrance pupil of the emmetropized Gullstrand schematic eye (E). After refraction, it continues on as if it had passed through the center of the exit pupil (D) to its retinal intercept P_n. Point 0 is located at the center of the retinal sphere. (Courtesy of C.V. Mosby Co., St. Louis, MO)

sumed to fall upon the fovea F' at the posterior pole of the eye. The distance d (or DO) was set equal to 9.312 mm, and the radius of the retinal sphere, r, was assumed to be equal to 11.4 mm. These values were based on ocular mensuration data and the requirements of the Gullstrand eye, i.e., the Gullstrand eye made emmetropic dictates the distance DF' = d + r = 20.712 mm. A paraxial chief ray trace allows one to relate θ_i (angle of incidence measured at the center of the entrance pupil of the eye) to θ_3 (angle of refraction originating at the center of the exit pupil of the eye). Here, θ_i and θ_3 are specified in relation to the optic axis of the schematic eye.

$$\theta_3 = 0.8234\theta_i \quad . \tag{4.1}$$

Other basic relationships are

$$\phi = \theta_1 + \theta_3 \tag{4.2}$$

and

$$d/\sin \theta_1 \quad \text{or} \quad d/\sin(\phi - \theta_3) = r/\sin \theta_3 = x/\sin \phi \tag{4.3}$$

and

$$\cot \theta_1 = r/d \csc \phi + \cot \phi \quad . \tag{4.4}$$

If small-angle conventions are used, from (4.3) it can be shown that

$$\phi = [(d + r)/r]\theta_3 = 1.817\theta_3 \quad . \tag{4.5}$$

Combining (4.5) and (4.1), we get

$$\phi = 1.496\theta_i \quad . \tag{4.6}$$

Although small-angle conventions may seem inappropriate at first glance, the peak of the Stiles-Crawford function would otherwise be translated out of the dilated entrance pupil of the human eye about 15° from the posterior pole of the eye for both the center of the retinal sphere and the parallel receptor pointing hypotheses. This point is readily appreciated by looking at Fig.4.6. If a receptor at P_n is aligned with point 0 (located at the center of the retinal sphere), its projection in the exit pupil of the eye (designated by its displacement from the center of the exit pupil, D) is shown as the distance a_k''. A tangent relationship allows evaluation of a_k'', since d is given and   may be determined for a given value θ_i (4.6). A ray trace allows one to determine the projection of a_k'' in the exit pupil to the plane of the entrance pupil a_k

$$a_k = 1.096a_k'' \quad . \tag{4.7}$$

138

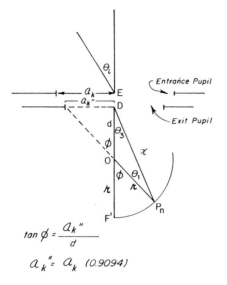

$$\tan \phi = \frac{a_k{''}}{d}$$

$$a_k{''} = a_k \ (0.9094)$$

Fig.4.6. For the center of the retinal sphere pointing hypothesis, the receptors are coaxial with the radii of the retinal sphere. With increase in angle ϕ, the forward projection of these photoreceptor axes to the pupillary plane results in marked displacements of that projection in the exit (and hence entrance) pupil. If the peak of the Stiles-Crawford function reflects the central tendency of receptor orientations, in the center of the retinal sphere pointing hypothesis one would expect maximum displacements of the Stiles-Crawford function in the entrance pupil, which would increase as θ_1 or ϕ becomes larger. In the parallel pointing case, all receptors are hypothesized as being parallel to the line DF'. The projection of such a parallel receptor located at P_n would intercept the exit pupil on the opposite side of the pupil as the projection in the center of the retinal sphere pointing hypothesis. The center of the exit pupil pointing hypothesis calls for the long axis of the receptor located at P_n to be coaxial with a ray having origin at point D (the center of the exit pupil of the eye). (Courtesy of C.V. Mosby Co., St. Louis, MO)

Here, a_k is the displacement in the entrance pupil of a ray projected from P_n relative to the center of the entrance pupil.

By a similar set of arguments one can determine the projection (to the plane of the exit pupil) of a receptor which is located at P_n but which is parallel to the line DF'. Using alternate interior angles, we get

$$\sin \theta_3 = a_k{''}/x \ . \tag{4.8}$$

Here, x can be determined from (4.3), and θ_3 is specified in (4.1). One solves for $a_k{''}$ and then determines its projection in the entrance pupil by using (4.7).

Since the anterior pointing or the center of the exit pupil pointing hypothesis requires the peak of the Stiles-Crawford function to remain centered in the exit (and hence entrance) pupil, let us allow that point to correspond to a_k = 0 mm (Fig.4.7). Then a displacement of the Stiles-Crawford peak indicating a center of the eye pointing model may be designated $+a_k$ and a displacement indicating the parallel pointing model may be designated $-a_k$. The predicted curves are shown in Fig.4.7. The computation

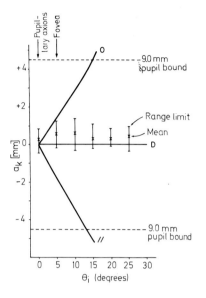

Fig.4.7. Data as well as predictions for each of the three hypotheses. Predictions are as follows: (0) center of the retinal sphere pointing hypothesis; (D) center of the exit pupil pointing hypothesis; ($||$) parallel pointing hypothesis. Here, a_k refers to position in the pupil of the eye (horizontal meridian). In this particular case, + a_k corresponds to the nasal side of the entrance pupil. It also indicates a prediction favoring the center of the retinal sphere hypothesis (predictor line 0). Also, θ_i is the angle of incidence of a chief ray measured at the center of the entrance pupil of the eye. Zero is defined as the pupillary axis. The values defined by angle θ_i can be translated into retinal test positions located on the horizontal meridian of the eye (as in Fig.4.1b). The total range of measured Stiles-Crawford peaks for three trained subjects and their combined mean positions in the entrance pupil are plotted for each test locus, θ_i. For comparison, the bounds of a centered, 9.0 mm, dilated pupil are shown by dashed lines above and below. The data clearly favor a center of the exit pupil pointing hypothesis (D) and indicate the presence of only one orientation law. Data for the same subjects in the vertical meridian are also available (see Fig.4.8)

of these predictive curves did not involve small-angle conventions, although paraxial ray traces were used.

The problem becomes complicated because the fovea is not on the optic axis of the eye, and a theory calling for anterior pointing logically requires the use of the posterior pole as the orientation reference. Asymmetries in the optics of the eye make specification of a unique optic axis difficult, and it is a problem to define an effective reference axis. In addition, there is essentially mirror symmetry in the two eyes (i.e., it is more meaningful to speak of nasal and temporal than left and right). Also, one needs to consider pupillary and retinal meridians other than the horizontal, and so on. Several of these questions have been treated in other studies, which may complicate sign conventions, etc., but does not meaningfully alter this abstracted analysis.

4.3.2 Experimental Results

The pupillary axis of the eye has been chosen as the reference axis in Fig. 4.7 as an estimate of the location of the posterior pole of the eye. It lies about 5° temporal in the visual field in relation to the point of fixation (fovea). The three predictive functions are indicated by D, anterior pointing; 0, center pointing; and ||, aligned parallel. Here, θ_i is a visual angle measured at the center of the entrance pupil of the eye (Figs.4.5,6), and a_k is the predicted or measured Stiles-Crawford peak position in the entrance pupil. The dashed lines give the bounds of an equivalent drug-induced, centered, dilated entrance pupil of 9 mm. Superimposed are plots of the combined mean locations of the peaks of the Stiles-Crawford function of three highly trained subjects [4.2,11]. The total range of measured peaks (three subjects) is designated by the vertical bars. Clearly, within the area tested, a single law of orientation pertains. Furthermore, there is little doubt that the measured central orientation tendency favors an anterior pointing hypothesis. Data taken in the vertical pupillary meridian corroborate this conclusion. The individual mean Stiles-Crawford peaks for each subject at the several test loci are plotted in Fig.4.8 within a hypothetical 9-mm entrance pupil. In addition, the group mean is plotted in relation to the pupil center. Note how the individual points cluster near the center of the pupil. On the basis of these data it would seem most probable that alignment of the receptors (central tendency) approximates the center of the exit pupil of the eye [4.11,12,15]. This study was considered to be so important that it was repeated on a second group of highly trained subjects with an identical result [4.15]. AGUILAR, PLAZA, and CRAWFORD have presented limited analogous data to those considered above [4.16,17]. However, their results do not allow one to draw meaningful conclusions on the subject.

There is a small nasal decentration of the Stiles-Crawford peak often found in Stiles-Crawford records. An attempt was made to determine whether the contracted pupil provided a better estimate of the locus of the Stiles-Crawford peaks than the dilated pupil did. The iris does not contract quite concentrically, and the drug-dilated state is not the photopic physiological norm. The contracted pupil proved to be about as good a reference center as the dilated pupil [4.12]. No intermediate point between the extremes of dilation and concentration proved to be a superior reference. Similarly, the corneal reflex did not provide a superior reference. The small nasal decentration of Stiles-Crawford peaks will be further commented on below.

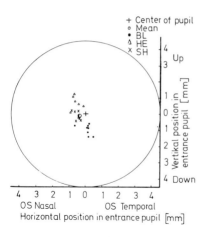

Fig.4.8. An arbitrary dilated, 9-mm entrance pupil is indicated. The small
cross indicates the center of the pupil
and corresponds to point E in Figs.4.5,6.
The data points for each observer correspond to the values of θ_i shown in Fig.4.7.
A grand mean across three observers, 0,
has been plotted and fell quite close to
the pupil center. (Courtesy of C.V. Mosby,
Co., St. Louis, MO)

When evaluating these data, please keep in mind that a ray displaced 1 mm
in the human entrance pupil of the Gullstrand schematic eye (a_k) is equal to
a 2.5° change in projected angle of incidence at the retina [4.18]. The receptor is tens of micrometers long, the eye is tens of millimeters long.
Consider the accuracy of pointing a 50-70 µm receptor at an aperture 20+ mm
away. How accurately can one point one's arm at an object several blocks
away? Is a 1-2° error possible? Although the small nasal displacement in
many sets of these data is probably real, the impressive point is the small
magnitude of the distribution of peaks.

As we look at the human eye, we see the entrance pupil (the image of the
iris aperture seen through the refractive cornea). Similarly, the receptor
does not "look" at the iris aperture, but rather it "sees" it through the
exit pupil of the eye (the image of the iris aperture as seen through the
eye lens). A ray traversing the center of the exit pupil of the eye traverses both the center of the true aperture of the eye and the center of
the entrance pupil of the eye.

Thus, if our initial assumptions are correct, namely, that the peak of
the Stiles-Crawford function reflects the central orientation tendency for
receptor alignment in the sampling area tested, then the point toward which
the receptors most probably tend to align approximates the center of the
exit pupil of the eye. This conclusion allows a finer localization of the
anterior pointing locus in man than is possible by histology. This conclusion does not state, as some authors have since implied, that all human
receptors are aligned with the pupil, but only that their central tendency
is to align with the pupil. We will consider this point further. This technique provides added evidence for anterior pointing, the approximate locus

for convergence of the axes of the photoreceptors has been defined, and human anterior pointing has been verified.

4.4 Retinal Receptor Orientation Studied by Means of X-Ray Diffraction

WEBB, using X-ray diffraction, has provided a third technique for the study of receptor orientation [4.19]. He demonstrated the presence of anterior pointing in the frog. Laties has provided added qualitative histological evidence for anterior pointing in the frog.[2]

4.5 Relationship Between Receptor and Tapetal Alignment

There is another bit of pertinent histological evidence. Some years ago, it was noticed that back-reflecting tapetal plates in certain elasmobranch fishes (sharks, rays, and so forth) were aligned essentially perpendicularly with lines connecting them to the center of the eye pupil [4.20-22]. In these fish there are also dark-pigment-containing cells which allow dark pigment to slide in front of the tapetal back-reflecting plates to prevent back reflection at high light levels. For simplicity, think of these dark-pigment-containing cells and the tapetal plates as a sort of venetian blind, the plates being the leaves of the venetian blind, and the pigment sliding in and out of spaces lying between the plates. Enoch called this work to the attention of Laties, who in turn enlisted S. Thorpe, then of Brown University and Woods Hole, to collect and fix the eyes of *Mustellus canis* (an elasmobranch). These retinas were analyzed in Laties's laboratory. Samples of the *M. canis* retina are seen in Fig.4.9. E. MACNICHOL has observed waveguide modes in this species [4.7]. Near the posterior pole (Fig.4.9a), the tapetal plates are essentially perpendicular to the receptors, and this perpendicular relationship also holds for a peripheral point (Fig.4.9b). In fact, if one places a draftsman's right angle on top of these drawings, the relationship between the receptor and the tapetum is almost a perfect right angle. This relationship between receptor orientation, tapetal plate alignment, and aperture cannot be a chance occurrence. A right-angle relationship provides maximum effective back-specular reflectance of

---

2 A.M. Laties (personal communication).

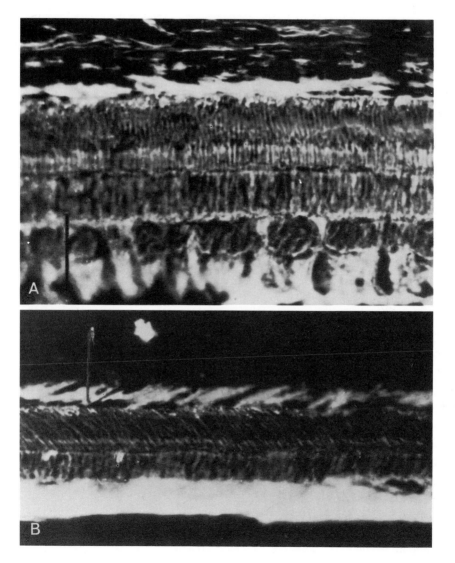

Fig.4.9a,b. Two views of a Mustellus canis (elasmobranch) retina. Waveguide modal patterns have been observed in this species by E. MACNICHOL (cited in [4.7]). (a) This section (light passing upward from the bottom) is from the posterior pole of M. canis. The receptors are nearly perpendicular to the (here highly fluorescent) tapetal reflecting surface. In a sense, this picture is not too different from the view near the posterior pole seen in Figs.42a,3a,4a. However, note the dark pigment that tends to slide laterally into the highly fluorescent zone at a few points. (b) The relation between the receptors, reflecting surfaces, and dark-pigment (shielding) cells is more readily seen in this picture taken from the peripheral retina of the same species. The receptors are aligned in approximation with the pupillary aperture, and the reflecting surfaces are angled at an almost perfect 90° from the receptor longitudinal axis. The black shielding pigment can be seen between some of the reflecting surfaces. (Courtesy of A. Laties, University of Pennsylvania, Philadelphia, and the Polytechnic Institute of New York, NY)

enery that has passed through the receptor. This reflectance enhances local
quantum-catch efficiency at low light levels, but results in the creation
of a form of partial integrating sphere at high light levels. This development would then act to reduce acuity and contrast sensitivity. Hence, the
need exists for the pigment to slide across and shield the back-reflecting
plates. One should not generalize this finding to all tapetal systems.

4.6 Realignment of Receptors Following Retinal Pathology

There are other pertinent indicators of the presence of a mechanism for
receptor alignment. These indicators also provide insight into the mechanisms for alignment. Namely, a small number of individuals who have had
their retina detached, then reattached, have been studied. We have observed
the redevelopment of receptor alignment as revealed by the study of the
Stiles-Crawford function. An early meaningful set of such data was published
by FANKHAUSER and ENOCH [4.23], although similar action was considered by
FANKHAUSER and associates [4.24]. ENOCH and associates [4.25] recently
published a data set demonstrating the recovery of orientation in the fovea
of a young man whose retina had been totally detached after a fight (Fig.
4.10). It is clear that in both cases the receptors had reoriented or, at
least, showed a clear tendency toward approximating normal orientation. We
are currently studying the rates and nature of recovery (and in a few patients nonrecovery) of visual response in almost 20 cases of patients who
have had surgical repair of retinal detachment.

Individuals whose retina had been literally lifted up by underlying
pathological processes have also been studied. For example, a 9-year-old
child with Best's vittelliform macular dystrophy was reported by BENSON
and associates [4.26]. The vittelliform material raises the retina up off
the pigment epithelial substrate by almost 1/3 mm in the macular area.
Amazingly, in patients exhibiting this rare congenital anomaly, quite good
vision is maintained for a remarkable period of time despite this lesion.
We found that this lifted-up section of retina had an essentially normal
(but a bit flat) Stiles-Crawford function approximately centered in the
pupil. In the child's second eye, following a recent hemorrhage, the Stiles-
Crawford function was greatly distorted by resultant scarring, and vision
was greatly reduced. Of particular interest in this disease is that in all
but two reported cases [4.27,28] the standing potential of the eye, the

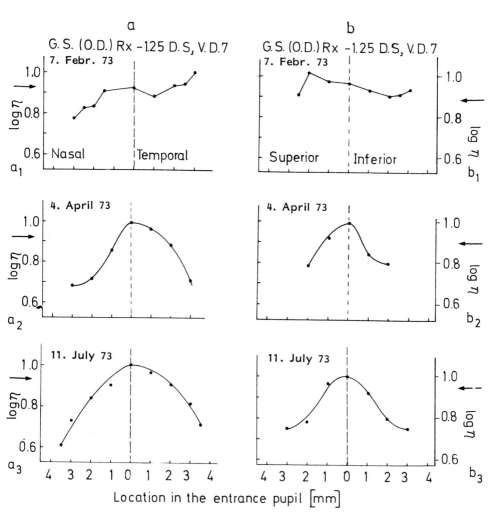

Fig.4.10a,b. Recovery of directional sensitivity of the retina following scleral buckling surgery for total retinal detachment. (a) Horizontal meridian, Stiles-Crawford function, and (b) vertical meridian, Stiles-Crawford function. The first pair of data (a and b) on the top were taken on 7 February, 1973, 15 days after surgery. The central pair of data were taken on 4 April, 1973, i.e., 2 months laters. The sets of data to the bottom were obtained on 11 July, 1973, nearly 6 months after surgery. The horizontal data set obtained on 11 July approaches the control data set obtained from the second unaffected eye of the patient. These data are hypothesized to reflect recovery of receptor alignment (in time) following the detachment and subsequent surgery. Visual acuity (resolution) and sensitivity improved with improvement in the Stiles-Crawford function. Fixation was always good. (Courtesy of C.V. Mosby Co., St. Louis, MO)

electrooculogram (EOG), has been anomalous. Unfortunately, the equipment to test that property was not available for the case mentioned.

Recently FITZGERALD et al. [4.29a] examined two cases of interest: one patient had a vitelliformlike lesion and another patient had pathology of the pigment epithelium-receptor interface with extinction of the electrooculogram (EOG). In the former case, identical results were obtained to those reportpd by BENSON, KOLKER and ENOCH [4.26]. The S-C curve was flattened but centered. In the second eye of this patient, the anomaly was very limited and a more normal looking S-C function was obtained. The egg-yolk-like lesion behind the neurosensory retina certainly could have contributed to the somewhat flattened S-C pattern. The EOG was reduced but not eliminated in this individual. In the second patient, in the area of active foveal lesion, the S-C function was anomalous, but at several other retinal points studied, the S-C function was absolutely normal, in spite of the fact that the light-induced portion of the EOG, a retina-wide function, was not measurable.

In another case reported by FITZGERALD et al. [4.29a], an individual with a serous neurosensory detachment above a serous pigment epithelial detachment was studied (Fig.4.11) [4.30]. It was significant that with remission of the serous neurosensory retinal detachment, but in the presence of a remaining pigment epithelial detachment, evidence for at least partial realignment (partial recovery of the S-C function) was present. This suggests that the pigment epithelium plays some role in the maintenance of receptor orientation.

We have known for some time that transretinal stress can alter receptor orientation. Such studies were conducted by FANKHAUSER and associates early on [4.24]. Sets of foveal Stiles-Crawford functions obtained just before and after light coagulation in the peripheral retina are shown in Fig.4.12. Here, unequivocally, the Stiles-Crawford function was altered by transretinal traction after light coagulation. FANKHAUSER, ENOCH and CIBIS presented fundus photograhs of a retina recorded at various times after a substantial light coagulation burn in the foveal area [4.24]. Some retinal vessels were translated across the retina, a distance roughly equivalent to a disk diameter (5-7° of the visual field), toward the burn area during the healing and scarring process. Lest these comments be misinterpreted, lesser burns cause lesser effects. Had we done this work today, these individuals would have been followed in time (about a year) in order to determine if the receptors in this area would partially or totally reorient themselves as they had in the young boy who had had the detachment (Fig.4.10).

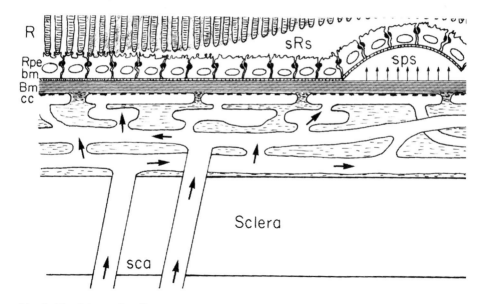

Fig.4.11. Schematic diagram summarizing changes occurring in many macular degenerations. There is both a focal serous detachment (sps) of the retinal pigment epithelium (Rpe) and of the neurosensory retina (sRs), i.e., the rods and cones (R). These detachments are secondary to a focal leak in the walls of the capillaries of the choriocapillaris (cc) in the central retinal area. Rpe cells are densely adherent to each other by a system of tight junctions (symbolized here by the small dark condensations shown at the interfaces of neighboring Rpe cells). These may be loosened in the presence of a Rpe detachment and its exudative process (sps). The basement membrane (bm + BM) is also affected in the area of the lesion. The vascular supply to the underlying choroid is indicated by sca. (From [4.30]). (Figure is reproduced with permission of the author and the publisher of the Am. J. Ophthalmol.)

CAMPOS et al. studied a most interesting young man who had been hit in the eye with a piece of rod which caused a rupture of the choroid (vascular layer external to the retina) and an overlying serous detachment of the neurosensory retina (Fig.4.11) with alteration in receptor orientation in the involved area and inner retinal neural involvement [4.31]. With remission of this traumatic lesion, receptor orientation and resolution recovered in the amazingly brief period of three weeks in the fovea. As the serous fluid was resorbed and scarring occurred, traction folds were seen on the inner surface of the retina. At a "control" point just $5°$ (visual angle) in the nasal field from the fovea, the S-C function went out of alignment (under a traction fold) just as foveal response returned to normal. Several weeks later the $5°$ eccentric point had returned near to its original alignment even though on overlying traction fold remained. The foveal distribution

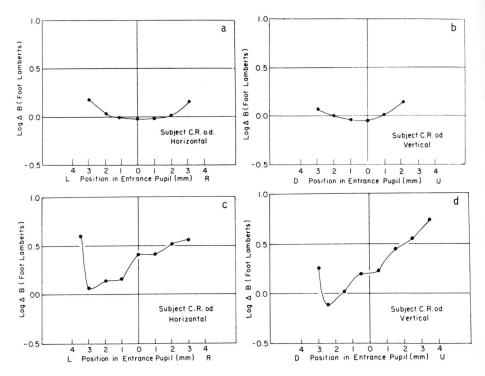

Fig.4.12a-d. Stiles-Crawford-type data plotted as threshold functions rather than sensitivity functions (threshold = 1/sensitivity). The relationships are simply inverted in relation to Fig.4.10. These Stiles-Crawford data represent traverses of a test beam in the entrance pupil of the eye in the horizontal (a,c) and vertical (b,d) meridians of the same foveal test point in the same patient. Curves (a) and (b) were determined just before and curves (c) and (d) were determined after light coagulation of the patient's peripheral retina. There is clear evidence of a shift in the Stiles-Crawford peak, presumably caused by transretinal stress relative to postcoagulation scar formation (Courtesy J. Opt. Soc. Am., Lancaster, PA)

remained unchanged. These data suggest that corrective alignment can occur on a very local basis and can occur in the presence of at least limited traction effects. Recently BRESNICK et al. reported somewhat similar results in a number of cases [4.32].

In some individuals, anomalous Stiles-Crawford functions have not exhibited recovery in time. If anomalies in receptor alignment are foveal, modest reduction in retinal resolution can occur [4.33,34]. These must be regarded as failures of the mechanism for alignment (or realignment).

In conclusion, in the presence of some conditions in which the retina is lifted off the pigment epithelium and then returns to proper contact (as in central serous choroidopathy [4.23]), senile macular degeneration [4.29b], or neurosensory retinal detachment) or is raised up on top of a lesion (as in vitelliform maculopathy), orientation reestablishment or maintenance of alignment is still possible. This property lasts into adulthood. Since the system can reestablish orientation (see also experimental studies of retinal detachment by MACHEMER [4.35,36]), the presence of a mechanism for orientation is implied.

4.7 A Unique Set of Directional Sensitivity Functions

A most interesting patient (female, white, mid-twenties) was recently reported by BEDELL and ENOCH [4.37]. This young lady, modestly amblyopic in one eye, was admitted to a study of transretinal alignment properties in amblyopia being conducted by BEDELL [4.38], in which he asked whether a disturbance in alignment in the central retina reflected a local event or a broadly based response characteristic. Except for this one young lady, no individual in his sample had a broadly based, i.e., transretinal, anomaly of directional sensitivity. In both eyes her Stiles-Crawford function was normal at fixation (assumed fovea). In her amblyopic right eye, when she was tested at various loci across the retina (Figs.4.13a, and 14a), the S-C peaks approximated the entrance pupil center at all loci. In her left eye which had normal visual acuity there was a systematic shift of alignment across the pupil as the visual field (retinal) test locus was shifted. Her peaks approached and followed the curve of the model but did not overlie the line describing the center-of-the-retinal sphere model defined in Fig. 4.7. An extensive set of measures was taken over a 6-month period at numerous horizontal and vertical field positions. The horizontal data for this eye are shown in Figs.4.13b,14b. Figures 13a,b are schematic plots showing the projected locations of her S-C peaks at several points along the horizontal meridian in both eyes. Note in the left eye that the lines cross at a locus between the pupil *and* the center of the eye. Her S-C functions (Figs.14a,b) approximated normal (perhaps slightly flatter OS).

This then is the first transretinal alignment anomaly-clearly her response system is following a different "law". As might be predicted, other response functions show decrements in sensitivity in the *periphery* of the field. For example, photopic dark adaptation is more affected than scotopic (rod) dark adaptation, etc.

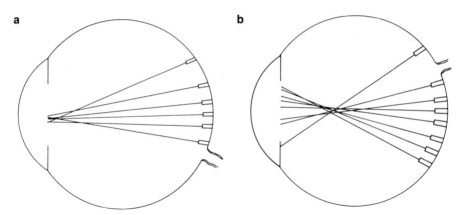

Fig.4.13a,b. This is a schematic representation of receptor alignment characteristics in the horizontal meridian of patient P.C. The retinal projections are inferred from measures of the S-C function at multiple points across the retina in the two eyes. (a) Right eye. (b) Left eye

Fig.4.14.
Figure caption see
opposite page

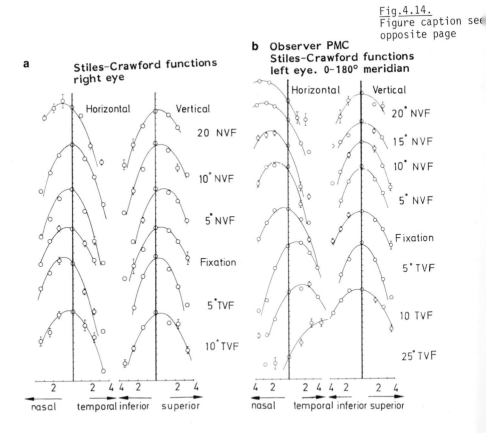

To determine whether this is a genetic variant both parents and one female sibling were studied. All showed normal transretinal S-C alignment in both eyes. It is not clear what mechanism can generate this retina-wide anomaly in alignment and how the distributive factors (which must influence the shape of the S-C function) are not (or only modestly) altered.

4.8 Distribution of Receptor Alignment

Are all receptors pointing at the same point? Obviously not. Although one may criticize the details of the measurements of the directionality of individual receptors, it would seem that individual receptors have relatively narrow directionality transmissivity. Directionality of individual rods and cones as estimated [4.3] is narrower than half angles of Stiles-Crawford functions measured to date. The Stiles-Crawford function represents some combination of the directionality of individual cells (or groups of cells), a distributive factor relating to orientation variation in the sampling area, and possibly some neural interactive factor related to how signals from the receptor cells are combined to produce the visual signal. RIPPS and WEALE demonstrated that the Stiles-Crawford function measured by fundus reflectometry is essentially the same as that measured by psychophysics [4.39]. Furthermore, studies of radiation patterns of populations of both rods and cones approach measured psychophysical functions (e.g., [4.40,41]). Thus, apparently, the neural interactive factor is a modest one; however, one cannot assume the distributive factor to be negligible. There is variance in all biological functions. Having viewed thousands of retinal preparations, I feel that the major variation in orientation occurs between receptor groups (e.g., [4.34,42]). Obviously, there must be some variation within groups as well. The studies of O'BRIEN and MILLER, who used a simple experimental technique, reveal the presence of subgroups having different orientations in any

---

Fig.4.14a,b. Patient P.C. Stiles-Crawford functions measured in (a) the right eye and (b) the left eye at several loci along the horizontal meridian in the visual field. Both horizontal and vertical traverses in the entrance pupils of her eyes were executed. Several replications exist. At all loci tested in the right eye, the S-C peaks approximate the pupil center in the horizontal and vertical meridians. In the left eye there is an orderly "march" of the peaks of the horizontally determined S-C functions across her entrance pupil, with the data obtained in the vertical pupillary meridian remaining aligned approximately with the pupil center

152

retinal area [4.43]. This experiment has been repeated in this laboratory and reaffirmed [4.33]. I have made comparable observations during studies of color mechanisms of the Stiles-Crawford function [4.44]. Furthermore, attempts at more quantitative measures of subgroups having a differing orientation within a sampling area have been reported separately by several investigators (e.g., [4.45-48]).

4.9 Rod and Cone Alignment Compared

The orienting mechanism must be somewhat similar for rods and cones on the basis of data recently presented by VAN LOO and ENOCH [4.49], wherein rod and cone directional sensitivity functions measured at the same retinal locus were symmetrical about the same point in the entrance pupil of the eye (Fig.4.15). Thus whatever differences exist in the directional sensitivity of rods and cones, the alignment characteristic has a common central tendency and locus. Furthermore, one can argue from Laties's fine histological studies (e.g., Figs.4.3,4) that rod-dominant species share the anterior pointing property. Also, rods do not seem to have orientation properties different from those of cones in obviously mixed cell populations (e.g., Fig.4.2). ENOCH has attempted to draw some of these concepts together in a recent publication [4.50].

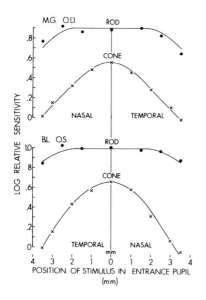

Fig.4.15. Photopic and scotopic Stiles-Crawford functions measured at the same retinal locus (6° from the fovea) in two different observers. A Kodak Wratten 23A (orange red) filter was used at the two adaptation levels. Photopic and scotopic functions exhibit common centration in both cases. (Courtesy of Pergamon Press, London)

One point evident in Figs.4.2,4 is that the inner segments of rods in the far retinal periphery in Laties's preparations seem to take on some of the morphological characteristics of cones, that is, there seems to be an increase in inner-segment diameter relative to outer-segment diameter. This observation needs to be assessed carefully by a highly qualified histologist and could be most important from the point of view of receptor optics. It suggests a unity in characteristics between rods and cones which has not been identified previously. This observation is presented as a suggestion for special study and should not be generalized at this time.

4.10 The Nature of the Orienting Mechanism

If there is an orientation mechanism, and this orientation mechanism can correct itself in the presence of certain anomalies, one must then ask about the nature of that mechanism. Is there some form of control system with an active feedback mechanism? What is the possible error signal, and what is the mechanism for alignment? In order to answer such questions, some means of inducing error in a nondamaging way is needed [4.51]. Furthermore, we can inquire as to how the orientation is originally achieved in an animal. For example, is anterior pointing present at birth? This question was asked in [4.9]. Rhesus macaque monkeys were delivered by cesarean section at the primate facility in Puerto Rico (this work was conducted by Laties), and it was shown that anterior pointing was present in the near-term rhesus fetus. Furthermore, comparable findings were observed in the chick retina [4.2]. One possible factor influencing orientation is radiant energy within the visible spectrum. Although it is doubtful that much light enters the womb of a rhesus monkey, light does penetrate the shell of a chicken.

BONDS and MACLEOD have reported studying a person who had a grossly displaced pupil and whose Stiles-Crawford peak was also displaced in a comparable manner (Fig.4.16) [4.52]. However, it should be pointed out that there have been several cases of retinal pathology with displaced Stiles-Crawford functions that are no longer centered in the pupil. It is necessary to rule out some form of scarring or tractional or other lesion.

Rather than study added pathological aberrants, ENOCH and associates reversed the question [4.53-55]. If light influences orientation, elimination of light should result in dispersal of alignment or other alteration in alignment.

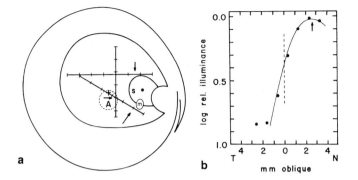

Fig.4.16a,b. Data obtained from the amblyopic eye of A.B. BONDS [4.52]. The normal and dilated pupil of iris of this eye are displaced. The Stiles-Crawford function is displaced in a near-similar manner. Several meridia were tested, and the peaks of the photopic S-C functions obtained on specific surveys of this pupil are indicated. (Courtesy A.B. Bonds, D.A. MacLeod, and the Publishers of Investigate Ophthalmology, C.V. Mosley, Co., St. Louis, MO)

Normal adult observers of both sexes aged 20-50 were occluded in one eye for a period of days with a black patch and full S-C functions were measured in the horizontal and vertical pupillary meridians at several loci across the retina in both eyes. Determinations were made before patching, at the end of patching and on recovery from patching (Fig.4.17). In the patched eye only, at all points tested out to 20° from fixation there was a profound flattening of the S-C function which reached its peak alteration in 3-5 days. The unpatched eye was not affected. Recovery took place in a comparable period of time after discontinuing patching. In order to better monitor the situation, a rapid method of peak and S-C function testing was developed. This required only 5 min of light exposure per retinal point tested per day for the necessary evaluations in two meridians.

It was also shown that the effect is due to light removal, and was not due to interference with form vision, i.e., the use of a white diffuser rather than black occluder had little effect on resultant S-C functions. Also, patching one eye did not alter measured S-C functions in the second eye.

The peak of the S-C function either did not shift during occlusion or shifted only slightly nasally.

With flattening of the pattern, overall ocular sensitivity was the same or increased (not decreased), and some individuals noted limited reduction in resolution (more so at lower luminances) and some reported modest alterations in perceived color [4.53-56].

20° NVF

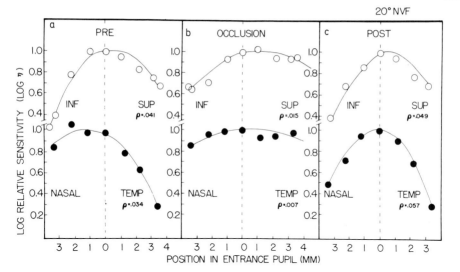

Fig.4.17a-c. The effect on Stiles-Crawford functions of monocularly occlud-
ing an eye with a black patch for a period of days. (a) Prepatch S-C test.
(b) Data obtained at the end of 7 days. (c) Data obtained 7 days after termination of patching. Both horizontal and vertical pupillary meridians were
tested, and the test point here was located 20° from fixation. The observer
was male and 29 years old. Note how flat the central area of the S-C function had become with patching. Testing was unequivocally photopic

It is not clear whether this represents a dispersal in receptor alignment
or alteration in receptor disk alignment. The latter seems to occur to some
extent with aging and can occur following intense exposure to light. The fact
that overal sensitivity increased is in keeping with recent findings of increasing receptor length, number of disks, rhodopsin concentration, and visual sensitivity in the dark. In the black occluder experiment, the S-C function flattened but showed little shift in peak, while just the opposite occurred in the case of the patient reported by BEDELL and ENOCH and BEDELL,
i.e., the peak shifted and there was only modest S-C pattern alteration
(Figs.4.13,14) [4.37,38]. Any theory of alignment properties must account
for both variants.

Another point is that the standing potential of the eye is a candidate
for control or modification of alignment as it provides an oriented potential
vector with an anterio-posterio axis in the eye. It was noted above [4.29a]
that the S-C could be maintained even though the light-induced portion of
the standing potential of the eye (as evaluated by the electrooculogram, EOG)
was extinguished. With dark patching, dispersal of aligned components took
place, i.e., the S-C function flattened. This suggests that the non-light-

induced portion of the EOG (that present in darkness) also has little influence on alignment [4.53-56]. Similarly, the measured EOG was normal in the eye with "center-of-the-retinal-sphere pointing" which was reported by BEDELL and ENOCH [4.57].

In a separate experiment the author tried wearing a specially fitted 15-mm-chord-diameter, black-painted HEMA soft-type contact lens with a 2-mm clear aperture centered 2 mm temporally. The lens had a double slab-off front surface designed to prevent rotation of the lens (Titmus-Eurocon, Aschaffenburg, Fed. Rep. of Germany). Homatropine cycloplegia was used. After a few days the S-C peak shifted in the nasal direction rather than temporally! At first this confounded ENOCH and BIRCH [4.55]. Removal of the lens resulted in a prompt (3-4 day) shift of the S-C function back towards the entrance pupil center (there is a slight nasal bias) at both the fovea and at a point 5° in the nasal visual field. These changes occurred during continued homatropine cycloplegia.

Careful study clarified the issue. In both eyes at 5° and 10° in the temporal visual field (towards the blind spot), alignment was markedly shifted in the nasal direction under normal conditions of illumination. On retest it is clear that alteration from 10° temporal field to 20° in the nasal field changes in a wavelike fashion spreading out from the blind spot area. This suggests some form of (tractional?) vector which has origin at the blind spot in Enoch's two eyes and which extends across the retina in the horizontal meridian. Light apparently counters this tractional force and provides a countervailing action which results in more normal alignment up to a point. Thus, normally the fovea and temporal retina (nasal field) align approximately with the pupil center. The 2-mm-aperture contact lens did not provide enough light (even though at least three hours a day was spent in the bright sunlight) to maintain alignment, but the general S-C pattern was retained.

Repeating the experiment using a black-painted contact lens with a 3-mm aperture displaced 2.5 mm from the center of the contact lens resulted in a realignment of receptors on the temporal retina towards the displaced pupil. The nasally aligned receptors near the blind spot hardly "budged." The point of fixation acted as a dividing line between response dominated by the possibly retinal-based tractional forces and the apparently light-driven response induced by the displaced pupil [4.55]. This experiment needs to be repeated on someone not subject to as strong a retinally based tractional factor influencing alignment.

Thus, here at least two factors were at play 1) a transretinal horizontal tractional vector of decreasing magnitude with its maximum at the optic nerve head, and 2) retinal illumination, presumably equally distributed (near the posterior pole — not in the far periphery). The (time-averaged?) amplitude of light passing the displaced 2-mm aperture was not sufficient to overcome this tractional force, while the energy admitted by the 3-mm aperture was adequate to achieve the desired goal, except where the tractional effect was greatest.

There is the suggestion that light acts in two modes. First, it seems to maintain or hold alignment, and a given alignment can be maintained over a period of low light (or even of withdrawal of light at night, during slepp, or in a dark environment, etc.). Second, light may act as a "driving" mechanism or signal to some error detecting feedback mechanism. This system must react to time-averaged luminance. It is this latter mechanism which apparently mediates alignment. Clearly, these are as yet inadequately defined mechanisms requiring further clarification and testing.

It is important to realize that wearing a black patch for a period of months is a routine training regime in amblyopia therapy, and patching for aperiod of time is common after ocular surgery and/or during medical therapy. Further, miotic (pupil contraction) therapy is common long-term therapy in glaucoma, and in some forms of accommodative-convergence anomaly, etc. Thus the maneuvers performed above are routinely used in ophthalmic practice.

4.10.1 Embryological State

There are several potential mechanisms capable of influencing orientation. MOSCONA and SHEFFIELD have removed embryonic chick retinal material and placed it in a tissue culture medium [4.58,59]. They then separate the retinal cells. These cells disperse and float in a single layer in random order in tissue culture medium. After cell separation, events are often recorded by time-lapse photography. In their studies one can readily observe the individual cells moving about and effectively identifying each other, then aggregating to form ringlike ("rosette") structures, and, finally, developing receptor segments aligned radially with the center of these rings. These segments are oriented as perfectly as spokes on a wheel. If liver and retinal cultures are mixed together, each group of cells separates and forms appropriate subunits. Clearly, the factors involved must influence initial receptor alignment in the embryo.

There are a number of other mechanisms, e.g., differential growth rates of pigment epithelium and retinal receptors (e.g., [4.60]) which can provide for development of primary orientation in the embryo prior to exposure to light or prior to development of postnatal ocular potentials. When analyzing a problem such as this, one cannot rule out the presence of multiple mechanisms.

4.10.2 Mechanisms Subserving Alignment

We know virtually nothing about possible mechanisms subserving alignment of photoreceptors. In many species of fish and amphibians marked photomechanical movements are recorded along the axis of the receptor [4.61]. In these species the inner segment, especially the myoid portion, is greatly altered in the light-adapted state, and these effects are influenced by temperature

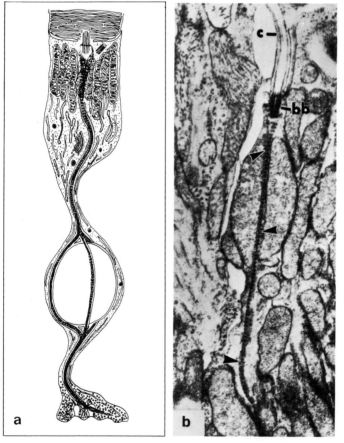

Fig.4.18a,b.
Figure caption
see opposite page

and circadian rhythms. However, such translations have never been noted in mammals, including primates.

Recently in a series of studies LATIES and BURNSIDE have noted the presence of actin filaments at various loci in the receptor [4.62,63]. Actin filaments are often associated with motile processes. Figure 4.18 (taken from [4.64]) shows organized filaments in the inner segment and deeper components of the guinea pig receptor. At this point in time it is not clear how such a mechanism might function or whether this is the critical mechanism.

The receptor outer segments are specializations originally derived from a primitive cilium. One question which might be asked is where does the cell bend? Does the rod or cone orient 1) at the junction of the inner and the outer segment near where the basal body of the cilium has its origin or 2) at or near the external limiting membrane, both the outer and the inner segment thus being included. From the vast majority of Laties's fine photographic records it would seem that the second possibility is correct.

Possibly, action of the myoid or the cilium or some sort of differential stretch or packing controls orientation on a local basis. Then, too, one has to consider the role of the interdigitating pigment epithelium fibrils. These fibrils and their relationship to the photoreceptor may take on elaborate forms, such as the structures STEINBERG and WOOD have described enveloping the outer segments of the cones of cats [4.65].

There is no question that the stretch or elastic characteristics of the retina and underlying choroid are different. For simplicity, think of the receptors as being attached to the retina by means of the external limiting membrane, and of the pigment epithelium and its processes as being attached to the underlying Bruch's membrane, which in turn is attached to the choroid. If one cuts a square block of retina, choroid, and sclera, the leatherlike sclera essentially lies there, the choroid retracts grossly from the cut scleral edge, and the retina tend to overhang the block a bit. Thus there are different elastic characteristics in each of the major layers. It is im-

---

Fig.4.18a,b. These figures are taken from a recent paper by SPIRA and MILMAN [4.64]. The schematic drawing (a) is the result of numerous records of photoreceptors, such as are seen in (b). The fibrile system of this guinea pig receptor is depicted. The arrows in (b) identify the fibrils, here passing through the mitochondria of the ellipsoid portion of the receptor and terminating at (near) the basal body (*bb*) of the receptor cilium (*c*). The cilium connects the inner and outer segments of the receptor. (Reproduced with the permission of the authors and of the publishers of the Am. J. Anat., Westar Press)

portant not to think of the receptors as essentially rigid bodies totally protected from the external environment. Over and above factors influencing their being held in place, receptors are not exempt from various physical forces, for example G forces [4.66,67], stretch effects [4.68-72], or shear [4.13,51]. RICHARDS has shown that following ordinary saccadic eye movements the peak of the Stiles-Crawford function is transiently displaced [4.73]. In recent work in this laboratory it was found that the act of accommodating causes a change in the position of the peak of the Stiles-Crawford function [4.13,51].

When the ciliary body contracts during accommodation, the retina advances anteriorly toward the ciliary body. Really, it is the choroid (which is continuous with the ciliary body) which advances; the retina is adherent to the choroid at its anterior edge, the ora serrata [4.68-72]. In man the advance can be as much as a half a millimeter [4.70]. This advance is enormous when one considers the total area of the retina [4.74]. There are several reported cases of tears in the peripheral retina (e.g., [4.75,76]) associated with the use of powerful drugs (such as phospholine iodide) causing contraction of the ciliary body and pupillary constriction. In this laboratory it has been demonstrated that the Stiles-Crawford peak is translated about a millimeter nasally in the pupil during marked accommodation (Fig.4.19). The direction of translation (Fig.4.20) implies that the underlying pigment epithelium is advancing more rapidly than the overlying external limiting membrane (retina proper). The presence of this induced nasal displacement may account for some part of the slight nasal orientation bias of the Stiles-Crawford function of the first type which is seen in several studies. All that would be needed for this to occur is a different causative and restorative rate constant. Similarly, a tractional force having origin at the optic nerve head (as recently revealed in the author's eyes) [4.53,54] could result in the same effect.

It is important to understand the translation seen in Fig.4.19. If the retina is drawn forward by an extensional stress from above and an extensional stress from below, a point lying on the horizontal meridian near the posterior pole (such as the fovea) should have placed upon it an equal force from above and below [4.72]. Of course, the balance in forces could be altered by some traction associated with retinal or chordoidal vasculature or the presence of adhesions or scars. However, in the horizontal retinal meridian the extensional stress caused by the forward advance of the nasal and temporal retina and translated back to the posterior pole or fovea cannot be the same as the extensional stress in the vertical meridian because of the interposition of the optic nerve head. The retina is attached at the optic

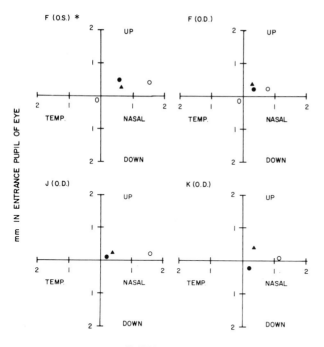

Fig.4.19. Mean locations of the peaks of Stiles-Crawford functions are presented in relation to the location of the corneal reflex as measured in the plane of the entrance pupil of the eye. (●) Test object optically projected to infinity, emmetropic young observers, no lens; (▲) -0.50-diopter soft contact lens placed over observer's eye; (○) -9.00-diopter soft contact lens placed over observer's eye. Note, nasal and temporal are reversed in direction O.S from O.D. The -0.50-diopter lens served as a near-plano control; the -9.00-diopter lens provided the accommodative demand for the test task. The observer and eye tested are indicated in each case. Up, down, nasal, and temporal refer to the entrance pupil of the eye tested. Separate determinations were made for the horizontal and vertical meridians. Marked accommodation causes a statistically significant nasal shift in the measured peak of the photopic Stiles-Crawford function. (Courtesy of Professional Press, Chicago)

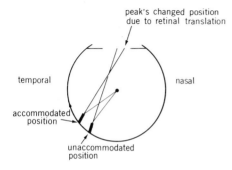

Fig.4.20. The outer ends of the receptors and associated structures (near the pigment epithelium) in the foveal sampling area tested are most probably advancing anteriorly at a rate faster than that of those parts of the receptors near the external limiting membrane. This process would result in a nasal Stiles-Crawford peak shift. (Courtesy of Professional Press, Chicago)

nerve head, and hence the force on the nasal retina cannot be effectively translated past the optic nerve head. However, the strain of the temporal retinal advance is translated to the foveal area, and the balance of forces assumed present in the vertical meridian is not duplicated. Thus, there is a possibility for a horizontal displacement and shear at the fovea associated with marked accommodation. A change in position of the peak of the Stiles-Crawford function implies a shearing-over rather than a simple translation [4.13], although most probably both are occurring. It is necessary to note that there may be added effects due to changes in the eye lens. There is evidence to suggest that the retinal stretch and shear effects just considered may still be present in elderly people. That is, though the eye lens can no longer change its shape, the ciliary body can still contract.[3] Thus the act of trying to accommodate may still cause extensional stresses and resulting strains in the elderly. These considerations may be important in cases in which the retina is poorly adherent to its substrate.

4.11 Summary

At this time, based on still limited studies, it would seem that a fundamental property of vertebrate receptor optics is anterior pointing by receptors. The anterior pointing locus is most probably a point near the center of the exit pupil of the eye. There is evidence for the recovery of orientation when that orientation is disturbed. One or more mechanisms mediate that orientation, and we must seek to define and understand their functional properties.

We must better define how disturbance in orientation influences vision. Briefly stated, substantially disturbed receptors have reduced light-guiding capability (and hence reduced sensitivity), reduced contrast sensitivity (caused by increased cross talk and so forth), and reduced resolution capability. These combine to create a detectable (but not necessarily common), modest form of amblyopia [4.33]. In a sense, these cases of poor acuity would seem to represent a failure of the photoreceptor alignment system. Similarly, receptor malalignment also will modestly influence perceived color because of factors relating to the Stiles-Crawford effect of the second kind.

An interesting set of theories has evolved in relation to receptor alignment in the neonate. There is considerable mechanical hydraulic stress on the newborn ocular vascular system at the time of delivery. Retinal hemor-

---

3 H. Goldmann (personal communication).

rhages and transient edema of the optic papilla following intracranial pressure rise and later rapid decompression during delivery commonly occur. These can readily cause disturbances in receptor alignment. In some instances the macula is involved. While hemorrhages are apparently rapidly absorbed, if foveal receptors remain misaligned during the critical period (not yet really defined) for the development of vision, central fine-resolution capability may fail to develop. That is, subsequent elaboration of the visual system is dependent upon the quality of the optically transmitted and neurally transformed retinal image during the critical period for visual development.

If the capability for realignment exists and occurs after part or all of the critical period, then signs of a prior receptor disarray may no longer be present in later years, but reduced resolution capability can persist. There is interesting literature on the subject [4.77-82]. K. Simons is currently reviewing this literature as part of his dissertation.[4] Simons has pointed out that the rupture of vessels may in fact partially serve as a safety-valve protective of the infant in the presence of these substantive forces, i.e., fine-vessel rupture may limit the potential for damage to the eye. At this stage of investigation it is dangerous to carry speculation too far.

Light clearly plays a role in maintenance of photoreceptor orientation. The standing potential of the eye, as sampled by routine clinical EOG techniques seems not to play a definitive role. Transretinal forces (shear, stretch), packing, physiological factors (accommodation, eye movements, exposure to gravitational forces), and pathology, can alter alignment, at least transiently. Clearly, maintenance of alignment is an active process and multiple factors are at play. Above all, one cannot but be impressed by the overall accuracy of alignment and its general stability [4.57].

Thus, not only must we consider the mechanism for orientation, the nature of a possible error signal, and the mechanisms for maintaining orientation and relating the retina to its substrate, but we must also analyze the consequences of disturbances in these mechanisms in relation to the individual's visual capability. Of equal importance is the need to understand the mechanisms and processes leading to failure of recovery of proper photoreceptor orientation. One must consider these studies taken as a whole as a pathology of receptor orientation.

Alignment of receptors is most probably related to the properties of the receptor as a waveguide. The alignment of the receptor with the pupillary

---

4 Personal communication.

aperture ensures optimal utilization of the pertinent signal necessary for survival of the animal.[5] The separation of the receptors, the presence of limiting receptor apertures, the directionality, the light-guiding and collecting properties, and so forth, help the photoreceptor to distinguish a pertinent signal passing through the pupil from intraocular noise and marked aberrations, and to distinguish the signal in one detector from that in its neighbors. The orientation and stacking of the transducing photolabile pigment in the outer segment enhance capture of quanta guided to that point by the photoreceptor. This proposed system is supplemented by added dark absorbing pigments, and, in some species, detection is aided by specular back reflections from a tapetum. If these are valid arguments, it becomes evident why the vertebrate and the invertebrate eye each had to evolve fiber optic elements that would be integral parts of the primary transduction mechanisms of the visual system. Many specialized waveguide properties have evolved to bias the particular signal presented to the transducers in different species. The specializations are fascinating in and of themselves. However, the key to the whole may well be the presence of the receptor waveguide and the utilization of the guide as part of the total aperture system of the eye [4.2].

Acknowledgement. This work was supported in part by a grant from the National Eye Institute No. EY-01418, NIH, Bethesda, MD.

References

4.1 N.A. Locket: Retinal anatomy in some scopelarchid deep-sea fishes. Proc. R. Soc. London B*178*, 161 (1971)
4.2 J.M. Enoch: Retinal receptor orientation and the role of fiber optics in vision. Am. J. Optom. Physiol. Opt. *49*, 455 (1972)
4.3 F.L. Tobey, Jr., J.M. Enoch, J.H. Scandrett: Experimentally determined optical properties of goldfish cones and rods. Invest. Ophthalmol. *14*, 7 (1975)
4.4 J.J. Gibson: *The Senses Considered as Perceptual Systems* (Houghton-Mifflin, Boston 1966) pp.163-165
4.5 A.M. Laties: Histological techniques for study of photoreceptor orientation. Tissue Cell *1*, 63 (1969)
4.6 A.M. Laties, J.M. Enoch: An analysis of retinal receptor orientation. I. Angular relationship of neighboring photoreceptors. Invest. Ophthalmol. *10*, 69 (1971)

---

5 I do not hold that all receptors are aligned with the center of the pupillary aperture, as MacLeod maintains.

4.7   J.M. Enoch, B.R. Horowitz: "The Vertebrate Retinal Receptor as a Wave-
      guid", in *Symposium on Optical and Acoustic Micro Electronics*, N.Y.
      Microwave Research Institute Symposium, Vol.23, ed. by J. Fox (Poly-
      technic, New York 1974) pp.133-159

4.8   A.M. Laties, P.A. Liebman, C. Campbell: Photoreceptor orientation in
      the primate eye. Nature *218*, 172 (1968)

4.9   J.M. Enoch, A.M. Laties: An analysis of retinal receptor orientation.
      II. Predictions for psychophysical tests. Invest. Ophthalmol. *10*, 959
      (1971)

4.10 W.S. Stiles, B.H. Crawford: Luminous efficiency of rays entering eye
      pupil at different points. Proc. R. Soc. London B*112*, 428 (1933)

4.11 J.M. Enoch, G.M. Hope: An analysis of retinal receptor orientation.
      III. Results of initial psychophysical tests. Invest. Ophthalmol. *11*,
      765 (1972a)

4.12 J.M. Enoch, G.M. Hope: An analysis of retinal receptor orientation.
      IV. Center of the entrance pupil and the center of convergence of
      orientation and directional sensitivity. Invest. Ophthalmol. *11*, 1017
      (1972b)

4.13 K. Blank, R.R. Provine, J.M. Enoch: Shift in the peak of the photopic
      Stiles-Crawford function with marked accommodation. Vision Res. *15*,
      499 (1975)

4.14 A. Gullstrand: cited by H. von Helmholtz: *Treatise on Physiological
      Optics*, Vol.1, ed. and transl. by J.P.C. Southall, 3rd. ed. (Dover,
      New York 1962) p.351

4.15 J.M. Enoch, G.M. Hope: Directional sensitivity of the foveal and para-
      foveal retina. Invest. Ophthalmol. *12*, 497 (1973)

4.16 M. Aguilar, A. Plaza: Efecto Stiles-Crawford en vision extrafouveal.
      An. R. Soc. Esp. Fis. Quim. A*50*, 119 (1954)

4.17 B.H. Crawford: Luminous efficiency of light entering eye pupil at dif-
      ferent points and its relation to brightness threshold measurements.
      Proc. R. Soc. London B*124*, 81 (1937)

4.18 J.M. Enoch: Nature of the transmission of energy in the retinal recep-
      tors. J. Opt. Soc. Am. *51*, 1122 (1961)

4.19 N.G. Webb: X-ray diffraction from outer segments of visual cells in
      intact eyes of the frog. Nature *235*, 44 (1972)

4.20 E.J. Denton: On the organization or reflecting surfaces in some marine
      animals. Philos. Trans. R. Soc. London B*258*, 285 (1970)

4.21 E.J. Denton, J.A.C. Nicol: The choroidal tapeta of some cartilagineous
      fishes. J. Mar. Biol. Assoc. U.K. *44*, 219 (1964)

4.22 J.A.C. Nicol: The tapetum in Scyliorhinus canicula. J. Mar. Biol.
      Assoc. U.K. *41*, 271 (1961)

4.23 F. Fankhauser, J.M. Enoch: The effects of blur upon perimetric thres-
      holds: A method for determining a quantitative estimate of retinal
      contour. Arch. Ophthalmol. *68*, 240 (1962)

4.24 F. Fankhauser, J.M. Enoch, P. Cibis: Receptor orientation in retinal
      pathology. Am. J. Ophthalmol. *52*, 770 (1961)

4.25 J.M. Enoch, J. Van Loo, E. Okun: Realignment of photoreceptors dis-
      turbed in orientation secondary to retinal detachment. Invest. Opthal-
      mol. *12*, 849 (1973)

4.26 W.E. Benson, A.E. Kolker, J.M. Enoch, J. van Loo, Y. Honda: Best's
      vitelliform macular dystrophy. Am. J. Ophthalmol. *79*, 59 (1975)

4.27 L.A. Birndorf, W.W. Dawson: A normal electrooculogr4m in a patient
      with a typical vitelliform macular lesion. Invest. Ophthalmol. *12*,
      830 (1973)

4.28 A.F. Deutman: Electroculography in families with vitelliform dystrophy
      of the fovea: Detection of the carrier state. Arch. Ophthalmol. *81*, 305
      (1969)

4.29a C.R. Fitzgerald, J.M. Enoch, D.G. Birch, M.S. Benedetto, L.A. Temme, W.W. Dawson: Anomalous pigment epithelial/photoreceptor relationships and receptor orientation. Invest. Ophthalmol. (submitted 1979)

4.29b C.R. Fitzgerald, J.M. Enoch, E.C. Campos, H.E. Bedell: Comparison of changes in visual functions in two cases of senile macular degeneration. Albrecht von Graefes Arch. Klin. Exp. Ophthalmol. *210*, 79-91 (1979)

4.30 J.D.M. Gass: Pathogenesis of disciform detachment of the neuroepithelium. Am. J. Ophthalmol. *63*, 573 (1967)

4.31 E.C. Campos, H.E. Bedell, J.M. Enoch, C.R. Fitzgerald: Retinal receptive field-like properties and Stiles-Crawford effect in a patient with a traumatic choroidal rupture. Doc. Ophthalmol. *45*, 381 (1978)

4.32 G.H. Bresnick, V.S. Pokorny, J. Pokorny: Visual function abnormalities in macular heterotopia due to proliferative diabetic retinopathy. Am. J. Ophthalmol. (submitted 1979)

4.33 J.M. Enoch: The current status of receptor amblyopia. Doc. Ophthalmol. *23*, 130 (1967)

4.34 H. Ohzu, J.M. Enoch: Optical modulation by the isolated human fovea. Vision Res. *12*, 245 (1972)

4.35 R. Machemer: Experimental retinal detachment in the owl monkey. IV. The reattached retina. Am. J. Ophthalmol. *66*, 1075 (1968)

4.36 J.M. Enoch: Photoreceptor orientation following retinal detachment. Am. J. Ophthalmol. *67*, 603 (1969)

4.37 H.E. Bedell, J.M. Enoch: An apparent failure of the photoreceptor alignment mechanism in a human observer. AMA Arch. Ophthalmol. *98*(11) 2023-2026 (1980)

4.38 H.E. Bedell: "Retinal Receptor Orientation in Amplyapic Eyes Assessed at Several Retinal Locations Using the Psychophysical Stiles-Crawford function"; Dissertation, University of Florida (Gainesville, 1978)

4.39 H. Ripps, R.A. Weale: Photo-labile changes and the directional sensitivity of the human fovea. J. Physiol. London *173*, 57 (1964)

4.40 J.M. Enoch, J.H. Scandrett: Human foveal far-field radiation pattern. Invest. Ophthalmol. *10*, 167 (1971)

4.41 F.L. Tobey, Jr., J.M. Enoch: Directionality and waveguide properties of optically isolated rat rods. Invest. Ophthalmol. *12*, 873 (1973)

4.42 J.M. Enoch, L.E. Glismann: Physical and optical changes in excised retinal tissue: Resolution of retinal receptors as a fiber optic bundle. Invest. Ophthalmol. *5*, 208 (1966)

4.43 B. O'Brien, N.D. Miller: " A Study of the Mechanisms of Visual Acuity in the Central Retina", WADC Tech. Rpt. 53-198, Wright Air Development Center, Wright-Patterson Air Force Base, Ohio (1953)

4.44 J.M. Enoch, W.S. Stiles: The colour change of monochromatic light with retinal angle of incidence. Opt. Acta *8*, 329 (1961)

4.45 W.L. Makous: A transient Stiles-Crawford effect. Vision Res. *8*, 1271 (1968)

4.46 G. Heath, P.L. Walraven: Receptor orientations in the central retina. J. Opt. Soc. Am. *60*, 733 (1970)

4.47 J.R. Coble, W.A.H. Rushton: Stiles-Crawford effect and the bleaching of cone pigments. J. Physiol. Londong *217*, 231 (1971)

4.48 D.I.A. MacLeod: Directionally selective light adaptation: A visual consequence of receptor disarray? Vision Res. *14*, 369 (1974)

4.49 F. Van Loo, J.M. Enoch: The scotopic Stiles-Crawford effect. Vision Res. *15*, 1005 (1975)

4.50 J.M. Enoch: "Vertibrate Rod Receptors are Directionaly Sensitive", in *Photoreceptor Optics*, ed. by A. Snyder, R. Menzel (Springer, Berlin, Heidelberg, New York 1975) pp.17-37

4.51 J.M. Enoch: Marked accommodation, retinal stretch, monocular space perception, and retinal receptor orientation. Am. J. Optom. Physiol. Opt. *52*, 376 (1975)

4.52 A.B. Bonds, D.I.A. MacLeod: A displaced Stiles-Crawford effect associated with an eccentric pupil. Invest. Ophthalmol. Vis. Sci. *17*, 754-761 (1978)

4.53 J.M. Enoch, D.G. Birch, E. Birch: Monocular light exclusion for a period of days reduces directional sensitivity of the human retina. Science *206*, 705-707 (1979)

4.54 J.M. Enoch, D.G. Birch, E. Birch, M. Benedetto: The effect of occlusion on some visual functions. Proc. Ninth Ophthalmological Symp. Amblyopia, Cambridge Univ., Sept. 9-11, 1979. Trans. Ophthalmol. Soc. U.K. *99*, 407-412 (1979)

4.55 J.M. Enoch, D.G. Birch: Evidence for alteration in photoreceptor orientation. Ophthalmology *87*, 821-834 (1980); Inferred positive phototropic activity in human photoreceptors. Philos. Trans. R. Soc. London B (in press, 1981)

4.56 D.J. Birch, E. Birch, J.M. Enoch: Visual sensitivity, resolution and Rayleigh matches following monocular occlusion for one week. J. Opt. Soc. Am. *70*, 954-958 (1980)

4.57 H. Bedell, J.M. Enoch: A study of the Stiles-Crawford (S-C) function at $35°$ in the temporal field and the stability of the foveal S-C function peak over time. J. Opt. Soc. Am. *69*, 435-442 (1979)

4.58 A.A. Moscona: In *Cells and Tissues in Culture*, Vol.1, ed. by E.N. Willmer (Academic, New York 1965) pp.489-529

4.59 J.B. Sheffield, A.A. Moscona: Electron Microscope analysis of aggregation of embryonic cells: The structure and differentiation of aggregates of neural retinal cells. Dev. Biol. *23*, 36 (1970)

4.60 R.M. Gaze, K. Strazicky: The growth of the retina in *Xenopus laevis*: An autoradiographic study. J. Embryol. Exp. Morphol. *26*, 67 (1971)

4.61 M.A. Ali: Les responses retinomotorices: Caracteres et mecanismes. Vision Res. *11*, 1225 (1971)

4.62 A.M. Laties, B. Burnside: "The Maintenance of Photoreceptor Orientation", in *Motility and Cell Function*, Proc. 1st John M. Marshall Symposium in Cell Biology, ed. by F.Pepe, V. Nachmias, J.W. Sawyer (Academic, New York 1978)

4.63 B. Burnside: Thin (actin) and thick (myosin-like) filaments in cone contraction in the teleost retina. J. Cell Biol. *78*, 227-246 (1978)

4.64 A.W. Spira, G.E. Milman: The structure and distribution of the cross-striated fibril and associated membranes in guinea pig photoreceptors. Am. J. Anat. *155*(3), 319-337 (1979)

4.65 R. Steinberg, I. Wood: Pigment epithelial cell ensheathment of cone outer segments in the retina of the domestic cat. Proc. R. Soc. London B*187*, 461 (1974)

4.66 W.J. White, W.R. Jarve: "The Effects of Gravitational Stress on Visual Acuity", WADC Tech. Rpt. No. 56-247, Wright Air Development Center, Wright-Patterson Air Force Base, Ohio 1956)

4.67 W.J. White: "Experimental Studies of the Effects of Accelerative Stress on Visual Performance", Ph.D. Dissertation, Ohio State University (1958)

4.68 P. Czermak: Cited by H. von Helmholtz: *Treatise on Physiological Optics*, Vol.1, ed. and transl. by J.P.C. Southall, 3rd ed. (Dover, New York 1962) p.401

4.69 W.H. Luedde: Hensen and Voelcker's experiments on mechanism of accommodation: An interpretation. Trans. Am. Ophthalmol. Soc. *25*, 250 (1927)

4.70 R.A. Moses: *Adler's Physiology of the Eye: Clinical Application*, 6th ed. (Mosby, St. Louis 1975) pp.306-312

4.71 G. Van Alphen: On emmetropia and ametropia. Supplementum ad Ophthalmologica *142*, 1 (1961)

4.72 K. Blank, J.M. Enoch: Monocular spatial distortions induced by marked accommodation. SCIENCE *182*, 393 (1973)

4.73 W. Richards: Saccadic suppression. J. Opt. Soc. Am. *59*, 617 (1969)

4.74 J.M. Enoch: Effect of substantial accommodation on total retinal area. J. Opt. Soc. Am. *63*, 899 (1973)

4.75 H.H. Lemcke, D.K. Pischel: Retinal detachments after the use of phospholine iodide. Trans. Pac. Coast Otoophthalmol. Soc. *47*, 157 (1966)

4.76 L.M. Spencer, B.R. Straatsma, R.Y. Foss: In *Symposia on Retina and Retinal Surgery*, Trans. New Orleans Acad. Ophthalmol. (Mosby, St. Louis, MO 1969)

4.77 H.M. Burian: The effect of twin flashes and of repetitive light stimuli on the human electroretinogram. Am. J. Ophthalmol. *48*, 274 (1959)

4.78 J.M. Chang: Panel D-15 test: Clinical practical application for congenital color anomaly. Trans. Soc. Ophthalmol. Sin. *13*, 21 (1974)

4.79 C.L. Giles: Retinal hemorrhages in the newborn. Am. J. Ophthalmol. *49*, 1006 (1960)

4.80 J.T. Planten, P.C. Von Der Schaaf: Retinal hemorrhage in the newborn: An attempt to indicate and explain its cause and significance. Ophthalmologica *162*, 213 (1971)

4.81 F. Sezen: Retinal hemorrhages in newborn infants. Br. J. Ophthalmol. *55*, 248 (1970)

4.82 G.K. Von Noorden, A. Khodadoust: Retinal hemorrhage in newborns and organic amblyopia. Arch. Ophthalmol. *89*, 19 (1973)

5. Waveguide Properties of Retinal Receptors: Techniques and Observations

J. M. Enoch and F. Tobey, Jr.

With 24 Figures

Historically the first direct proof that the vertebrate receptor outer segments were optical waveguides was Enoch's observation of the characteristic modal patterns in excised retinas (Figs.5.1,2) [5.1]. Previously TORALDO DI FRANCIA had predicted on theoretical grounds that outer segments should be waveguides [5.2]. JEAN and O'BRIEN and ENOCH and FRY had shown that scaled up polystyrene foam models of receptors acted as waveguides at appropriate microwave frequencies [5.3,4].

Much earlier (1843), BRÜCKE had pointed out that the high-index outer segments should cause light incident at small angles to the receptor axis to undergo total internal reflection and hence be confined to the receptor [5.5]. Such "light pipes" are the ray optics analogue of waveguides. In the same year HANNOVER described complex geometrical figures at the terminals of frog receptors when viewed end-on (Chap.1) [5.6]. This was probably the first reported observation of waveguide modal patterns, although he interpreted them as fine structures since waveguide phenomena were totally unknown at the time.

The modal patterns remain the most characteristic feature of optical waveguides, at least in the visible region where they are readily apprehended by the eye. The patterns observed at the terminals of the guides, where the latter act as antennas, are a consequence of the discrete modes in which electromagnetic energy is transmitted along dielectric waveguides. These modes are an interference phenomenon and result in maxima and minima of energy density within the guide volume. The pattern of the maxima and minima is determined by the particular mode, while the scale of a particular pattern depends also on the guide diameter and the refractive indexes of the interior of the guide and its surrounding medium ("core" and "cladding").

In this chapter we describe some of the techniques for making optical observations on retinas and single receptors and some of the results obtained to date.

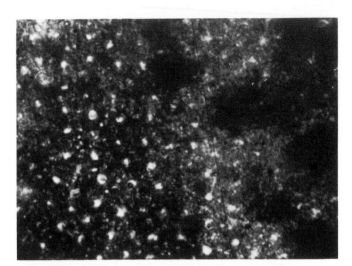

Fig.5.1

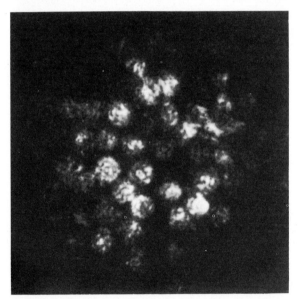

Fig.5.2

Fig.5.1. Receptor terminations in the parafovea of a squirrel monkey under narrow-angle illumination in the normal direction. Note the relatively simple modal patterns in these small-diameter receptors (ca.1-1.5 μm). [5.7]

Fig.5.2. Receptor terminations in a frog retina ("red" rods). Narrow-angle illumination in the normal direction. The patterns are much more complex because of the larger size of the frog outer segments (ca.6μm). [5.8]

5.1 Sample Preparation

It should be evident that any treatment which alters the dimensions, alignment, refractive indices, or pigment content of the retinal structures from
their normal conditions will alter the measured optical properties. Because
the retinal receptors form a closely spaced and closely aligned bundle of
optical waveguides, even small changes in the above properties can have
noticeable effects. Furthermore, changes in the material surrounding the receptor waveguides, or preceding them in the light beam, or even changes in
neighboring receptors can affect the optical properties of the receptors
under investigation. Ideally, one would like to study the retina in the living eye. While a great deal of information has been obtained thus by slit
lamp studies, fundus reflectometry, and especially by retinal spectroden-
sitometry, the long working distances required and the consequent limitation
on optical magnification have prevented observations of the living retina
at the cellular level, with few exceptions [5.9-12].

Much more can be learned about retinal receptor optics from undamaged
samples free of the intervening media. Sample preparation may be resolved
into two problems: that of 1) obtaining an undamaged sample initially and
2) preserving it long enough in a form suitable for making measurements. The
receptor cells are mechanically fragile, with their long, thin, inner and
outer segments projecting beyond the external limiting membrane toward the
pigment epithelium (Chap.2). It is this same structure coupled with the
lower refractive index in the interstitial medium which imparts their waveguide character. Separating the retina from the pigment epithelium with the
receptors intact in their matrix requires careful work.

5.1.1 Species Employed

While a large variety of animals have been subjects of retinal investigations over the years, for most workers interested in optical properties
of the retina, the number of species which are both suitable and available
will be much more limited. Particularly if substantial pieces of undamaged
retina are required, both initial cost and the logistics of maintenance become serious considerations. We have found that quality preparations of this
type demand practice on several animals daily over a period of many weeks.
The larger and more expensive the animal, the more difficult it is to maintain such a schedule. It is advisable to inquire about availability over
the projected duration of the experiment. Some species may be abundant at
one season and quite unobtainable at another. The stage of the life cycle

and past history may be important considerations. Animals undergoing hibernation may show a very different retinal condition than at a peak of activity. Animals collected from the wild often come from widely different locations and are shipped to a central point for distribution (e.g., *rana pipiens*). In the process some species may go for long periods without being fed or with inadequate diets, which can seriously affect the condition of the retina. Retinas of the albino rat are effectively destroyed by even brief exposure of the animal to bright light, e.g., by placing them in sunlight. There is no "perfect" animal for this work and each has its own particular advantages and disadvantages depending on the experiment on view.

Albino rats provide the cheapest, readily dissectable mammalian retinas. The receptors are uniform and rodlike in appearance although there is evidence for cones and cone function [5.13-15]. The outer segments are a reasonable match in dimensions to the rods of primates and of man. Variations across the retina appear minor. Since they are supplied by breeders, the genetic heritage may be somewhat more consistent than animals collected from the wild. Albinism simplifies dissection and avoids other problems associated with dense pigment epithelial melanin. If one obtains weanlings and enhances their diet for about three weeks with Vitamin A, stable, rhodopsin rich retinas are obtained. Animals are sacreficed at a weight of about 160-190 g. Albino rats must never be placed, even briefly, in direct sunlight or other bright light as the retinas deteriorate rapidly under these conditions.

Rabbits have retinas rather similar to rats but are more expensive (though not prohibitively so), and require more space per animal. Rabbits are particularly suitable for "choroidal block" preparations (see below). Mouse eyes are too small for easy handling, although genetic variants with displaced pupils may be valuable for studying factors affecting orientation of receptors.

The small size of mammalian receptors will push the limits of any optical system where single receptors are to be studied end-on. Frog red (transmitting) rods are 6 µm in diameter and, consequently, much easier to study in isolation. By the same token, however, the optical properties must differ in important respects from mammalian retinas since they support higher order modal patterns (Fig.5.2). The frog retina is fairly complex, containing besides the red rods, the "green" rods and small single and double cones. The latter are difficult to identify looking at the receptors end-on, but the radiation transmitted by the retina appears to be dominated by the red rods. As the frog retina deteriorates, the red rods appear to separate along longitudinal folds resulting in a "lobulated" cross section (Fig.5.3). This changes

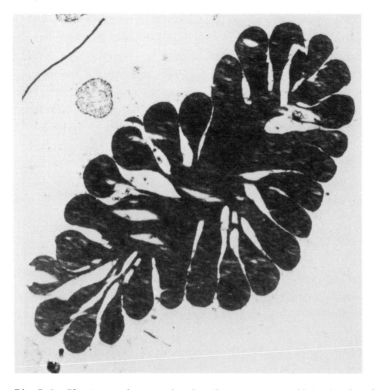

Fig.5.3. Electron micrograph of a frog receptor disk showing lobulations.
The disk is grossly swollen and the outer cell membrane has been removed.
[5.8] (Courtesy of Dr. Adolph Cohen)

the waveguide characteristics considerably. The receptor myoids and the melanin pigment show "photomechanical" shifts with light level.

Goldfish are among the easiest animals to maintain. Commercial supplies
are plentiful and the equipment is available from most pet shops. Fish obtained from local suppliers may sometimes be blind, however. The receptors
are much smaller in diameter than frog, but enough larger than mammalian
receptors to ease optical problems substantially. Rods and cones are easily
differentiated in end-on observations, and both contribute significantly to
transmitted radiation. The retina is fairly complex, containing single and
double cones as well as the rods. The retinal mosaic varies considerably in
different regions of the retina. The goldfish also shows pronounced photomechanical effects (e.g., [5.16,17]).

Ground squirrels are reported to have an all-cone retina. The supplies
are somewhat uncertain, however, and we have found the retina difficult to
remove without shearing off the receptor outer segments.

The squirrel monkey has been the cheapest primate obtainable and has a well-developed fovea. Cone dimensions change from the fovea outwards more rapidly than in humans. Macaque retinas are closer to human in dimensions than squirrel monkeys, and their color vision also seems closer to human [5.18,19]. Monkey retinas tend to adhere to the underlying pigment epithelium.

If one has access to a medical institution doing enucleations, excellent fresh foveal preparations are obtainable from human eyes. In older humans adherence between the retina and its pigment epithelial substrate tends to be weak and it takes little effort to separate the retina proper from the pigment epithelium. Orientation may be virtually undisturbed under these conditions, but preparations must be obtained within minutes of enucleation.

5.1.2 Methods of Preparation and Observation

DARTNALL discussed techniques for obtaining whole retinas from small ex-perimental animals and from excised eyes of larger species [5.20]. Exper-imental animals are dark adapted, to maximize the amount of photopigment if this is to be retained for study, and also because removal of the retina is often easier.

The easiest preparations to make are those in which an isolated receptor cell is to be observed side-on. This type of preparation is useful for de-termining dimensions, index of refraction, linear dichroism, and birefrin-gence of single cells. It has also been used for most microspectrophotometry of photopigments in single cells performed to date. The excised retina is teased out, or agitated gently in a suitable medium, creating a suspension of fractured cells. A drop of the suspension is placed between microscope cover slips and the edges sealed to prevent evaporation. The preparation is then scanned to locate relatively undamaged receptor fragments for study. Cells will be fractured either at the myoid of the inner segment or at the cilium joining the inner and outer segments. It is preferable to select a cell fractured at the inner segment, as breaks in the outer segment membrane probably result in faster fluid transport into or out of the cell, as well as faster changes in dimensions and index of refraction. For the same reason, the immersion medium should be kept as nearly isotonic as possible (see be-low). Some workers use a 6-8% solution of purified gelatin to immobilize the cells under study [5.21,22]. DENTON measured dichroism of receptors in a preparation made by folding flat an excised retina with receptors on the out-side [5.23]. At the edge of the fold the receptors point out into the medium and can be observed side-on. This should alleviate the problem of dealing with fractured receptors. However, there might be difficulties in ensuring that the measuring beam traversed only a single receptor.

If the integrated optical properties of a group of receptor waveguides are to be studied, a correspondingly sizable piece of undamaged retina must be prepared in a flat configuration (see below). The same holds for single cells where the light must traverse them axially, since there is no other way of supporting the receptor axis parallel to the beam. The arrangement and type of receptor vary considerably with location in many species and procedures should be standardized so that samples are consistently obtained from the same region of the retina.

For flat preparations dark adaptation is essential for species in which melanin pigment migrates from the pigment epithelium into the region between the receptors under light adaptation (e.g., frogs, goldfish). A light-adapted goldfish retinal preparation is practically opaque and essentially useless for optical observation. Melanin migration is to some extent subject to diurnal rhythms and better preparations can often be made by judicious selection of the time of day.

Temperature and trauma also have an effect on melanin migration [5.16,24]. Even for species not known to be subject to melanin migration, dissection is often easier after dark adaptation.

Species showing melanin migration also show marked extensions and contractions of the receptor myoids (inner segments) with the stage of light adaptation. Under light adaptation cones contract toward the outer limiting membrane while rods extend away from it. Under dark adaptation the converse movements take place. The associated changes in conformation must be considered in interpreting measurements of optical properties of such retinas. Such photomechanical changes are common in fish and amphibians but are not evident in mammals (Chap.4).

Procedures for removing intact areas of retina seem to be very much an art and, in any case, must be adapted to the species and the needs of the experiment. With small experimental animals we prefer to leave the eye cup intact. Before opening the eye, the blood supply to the retina must be stopped to prevent bleeding. This may be done by decapitation (frog, goldfish) or by clamping the optic nerve sheath with a small curved hemostat (rat). The eye is advanced forward in the orbit and the anterior part cut away with one slice of a new razor blade. The cut is made posterior to the *ora serrata* because the retina adheres to the underlying pigment epithelium at its margin. In frog or rat, the retina can then be removed by gently grasping the edge with a pair of fine iris forceps with teeth ground smooth and gently peeling or lifting it out. The optic nerve is cut as it becomes accessible. The forceps should be oriented tangentially to the cut edge of retina, not radially, in order to minimize the area of damaged receptors. In goldfish much of the remaining vitreous adheres strongly to the retina and the two must be removed together. In larger species, retinas are dissected from freshly enucleated eyes. In enucleated human eyes, the fovea is easily located relative to the line joining the optic nerve and the insertion of the *lateral rectus* muscle. Location of the fovea is facilitated if the surgeon places a suture in the lateral rectus tendon stump during enucleation. Even so, it is difficult to obtain a fovea for observation. If one dissects out a 2-mm square block of rat retina, the sample contains a few million receptors and the probability that somewhere a group of a few hundred will be well-oriented relative to the microscope and to each other is rather good. However, to look at the fovea, a specific few thousand receptors must be well oriented and the probability of success is greatly reduced.

Pieces of retina are usually mounted in a shallow chamber or well ("flat" preparation). If the piece of retina is small relative to the radius of the eye it will lie nearly flat in a chamber slightly deeper than the thickness of the retina. A useful technique is to drill holes in selected microscope cover slips and cement them to whole cover slips to form the sides and bottom of the chambers. The depth of the chamber should be slightly greater than the retinal thickness which seems to be quite uniform within a species. The chamber must be deep enough so that the retina is not crushed when the chamber is closed, but shallow enough to inhibit drifting of the sample during observation (approximately 400+ μm).

The retina is placed in the chamber with receptors down if the light is to traverse the sample in the normal direction. The immersion medium is either present before the sample is placed in the chamber or is added drop-by-drop to one side of the retina. The chamber is closed with a microscope slide and the whole is rapidly inverted for use. In the case of goldfish, deeper chambers are required to accommodate the nonseparable blob of vitreous (1.5-2.5 mm). Such chambers can be made from slices of plexiglass tubing machined flat and cemented to a microscope slide.

Enoch has used a preparation in which a piece of sclera is removed from the lateral aspect of the proptosed living eye of an albino rabbit afer dissecting away intervening bone, eye muscles, and other tissue. The *venae vorticosa* and the ciliary ring of vessels about the optic nerve cup must be carefully avoided to prevent bleeding. The eye is sutured to a ring about the limbus for support. This arrangement allows viewing waveguide properties in vivo, in situ, by normal light path. The eye cornea and lens need not be removed. The observation is made through the transparent choroid body. Spaces exist between blood vessels for ready observation. For reverse path viewing, the same preparation can be used with cornea and eye lens removed and by insertion of narrow-diameter, high-power water-immersion objectives into the vitreous cavity. Single cells, modal patterns, and wavelength transmission properties are readily observed. However, pulse and respiration introduce movement in in vivo preparation, so that photomicrographic records are diffi- cult to obtain. The entire procedure is both exacting and time-consuming (3-4 h) and hence has received little use. The most important results were that modal patterns occurred in vivo and that excised retinal preparations could be obtained at a comparable level of quality. In fresh specimens the properties described in the excised rabbit preparations, the in vivo prep- arations, and the block preparations were essentially the same.

Where only a choroidal block is required, the elaborate surgical proce- dures are unnecessary. The eye is rotated to its extreme position, a scleral window is cut, and the choroidal block removed after clamping the blood sup- ply in the optic nerve and severing the muscles (containing blood vessels). This provides a sample which has been subjected to much less direct handling than is the case for the dissected retinas described above. The albino rabbit is particularly suitable for this procedure because the retina adheres strong- ly to the choroid but the latter separates easily from the sclera.

Immersion media should be as nearly isotonic as possible to minimize changes in dimension and refractive index of the cells in the retina. Un- fortunately, the precise composition of the retinal fluids is not known and probably varies from one species to another. We have used a standard cold- blooded Ringer's for frog and commercially available tissue culture medium for rat, monkey, and human retinas (TC 199 without indicator, Difco, Detroit). Silicone oil was used to cover the goldfish and rabbit preparations described (Dow Corning Silicone 200 fluid, 50 centistokes). It is immiscible with water solutions and tends to retain natural fluids within an isolated "cell" while allowing gaseous exchange. (This material has been used as an artificial vitreous). Goldfish preparations using this medium have been among the most durable we have obtained — in some cases permitting measurements for up to an hour before clouding sets in. Comparable durations were not obtained with the choroidal block from rabbits, however. Silicone oil is difficult to use for floating isolated retinas as these tend to ball up in this medium.

However, we have successfully obtained flat mounts of excised rat retina with
silicone oil. Enoch has used aqueous from the second eye as well as from the
eyes of "black mollies" (a large eyed strain of goldfish) as the medium for
frog and goldfish receptors [5.25a].

With experience, the quality of a preparation can be judged by its high
transmittance and by the brilliance and sharpness of the mode patterns observed in a microscope under proper illumination at the receptor terminations.
In a good preparation, as the microscope is focused down along the receptor
waveguide, the modal patterns will shift laterally very little. If they all
move laterally in the same direction, the retina is probably intact but
tilted relative to the axis of the instrument. If they move off in all directions, the receptors are scrambled. If an oil-drop-like appearance is seen
at cell termination and the guide length seems to be very short, the outer
segments are probably sheared off.

If the presence of photopigment is necessary for the experiment, preparation must be carried out under conditions such that bleaching is minimized.
Depending on the experiment contemplated, the photopigment present, and the
species, one may use either deep red light or infrared illumination. One must
keep in mind, however, that significant bleaching can occur well outside the
nominal wavelength limits of pigment sensitivity if the exposure is broad
band and lengthy and the retina is in a condition such that pigment regeneration is impaired. One must also resist the temptation to increase the lamp
voltage or move it close to the preparation in order to improve visibility.
The situation is aggravated, of course, if some of the more red-sensitive
receptors are being studied. If red light is to be used, preliminary test
for bleaching should be made using the same filter, lamp, voltage, distance,
and exposure time which will pertain in the actual experiment. If this is not
practical, at least some estimate of bleaching should be made from known
properties of the pigment and of the light source to be used. A preferable
solution is to use a filtered source of near infrared radiation and one of
the infrared image converters now widely available.

Time is an essential element in obtaining preparations usable for optical
measurement, not only because of pigment bleaching but also because of the
rapid development of clouding in excised retinas [5.7]. In the typical preparation of a large piece of retina, only a very few minutes are available
for obtaining optical data before light scattering makes further measurements
useless. We have found it necessary to put considerable effort into repetitive
practice and improving the efficiency of the entire operation in order to
obtain data in the few minutes available. Timing the operation from the moment

Table 5.1. Order of validity of preparations

In situ preparations

1. In vivo, in situ, scleral dissection, albino animal
2. In situ, carotid perfusion, scleral dissection, albino animal
3. In situ, scleral dissection, albino animal

In vitro preparations

4. In vitro, flat retinal-choroidal block mount, albino animal; nonmiscible, nontoxic mountant medium
5. In vitro, flat retinal mount (less pigment epithelium); nonmiscible mountant medium
6. In vitro, flat retinal mount (less pigment epithelium) miscible compatible mountant medium (index of refraction n_m) circulated with gaseous constituents
7. Same as 6, but noncirculated medium

---

blood supply to the retina is cut off is essential. It is possible to have a completed preparation on the microscope stage within 2 min. The development of clouding is highly unpredictable from one sample to the next under apparently identical conditions.

It is informative to consider the types of preparations available for studying the physical properties of through-the-retina single-cell waveguide modal patterns. Table 5.1 assigns a descending order of validity to preparations useful for direct observation of physical properties of receptor waveguides [5.25b].

Depending on the problem considered, each preparation has its advantages and disadvantages. In preparation 1, retinas studied in life following major surgery, one has to contend with tissue movement caused by breathing, pulse, tonic muscle contractions, etc. At 1000 × magnification these create a major problem. In addition, only albino preparations can be used for preparations 1-4. Dissection offers the potential for tissue damage, artificial media cause undefined alterations in physical properties, flow of medium interferes with stable record taking, and post mortem changes are inadequately defined. Albino rabbits have been used for preparations 1-4. Only techniques 5-7 are applicable to human preparations.

With some of the more exacting techniques described in the following sections, it is usually advisable to start with cheaper animals and simpler preparations and work upwards in Table 5.1 as experience is gained. With human or primate material in particular, procedures must be well worked out and practiced in advance because of cost and limited availability of specimens.

5.2 Qualitative Observations

5.2.1 Techniques for Observing Waveguide Behavior

Given a suitable sample, qualitative observations of waveguide behavior are relatively easy. The essential requirements are a light microscope with a level stage, a reasonably good, *dry* objective of moderately high power (40-60 ×) and *adequate working distance*, and provision for transilluminating the specimen with either monochromatic or white light over a very narrow cone angle (a few degrees at most). Lower power objectives are useful for locating promising areas in the sample.

The most critical element is narrow-angle illumination: the receptor waveguide has a small effective aperture angle, and light is transmitted in the waveguide modes only if it is incident at small angles to the guide axis. At higher angles of incidence the light is transmitted through and between the guide walls. In particular the usual large aperture microscope condenser is unsuitable because the major part of the light is transmitted in unguided modes which completely swamp the appearance of the typical guided mode patterns.

A simple illumination arrangement is to replace the condenser with a plane mirror and place an intense point source of light a couple of meters distant. A more efficient arrangement is to mount a short-focus camera lens equipped with an iris diaphragm in place of the microscope condenser and focus the light source on the ample. Some metallurgical objectives which have long working distances and iris diaphragms can also be used as condensers.

An interference filter of 10-nm bandwidth can be used to observe the mode patterns although a monochromator with its narrower band pass will give sharper patterns and permits quick changes in wavelength.

It is also useful to observe the back focal plane (BFP) of the microscope objective. All rays which emerge from the sample parallel to a given direction are brought to a point focus in the BFP. Hence the light distribution in the BFP provides a map of the angular dependence of light emerging from the sample. The BFP is readily observed with one of the telescopic eye pieces provided with phase microscopes for adjusting the phase annuli.

While the modal patterns remain the most characteristic evidence, other aspects of waveguide behavior can be demonstrated with the same equipment.

a) Wavelength Sensitivity of Transmission

The most spectacular demonstration is to illuminate the specimen with white light. The terminals of the receptors light up with overlapping colored patterns, even with fully bleached retinas [5.32]. This may occur because different transmission modes are favored at different wavelengths so that two or more modal patterns may be produced at the terminals, each by a different wavelength band. Where only a single modal pattern is excited, the scale of the pattern nevertheless varies, resulting in an intensity distribution that varies with wavelength and hence the appearance of color.

Another demonstration requires a monochromator in the illuminating beam. If the wavelength is changed rapidly the relative intensity between receptors will change noticeably, and a number will change mode patterns abruptly at a well-defined wavelength. The latter effect can be quite stable in a good preparation, occurring reversably for minutes as the wavelength is shifted back and forth.

b) Directional Sensitivity of Transmission

The original proposal that receptors might be waveguides was offered as an explanation of the Stiles-Crawford effect, i.e., the fact that any given point of the retina varies in sensitivity with the angle of incidence of impinging light [5.26,27]. Maximum sensitivity occurs for light arriving from near the pupil center in the normal eye. An appreciation of the directionality of transmission of a group of receptors can be obtained by observing the back focal plane (BFP) of the objective using a uniformly illuminated and well-oriented retinal preparation. The BFP image of a fresh preparation will be most intense near the center of the field, tapering off rapidly in all directions. By comparison, a perfect diffuser should give a uniform BFP image. In fact, if the retinal sample is watched over a period of time, the BFP image will approach more and more closely that of a diffuser as light scattering increases in the dying retina [5.7].

An alternative method is to provide an arrangement for varying the angle of incidence of the illumination at the retina while maintaining intensity constant and uniform. This can be done by rotating the condenser lens about an axis lying in the sample plane and simultaneously swinging the input mirror through half the angle. Observing the receptor terminals (not the BFP) a change of only 2-3° produces drastic changes in the radiation intensity from individual cells. Those which were initially very bright drop off rapidly;

some which were dim may suddenly light up, showing the strong sensitivity to angle of incidence typical of waveguides with small refractive index differences [5.1].

c) Image Transfer: The Retina as a Fiber Optic Bundle

Since the receptors are parallel, closely packed waveguides, they should act as fiber optic bundles and show the phenomena of image transfer, i.e., an image focused at the near end of the bundle should appear at the far end as a mosaic with the bright and dark terminals of the fibers as elements. This can, in fact, be observed in retinal preparations made as described. The reduced image of a pattern such as a resolution chart is focused at the input end of the receptor cells near the myoid, using the condenser lens. The pattern can then be observed by focusing a low-power objective on the output terminals of the fiber optic bundle (Fig.5.4). Optimum setting of the condenser is found by focusing for maximum brightness of the output mosaic [5.7].

Fig.5.4. Demonstration of image transfer in a retinal fiber optic bundle. This is a low-power photo of the output terminals (outer segments) of a rat retina, with an NBS resolution grating focussed at the input terminals [5.7]

5.2.2 Waveguide Behavior as an Aid in Observing Morphology

There is great variation in receptor dimensions between species and often within species. Since waveguide diameter strongly affects its behavior as a collector and transmitter of light, it is not surprising that marked differences in waveguide behavior are observed between receptor types.

The waveguide behavior can be very useful in elucidating the morphologies and arrangements of receptors in retinas of different species. This is aided considerably by the ability to focus on the radiation field within the receptor along a good part of its length. By thus focusing along the receptor one can note deflections in its path through the retina and variations in the mode pattern, guide cross section, and intensity. The level of highest intensity marks the effective radiating terminal of the waveguide.

When the excised retina is mounted receptors-up on the stage and illuminated as described above, the optical arrangement approximates that of the situation in vivo: light is incident over a narrow angle and traverses the inner and outer segments in that order. (In humans a typical 4-mm pupil produces an illuminating cone half-angle of about 5°). Under these conditions variations in intensity at the receptor terminals indicate the relative collection efficiency and transmissivity of the receptor waveguides. Differences in intensity between receptor types in the same retina emphasize the regular geometric ordering of the retinal receptor mosaic which is a pronounced feature in many species (e.g., [5.28]).

If the retina is mounted receptors down and illuminated optimally, the angular variation in light intensity emerging from the radiating terminal on the inner segment side is proportional to the ability of the receptor waveguide to accept radiation incident in the opposite direction, for excitation of the particular mode observed. This is a consequence of Helmholtz's reciprocity theorem of optics [5.29]. Similarly, the height of the radiating terminal defines the level for optimum radiation input, again for excitation of the observed mode. Unfortunately, it is difficult to locate the radiation terminal precisely with respect to the known anatomical structure of the retina. However, differences in observed height between the radiation terminals of nearby receptors do signify a corresponding difference in level of optimum radiation input. In species showing photomechanical behavior, changes in relative level of the input terminals and in inner segment diameter can be followed as a function of the state of light — or dark — adaptation.

a) Goldfish

The goldfish retina provides particularly clear illustrations of many of the above phenomena. The retina is complex, containing six reported types of cones, including double cones as well as rods [5.30]. Rods and some cones undergo complementary photomechanical contraction or extension in response to light- or dark-adapting conditions.

Double cones are readily distinguished from rods because of their much greater brightness, both by natural path illumination and by reverse path. Single cones present more of a problem, especially by reverse path, as their brightness level is less than that of the double cones. The double-cone receptors are arranged in a well-defined rectangular mosaic in some parts of the retina. The single cones regularly occur near the center of the rectangles defined by the double cones and display much larger and more complex mode patterns than the rods.

Under reverse path illumination the cone mosaic appears less prominent in the dark-adapted samples than in the light-adapted ones. In the light-adapted retinas, cone mode patterns also appear larger, and the more complex mode patterns seem to occur more frequently.

These observations are consistent with photomechanical effects reported in the literature. The cones should respond first with the onset of light adaptation by contracting toward the vitreous. As the myoid contracts, it increases in diameter, the mode patterns become larger, and higher order mode patterns occur more readily. Simultaneously, the plane of maximum cross section moves toward the vitreous. On the other hand, the photomechanical response of the rods and of the pigment epithelium occur more slowly for light adaptation [5.16].

At the level of maximum brightness, the most common mode pattern observed in cones is the bilobe, but the dot and ring or the linear trilobe patterns are also frequently observed. The more complex dot and double ring occurs regularly. Occasionally, still higher order mode patterns are detected. The lowest order single-lobe pattern is rare. For the rods, the bilobe and annulus are overwhelmingly the most common, but the dot and ring pattern occurs regularly, as does the single lobe.

In light-adapted samples under reverse-path illumination one can follow the trace of the guided radiation over a considerable range in the double cones by starting with the plane of focus well toward the vitreous from the receptor cells and then moving slowly toward the sclera. One first encounters two very small, weak, separated mode patterns of low order (single lobe or

bilobe). The two patterns enlarge and intensify progressively, sometime forming higher order patterns which may be different in the two cells. Beyond the
point where the patterns are in near contact, the following four types of behavior may occur: 1) the patterns proceed parallel to each other; some interaction between the patterns may be apparent, but not always; 2) the patterns
may merge to form a larger pattern, sometimes of higher order (dot and ring
or dot and double ring). Details of the central part of the pattern often become indistinct; 3) an abrupt jump of radiation may occur from one guide to
the other; and 4) occasionally the radiation jumps to what appears to be a
different waveguide than the two we have been observing. More than one of the
above phenomena may occur in the same pair of guides at different levels. It
is over this range of levels that the mode patterns are most intense.

Proceeding further along the double cones, the patterns decrease in intensity, the traces can be observed to separate slightly and form smaller, lower
order mode patterns which eventually disappear, not necessarily at the same
level. There is an extended region of close apposition including the nucleus
and part of the ellipsoid levels. In this region strong interaction between
the electromagnetic radiation traversing the two guides would seem likely.
The observed separation of the mode patterns scleral to the region of interaction also fits with morphologic separation of the double-cone outer segments.

The small separated mode patterns initially observed show that waveguide
behavior is already occurring in the axons vitread to the point where the
double cones come into apposition within the outer nuclear layer. Thus the
guide must be surrounded by a medium of lower refractive index at that level.

b) Frog

Frog red-rod outer segments are much larger than goldfish, typically 6 μm,
and the mode patterns observed with receptors up are more complex, corresponding to the higher order modes which are permitted. The patterns are
strikingly similar to the elaborate geometrical patterns reported by HANNOVER
over a century ago (Chap.1) [5.6]. As the receptors age, the mode patterns
tend to break up into perhaps six or more subelements composed of a simple
mode pattern (about 6-9 min after blood cut-off). This change may occur because of the special nature of the fine structure of the individual membraneous disks of which the frog outer segment is composed. They are not simple
flat pancakes; each is lobulated and the lobules of successive disks are
aligned axially (Fig.5.3) [5.31]. Apparently, columns of lobules separate and
act as individual fiber optics elements.

By reverse path, although the modal-pattern separations are about right for rod receptors, the pattern diameters are small relative to the diameters of dark-adapted rod inner segments, and the modes are typically of lower order. Changes comparable to the apparent pattern break-up recorded with normal-path illumination are not observed. This indicates that under reverse-path illumination the radiating terminal does not lie in the outer segment. There appear to be two possibilities: a) the terminal may lie at a necked-down region of the myoid; b) alternatively, the terminal may occur in the axon vitread to the cell nucleus. This question is still unresolved.

c) Mammalian Retinas

Because of the small diameter of mammalian receptors (1-1.5 μm) which pushes the resolution of light microscopes, deductions about their morphology from end on observations have been limited. Nevertheless, modal patterns are readily seen on all mammalian species observed to date, including albino rat, albino rabbit, Rhesus macaque, squirrel monkey, baboon, and humans [5.1,7,32]. In primate, cones are readily distinguished from rods and the mosaic pattern is clearly evident [5.32]. In albino rat retinas one does see an occasional oversized modal pattern of higher order but there has been no systematic attempt to relate these to the cones which have been reported in rat retinas [5.13].

5.3 Quantitative Observations: The Fiber Optic Bundle

It is possible to make quantitative measurements of the resolution and of the modulation transfer function (MTF) of a piece of retina acting as a fiber optic bundle. It must be recognized from the outset that such measurements of a transmitted light signal are only indirectly related to the contributions of the retinal receptor bundle to the resolution and MTF of the visual system. This is because the latter is determined by the processes of light adsorption and signal transduction within the outer segments rather than just by the transmission of light along the whole length of the receptor waveguides. Nevertheless, such measurements provide useful information about the optical properties of the retina from which an elucidation of its overall contribution to the functioning of the visual system must start.

5.3.1 Resolution

For the retina, resolution is conveniently specified as cycles per micrometer or its inverse. ENOCH and GLISMANN have made resolution measurements on excised retinas of albino rats, albino rabbits, and squirrel monkeys [5.7]. Their appartus is shown schematically in Fig.5.5. Using monochromatic light from an interference filter (IF, 498 nm), a transparent grating (G) was imaged onto the near end of the retinal fiber optic bundle (R) by lens (L), which was independent of the observing microscope. The latter was focused on the far end of the bundle. The energy level was monitored, and wavelength, polarization, and angle of incidence at the retina was varied. The iris aperture (I) was set for an incident convergent cone of light of 7° at the retina.

As the grating spacing approaches the separation of the individual fibers, measured resolution is adversely affected by the superimposed pattern of the bundle itself. In this condition static and dynamic scanning of the image will yield different results. In static scanning all elements of the optical system are held stationary. In dynamic scanning the fiber optic bundle is moved about while the imaged grating is held fixed. The interfering mosaic pattern is effectively integrated out, enhancing the measured resolution [5.33].

In the work described, 71 measurements of resolution were made on well-oriented receptor samples, using dynamic scanning. Since mixed rod and cone retinas would be difficult to interpret, rat and rabbit retinas were used for rods and squirrel monkey foveas for cones.

Initially the excised rod retinas showed good resolution as measured, but resolution fell rapidly as the preparations aged. Figure 5.6 shows a loss in resolution by a factor of two, between 4 and 8 min. Thereafter, resolution decreased more slowly with time.

The maximum nonvernier resolution capability of the retinal fiber optic bundle can be related to interreceptor separation. The smallest separation, high-contrast, grating-resolution target resolved with an albino rat retina was 5.0 µm per grating cycle. In some cases, high-power photographs of the area were obtained. Mean intercell separation on several different preparations were determined, and ranged between 1.5-2.2 µm. In single preparations, the smallest recorded ratio of cycle length to (twice mean intercell separation) was 1.18. A more typical value, taking both measurements into consideration, was 1.45.

188

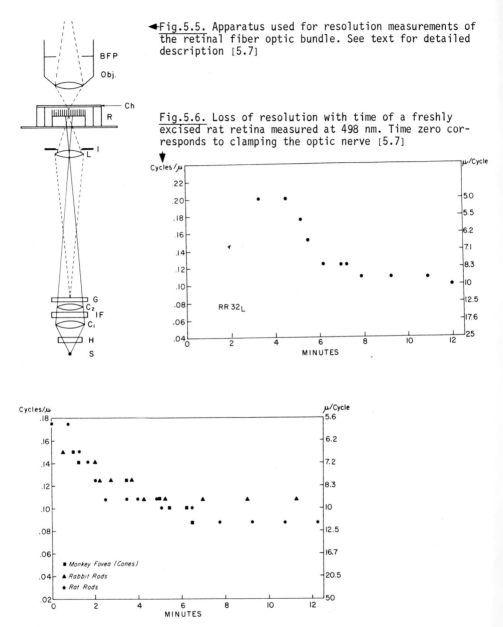

◄Fig.5.5. Apparatus used for resolution measurements of the retinal fiber optic bundle. See text for detailed description [5.7]

Fig.5.6. Loss of resolution with time of a freshly excised rat retina measured at 498 nm. Time zero corresponds to clamping the optic nerve [5.7]

Fig.5.7. Comparison of resolution loss in excised retinas of three different species. Different anesthesia and mounting techniques were used with each species. Time zero has been adjusted to bring the three sets of data into line [5.7]

In Fig.5.7 the three species are compared. The time scale has been adjusted
to an arbitrary zero in order to bring the three sets of data into line. All
three species showed essentially the same type of decrement in resolution as
a function of time, in spite of differences in mounting techniques and immersion media.

5.3.2 The Modulation Transfer Function of the Retina

The optical transfer function (OTF) has found increasing use for evaluating
the image- and information-processing capability of optical systems [5.29,34,
35]. It is not limited to optical and electrooptical systems, but provides
a valuable tool for vision research as well (e.g., [5.36-42]).

The optical transfer function of an optical system measures, as a function
of spatial frequency, the changes in modulation (or contrast) and the shifts
in phase introduced into the image relative to the object by the optical
system (see below for definitions). In general, for a given spatial frequency,
the OTF is a two-dimensional function of the transverse coordinates in the
object plane and hence has different values at different positions in the
object plane; it may also vary with orientation of the "wave" normal of the
spatial frequency. In image-forming (lens and mirror) systems, the OTF is
critically sensitive to the aperture of the optical system, to aberrations,
and to focus. For a given optical system it will depend on the wavelength of
the light used to form the image.

The OTF can be separated into two factors: the modulation transfer function (MTF), describing the change in modulation of the image, and the phase
transfer function, which describes the shift in phase. It can be shown that
for a diffraction-limited image-forming system there is no phase shift and
the phase transfer function is a constant.

Fortunately, for optical systems which are circularly symmetrical and
well centered, it is possible to obtain meaningful measurements of the MTF
by using a transilluminated one-dimensional grating of known modulation as
object, and measuring the modulation in the image.

Figure 5.8 is a schematic plot of radiance (or luminance) along a transverse
(X) direction for a transilluminated cosine grating object and its image.
To allow comparison, X' is measured in units of (AX) where A is the magnification of the system. Noncoherent illumination is assumed. The variation of
radiance across the object plane may be represented by

$$E = E_{av} + E_0 \cos[2\pi(X/d)] \tag{5.1}$$

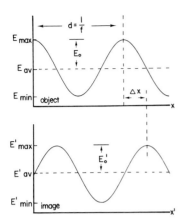

Fig.5.8. Object-image relationships of a sinusoidal grating used to define the optical transfer function and the modulation transfer function [5.37]

where d is the grating spacing.

Similarly for the image

$$E' = E'_{av} + E'_0 \cos\left[2\pi\left(\frac{X'}{d'} + \frac{\Delta X'}{d'}\right)\right] . \tag{5.2}$$

The modulation of the object grating is defined by

$$M = \frac{E_{max} - E_{min}}{E_{max} + E_{min}} = \frac{E_0}{E_{av}} . \tag{5.3}$$

Similarly for the image

$$M' = \frac{E'_0}{E'_{av}} , \tag{5.4}$$

the phase shift is given by

$$\phi = 2\pi(\Delta X'/d') , \quad d' = Ad . \tag{5.5}$$

Then the value of the MTF at the frequency, f, is given by the ratio of the modulation

$$R_1(f) = \frac{M'}{M} = \frac{E'_0/E'_{av}}{E_0/E_{av}} . \tag{5.6}$$

It is customary to normalize R to 1 at zero frequency.

If the image produced by one optical system is used as the object for a second optical system, then the modulation of the final image can be expressed as

$$M'' = R_2 M' = R_2 \cdot R_1 M$$

and

$$R_{1,2} = \frac{M''}{M} = R_1 \cdot R_2 \quad . \tag{5.7}$$

Thus the MTF of a complex optical system is given by the product of the MTFs of the components.

CAMPBELL and GREEN used this latter property to evaluate the optical system of the human eye [5.39]. They used a divided laser beam to produce a high-contrast cosine grating on the retina. By diluting the pattern with uniform illumination of the same color, while keeping luminance constant, they were able to determine contrast sensitivity (threshold^{-1}) as a function of spatial frequency for the combined retina and neural system. They then determined contrast sensitivity for the total system by displaying to the observer a cosine grating pattern generated on an oscilloscope. From the ratio of the two functions they could determine the MTF of the intervening optical system of the eye.

The same approach should permit a separation of the MTFs of the retinal receptors and the visual neural system if either could be determined independently. A theoretical treatment of the MTF of the retinal fiber optic bundle has been published by OHZU [5.36].

a) *Experimental Determination of Retinal MTF*

OHZU, ENOCH, and O'HAIR undertook measurements of the retinal MTF of albino rat and squirrel monkey foveas and OHZU and ENOCH made similar measurements on three human retinas using the same equipment [5.37,43].

A schematic diagram of the apparatus is shown in Fig.5.9. A square wave grating (G) of varying spatial frequency was focused on the input side of the retinal fiber optic bundle (R). This grating was mounted on a rotating drum (D) driven at a constant 4 rpm. Large transparent and "opaque" areas provided calibration of the dynamic range of the device. Complete calibration and MTF measurements could be obtained every 15 s. The image on the retina was reduced optically so that the range of spatial frequencies applied to the retina was 18-230 c/mm.

The microscope objective (L_4) was focused on the grating image after it had been transferred through the retinal fiber optic bundle to the photoreceptor terminations. The primary image of the objective was then projected onto the plane of the scanning slit (S) by an eye-piece (L_5) at a magnification of 135 times that of the specimen. The final image of the grating was parallel to the scanning slit and moved perpendicularly to it. The scanning slit projected backwards to a 39 by 0.15 μm test area at the specimen plane.

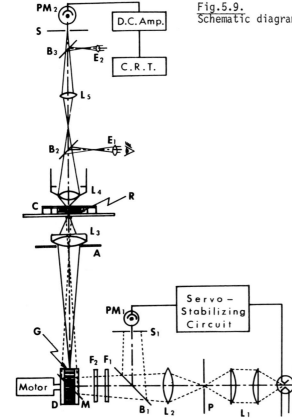

Fig.5.9.
Schematic diagram of the MTF apparatus [5.37]

Since the highest frequency of measurement was 230 c/mm, and the receptor spacing was of the order of 2 μm, the effect of the slit width on modulation measurement was negligible. Assuming the cells were packed at random, the conditions for dynamic scanning of a fiber optics bundle were also satisfied by the selected slit length. The difference in measurement caused by a location change of the slit relative to the specimen was virtually negligible when an area of well-oriented retina having reasonably uniform packing was tested.

The light signal passing through S was detected by photomultiplier PM_2. The output photocurrent of PM_2 was amplified by a dc amplifier and displayed on an oscilloscope fitted with a camera. The horizontal sweep of the oscilloscope was synchronized with the rotation of the drum.

The xenon-arc light source was stabilized by an optical feedback loop, and the entire apparatus (except the source) was mounted on a vibration-

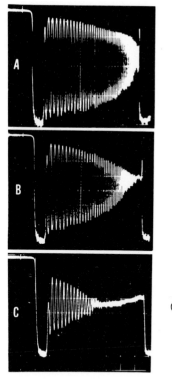

EXAMPLE I

A. MODULATION
   THROUGH THE
   OPTICAL SYSTEM
   OF THE INSTRU-
   MENT WITHOUT
   RETINA

B. MODULATION WITH
   A WELL ORIENTED
   RETINA IN PLACE

C. MODULATION WITH
   A POORLY
   ORIENTED RETINA

Fig.5.10a-c. CRT traces from the MTF measurements of OHZU, ENOCH, and O'HAIR. Ordinate corresponds to power at the detector. Spatial frequency varies along the abscissa. The initial flat region corresponds to zero transmission. The flat regions bracketing the oscillatory signal correspond to maximum unmodulated power, see text for details [5.37]

damped table. Typical oscilloscope records are shown in Fig.5.10. All of the MTF data were obtained by passing light through the retina in the same direction as in vivo.

b) MTF of the Albino Rat Retina

Seventy-three rat retinal specimens were studied. Receptor orientation was checked by varying the focal plane of the microscope objective. If lateral translation of the cell trace occurred, the orientation of the cell was less than perfect. The area of best orientation of photoreceptors was always selected for measurement.

The quality of the MTF decreased as a function of time after dissection. Well-oriented retinal specimens were divided into three classes based on behavior during the first 10 min of recording.

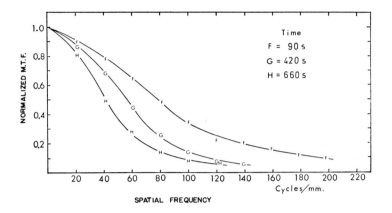

SPATIAL FREQUENCY

Fig.5.11. MTF of a well-oriented albino rat retina at various times after clamping the optic nerve [5.37]

Class 1: The retina showed only minor change in transmission of light at zero frequency and in the MTF during measurement (ca. 10% of samples).

Class 2: The specimens showed only small changes in transmission at zero frequency but a large decrease in the MTF at higher frequencies (ca. 60%).

Class 3: The retinas showed alterations in both transmission at zero frequency and in the MTF at all frequencies (ca. 30%).

All samples exhibited relatively stable measurement characteristics up to about 4 min after clamping of the optic nerve during dissection (time zero). Representative sets of data taken from a well-oriented rat retina of the first class are shown in Fig.5.11. Total transmissivity did not change markedly in groups one and two during test sessions.

c) MTF of Squirrel Monkey Foveas

Nine well-prepared squirrel monkey foveal specimens were tested. Preparation of the test specimen prior to observation took 5 min. The preparation was well illuminated in the normal manner, i.e., light passed through the retina and emerged at the terminations of the outer segments. Light-conductive behavior in these foveal cone receptors was more complex than that observed in rat rods. Figure 5.12 shows the mean value of the MTF.

The same type of MTF degradation in time was noted with squirrel monkey foveas as with rat (rod) retinas, but the changes occurred more slowly in the former. The thinner neural layer overlying the monkey fovea is the likely reason for the difference.

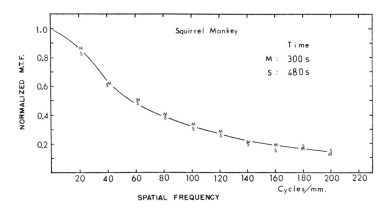

Fig.5.12. MTF of well-oriented squirrel monkey foveas at two different times after clamping optic nerve. Peak wavelength 530 n [5.37]

d) MTF of Human Foveas

While the preceding experiments were being carried out, three high-quality human foveas became available for study. During the enucleation procedure, the surgeons tied one or more sutures to the eye in order to help localize the fovea and to indicate the region of optimal incision. This greatly speeded up specimen dissection and allowed minimal disturbance of the globe pending future analysis in the ocular pathology laboratory. The removal of the retina (fovea) from its pigment epithelial bed was far more easily accomplished in the human than in the squirrel monkey.

The times between clamping of the optic nerve during enucleation in the operating room and the initiation of data collection in the three preparations were

A) 9 min, 45 s
B) 14 min, 30 s
C) 10 min, 0 s

A central foveal area with nearly perfect orientation of receptors was found in each specimen. Little residual melanin pigment was observed between the receptors compared to the squirrel monkey and the center of the human fovea appeared grossly and microscopically thinner than the squirrel monkey foveas previously studied. The latter observation was confirmed histologically.

Figure 5.13 is the first record obtained from the fovea of specimen A. Compared with Fig.5.10, it is evident that high-frequency response in humans

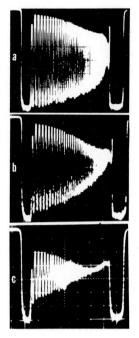

a.

Modulation with
the instrumental
optical system
(no specimen)

b.

Modulation with a
human fovea
(receptors well
oriented)

c.

Modulation with a
human para-
fovea (see text)

Fig.5.13a-c. CRT readout using a
human fovea in the MTF apparatus.
Compare with Fig.5.10 [5.43]

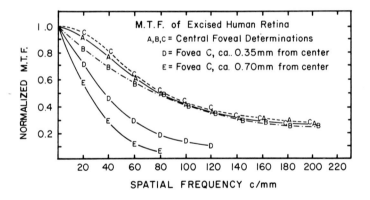

Fig.5.14. MTF curves of three human foveas, A, B, and C (see text). Receptors were well oriented [5.43]

is superior to that determined on much fresher rat retinal specimens having
well-oriented receptors. Normalized data from the three human foveal preparations are presented in Fig.5.14. Each point is a mean of data taken from
a minimum of four records.

In the parafoveal region there was a rapid falloff in the quality of the
modulation transfer function and in the resolution capability of the preparation. Curves D and E show the effect clearly. As with the other preparations, human foveal MTFs decreased in time, but the rate was much slower
then with rat retinas or even monkey foveas. In general, parafoveal curves
are more similar to rat than are the foveal curves, as might be expected
from the higher proportion of rods and the thicker neural layers of the
latter. MTFs of the three mammalian species are compared in Fig.5.15.

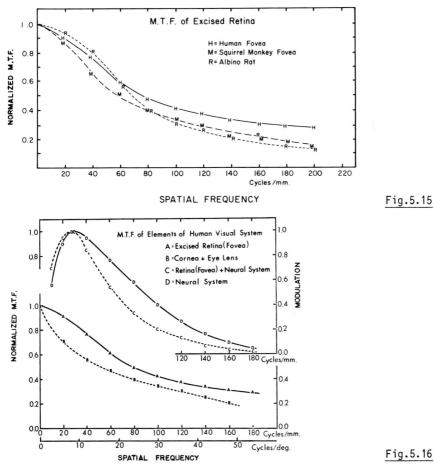

Fig.5.15

Fig.5.16

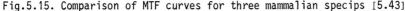

Fig.5.15. Comparison of MTF curves for three mammalian specips [5.43]

Fig.5.16. MTF of the human foveal-neural response system (curve D) estimated
from measured MTFs of the human retina-plus-visual-response system (C) and
of the retina alone (A), see text for details [5.43]

Insofar as the retinal *transmission* MTF just described can be taken as representative of a retinal *response* MTF, the data of CAMPBELL and GREEN [5.39] can be combined with that of OHZU and ENOCH to obtain an estimate of the MTF of the neural system, i.e., by dividing curve C by curve A to get curve D, Fig.5.16. The result suggests that the low-frequency peak observed by CAMP-BELL and GREEN is introduced by the neural processing system and not by the retina. Note also that the optical properties of the foveal retina are well matched to those of the remainder of the optical system of the eye and should not significantly degrade the input image.

5.3.3 Directional Transmissive Properties of the Fiber Bundle

As pointed out in Sect.5.2.1, the light distribution in the BFP of a microscope objective maps the directional properties of light emerging from the specimen. It should be possible to measure the irradiance distribution at the BFP and determine the directional transmissive properties of the retinal fiber bundle under the illumination conditions used.

Semiquantitative experiments of this type have been done by ENOCH and SCANDRETT on a human fovea (cones) and by TOBEY and ENOCH on rat retinas (rods) [5.44,45a]. The results on human fovea showed a relatively peaked directionality (Fig.5.17) while the rat retinas showed a nearly flat central portion with a sharp drop toward the edges (Fig.5.18) (see also [5.45b]).

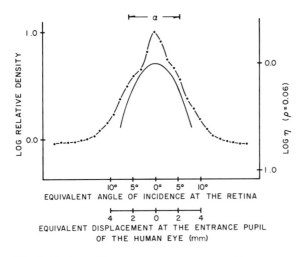

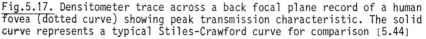

Fig.5.17. Densitometer trace across a back focal plane record of a human fovea (dotted curve) showing peak transmission characteristic. The solid curve represents a typical Stiles-Crawford curve for comparison [5.44]

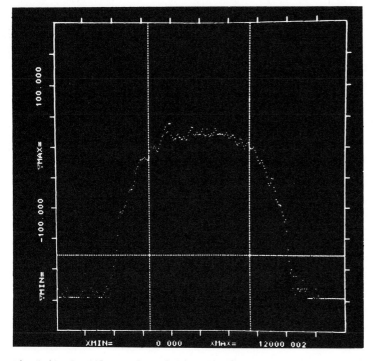

Fig.5.18. Profile of far-field radiation pattern of rat retina under reverse path illumination. Ordinate: Log relative irradiance × 100. Abscissa: distance on this axis corresponds to angular separation at the specimen plane [5.45a]

5.4 Quantitative Observations of Single Receptors

5.4.1 Techniques

In order to fully understand the part that the optical properties of the retina play in the overall functioning of the visual system it is finally necessary to study the optics of individual receptors. This is difficult experimentally because of the need to isolate and stabilize single receptor cells for measurement. Three techniques have been used to achieve this isolation. a) Bodily separating the cell from its surroundings as described in Sect.5.1. b) Optical isolation, a technique that is required in most cases where the waveguide properties of the receptors are intimately involved in the results. This is because an individual receptor must be observed end-on, and the present state of the art permits doing so only by leaving the receptor in a large supporting matrix of neighboring receptors. This technique is described in Sect.5.4. c) A third technique makes use of an observer's ability to fix his attention on a single cell among a large group of surrounding cells. The technique is useful where a readily apprehended change can be induced in the experimental cell without deflecting the observer's attention. It has been used earlier in semiquantitative estimates of the effect of a change of angle of incidence of illumination by ENOCH and recently in a direct measurement of the waveguide parameter, V, of a receptor cell [5.1,25b].

All of the reliable spectrophotometry of individual receptors has been done side on with cells separated from the retina, and hence does not consider waveguide effects (e.g., [5.30,46-49]). This work is described in Chap.9 of this book and will not be considered here. Excellent reviews of earlier work are available [5.22,50].

5.4.2 V and Its Measurement

In order to treat the optics of a single receptor theoretically, it is modeled as a cylindrical dielectric waveguide of circular symmetry, with two fixed indices of refraction inside and outside the guide (Chap.6). Admittedly this is an idealization, but a comparison of results expected from the model with the behavior of real guides should tell us much about the latter.

Light is propagated along the guide in one or more of a set of "modes", each associated with a characteristic three-dimensional distribution of energy density in the guide. In the ideal guide their distributions may be very nonuniform but will show a high degree of symmetry. Modes are grouped into "orders", related to the number of distinct nodal surfaces on which the electromagnetic fields go to zero. Thus the lowest (first-) order mode (designated HE_{11}) goes to zero only at an infinite distance from the guide axis. Note that for all modes the fields extend to infinity in principle. At the terminus of the guide it becomes an antenna, radiating light energy into space. If one observes the end of an optical waveguide with a microscope, one sees a more or less complex pattern of light which roughly approximates a cross section through the modes being transmitted.

The optical properties of the ideal guide are completely determined by a parameter, V, defined by

$$V = \frac{\pi d}{\lambda} \sqrt{n_1^2 - n_0^2} \tag{5.8}$$

where d is the diameter of the guide, λ is wavelength in vacuum, and n_1 and n_0 are the indexes of refraction of the inside (core) and outside (cladding) of the fiber, respectively. For a given guide, V is readily changed by changing λ. With one exception, each of the various modes has associated with it a cutoff value, V_c, below which light cannot propagate indefinitely along the fiber in that mode. The lowest order, HE_{11}, mode does not have a cutoff value (e.g., [5.51]). V may be regarded as setting the radial scale of the energy distribution; for $V \gg V_c$, the energy is concentrated near the axis of the guide with relatively little transmitted outside the core. As V approaches V_c, the energy spreads out with proportionally more being transmitted outside the guide wall. SNYDER has defined a quantity η which represents the fraction of transmitted energy which propagates inside the guide wall [5.52]. For a given mode or combination of modes and given V, η can be evaluated from waveguide theory. Figure 5.19 shows schematic representations of the mode patterns commonly observed in vertebrate receptors together with V_c for the corresponding modes [5.1]. Figure 5.20 shows values of η as a function of V for some of the lower order modes [5.53].

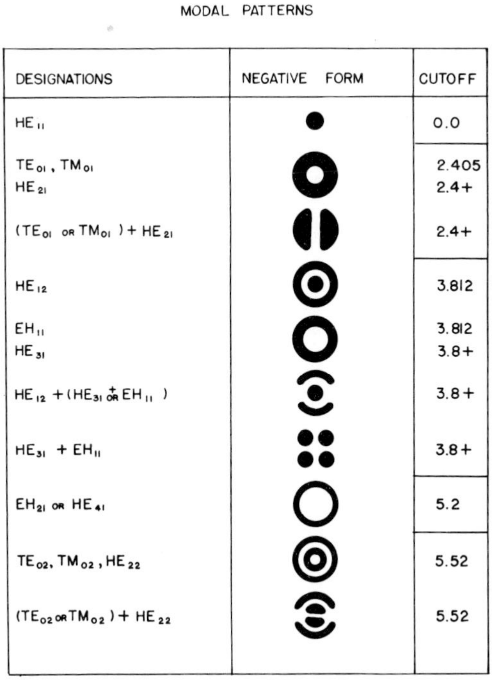

MODAL PATTERNS

DESIGNATIONS	NEGATIVE FORM	CUTOFF
HE_{11}		0.0
TE_{01}, TM_{01} HE_{21}		2.405 2.4+
$(TE_{01}$ or $TM_{01})$ + HE_{21}		2.4+
HE_{12}		3.812
EH_{11} HE_{31}		3.812 3.8+
HE_{12} + $(HE_{31}$ or $EH_{11})$		3.8+
HE_{31} + EH_{11}		3.8+
EH_{21} or HE_{41}		5.2
TE_{02}, TM_{02}, HE_{22}		5.52
$(TE_{02}$ or $TM_{02})$ + HE_{22}		5.52

Fig.5.19. Schematic representation of low-order modal patterns with their cutoff values of V [5.1]

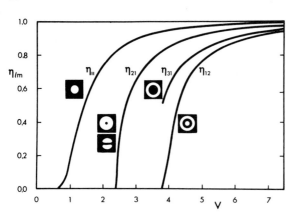

Fig.5.20. Theoretical values of η, the fraction of guided power carried within a waveguide as a function of V [5.53]

It is evident that an understanding of receptor waveguide behavior requires a determination of V for a range of typical receptors. An obvious straightforward approach is to measure the two indexes n_1 and n_0, determine d, and calculate V for different wavelengths. This is difficult experimentally; mammalian receptors are often only two to three wavelengths in diameter, with even smaller separations between them; they are also fragile and subject to osmotic changes in dimension and refractive index when removed from the eye, especially if the cell membrane has been damaged.

A number of measurements have been made of the refractive indexes of receptor cells (Sect.5.4.3), but determination of V requires a measurement of the index for the interstitial material as well and no means has yet been devised to do this directly. In addition, an accurate measurement of the diameter is required on the same cell.

An alternate approach is suggested by examination of (5.8). If V could be found at any one wavelength, it could be found for other wavelengths by simple calculation. For ultimate precision it might be necessary to include a correction for change in the refractive indices with η, but one would expect this to be small.

If one observes a "flat" retinal preparation illuminated by a monochromator so that the mode patterns are sharp, then by rapidly changing wavelength throughout the visible some of the receptors will show an abrupt change of the pattern from one mode type to another. This behavior suggests that for these receptors V is moving through the value of V_c for the more complex mode resulting in a change in the dominant mode being excited.

To be valid this approach requires that the fraction of energy within the guide ($n_{\ell m}$) should fall abruptly to zero at cutoff. In fact, theoretical work of KIRSCHFELD and SNYDER and of SAMMUT and SNYDER showed that this condition should hold for a very few low-order modes, although not for most others (the latter are the so-called "leaky" modes) [5.53,54]. In particular, they predict that η should go to zero at cutoff for the *second-order modes* characterized by bilobed or thick annular patterns (Fig.5.20). These modes include HE_{12}, TE_{01}, $TM_{01}(TE_{01} + HE_{21})$, and $(TM_{01} + HE_{21})$. In our experience these modal patterns are by far the most frequently observed in small-diameter mammalian receptors.

ENOCH and TOBEY have carried out such an experiment on albino rats using a microscope fitted with a 35-mm focal length camera lens in place of the condenser [5.25a]. Total magnification used was about 1000. The 1.5-mm slit of a monochromator was focussed near the external limiting membrane (vitread end of the receptor inner segments). Because the cell images were much smaller than the exit slit image, the spectral bandwidth of illumination was determined essentially by the entrance slit of the monochromator alone which was set for 2-nm spectral slit width. The wavelength could be varied rapidly by the observer using a servo motor. A 1000-W xenon arc was used as the source of illumination.

Cells were illuminated with aperture settings of f/22 to f/18 corresponding to full illuminating cones in the aqueous medium of approximately 2 to 2.5°. Candidate cells were selected if they transmitted a strong HE_{11} (single lobe) mode in the red and were well aligned with the instrument axis. They were tested for transition from single lobe to bilobe or broad annulus by rapidly shifting to shorter wavelengths. If a transition occurred, the observer shifted back to the red, then moved slowly through the transition region from long to short wavelengths, stopping when the higher order pattern just began to be visible. Wavelength was then read from the monochromator drum. At least four readings were taken after which the wavelength was shifted away from the mode transition and a Polaroid record was made for use in estimating cell diameters.

Twenty-nine sets of measurements were made using TC199 as mountant medium and 13 sets using silicone oil as a check on possible changes in the interstitial matrix index (n_0). No significant difference was noted between the two samples.

The value of V_c associated with the bilobe or thick annular modal patterns is 2.40 and the measurements determined the wavelength at which V passed through this value in the test receptors. The transitions observed occurred in a relatively limited wavelength band about 600-540 nm. However, these limits were set by the observer's visual system rather than any property of rat receptors.

The vast majority of modal patterns observed in rat photoreceptors were annular or bilobe patterns and did not undergo transition within the range accessible to measurement. This implies that most rat receptors have a V exceeding 2.4 in the visible region. If one postulates approximately constant concentrations of solids and, hence, of refractive indices within rat receptors and throughout the interstitial matrix, then the majority of rat receptors must have been larger than the population tested. This is in agreement with SIDMAN's value of 1.4-μm average diameter for unfixed rat rod receptors [5.55], while the mean value for tested cells in this experiment was approximately 1.1 μm. The method tends to underestimate λ at cutoff or to overestimate V at the reported cutoff wavelengths because the emerging bilobe radiation pattern can only be detected after η has been brought to some indeterminate level above zero. Since the falloff in η_{21} is fairly sharp (Fig.5.20), this effect should be small. The wavelength for the Polaroid record was chosen to get as high a value of η as possible by shifting away from the transition wavelength so that the radiation would be more strongly concentrated within the cell. The amount of shift was limited by film sensitivity on the red side and by the observer's inability to identify the test cell in the blue region where scattering became severe. The edge of the photographic pattern was assumed to be a reasonable match to the boundary of the test cell at the higher values of η. Maximum and minimum diameters were taken from the photographic modal pattern record and averaged.

By obtaining measurements of diameter on the same cells in which mode transitions were observed, it was possible to calculate $\Delta n = n_1 - n_0$, the difference in indices between the receptor waveguide and the surrounding matrix. We can write

$$n_1^2 - n_0^2 = \Delta n(2n_0 + \Delta n) \quad . \tag{5.9}$$

Since

$$\Delta n \ll 2n_0 \quad , \quad \text{then}$$

$$V \cong \frac{\pi d}{\lambda} \sqrt{2n_0 \Delta n} \tag{5.10}$$

$$\Delta n = \frac{1}{2n_0} \left(\frac{V}{\pi}\right)^2 \left(\frac{\lambda}{d}\right)^2 \quad . \tag{5.11}$$

Assuming that n_0 = 1.34 and that V = 2.40 at the specified transition, then

$$\Delta n = 0.218 \left(\frac{\lambda}{d}\right)^2 . \tag{5.12}$$

Note that the values obtained for Δn by this method are relatively insensitive to the value assumed for n_0 because even the extreme range of possible values for the latter is quite small. The situation is quite different when n_1 is measured directly and n_0 has to be assumed. In the latter case, n becomes very sensitive to the value assumed for n_0 as well as any errors in n_1. The observed values of Δn, diameter, and transition wavelengths are summarized in Table 5.2. The mean value of Δn = 0.06 compared favorably with BARER's earlier estimate of 0.0606 [5.56].

Table 5.2.

Mountant medium	Mean[a] Δn	Standard deviation	N
TC199	0.0590	0.0196	29
Silicone oil	0.0603	0.0199	13

Mountant medium	Diameter	Standard deviation	N
TC199	1.16 μm	0.201 μm	29
Silicone oil	1.13 μm	0.173 μm	13

Mountant medium	λ at modal pattern transition	Standard deviation	N
TC199	0.581 μm	0.0170 μm	29
Silicone oil	0.574 μm	0.0228 μm	13

[a]Means are based on individual (not pooled) data sets.

5.4.3 Refractive Index and Cell Dimensions

Two methods have been used for measuring the refractive index of living receptor cells: interferometry with a two-beam interference microscope and refractometry using a phase-contrast microscope. With the interference microscope one measures the difference in the optical path length between the sample and an equal depth of the immersion medium. Determining the refractive index then requires measuring the depth of the sample and the refractive index of the medium. The latter is easily obtained with one of the standard precision refractometers on the market. Depth measurements can be difficult. With receptors it is usual to measure the transverse diameter and assume they are circularly symmetric.

BARER has developed a very elegant technique for refractometry of living cells with a phase microscope [5.56]. It depends on the fact that with positive phase contrast, a higher index sample in a lower index medium appears darker than the background and conversely the reverse is true of negative phase contrast). By adjusting the index of the medium until the sample contrast is a minimum, the index of the sample equals that of the medium. Similar techniques had long been in use for mineral samples but were not successful for cells because of the conflicting needs to vary the refractive index of the medium and maintain isotonicity between it and the interior of the cell. BARER used bovine plasma albumin to adjust the refractive index. Because of its very large molecular weight (about 70,000), it contributes substantially to the refractive index while affecting the tonicity very little.

SIDMAN used Barer's technique to measure refractive indices of receptor cells of frogs, salamanders, turtles, chicks, pigeons, rats, mice, and monkeys (*Macaca macaca*) [5.55]. He found values for rod outer segment ranging about 1.41, for cone outer segment about 1.39, and for rod and cone myoids about 1.36

The refractive index can be used to estimate the concentration of solids in biological material. The refractive index of a solution may be represented by

$$n = n_s + \alpha C$$

where n_s is the index of the solvent, C is the concentration expressed as mass per unit volume, and α is termed the *specific refractive increment* and depends on the nature of the solute.

SIDMAN reported values of C (gm per 100 ml) of 40-43% for rod outer segments, approximately 29-34% for cone outer segments, and 15-17% for myoids. He used a value of α of 0.0018 as typical for protoplasm. He also reported values and distributions for diameters and lengths of receptor outer segments.

A detailed understanding of receptor-waveguide behavior requires accurate values for guide diameters. Because of the low contrast in "live" material, most earlier measurements of receptor diameters were made on fixed and stained preparations and required a rather large and uncertain correction for shrinkage. With the development of phase and interference contrast, microscope measurements could be made directly on live receptors. There still remain potential pitfalls due to swelling or shrinkage in media of improper tonicity, and optical artifacts sometimes introduced at the edges of specimens by these instruments.

While the measurements of bleached receptor diameters may now be relatively straightforward, a number of workers have reported changes in diameter due to the bleaching itself [5.57-60].

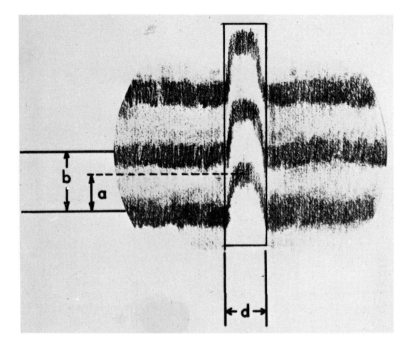

Fig.5.21. Schematic drawing of fringes crossing a receptor outer segment as viewed through the interference microscope. a: Fringe shift; b: fringe separation; d: receptor diameter [5.25a]

ENOCH et al. undertook measurements of the diameters and refractive indices of frog (red) rod outer segments before and after a total bleach [5.25a]. In order to preserve the photopigment the experiments were done in the infrared.

A Zeiss interference microscope (Jamin-Lebedeff type) was adapted for infrared by introducing a 150-W xenon arc source, an infrared interference filter (820-832 nm) to provide the monochromatic beam, and a calcite polarizer to improve infrared transmittance. A special quarter wave plate was designed for 826 nm. An analyzer eyepiece with special Wollaston prism elements was used to introduce the fringes into the field (Fig.5.21). The 40 × condenser and objective units were retained.

The fringe field was imaged onto an infrared image converter and a camera was used to record the phosphor display.

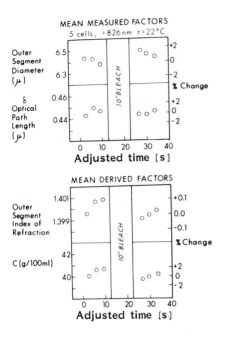

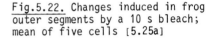

Fig.5.22. Changes induced in frog outer segments by a 10 s bleach; mean of five cells [5.25a]

The E vector of the measuring light was set perpendicular to the long axis of the outer segment and parallel to the fringe pattern. Thus the E vector lay in the plane containing the absorption axes of the visual pigment, but not in the adsorption plane of certain photoproducts (e.g., [5.49]).

Only outer segments with attached ellipsoids were used, and measurements were limited to the outer segment. All dissection was conducted in total darkness with the aid of a second infrared image converter and infrared sources. Fresh aqueous humor from the eye of a large goldfish was used as immersion medium. The index of refraction of the immersion medium was measured with a refractometer.

The optical path difference δ was computed from the evaluated fringe shift, a, and fringe separation, b (Fig.5.21),

$$\delta = \frac{a}{b} \lambda \quad (\lambda = 0.826 \; \mu m) \; . \tag{5.13}$$

The thickness (d) was assumed equal to the width which was derived from a photograph taken at 826 nm. The index of the outer segment (n_1) is given by

$$\delta = (n_1 - n_0)d \; , \tag{5.14}$$

$$n_1 = \frac{\delta}{d} + n_0 \qquad\qquad (5.15)$$

where n_0 is the index of the immersion medium.

A computer-controlled microdensitometer was used to determine fringe shift, fringe displacement, outer segment diameter, and taper from the photographic records. The algorithms used for these redcutions are described in more detail by ENOCH, SCANDRETT and TOBEY [5.25].

Figure 5.22 shows the results for the mean measurements on five rod outer segments. The diameters tended to show a small decrease just before bleach, a small jump (about 2%) with the bleach followed by a continued decrease. The refractive index, n_1, showed the reverse behavior with a drop of about 0.15% in total index over the bleach period. Recall, however, that this corresponds to a change of about 3% in the index *difference*. These changes are small enough that no pronounced change in waveguide behavior would be expected.

5.4.4 Optical Isolation of Single Receptors

A receptor waveguide which is to be viewed end-on cannot be bodily removed from its supporting matrix of interstitial material and neighboring cells. Hence, studies of waveguide behavior of single receptors require a means of optically isolating light which has traversed that receptor only. This is a requirement both for studying the directional dependence of light emerging from a single receptor waveguide and for obtaining its end-on spectral transmission.

Direct masking at the sample level is impractical both because of the small size of vertebrate receptors (1-6 μm diameters cover most species) and the need to avoid any damaging mechanical contact with the receptors.

A number of early workers attempted to isolate receptors for end-on microspectrophotometry by focusing a small spot on the end of the receptor to be studied. However, this requires a fairly large numerical aperture in the condenser optics and the major part of the light will be scattered outside of the receptor waveguide and through neighboring structures. As an example, to produce an Airy disk 1 μm in diameter with 600-nm (orange) light requires a numerical aperture of 0.59, corresponding to a half angle of the illumination cone of 36°. The latter substantially exceeds the half-power acceptance angle of receptor waveguides, especially for the inner segments which are probably the relevant structure for light collection in vertebrate receptors. (By comparison a typical 4-mm diameter human pupil produces an illuminating half angle of only 5°). Thus end-on study of the optical properties of the receptor waveguides requires illumination over narrow cone angles comparable to their numerical aperture.

A more practical approach to optical isolation is to produce a real enlarged image of the retina and mask off the image of a single receptor. Because the output from a receptor waveguide is confined to a relatively narrow cone, there is an additional advantage to be gained if the collecting optics accept only light arriving within that cone. This can be accomplished by masking off the far-field pattern at the BFP of the objective or equivalently at a real image of the BFP.

ENOCH has designed an instrument incorporating all three of these principles which can be used both to study the directional dependence of light from a single receptor waveguide and for end-on microsppectrophotometry [5.8, 61]. The instrument is sufficiently versatile that it can also be used for side-on single-receptor microspectrophotometry. Figure 5.23 is a schematic diagram of this special microscope-microspectrophotometer. A mirror chopper directs the beam alternately to the sample and reference paths. A reduced image of the monochromator exit slit is imaged on the specimen by the condenser lens, using a small aperture. No attempt is made to restrict the spot to a single receptor. The condenser lens has a variable iris diaphragm that provides illuminating-cone half-angles ranging from about 5° to over 20° (in air). Demagnification is $32 \times$ and the working distance is about 6 mm. The condenser mount is fixed vertically and the condenser is focused by translating the stage and all succeeding components up to and including the eyepieces as a unit. Similarly, the objective is focused by translating the objective mount and all components beyond it as a unit relative to the stage.

The lower low-power (LP) objective produces a real image of the objective back focal plane (BFP), which is accessible for masking. At the same time, the specimen image plane is shifted to a point about 27 mm ahead of the BFP image plane. Apertures of known diameters can be placed at any desired position in the specimen or BFP images. Magnification of the specimen image is about $10 \times$. An upper low-power microscope can be focused onto either the specimen image or the BFP image.

A removable beam splitter feeds a horizontal viewing port for an infrared image converter. A trinocular microscope body with removable prism is used for viewing in experiments not requiring infrared. Telescopic and regular eyepieces in the binocular viewing ports allow simultaneous evaluation of specimen (near-field) and BFP (far-field) light distributions. The vertically mounted eyepiece projects the beam either to the photomultiplier or to a removable camera. Background illumination (infrared or visible) can

SCHEMATIC DIAGRAM OF MSP OPTICS

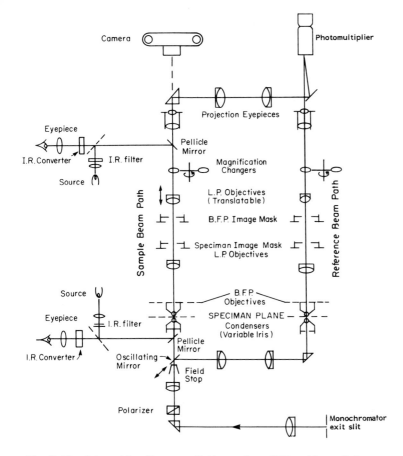

Fig.5.23. Schematic diagram of the end-on MSP optics [5.8]

be switched into the path below the condenser by means of a mirror and pellicle; this is used for locating cells in the field suitable for study. This source is switched off during recording.

The sample can be viewed in the reverse direction using the condenser as objective and the optics above the stage as an illumination system. This arrangement permits focusing the condenser independently on the input plane of the fiber optic bundle which is displaced from the exit plane by the length of the receptor waveguides. A beam splitter is placed in the path below the condenser, and either another image converter or a standard eyepiece may be used.

The light source is an Osram 900-W xenon arc with screw drives for translation in the x, y, and z directions. The arc is powered by its own dc generator and is stabilized by a feedback loop that samples light output from a beam splitter placed behind the entrance slit of the monochromator but ahead of the grating. The same portion of the arc is sampled as irradiates the specimen.

The arc is focused on the entrance slit of a Bausch and Lomb 250-mm grating monochromator. The 5-cm square grating is ruled with 600 lines/mm and is blazed for 500 nm in the first order. Dispersion is 3.3 nm/mm. When the instrument is used as a microspectrophotometer, the wavelength drive can be stepped automatically through a wide selection of previously chosen wavelengths.

The monochromator output is collimated and directed into an automatically sequenced filter bank to reduce stray light. The collimated beam is then sent into the special microscope-microspectrophotometer via a right-angle prism. Alternately, the beam (white light or infrared) can be bypassed around the monochromator and sent directly into the instrument.

5.4.5 Directional Properties of Single Receptors

The instrument described in the preceding section has been used to measure single-receptor directional transmission properties of frog (red rods), albino rats (rods), and goldfish (cones and rods) [5.8,45a,62].

For this purpose retinas were mounted receptors-down on the stage so that light traversed them in the reverse of the normal direction. Thus the cone of light emerging from the inner segments corresponded exactly to its acceptance cone for light arriving from the physiological direction when the same mode was excited.

Single receptors were masked off at the specimen image plane and the image of the back focal plane was recorded photographically. Relative intensity and angle calibrations were taken and a computer-controlled microdensitometer was used to determine the angular spread at which the irradiance in the BFP image fell to one half its peak value. Figure 5.24 illustrates the nature of

Fig.5.24a-d. Determination of the directional properties of a goldfish cone radiating a trilobed modal pattern: a) near-field pattern at the radiating terminal of the receptor waveguide; b) far-field pattern from the same receptor; c) computer-generated contour map of the far-field pattern, used to select the trace along which the profile is measured; d) profile of the far-field pattern, log relative irradiance on the ordinate, abscissa is proportional to angle [5.62]

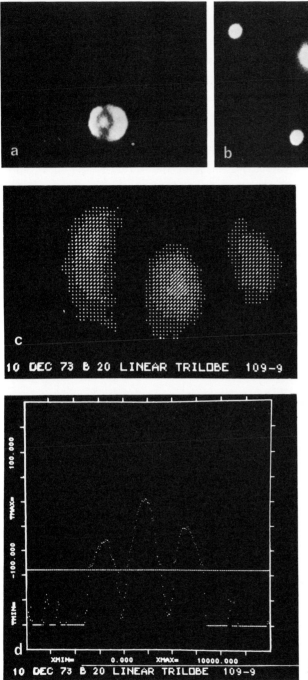

10 DEC 73 B 20 LINEAR TRILOBE   109-9

10 DEC 73 B 20 LINEAR TRILOBE   109-9

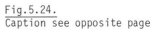

Fig.5.24.
Caption see opposite page

the data records obtained for a goldfish cone radiating a trilobed modal
pattern.

The measurements were complicated by diffraction introduced by the very
small masking aperture at the specimen image plane. This resulted in spreading of the pattern in the BFP image beyond what would have occurred without
the mask. An empirical correction procedure was devised by recording BFP
images from a glass fiber using a series of masking apertures [Ref.5.62,
App.II]. The angular measurements must also be corrected for the refractive
index of the medium. For small angles this is nearly equivalent to dividing
the observed angles by the refractive index of the medium (taken to be 1.34).

Table 5.3. Half-angles for the light cone emerging from the inner segments of
single receptors under reverse-path illumination. All measurements were taken
at the halfpower point and are corrected for the refractive index of the immersion medium. α: Corrected for diffraction; α': uncorrected

Receptor type	Mode	α [deg]	[α' deg]
goldfish cone	bilobe	2.4	10.8
		2.2	8.9
		3.2	12.7
		1.2 [a]	4.6
		2.8	11.4
		2.3	9.4
goldfish cone	dot and ring	2.3 [a]	9.1
		2.7	10.8
		2.2 [a]	8.9
		2.2 [a]	8.9
		2.7 [a]	10.8
		2.2 [a]	8.7
		2.0 [a]	8.2
		2.7	8.7
goldfish rod	bilobe	2.8	16.9
		0.9 [a]	5.7
		2.5	15.8
		2.3	14.1
rat rod	HE_{11}	1.3	6.4
rat rod	HE_{11}	1.6	6.4

[a] One lobe fell below the half-power point criterion.

Table 5.3 summarizes the results for goldfish cones and rods and for rat
rods. Theoretical studies of WIJNGAARD and of HOROWITZ indicate that the diffraction correction procedure used probably overcorrects the effect so that
α represents a lower limit to the half-angles. On the other hand α' certainly
represents a high upper limit since the spreading induced by the small masking apertures is substantial. Thus, there can be no doubt that light is emerg-

ing from the inner segment waveguides into very tight cone angles. Conservatively, one can conclude that for receptors measured to date and for excitation of the modes observed, acceptance half-angles at the half-power point are of the order of $5°$ with an uncertainty factor of less than 2. This applies to both the cones and the rods that have been measured.

5.5 Summary and Conclusions

The work described in this chapter has shown conclusively that vertebrate retinal receptors are optical waveguides and that the vertebrate retina is an optical fiber bundle. This is shown directly by the observation of waveguide modal patterns in individual receptors and by the transfer of images across the retinal receptor layer. There is also indirect evidence provided by the physical properties to be expected from the known morphology of retinal receptors (Chap.2) coupled with the refractive index measurements, i.e., a high-index cylinder in a low-index surround must act as a waveguide.

Two important advantages are immediately evident for the waveguide property.

1) Because the light impinging on the receptor terminal is trapped in the guide, the greater sensitivity of a thicker layer of pigment can be utilized while retaining the resolution inherent in the receptor spacing. Without light guiding these two properties would tend to be mutually exclusive.

2) Because the guide preferentially transmits and traps light arriving within a small solid angle centered on the guide axis, the receptor is preferentially senstive to light arriving along the axis and discriminates against stray and scattered light from elsewhere in the eye (Chaps.3,4).

A beginning has been made at quantitative assessment of the retinal optical properties. First measurements of individual receptor directivities tend to show that these are more narrow than the observed Stiles-Crawford effect and, hence, could account for the latter, but a detailed synthesis is still in the future. Measurements of resolution and the MTF of the retina acting as a fiber optic bundle suggest that it is well matched to the capabilities of the rest of the visual system. There is now a need, however, for a good theoretical analysis of how the *detection* properties of the retina would be expected to relate to the optical *image transfer* properties.

Measurements of the waveguide parameter V have yielded data in agreement with what would be predicted from earlier measurements of refractive index, at least for albino rat.

The effect of bleaching on diameter and refractive index of frog rods has been studied and shows that effects are minor *insofar as waveguide properties are concerned*. Data on other species should soon be available.

Good end-on microspectrophotometry of single receptors has not yet been achieved but, with the aid of the new knowledge about their waveguide properties, this goal is being actively pursued in Enoch's laboratory.

References

5.1   J.M. Enoch: Visualization of waveguide modes in retinal receptors. Am. J. Ophthalmol. *51*, 1107/235-1118/246 (1961)

5.2   G. Toraldo di Francia: Retinal cones as dielectric antennas. J. Opt. Soc. Am. *39*, 324 (1949)

5.3   J. Jean, B. O'Brien: Microwave test of a theory of the Stiles-Crawford effect. J. Opt. Soc. Am. *39*, 1057 (1949)

5.4   J.M. Enoch, G. Fry: Characteristics of a model retinal receptor studied at microwave frequencies. J. Opt. Soc. Am. *48*, 899-911 (1958)

5.5   E. v. Brücke: cited in H. von Helmholtz: *Treatise on Physiological Optics*, Vol.2, ed. and transl. by J.P.C. Southall (Dover, New York 1962) p.229

5.6   A. Hannover: Vid. Sel. Naturv. og Math. Sk. *X* (1843) [see Ref.5.8]

5.7   J.M. Enoch, L.E. Glismann: Physical and optical changes in excised retinal tissue: resolution of retinal receptors as a fiber optics bundle. Invest. Ophthalmol. *5*, 208-221 (1966)

5.8   J.M. Enoch, F.L. Tobey: A special microscope-microspectrophotometer: optical design and application to the determination of waveguide properties of frog rods. J. Opt. Soc. Am. *63*, 1345-1356 (1973)

5.9   W.A.H. Rushton: The difference spectrum and photo sensitivity of rhodopsin in the living human eye. J. Physiol. London *134*, 11-29 (1956)

5.10 W.A.H. Rushton: The rhodopsin density in the human rods. J. Physiol. London *134*, 30-46 (1956)

5.11 H. Ripps, R.A. Weale: Cone pigments in the normal human fovea. Vision Res. *3*, 531-543 (1963)

5.12 J.M. Enoch: An approach toward the study of retinal receptor optics. Am. J. Optom. Arch. Am. Acad. Optom. Monograph 336 (1965)

5.13 R.L. Sidman: Histochemical studies on photoreceptor cells. Ann. N.Y. Acad. Sci. *74*, 182-195 (1958)

5.14 R.W. Massof, A.E. Jones: Electroretinographic evidence for a photopic system in the rat. Vision Res. *12*, 1231-1239 (1972)

5.15 D. Birch, G.H. Jacobs: Behavioral measurements of rat spectral sensitivity. Vision Res. *15*, 687-691 (1975)

5.16 M.A. Ali: Les responses retinomotrices: caractères et mécanismes. Vision Res. *11*, 1225-1288 (1971)

5.17 L. Arey: The movements in the visual cells and retinal pigment of the lower vertebrates. J. Comp. Neurol. *26*, 121-201 (1916)

5.18 R.L. Devalois, H.C. Morgan, M.C. Polson, W.R. Mead, E.M. Hull: Psychophysical studies of monkey vision I. Macaque luminosity and color vision tests. Vision Res. *14*, 53-67 (1974)

5.19 R.L. Devalois, H.C. Morgan: Psychophysical studies of monkey vision II. Squirrel monkey wavelength and saturation discrimination. Vision Res. *14*, 69-73 (1974)

5.20 H.J.A. Dartnall: *The Visual Pigments* (Methuen, London; John Wiley, New York 1957)

5.21 W.H. Dobelle, W.B. Marks, E.F. MacNichol, Jr.: Visual pigment density in single primate foveal cones. Science *166*, 1508-1510 (1969)

5.22 P.A. Liebman: "Microspectrophotometry of Photoreceptors", in *Photo-chemistry of Vision*, ed. by H.J.A. Dartnall, Handbook of Sensory Physiology, Vol.7/1 (Springer, Berlin, Heidelberg, New York 1972) pp.481-528

5.23 E.J. Denton: A method of easily observing the dichroism of the visual rods. J. Physiol. London *124*, 16-17 (1954)

5.24 S.A. Thorpe: The Effect of Chromatic Adaptation and Temperature on the Spectral Sensitivity of the Goldfish, *Carassius auratus*, Ph.D. Dissertation (Brown University, Providence, RI 1972)

5.25a J.M. Enoch, J. Scandrett, F.L. Tobey: A study of the effects of bleaching on the width and index of refraction of frog rod outer segments. Vision Res. *13*, 171-183 (1973)

5.25b J.M. Enoch, F.L. Tobey: Use of the waveguide parameter V to determine the difference in the index of refraction between the rat rod outer segment and the interstitial matrix. J. Opt. Soc. Am. *68*, 1130-1134 (1978)

5.26 B. O'Brien: Vision and resolution in the central retina. J. Opt. Soc. Am. *41*, 882-894 (1951)

5.27 W.S. Stiles, B.H. Crawford: The luminous efficiency of rays entering the eye pupil at different points. Proc. Roy. Soc. London B*112*, 428-450 (1933)

5.28 K. Engstrom: Cone types and cone arrangements in the retina of some cyprinids. Acta Zool. Stockholm *41*, 277-295 (1960)

5.29 M. Born, E. Wolf: *Principles of Optics*, 2nd ed. (Pergamon, Oxford 1964) p.381

5.30 F.I. Harosi, E.F. MacNichol, Jr.: Visual pigments of goldfish cones, spectral properties and dichroism. J. Gen. Physiol. *63*, 279-304 (1974)

5.31 S. Nilsson: Receptor cell outer segment development and ultrastructure of the disk membranes in the retina of the tadpole (Rana pipiens). J. Ultrastruct. Res. *11*, 581-620 (1964)

5.32 J.M. Enoch: Nature of transmission of energy in the retinal receptors. J. Opt. Soc. Am. *51*, 1122-1126 (1961)

5.33 N.S. Kapany, J.A. Eyer, R.E. Keim: Fiber optics. Part II. Image transfer on static and dynamic scanning with fiber bundles. J. Opt. Soc. Am. *47*, 423-427 (1957)

5.34 C.S. Williams, O.A. Beckland: *Optics: A Short Course for Engineers and Scientists* (Wiley, New York 1972)

5.35 W.J. Smith: *Modern Optical Engineering* (McGraw-Hill, New York 1966) Chap.11

5.36 H. Ohzu: Comments on the application of fiber optics to retinal studies. Part I. Image processing by retinal receptors. Jpn. J. Clin. Ophthalmol. *20*, 1031-1034 (1966)

5.37 H. Ohzu, J.M. Enoch, J. O'Hair: Optical modulation by the isolated retina and retinal receptors. Vision Res. *12*, 231-244 (1972)

5.38 F.W. Campbell: The human eye as an optical filter. Proc. IEEE *56*, 1009 (1968)

5.39 F.W. Campbell, D.G. Green: Optical and retinal factors affecting visual resolution. J. Physiol. London *181*, 576-593 (1965)

5.40 G. Westheimer: Modulation thresholds for sinusoidal light distribution on the retina. J. Physiol. *152*, 67-74 (1960)

5.41 G. Westheimer: Pupil size and visual resolution. Vision Res. *4*, 39-45 (1964)

5.42 G. Westheimer: The Maxwellian view. Vision Res. *6*, 669-682 (1966)

5.43 H. Ohzu, J.M. Enoch: Optical modulation by the isolated human fovea. Vision Res. *12*, 245-251 (1972)

5.44 J.M. Enoch, J. Scandrett: Human foveal far-field radiation pattern. Invest. Ophthalmol. *10*, 167-170 (1971)

5.45a F.L. Tobey, J.M. Enoch: Directionality and waveguide properties of optically isolated rat rods. Invest. Ophthalmol. *12*, 873-880 (1973)

5.45b J.M. Enoch: "The Retina as a Fiber Optics Bundle", App. B, in *Fiber Optics Principles and Applications*, ed. by N.S. Kapany (Academic, New York 1967)

5.46 W.B. Marks: "Difference Spectra of the Visual Pigments in Single Gold-fish Cones", Ph. D. Thesis (The Johns Hopkins University, Baltimore, MD 1963)

5.47 W.B. Marks: Visual pigments of single goldfish cones. J. Physiol. London *178*, 14-32 (1965)

5.48 P.A. Liebman, G. Entine: Sensitive low-light-level microspectrophoto-meter: detection of photosensitive pigments of retinal cones. J. Opt. Soc. Am. *54*, 1451-1459 (1964)

5.49 F.I. Harosi: "Frog Rhodopsin in situ: Orientational and Spectral Changes in the Chromophores of Isolated Retinal Rod Cells", Ph.D. Thesis (The John Hopkins University, Baltimore, MD 1971

5.50 S.D. Carlson: Microspectrophotometry of visual pigments. Q. Rev. Biophys. *5*, 349-393 (1972)

5.51 E. Snitzer: Cylindrical dielectric waveguide modes. J. Opt. Soc. Am. *51*, 491-498 (1961)

5.52 A. Snyder: Power loss on optical fibers. Proc. IEEE *60*, 757-758 (1972)

5.53 K. Kirschfeld, A.W. Snyder: Measurement of a Photoreceptor's Charac-teristic Waveguide Parameter. Vision Res. *16*, 775-778 (1976)

5.54 R. Sammut, A.W. Snyder: Ambiguity in Determination of Waveguide Par-ameter V from Mode Cutoffs. Vision Res. *16*, 881-882 (1976)

5.55 R.L. Sidman: The structure and concentration of solids in photorecep-tor cells studied by refractometry and interference microscopy. J. Biophys. Biochem. Cytol. *3*, 15-30 (1957)

5.56 R. Barer: Refractometry and interferometry of living cells. J. Opt. Soc. Am. *47*, 545-556 (1957)

5.57 F. von Hornbostel, cited by W. Kühne: Veränderungen der Stäbchen durch Licht. Unters. Physiol. Inst. Univ. Heidelberg *1*, Heft 1, 409-411 (1877)

5.58 J. Wolken: Structure and molecular organization of retinal photorecep-tor. J. Opt. Soc. Am. *53*, 1-19 (1963)

5.59 E. Friedman, T. Kuwubara: The retinal pigment epithelium. IV. The damag-ing effect of radiant energy. AMA Arch. Ophthalmol. *80*, 265-279 (1968)

5.60 T. Kuwubara: Retinal recovery from exposure to light. Am. J. Ophthal. *70*, 187-198 (1970)

5.61 J.M. Enoch: Retinal microspectrophotometry. J. Opt. Soc. Am. *56*, 833-835 (1966)

5.62 F.L. Tobey, J.M. Enoch, J.H. Scandrett: Experimentally determined optical properties of goldfish cones and rods. Invest. Ophthalmol. *14*, 7-23 (1975)

6. Theoretical Considerations of the Retinal Receptor as a Waveguide

B. R. Horowitz

With 22 Figures

The preceding chapters and the references contained therein have firmly established that the photoreceptor layer of the retina acts as a fiber optics bundle, and that the individual photoreceptors act as fiber optics elements or dielectric optical waveguides. This behavior occurs because of the physical properties of the retina and is now routinely observed in experimental studies of retinal photoreceptors. Within a given retinal neighborhood, the photoreceptor cells are, roughly speaking, parallel cylindrical dielectric structures with diameters of the order of the wavelength of light, which are embedded in a surrounding medium of lower refractive index. Moreover, these receptors are aligned with the entrance pupil of the eye and are thus able to be maximally stimulated by the relevant illumination. Recent reviews of waveguide properties of retinal photoreceptors may be found in the proceedings of a workshop dedicated to photoreceptor optics [6.1] and in a paper presented at a symposium [6.2].

Waveguide properties must be given due cognizance when considering the retina as an optical element, and when attempting to draw inferences requiring knowledge of the optical excitation of individual photoreceptors. Application of the waveguide properties of the retinal photoreceptors can be separated into two classes. 1) There are studies of the manner in which the optical or waveguide properties of the individual photoreceptors, or of the entire photoreceptor layer, influence visual function. Here, the object is to relate visual function to the optical stimulation (light collection plus absorption) of individual or groups of photoreceptors. An example is the theoretical calculation of retinal direction (Stiles-Crawford) effects. 2) There are studies that employ optical measurements of the photoreceptors in order to determine the optical and structural properties of those receptors. Indeed, this method which we may refer to as modal microscopy offers the advantage of a novel histological technique for working with relatively fresh, unfixed, and unstained biological material (Chap.5). Here, optical measurements are used to draw inferences regarding photoreceptor index of refraction (including birefringence), directionality, pigment properties, and receptor orientation and packing. In both classes of studies, theoretical calculations based upon the analysis of an appropriate photoreceptor model are correlated with experimental observations. The modeling and subsequent theoretical calculations can thus serve as useful tools for the design, analysis, and interpretation of experiments.

In this chapter we review the theoretical aspects of the waveguide nature of individual vertebrate retinal photoreceptors; arrays of photoreceptors are treated in Chap.7, and invertebrate photoreceptors are discussed in Chap.11. Modeling of the retinal photoreceptors and a discussion of the analysis approach appear in Sect.6.1. Mode properties of the dielectric waveguide as well as the specialization of those modes to the weakly guiding retinal photoreceptors are presented in Sect.6.2. Section 6.3 treats optical excitation of the model waveguides, and in Sect.6.4 we discuss further properties of the photoreceptors, namely, effective aperture, directionality, and spectral sensitivity.

In selecting material for this chapter, attempts were made to review a wide area, and still allow for an excursion giving a description of the origin and formulation of the waveguide modes. Topics of interest to both those with and without prior backgrounds in electromagnetics are included. Mathematical expressions that are presented in the text provide our notation and serve as anchor points to work appearing in the literature. Attempts have been made to organize the text in the form of a guided tour, with discussions both in general and about the origins, limitations, and uses of the mathematical relations. In referencing the various mathematical expressions, it was not always possible to give sources in which the expressions appear in exactly the form given herein; some equivalent forms are used, and some differences in sign will be experienced when references are cited in which the time variation $\exp(-i\omega t)$ is used. (Note, for example, expressions using this time variation may be converted by replacing $-i$ by j, where $\sqrt{-1} = j$.)

6.1 The Model

The retinal layer of rods and cones (photoreceptors) is defined as beginning at the external limiting membrane (ELM) [6.3-5], and it is from that boundary that the photoreceptors are believed to take their orientation relative to the pupillary aperture [6.6-8]. In any local area within that layer, the photo-

receptors are roughly parallel [6.6,7] cylindrical structures of varying degrees of taper, which have diameters of the order of magnitude of the wavelength of light [6.3-5,9-14], and which are separated laterally by an organized lower refractive index substance, called the interphotoreceptor
matrix (IPM), which is identified as an acid-mucopolysaccharide-like complex
[6.15-20]. The photoreceptors thus form a highly organized array of optical
waveguides, whose longitudinal axes are, beginning at the ELM, aligned with
the central direction of the pupillary illumination [6.2,9-12,21-28]. The
IPM serves as the cladding for those waveguides.

We will assume, as was done in the past, that the ELM approximates the
inner bound of the receptor fiber bundle for light that is incident upon
the retina in the physiological direction [6.11,21,29]. Portions of the rod
and cone cells that lie anterior to the ELM are not necessarily aligned
with the pupillary aperture, and might thus be struck on the side, rather
than on axis,.by the incident illumination. They would therefore be only
weakly excited by the incoming illumination [6.2,10,11,21,23].

Although the retinal receptors do exist in the form of an array, it is
useful to perform an analysis of the optical properties of an otherwise isolated, single retinal photoreceptor model situated in a homogeneous surround. This might, at first glance, appear to be a brash step. It is done
for a number of reasons. 1) The individual photoreceptor model is considerably more analytically tractable than is the array; 2) experimental observations of photoreceptors in situ within excised retinae appear to yield
first-order agreement with calculations performed upon such a model [6.11,
20,30]; and 3) modes on arrays can be constructed from combinations of the
modes of the individual receptors [6.31,32] (Chap.7). Calculations performed
upon a model array consisting of hexagonally arranged dielectric cylinders
show that propagation along that array is governed by a dispersion relation
that reduces to that of the isolated fiber when the fibers are tightly packed
[6.33].

6.1.1 Three-Segment Tapered Model

The idealized three-segment model photoreceptor is shown in Fig.6.1. All three
segments are assumed to have circular cross section, with uniform myoid and
outer segment and a smoothly tapered ellipsoid. For receptors with equal
myoid and outer segment radii, the ellipsoid is untapered. In the model, the
sections are taken to be homogeneous, isotropic, and of higher refractive
index than the surrounding homogenous isotropic medium, which is identified
with the IPM.

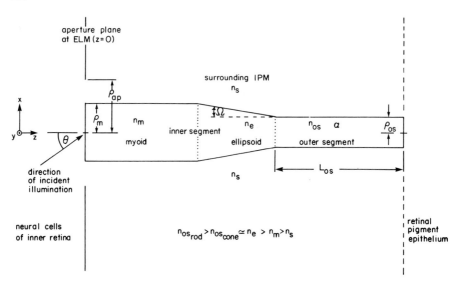

Fig.6.1. Longitudinal view of three-stage photoreceptor model with circular cross section. The physiological direction for incident illumination is from the left. The aperture plane at the external limiting membrane (ELM) is imposed upon the geometry for computational purposes; n = refractive index, α = naperian absorption coefficient. In an actual retina, receptors may or may not extend to the surface of the retinal pigment epithelium (RPE)

Ranges of refractive index, measured across species by SIDMAN with BARER, are shown in Table 6.1 [6.15,17]. Measurements were carried out using immersion refractometry in unpolarized light and, thus, represent an average figure for the receptors. The ranges in Table 6.1 are fairly tight, reflecting the fact that the data show a high degree of uniformity when viewed across species. A source of error in the data is that measurements were made upon cells that were teased out from the retina and which were often fractured along the inner segment, thus opening the way for possible swelling. An example of swelling was presented by ENOCH [6.22], who measured a 2.2-µm variation in the diameter of a (6-µm nominal diameter) frog rod outer segment during the first 5 min after stoppage of blood flow during dissection, and a variation of 3.1 µm during the first 35 min. The inner segment of the cell was described as being "largely intact", and the cell was observed while immersed in aqueous humor from the same eye. Variations in measured refractive index were described as paralleling the diameter changes.

Both the myoid and the ellipsoid are assumed to be lossless, or nonabsorbing; cytochrome pigments in the mitochondria of the ellipsoids are not considered in this model. Absorption in the outer segment by the photolabile

Table 6.1. Measured refractive indices[a,b,c]

			Refractive Index	
			Rod	Cone
Outer segment (n_{os})			1.405 – 1.411	1.387 – 1.396
Inner Segment	Ellipsoid	(n_e)	1.398	1.390 – 1.398
	Myoid	(n_m)	1.361	1.361 – 1.364
	Paraboloid[d]	(n_p)	–	1.378

[a] Determinations across species (frog, salamander, turtle, chick, pigeon, rat, mouse, monkey) from refractometery measurements in [6.15,17].

[b] The IPM was not measured; BARER estimated its refractive index as 1.347 [6.17]. In recent work, ENOCH and TOBEY reported a value of 1.348 from measurements on albino rat [6.20].

[c] We note that universal agreement does not exist regarding the applicability of this data to the human retina, as human receptors were not measured. MILLER and SNYDER [6.34] and SNYDER and PASK [6.29] used refractive index estimates obtained by interpreting the density of staining in electron micrographs, in consultation with this data. Among their results is the estimate that human foveal cones have an outer segment index of approximately 1.419 [6.34].

[d] Present only in cones of some species; not included in model.

pigment molecules aligned within the closely packed disk membranes [6.35-38] is described by naperian absorption coefficient α. The aperture imposed at the ELM facilitates calculation of the equivalent illumination incident upon the receptor.

A number of assumptions and approximations have been made in constructing the model; these can be thought of in two overlapping groups: assumptions and approximations regarding structure, dimensions, and optical constants, and those pertaining to the electromagnetic validity of the model. Approximations and assumptions belonging to the first group are:

1) The receptor cross section is approximately circular and its axis is a straight line. This appears to be a reasonable assumption for the intact eye or for fresh excised tissue that is free of pathology. Deviation from circularity will produce form birefringence.

2) Ellipsoid taper is assumed to be smooth and gradual. Tapers that deviate from these conditions may introduce strong mode coupling and radiation loss.

3) Individual segments are assumed to be homogeneous. This is a first-order approximation, as each of the photoreceptor's sections has different inclusions [6.14,39] which produce local inhomogeneities. In addition,

because only average values of refractive index have been measured for those sections, the existence of a slow refractive-index gradient cannot be ruled out. Although a step profile in core-cladding (photoreceptor-IPM) refractive index is employed in our model, that is not the only profile which would predict mode behavior that would provide a reasonable match with experimental observations. It has been suggested that the modes of a fiber having a circular parabolic profile also provide a good match for those modes observed in retinal photoreceptors [6.40].

4) Individual segments are assumed to be isotropic as a first-order approximation. The ellipsoid is packed with highly membraneous mitochondria, whose dimensions approach the wavelength of light [6.14], and which are shaped differently in different species. Mitochondria introduce local inhomogeneity and thus produce scattering, and may introduce anisotropy. The outer segment is composed of transversely stacked disks, having a repetition period an order of magnitude smaller than the wavelength of light [6.13,14]; these disks contain the photolabile pigment [6.14,35-38]. Form birefringence is produced by the laminated structure of the outer segment, and intrinsic birefringence is produced by the molecular alignment ([6.35,38,42-44] and [Ref.6.41, pp.705-708]). In addition, circular and linear dichroism are contributed by the configuration and alignment of the photopigment molecules [6.35,38,42,45-52]. Those topics are pursued further in Chaps.8 and 9.

5) Linear media are assumed. Nonlinearities may arise from a number of sources. a) Absorption of light is accompanied by bleaching of the photopigments and the production of absorbing photoproducts that have different absorption spectra from those of the pigments. Bleaching is energy and wavelength dependent and the bleached pigment is removed from the pool of available pigment, thus affecting the absorption spectrum. Under physiological conditions in the intact organism where pigment is constantly renewed, the relatively low illumination levels that result in a small percent bleach can be expected to be consistent with the assumed linearity. However, for either the high illumination levels used in retinal reflectometry (for discussions, see [6.53,54]) or the spectral measurements performed upon excised retinae or isolated receptors, where renewal is virtually nonexistent, · workers must utilize special procedures to avoid obtaining distorted spectra [6.46,55-59]. Fully bleached samples of retina can be expected to behave fairly linearly, provided that illumination-dependent heat buildup does not occur. b) Nonlinearity may arise in the ellipsoid owing to illumination-dependent activity of the mitochondria. It is known that isolated liver mitochondria exhibit metabolically linked mechanical effects [6.60]. If such

effects were to occur in the retina, then illumination-driven activity could alter inner segment scattering. c) Some species exhibit illumination-dependent photomechanical responses that result in gross movement of retinal photoreceptors [6.61,62]. That movement is accompanied by an elongation or contraction of the myoid and has been analyzed by MILLER and SNYDER in terms of its action as an optical switch [6.63].

6) Refractive indices of the segments are assumed to be independent of wavelength. This neglects normal dispersion as well as anomalous dispersion, which is the name given to the variation in refractive index that occurs in the vicinity of an absorption peak. Discussion of the effects of anomalous dispersion are given by SNYDER and RICHMOND [6.64,65] and STAVENGA and VAN BARNEVELD [6.66].

7) The medium to the left of the external limiting membrane (ELM) in Fig. 6.1 is assumed to be homogeneous. Again, this is an approximation. Although the ocular media and inner (neural) retina are largely transparent, scattering does occur within the cornea, aqueous, lens, vitreous, and inner retina. Preparations of excised retinae will be devoid of the ocular media, but the inner retina, and possibly some adhesions from the vitreous, will be present. A description of the scattering, however, would allow its effect upon the incident illumination to be accounted for. A discussion of intraocular scattering and an extensive list of references are found in the book by MILLER and BENEDEK [6.67].

8) It is assumed that the medium surrounding the photoreceptors is homogeneous and nonabsorbing. This, too, is an approximation. Microvilli originating in the Mueller cells extend into the space between the inner segments [6.3,4,14], and microfibrils of the retinal pigment epithelium (RPE) surround portions of the outer segments [6.14,68]. In some species, the microfibrils contain an absorbing melanin-based pigment. The assumption of homogeneity of the surround also neglects the reflection and backscatter produced by the RPE [6.69], choroid, and sclera; without that reflection, retinal reflectometry would not be possible [6.53,54]. We note that preparations of excised retinae that are used for observing waveguiding are usually devoid of the RPE [6.10-12]. Moreover, because the RPE and choroid are highly absorbing, the reflected light will be a small fraction of the incident light, except in the case of albinos and species having tapeta; the tapeta produce back-reflected light and are responsible for the reflected light seen at night from the eye of the cat [6.2,23] (Chap.10).

9) In some species the outer segment is tapered [6.3-5,37]. Gradual taper can be accommodated in the model. Sharp tapers, on the other hand, must be dealt with separately [6.70].

10) The exact location of the effective input plane to the retinal fiber optics bundle is not known. In early experimental work, ENOCH [6.11,21] assumed that the ELM marks the inner bound of that fiber bundle, and this assumption still appears to be the most reasonable. For this assumption to be valid, the receptor myoids must be separated by a large enough space of sufficiently low refractive index, so that this portion of the retina appears optically as a fiber bundle, rather than as a largely homogeneous medium. In the absence of such a separation, only weak waveguide excitation could occur within the myoid region; the next most sclerad candidate for the input plane would be the level of the ellipsoids, which contain densely packed mitochondria, and whose refractive index is rivaled only by that of the outer segments (see Table 6.1). Different locations for the input plane would interface the incoming illumination with fiber bundles of different optical characteristics, and would thus influence the optical excitation of the retinal photoreceptors.

At the the ELM, microvilli originating in the Mueller cells extend into the space between the myoids [6.3,4,14]. Those processes are of unknown refractive index; however, they do not extend far along the myoid, and the remaining space appears to be occupied by the lower refractive index IPM. Support for locating the input plane in the vicinity of the ELM comes from a) histochemistry (e.g., [6.18,19]), which indicates the presence of acid mucopolysaccharides between the inner segments and extending vitreally to the microvillous surface of the Mueller cells; and b) observations of waveguiding in the myoids of goldfish double cones [6.28] and human and monkey cones (personal communication from Enoch) that were illuminated in the physiological direction, and observations of waveguiding in the myoids, and in some of the thinner inner segment processes lying anterior to the ELM, of rod and cone receptors of a number of species, that were illuminated in a direction opposite to the physiological direction [6.27,28,71]. Moreover, the ELM is the plane from which the receptors appear to assume their orientation relative to the pupillary aperture [6.6-8] and may therefore be the plane at which the receptors can be optimally excited. We note that universal agreement does not exist regarding the location of the effective input plane; some investigators assumed it to be located at the outer segment level [6.72].

Photoreceptor models are constructed from available data. Whereas these data do include some information on receptor mosaics as well as dimensions, refractive index, and ultrastructure of representative photoreceptors, systematic characterization of local as well as across-the-retina variations in receptor parameters and mosaic populations and geometries is rare (Chap.2). Such data are needed for more complete analyses of the optics of photoreceptors.

Different fixation techniques are known to alter both structure and dimension, and to preserve certain features for observation at the expense of others [6.19,73-75]. In addition, the process of excising and cutting the delicate tissue can disturb cell orientation and introduce artifacts. The histologist chooses those preparation techniques that are best suited to the objectives of a particular investigation. It is rare indeed that the objectives are to construct a viable model for the pursuit of photoreceptor optics. Clearly, there is ample room for investigations that would aim to provide the requisite parameters for that pursuit.

The electromagnetic validity of the model is based upon the supposition that first-order effects can be calculated in the presence of the assumptions (1-10) given above. The additional assumption is that meaningful results relating to the in situ properties of a retinal photoreceptor can be obtained by means of calculations performed upon a single photoreceptor model. Involved here are considerations regarding optical excitation, optical coupling between neighboring receptors, and the role of scattered light within the retina. These assumptions may be summarized as follows:

1) The illumination incident upon the single photoreceptor through the aperture of Fig.6.1 is to be representative of that illumination available to an individual receptor located within the retinal mosaic. Figure 6.2 illustrates the cross section of a uniform receptor array as seen from the ELM. For a uniformly illuminated, infinitely extended array, the total illumination available to each receptor would be that falling on a single hexagonal area (shown with broken lines in Fig.6.2). Thus for modeling purposes, it would appear reasonable to interpose a truncating aperture at the ELM between the single receptor model and the incident illumination. Circular truncation apertures have been employed in the photoreceptor optics literature [6.29,30,76-78]. Such apertures do serve to delimit the incoming illumination, but they also produce diffraction effects owing to their edges. The assumption here is that such effects can be tolerated in the model.

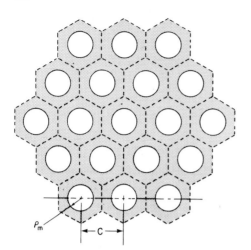

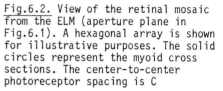

Fig.6.2. View of the retinal mosaic from the ELM (aperture plane in Fig.6.1). A hexagonal array is shown for illustrative purposes. The solid circles represent the myoid cross sections. The center-to-center photoreceptor spacing is C

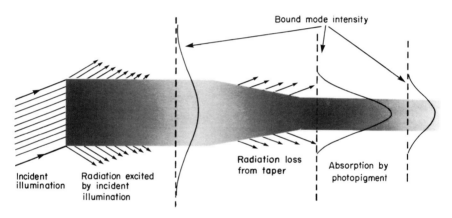

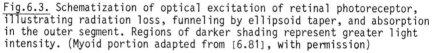

Fig.6.3. Schematization of optical excitation of retinal photoreceptor, illustrating radiation loss, funneling by ellipsoid taper, and absorption in the outer segment. Regions of darker shading represent greater light intensity. (Myoid portion adapted from [6.81], with permission)

2) As required by the boundary conditions, the incident illumination excites both guided modes that are bound to the receptor and carry power along the receptor axis, and fields that carry power away from the receptor into the radiation field (see Fig.6.3) [6.79,81]. In addition, the tapered ellipsoid also radiates power array from the receptor by causing a fraction of the bound-mode power to be coupled into the radiation field. The mode coupling on receptors having smooth gradual tapers may be neglected in first-order calculations, except in the vicinity of a mode's cutoff where

strong radiation may occur [6.70]. Any such radiated power would represent additional unguided light within the retina. The bound modes carry power within the receptor cross section, as well as in the surrounding medium outside the receptor boundary. Coupling to neighboring receptors can occur due to both the unguided light and that portion of the bound mode power that resides within the IPM and extends to a neighboring receptor. These effects are treated in Chap.7. In the single receptor model, coupling effects are neglected for a first-order approximation to the fields of receptors that are directly illuminated. This approximation is strenthened by the dichroism exhibited by the photopigment, which favors absorption of light traveling in the axial direction.

3) Coherent illumination of the photoreceptors is assumed. This is employed as a first-order approximation for many experimental situations and for illumination of the photoreceptors in a given retinal neighborhood through the eye pupil under physiological conditions.

An estimate of some physiologically compatible truncation apertures can be made by considering the data obtained from a specimen of squirrel monkey retina which show that the intercellular distance at the ELM (measured as the distance from the edge of one receptor to the edge of its nearest neighbor) remains constant at 0.83 μm, and that the inner segment diameter varies from 1.07 μm at the foveal center to 5.9 μm at a distance of 1 mm from that center [6.11]. Assuming a hexagonal array of receptors, these dimensions translate to effective circular illumination diameters 1.9 and 1.2 times larger than those of the receptor inner segments, for receptors located at the foveal center and 1 mm distant from that location, respectively. In making this comparison, we defined the equivalent aperture as a circle whose area is equal to that of the hexagon (see Fig.6.2) specified by the appropriate intercellular spacing and inner segment diameter. We note that a human radial shrinkage correction factor based on average bulk tissue shrinkage was applied to obtain those histological data [6.11]. Such corrections do not account for differential shrinkage within the sample, and correction factors of the order of 20% are not unusual for the preparation techniques employed [6.11,73].

6.1.2 Approaches to Analysis of the Model

The earliest analyses of light propagation within the retinal photoreceptors employed geometric optics. As early as 1843, BRÜCKE discussed the photoreceptor's trapping of light via total internal reflection, but stopped just short of suggesting a directionality of vision [6.82]. Interest resumed after the discovery by STILES and CRAWFORD of a directionality effect in human vision, now called the Stiles-Crawford effect of the first kind (SC-I) [6.83]. The essence of that effect is that for beams of light, incident upon the same retinal area, sensitivity is greater for the beam passing through the pupil near its center (along the receptor's axis) than for light passing through other pupil locations (off the receptor's axis). Furthermore, the magnitude of the effect is strongly wavelength dependent [6.84]. Discussions of the Stiles-Crawford effect are given by STILES [6.85], by CRAWFORD [6.86], and in Chap.3.

Following discovery of the Stiles-Crawford effect, WRIGHT and NELSON [6.87] and O'BRIEN [6.88,89] applied geometric ray tracing to explain it. O'BRIEN [6.88] suggested a light-funneling effect produced by the tapered ellipsoids, and WINSTON and ENOCH [6.90] employed geometric optics to investigate the light-collecting properties of retinal photoreceptors. The further application of geometric optics to general fibers can be found in a book by KAPANY [6.91]; and a discussion of light collection in fibers is given in Chap.8. We present below the rudimentary concepts from geometric optics that we shall refer to in subsequent sections.

a) Geometric Optics Considerations

Recall the behavior of rays incident upon the interface separating two dielectric media (see Fig.6.4). Rays incident from medium 1 at an angle θ_1 to the interface normal, will be refracted at an angle θ_2 in medium 2; the angles are related by Snell's Law

$$n_1\sin\theta_1 = n_2\sin\theta_2 \quad , \tag{6.1}$$

where n_1 and n_2 are the refractive indices of media 1 and 2, respectively. As shown in Fig.6.4, the refracted rays are bent toward the interface, for incidence from a medium of greater refractive index to a medium of lesser refractive index. For this situation a limiting angle of incidence $\theta_1 = \theta_c$, called the *critical angle of total internal reflection*[1], exists, for which

---

1 Note that in some treatments θ_c is defined as the complementary angle, i.e., the angle between the ray and the surface. In that case $\theta_c = \arccos(n_2/n_1)$.

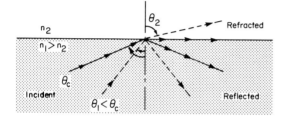

n_2

$n_1 > n_2$

θ_2 --- → Refracted

θ_c

Incident

$\theta_1 < \theta_c$

Reflected

Fig.6.4. Reflection and refraction at a dielectric interface. The dashed ray is incident at an angle smaller than the critical angle; and the solid ray is at critical incidence

the refracted ray travels along the interface ($\theta_2 = 90°$); from (6.1).

$$\theta_c = \arcsin(n_2/n_1) \quad . \tag{6.2}$$

Rays incident at angles smaller than θ_c will undergo partial reflection and refraction; rays incident at angles equal to or larger than θ_c will undergo total internal reflection.

Let us now consider a dielectric cylinder; the pertinent geometry is shown in Fig.6.5. A circular dielectric cylinder of refractive index n_1 is embedded in a lower-refractive-index surround n_2. Light is incident upon the fiber end face at an angle θ to the z or cylinder axis from a third medium of refractive index $n_3 \le n_1$. We consider *meridional rays*, i.e., rays which, in the course of their travel, intersect the fiber axis. There are three possibilities for meridional rays, shown by rays 1 through 3. Once inside the cylinder, ray 1 is incident upon the side wall at an angle smaller than θ_c, is only partially reflected and produces, at each reflection, a refracted ray that carries power away from the fiber. Ray 2 is critically incident and is thus totally reflected; this ray forms the boundary separating meridional rays that are refracted out of the fiber from meridional rays that are trapped within the fiber. Ray 3 is incident upon the side wall at an angle greater than θ_c, and is trapped within the fiber by total internal reflection.

Ray optics thus predicts a limiting incidence angle θ_L, related to θ_c via Snell's Law, given by

$$\sin\theta_L = (n_1/n_3) \cos\theta_c = \frac{\sqrt{n_1^2 - n_2^2}}{n_3} \quad , \tag{6.3}$$

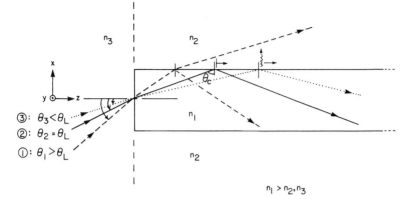

Fig.6.5. Ray optics view of light trapping showing incident meridional rays that form angles with the normal to the fiber side wall that are 1) less than θ_C, producing a partially reflected ray within the fiber and a refracted ray outside the fiber; 2) equal to θ_C, producing a totally reflected ray within the fiber; and 3) greater than θ_C, also producing a totally reflected ray within the fiber. The fields outside the fiber associated with the totally reflected rays are indicated; both propagate parallel to the cylinder axis, and that produced by ray 3 is evanescent. The reflected rays at the fiber entrance are not shown

for which incident meridional rays may be trapped within the fiber. Only those rays incident upon the fiber end face at angles $\theta \leq \theta_L$ are accepted; those incident at greater angles are quickly attenuated owing to refraction. As an example of the numbers involved, we identify the fiber in Fig.6.5 with a photoreceptor myoid. Using the likely refractive index estimates suggested by SIDMAN [6.15] and BARER [6.17], $n_1 = n_m = 1.361$ and $n_2 = n_s = 1.347$, we find for θ_L, 8.31° and 8.23° if we take n_3 equal to n_s and n_1, respectively. The quantity $n_3 \sin\theta_L$ is referred to as the *numerical aperture* (N.A.) of the fiber [6.91] and is given by

$$\text{N.A.} = n_3\sin\theta_L = \sqrt{n_1^2 - n_2^2} \quad . \tag{6.4}$$

Application of meridional ray optics thus yields an acceptance property wherein only rays incident at $\theta \leq \theta_L$ are trapped within the fiber. An improved approximation that accounts for skew rays, predicts a decreasing, but nonzero, trapping of rays incident at $\theta < \theta_L$. Details of that analysis are given in the literature [6.77,91,92].

If we associate plane waves with the rays of Figs.6.4 and 6.5, we will find an evanescent field that propagates along the interface in the rarer medium and which decays exponentially transverse to the interface. The greater the angle of incidence is, the faster the attenuation will be [Ref.6.41,

Chap.1]. This result is consistent with results obtained from modal analysis, which show that the further away from cutoff a mode is, the more tightly bound to the cylinder it is, and the closer its central ray is to the cylinder axis [6.93]. Whereas the rays of geometric optics, representing attendant plane waves, work fine for planar interfaces, the curved side walls of the optical fiber introduce diffraction effects. Thus, questions have arisen regarding a reinterpretation of the classically defined critical angle [6.94,95] and the Fresnel reflection coefficients [6.96] from a quasi-optics vantage point.

Geometric optics, as it is often employed, does predict that an optical fiber can trap and guide light; it does not, however, provide an accurate quantitative description of phenomena on fibers having diameters of the order of the wavelength of light, where diffraction effects are important. Among these effects are considerations of angular directionality, effective receptor aperture, spectral sensitivity, and other wavelength-dependent effects. This was recognized by TORALDO Di FRANCIA who suggested viewing the photoreceptors as dielectric antennas, so that account might be taken of diffraction [6.97,98].

Before we move on to the wave optics approach, it behooves us to point out that ray optics can be applied to account for aspects of the modal fields that are commonly analyzed by wave optics. In this regard, a brief discussion [6.99] and a review [6.80] are relevant.

b) Electromagnetic Approach

In the remainder of this chapter, a simplified electromagnetic analysis of the photoreceptor waveguide model is reviewed and various applications of that analysis are discussed. Propagation along the waveguide is analyzed in terms of the natural modes of the structure. Illumination is assumed to be monochromatic and the Kirchhoff, or first Born, approximation, which ignores the contribution of waves that are scattered from the aperture, is used to match the incident illumination in the equivalent truncating aperture to the modes of the lossless guide that represents the myoid. Modes excited on the myoid are then guided to the tapered ellipsoid; some of the initially excited radiation field will have leaked away before the ellipsoid is reached.

Each segment is dealt with separately. Thus, the ellipsoid will guide or funnel the power received from the myoid to the outer segment. For a gradually tapered ellipsoid, only small amounts of mode coupling will occur [6.70], and that coupling, including the backward waves produced by reflec-

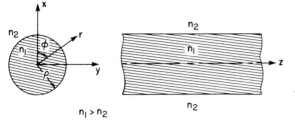

Fig.6.6.
Geometry for analysis of
infinitely extended fiber

tion along the taper, is neglected. Once the mode power reaches the outer
segment it may be absorbed by the photopigment. In most discussions only
bound modes are accounted for; modes that become cutoff on the structure
are not considered further. A perturbation result is employed to calculate
absorption. Absorption-related mode coupling and reflection due to index
differences between segments are neglected.

The end result of the simplified analysis is a series of approximations.
Although error estimates have appeared in the literature for many of those
approximations, estimates of the error incurred by combining those approximations are unknown. In spite of this deficit, the approximate analysis appears to provide reliable results for weakly guiding fibers such as those
found in the visual system. That analysis is also considerably less complicated than an exact electromagnetic analysis, which would require the full
solution to the boundary value problem, in which the illumination in the
aperture is not known a priori, but which must itself be obtained from a
solution to that boundary value problem. Furthermore, an exact solution to
the required boundary value problem is not known.

Both the myoid and the outer segments of the model are uniform cylinders.
Because each segment will be dealt with separately, and because the modes
of those segments are identical to those of an infinitely extended cylinder
of the same radius and refractive index, we may confine much of our initial
attention to the dielectric structure of Fig.6.6, which shows an infinite
uniform dielectric cylinder of radius ρ and refractive index n_1, embedded in
a lower-refractive-index surround n_2.

6.2 Modes of the Dielectric Waveguide

A *mode* of a dielectric waveguide is an electromagnetic field that propagates along the waveguide axis and has the same shape in any arbitrary transverse plane along that axis. Waveguide modes are obtained as source-free solutions of Maxwell's equations for the particular waveguide geometry (sources that produce the modes may be viewed as located at infinity). The total collection of such modes is called the waveguide's *mode spectrum*.

The open, or unbounded, transverse geometry of the dielectric waveguide supports two classes of modes, referred to as bound, guided, or trapped modes, and unbound, unguided, or radiation modes. Different behavior is exhibited by the two mode species. As their names suggest, the *bound modes* carry power along the waveguide, whereas the *unbound* or *radiation modes* carry power away from the guide into the radiation field. From a ray viewpoint, the bound modes are associated with rays that undergo total reflection and are trapped within the cylinder; unbound modes are associated with rays that refract or tunnel out of the cylinder.

For each value of wavelength, every mode has associated with it a transverse wave number that serves to scale the mode along the space coordinate in the guide cross section. Wave numbers belonging to bound modes occur at discrete values in wave-number space, whereas those belonging to unbound modes form a continuum. The subsets of bound and unbound modes are called the *discrete spectrum* and the *continuous spectrum*, respectively.

Bound modes exist as surface waves and may be excited individually. Unbound modes, on the other hand, do not exist individually (a single unbound mode can be excited only by an infinitely extended source located at infinity), but rather as a continuum whose integrated resultant is a clyindrical wave. Although we shall speak of individual unbound modes, only their collective effect has physical significance.

The modes are orthogonal over the infinite cross section, i.e., they do not exchange power with one another on the ideal guide and their excitations can be calculated independently. Both sets of modes taken together form a complete set. Completeness ensures that any incident illumination and any wave that propagates along the waveguide can be represented as a superposition of modes, comprising a finite sum of bound modes (for any given wavelength only a finite number of bound modes may propagate) and an integration (in wave-number space) over the unbound modes. That the bound modes themselves do not constitute a complete set can be seen by considering illumination in the form of incoming waves incident upon the open geometry. Such illumination could not be represented solely in terms of bound modes, as those modes are evanescent in the guide cross section and a sum of them could not represent an arbitrary incoming wave.

The optics of the eye, including photoreceptor alignment, favors the launching of bound modes. Unbound modes appear owing to scattering by in-homogeneities in the ocular media and retina (including the photoreceptor layer), and as a spillover from pupillary illumination that is incident upon the photoreceptors at angles that exceed their effective numerical apertures. The photoreceptor optics literature has dealt mainly with the bound modes. Those modes often provide the major contribution to light absorption in the outer segments, and we can expect that such treatments will provide a reasonable picture of light collection and absorption for receptors illuminated under physiological conditions. However, we should also expect that even under these conditions there may be observable phenomena whose existence depends upon the unbound light. In addition, an accounting of only the bound modes may predict, for certain variables, a discontinuous behavior which will be smoothed out upon inclusion of the unbound mode contributions.

In this section we outline the formulation of the modes and discuss their behavior. Major emphasis is placed on the bound modes, and approximate expressions are given for them.

6.2.1 Formulation and Basic Behavior of Exact Mode Solutions

The formulation and properties of the modes of the dielectric waveguide are well documented. In this section we draw considerably upon material presented in books on fiber optics by KAPANY [6.91], KAPANY and BURKE [6.100], and MARCUSE [6.101,102], articles on fiber optics by SNITZER [6.103], BIERNSON and KINSLEY [6.104], and SNYDER [6.79,80,93], a book on electromagnetic theory by STRATTON [6.105] and a book on guided waves by COLLIN [6.106].

It is most convenient to consider monochromatic light of radian frequency ω and to suppress the time variation $\exp(j\omega t)$. The time-varying *electric field intensity* $\underline{E}(r,\phi,z,t)$ is then represented in terms of the complex function $\underline{E}(r,\phi,z)$, and is obtained as

$$\underline{E}(r,\phi,z,t) = \mathrm{Re}\{\underline{E}(r,\phi,z)\exp(j\omega t)\} \quad , \tag{6.5}$$

with a similar expression for the *magnetic field intensity* $\underline{H}(r,\phi,z,t)$. Re{ } stands for the real part of { }, $j = \sqrt{-1}$, and the underline denotes a vector quantity.

Owing to the invariance of mode shape with position along the waveguide axis, $\underline{E}(r,\phi,z)$ and $\underline{H}(r,\phi,z)$ may be written in terms of variations $\underline{e}(r,\phi)$ and $\underline{h}(r,\phi)$ in the guide cross section multiplied by the *propagation factor* $\exp(-j\beta z)$, for propagation in the +z direction[2]. For a given mode, the fields within the fiber and surround travel along the guide together with identical phase velocities, hence with identical propagation factors in the two regions. The fields may therefore be written as

$$\begin{bmatrix} \underline{E}(r,\phi,z) \\ \underline{H}(r,\phi,z) \end{bmatrix} = \begin{bmatrix} \underline{e}(r,\phi) \\ \underline{h}(r,\phi) \end{bmatrix} \exp(-j\beta z) \quad , \tag{6.6}$$

where the axial *phase constant* β is related to the *phase velocity* v_p and the mode's *effective refractive index* n^{eff} by

$$\beta = \omega/v_p = n^{eff} k_0 \quad ; \tag{6.7}$$

and k_0 is the *wave number* of free space and is related to the free space wavelength λ_0 by

$$k_0 = 2\pi/\lambda_0 \quad . \tag{6.8}$$

Because of the cylindrical geometry, the mode fields are completely specified in terms of the z components E_z (or e_z) of the electric field and H_z (or h_z) of the magnetic field via Maxwell's equations, from which the transverse portions $\underline{E}_T$ (or $\underline{e}_T$) and $\underline{H}_T$ (or $\underline{h}_T$) may be obtained, as shown by (6.13) in Table 6.2. Once the z-directed fields are known, the transverse components are calculated via (6.13) and the total fields are formed by summing the components

$$\underline{E}(r,\phi,z) = \underline{E}_T(r,\phi,z) + \underline{z}_0 E_z(r,\phi,z) \tag{6.9}$$

$$\underline{e}(r,\phi) = \underline{e}_T(r,\phi) + \underline{z}_0 e_z(r,\phi) \quad , \tag{6.10}$$

where the components are related via (6.6), by

$$\underline{E}_T(r,\phi,z) = \underline{e}_T(r,\phi)\exp(-j\beta z) \tag{6.11}$$

$$E_z(r,\phi,z) = e_z(r,\phi)\exp(-j\beta z) \quad , \tag{6.12}$$

---

2 Note that for a time variation of the form $\exp(-j\omega t)$, the +z traveling wave would have a propagation factor $\exp(+j\beta z)$, which corresponds to a -z traveling wave for our time variation.

Table 6.2. Relationships pertaining to the formulation of modes[a]

	Inside Fiber (r<ρ)	Surround Region (r>ρ)	
Relationships for transverse field components[b]	$\underline{E}_T=(-j\beta\underline{\nabla}_T E_z+j\omega\mu\underline{z}_0\times\underline{\nabla}_T H_z)/(k^2-\beta^2)$ $\underline{H}_T=(j\beta\underline{\nabla}_T H_z+j\omega\epsilon\underline{z}_0\times\underline{\nabla}_T E_z)/(\beta^2-k^2)$		(6.13)
Transverse part of gradient operator	$\underline{\nabla}_T=\underline{\nabla}-\underline{z}_0\,\partial/\partial z$		
Reduced scalar wave equation	$(\nabla^2+k_1^2)\begin{bmatrix} E_z(r,\phi,z) \\ H_z(r,\phi,z) \end{bmatrix}=0$	$(\nabla^2+k_2^2)\begin{bmatrix} E_z(r,\phi,z) \\ H_z(r,\phi,z) \end{bmatrix}=0$	(6.14)
Wave numbers for plane waves	$k_1=2\pi n_1/\lambda_0$	$k_2=2\pi n_2/\lambda_0$	(6.15)
Transverse form of wave equation	$(\rho^2\nabla_T^2+U^2)\begin{bmatrix} e_z(r,\phi) \\ h_z(r,\phi) \end{bmatrix}=0$	$(\rho^2\nabla_T^2-W^2)\begin{bmatrix} e_z(r,\phi) \\ h_z(r,\phi) \end{bmatrix}=0$ or $(\rho^2\nabla_T^2+Q^2)\begin{bmatrix} e_z(r,\phi) \\ h_z(r,\phi) \end{bmatrix}=0$	(6.16)
Transverse part of laplacian operator	$\nabla_T^2=\nabla^2-\partial^2/\partial z^2$		
Eigenvalues	$U^2=\rho^2(k_1^2-\beta^2)$	$W^2=-Q^2=\rho^2(\beta^2-k_2^2)$	(6.17)
Normalized frequency, V	$V^2=U^2+W^2=U^2-Q^2=\rho^2(k_1^2-k_2^2)$ $=(\rho k_1)^2\delta=(2\pi n_1\rho/\lambda_0)^2\delta=(2\pi n_1\rho/\lambda_0)^2\cos^2\theta_c=(2\pi/\lambda_0)^2(n_1^2-n_2^2)$		(6.18)
Refractive index parameter	$\delta=1-\epsilon_2/\epsilon_1=1-(n_2/n_1)^2$		(6.19)

Table 6.2 (continued)

Eigenvalue equation[c]
(for bound modes only)

$$\left[\frac{J_n'(U)}{UJ_n(U)} + \frac{K_n'(W)}{WK_n(W)}\right]\left[\frac{J_n'(U)}{UJ_n(U)} + (1-\delta)\frac{K_n'(W)}{WK_n(W)}\right] = n^2\left[1-\delta\left(\frac{U}{V}\right)^2\right]\left[\frac{1}{U^2} + \frac{1}{W^2}\right]^2 \qquad (6.20)$$

[a] The particular form of (6.13) is from [6.107]; a closely related form is given in [6.80]. The remaining relations can be found in a number of different forms in [6.105,103,104,93]; the notation in the last two references is the closest.

[b] k and ε take the subscripts 1 and 2 in regions r<ρ and r>ρ, respectively; in dielectrics, the permeability $\mu=\mu_0$, the free space permeability; x is the vector cross product. Owing to (6.6), these relationships also hold if $\underline{E}_T$, $\underline{H}_T$, E_z, and H_z are replaced by $\underline{e}_T$, $\underline{h}_T$, e_z, and h_z, respectively.

[c] Prime (') denotes differentiation with respect to argument. U and W represent U_{nm} and W_{nm}. For the renumbering scheme where the modes are counted by the index ℓ and labeled as HE$_{\ell m}$ and EH$_{\ell-2,m}$, U and W represent $U_{\ell m}$ and $W_{\ell m}$, and are obtained by using n=ℓ for the HE$_{\ell m}$ modes and n=ℓ-2 for the EH$_{\ell-2,m}$ modes.

with similar expressions for the magnetic field quantities, and where $\underline{z}_0$ is the unit vector in the +z direction. $E_z(r,\phi,z)$ and $H_z(r,\phi,z)$ are solutions to the *reduced scalar wave equations* [(6.14) in Table 6.2], which by use of (6.12) may be cast into the transverse form of the wave equations [(6.16) in Table 6.2] governing the behavior of $e_z(r,\phi)$ and $h_z(r,\phi)$.

Solutions for $e_z(r,\phi)$ and $h_z(r,\phi)$ are obtained from the transverse wave equations (6.16) as the products of radial and azimuthal variations. The radial dependence is given in terms of cylinder (Bessel, Neumann, Hankel) functions and the azimuthal dependence is given in terms of the circular trigonometric functions.

Boundary conditions at the fiber-surround interface require continuity of those components of $\underline{E}$ and $\underline{H}$ that are tangential to the interface, i.e., the ϕ- and z-directed components. The condition at infinity requires that the bound modes decay outside the cylinder for large r; an oscillatory behavior is permitted for the radiation modes. In addition, the fields must be periodic in ϕ with period 2π, and fields on the fiber axis (r = 0) must be finite.

a) Bound Modes

Bound-mode solutions are of the form (see, for example [6.106])

$$\begin{bmatrix} e_z(r,\phi) \\ h_z(r,\phi) \end{bmatrix} = \begin{bmatrix} b_n \\ c_n \end{bmatrix} J_n(Ur/\rho)\exp(-jn\phi) \tag{6.21a}$$

for $r < \rho$, and

$$\begin{bmatrix} e_z(r,\phi) \\ h_z(r,\phi) \end{bmatrix} = \begin{bmatrix} d_n \\ f_n \end{bmatrix} K_n(Wr/\rho)\exp(-jn\phi) \tag{6.21b}$$

for $r > \rho$, where $J_n(\ )$ is the Bessel function of first kind and $K_n(\ )$ is the modified Hankel function, both of integer order n. U and W serve to scale the mode in the guide cross section within the fiber and surround, respectively, and are related to the axial phase constant β via (6.17) in Table 6.2[3].

The radial dependence for $r > \rho$ could just as well have been written in the form $H_n^{(2)}(Qr/\rho)$, where $H_n^{(2)}(\ )$ is the Hankel function of second kind[4] and

---

3 Use of n as a counting index for the modes should not cause confusion with its use, in subscripted form, for refractive index.

4 $K_n(\sigma) = -j(\pi/2)\exp(-jn\pi/2)H_n^{(2)}(-j\sigma)$, $-\pi/2 < \arg\sigma \le \pi$ [6.108].

$$Q = -jW \ . \tag{6.22}$$

For large argument, $K_n(\sigma)$ decays exponentially for large positive σ, and $H_n^{(2)}(\sigma)$ decays exponentially for large negative imaginary σ and exhibits an outgoing wave behavior for large positive σ. In addition, the azimuthal dependence could have been written as $\exp(jn\phi)$ or, as it often is, in terms of $\sin(n\phi)$ and $\cos(n\phi)$ instead of the complex exponential. Indeed, the solutions for $\pm n$ may be combined to form that solution, as the cylinder functions of orders $\pm n$ are linearly related [6.108], and a linear combination of solutions is also a solution. A discussion of the relative merits of the two azimuthal representations is presented by KAPANY and BURKE [Ref.6.100, p.97].

Application of (6.13) to solutions (6.21) yields the transverse fields $\underline{e}_T$ and $\underline{h}_T$, and satisfying the boundary conditions at the fiber-surround interface ($r = \rho$) yields four homogeneous algebraic equations involving the amplitudes b_n, c_n, d_n, and f_n. In order that these equations have a nontrivial solution, the determinant of this set of equations must vanish. This leads to a transcendental equation involving U and W (or, alternatively, U and Q) known as the *eigenvalue equation* (6.20). By use of this equation, along with (6.17) and (6.18) in Table 6.2, the allowed discrete values U_{nm}, W_{nm}, and β_{nm} of U, W, and β may be determined. The four equations obtained as a result of matching the boundary conditions determine the ratios of three of the amplitudes to the fourth, which must be specified by additional information, such as the power of a mode.

In general, the modes of the optical fiber are neither TE (transverse electric, also called H modes as they are completely determined by specification of the longitudinal magnetic field) nor TM (transverse magnetic, also called E modes as they are completely determined by specification of the longitudinal electric field), but are *hybrid modes* containing both longitudinal components E_z and H_z. The exact solutions for the circular fiber are two doubly infinite sets of hybrid modes, as can be seen from the quadratic nature of the eigenvalue equation [(6.20) in Table 6.2]. The exact solutions are often counted by the indices n and m [n = 0, 1, 2, ...; m = 1, 2, 3, ..., for the $\sin(n\phi)$ and $\cos(n\phi)$ variations; and n = 0, ± 1, ± 2, ...; m = 1, 2, 3, ..., for the $\exp(jn\phi)$ variation], and denoted either HE_{nm} or EH_{nm}, depending on the relative strengths of H_z and E_z [6.103]. Each of those modes, the n = 0 set excepted, is doubly degenerate, as evidenced by the azimuthal variations of $\sin(n\phi)$ or $\cos(n\phi)$ [or, $\exp(\pm j|n|\phi)$]. That degeneracy is a consequence of the cylindrical symmetry of the structure. The azimuthally invariant $n = 0$ modes degenerate into the TE_{0m} and TM_{0m} modes, which correspond to the odd and even EH_{0m} modes, respectively.

In anticipation of the natural groupings that are formed by solutions for small differences in refractive index, it is convenient for us to renumber the modes by using the index ℓ and making the substitutions $n \to \ell$ for the HE modes and $n \to \ell - 2$ for the EH modes. In this notation

$$HE_{nm} \to HE_{\ell m}$$
$$EH_{nm} \to EH_{\ell-2,m} \quad ;$$

thus, for example, $\ell = 2$ specifies U_{2m} and refers to both the HE_{2m} and EH_{0m} (TE_{0m} and TM_{0m}) modes.

The key to the behavior of a specific mode lies in finding its *eigenvalue* $U_{\ell m}$ from the eigenvalue equation for a given value of *normalized frequency*

$$V = \sqrt{U^2 + W^2} = (2\pi\rho/\lambda_0)\sqrt{n_1^2 - n_2^2} \quad ; \tag{6.23}$$

alternate forms for V are shown in (6.18) in Table 6.2. Once $U_{\ell m}$ is known, the corresponding values of $W_{\ell m}$ and $\beta_{\ell m}$ may be obtained. In general, explicit expressions for $U_{\ell m}$ cannot be found and the exact eigenvalue equation must be solved either numerically or graphically [6.100,101]. Solutions, plotted as functions of the parameters V and δ, are given for the twelve lowest order bound modes by BIERNSON and KINSLEY [6.104].

Bound modes propagate along the guide axis and attenuate exponentially transverse to that axis in the region $r > \rho$. This can be seen by examining the field in the surround region; for large enough values of Wr/ρ [6.108],

$$K_\ell(Wr/\rho) \sim \sqrt{(\pi\rho)/(Wr)}\exp(-Wr/\rho) \quad ,$$

where $\sim$ is read "is asymptotic to". The fields of (6.21) will be bound to the cylinder for $W_{\ell m} > 0$; larger values of $W_{\ell m}$ correspond to fields that decay more rapidly with increased r and are thus more tightly bound to the cylinder than are fields having smaller values of $W_{\ell m}$. As $W_{\ell m}$ decreases, the fields become more loosely bound to the cylinder and become unbound for $W_{\ell m} = 0$ (the asymptotic form is not valid for $W = 0$). This condition is known as *cutoff*, and occurs for a different value of $V = V_{\ell m}^c$ for each mode. $W_{\ell m} = 0$ ($Q_{\ell m} = 0$) is the transition point between real values of W (imaginary values of Q) which correspond to fields that are bound to the cylinder, and imaginary values of W (real values of Q) which result in fields that carry power away from the cylinder into the radiation field. By use of (6.23) the free space wavelength at which mode ℓm becomes cutoff is

$$\lambda^c_{\ell m} = (2\pi\rho/V^c_{\ell m}) \sqrt{n^2_1 - n^2_2} \quad . \tag{6.24}$$

For a given bound mode (ℓ, m fixed), the solution to the eigenvalue equation (6.20) for that mode's values of $U_{\ell m}$ begins at a lowest value $U_{\ell m} = U^c_{\ell m} = V = V^c_{\ell m}$, corresponding to $W_{\ell m} = 0$ and $\beta_{\ell m} = k_2$ at cutoff, and increases monotonically with frequency V to the asymptotic value $U^\infty_{\ell m} \equiv \lim\limits_{V \to \infty} U_{\ell m}$, corresponding to $W_{\ell m} \lesssim V$ and $\beta_{\ell m} \approx k_1$, the condition far from cutoff. Cutoff thus specifies values of $V = V^c_{\ell m}$ above which (values of λ_0 below which) individual bound modes may exist on the fiber. When V decreases and a bound mode passes through cutoff, its remaining power becomes part of the continuous spectrum and radiates from the guide. At any given value of V only a finite number of bound modes may propagate. The normalized frequency V thus defines which bound modes may propagate for a given geometry and wavelength, but does not itself specify which modes will actually be present, nor does it specify their amplitudes. In order to obtain that information we must also consider the illumination conditions, which are treated in Sect.6.3. Conditions at cutoff and far from cutoff are summarized in Table 6.3; the groupings that occur far from cutoff for all conditions and at cutoff for small refractive index difference are directly evident.

By using (6.7) and (6.15), we can express the range of the phase constant $\beta_{\ell m}$ for bound modes,

$$k_2 \leqq \beta_{\ell m} \leqq k_1 \tag{6.25}$$

as an equivalent range in terms of the mode's effective refractive index $n^{eff}_{\ell m}$, where

$$n_2 \leqq n^{eff}_{\ell m} \leqq n_1 \quad . \tag{6.26}$$

We thus see that the operating ranges for both the phase constant and effective refractive index lie between the values for the surround and fiber media, reflecting the presence of fields in both materials. Near cutoff, the fields are predominantly in the surround and that material's properties dominate the behavior, whereas far above cutoff the fields are tightly bound to the guide and its material properties dominate. Methods for measuring n^{eff} have been discussed by STAVENGA [6.109,110], and SNYDER [6.30]. Exact expressions for the bound modes are given by SNITZER [6.103], KAPANY and BURKE [6.100], and MARCUSE [6.101]; in Sect.6.2.2 we present approximate expressions for the bound modes.

Table 6.3. Exact and approximate cutoff conditions and far-from-cutoff conditons for bound modes[a]

Modes	Exact	$n_1-n_2 \to 0(\delta \to 0)$	Far-from-Cutoff Condition $(V \gg V^c_{\ell m})$ Exact and $n_1-n_2 \to 0(\delta \to 0)$
$HE_{1m}(\ell=1)$[b]	$J_1(U^c_{1m}) = 0$	$J_1(U^c_{1m}) = 0$	$J_{\ell-1}(U^c_{\ell m}) = 0$
$HE_{\ell m}\ (\ell \geq 2)$	$U^c_{\ell m} J_{\ell-2}(U^c_{\ell m})/J_{\ell-1}(U^c_{\ell m}) = -(\ell-1)\delta/(1-\delta)$	$J_{\ell-2}(U^c_{\ell m}) = 0$	$J_{\ell-1}(U^c_{\ell m}) = 0$
$EH_{\ell-2,m}(\ell \geq 2)$[c]	$J_{\ell-2}(U^c_{\ell m}) = 0$		

[a]This material appears in [6.103] in the form with the label n; here $n = \ell$ for HE modes and $n = \ell-2$ for EH modes.

[b]Root at $U^c_{11} = 0$ belongs to the HE_{11} mode, which does not have a cutoff.

[c]For $\ell=2$, TM_{0m} mode is the EH_{0m} even mode; TE_{0m} mode is the EH_{0m} odd mode.

b) Unbound Modes

Solutions for the unbound modes are of the form [6.70]

$$\begin{bmatrix} e_z(r,\phi) \\ h_z(r,\phi) \end{bmatrix} = \begin{bmatrix} g_\ell(Q) \\ o_\ell(Q) \end{bmatrix} J_\ell(Ur/\rho)\exp(-j\ell\phi) \tag{6.27a}$$

for $r < \rho$, and

$$\begin{bmatrix} e_z(r,\phi) \\ h_z(r,\phi) \end{bmatrix} = \left\{ \begin{bmatrix} s_\ell(Q) \\ v_\ell(Q) \end{bmatrix} J_\ell(Qr/\rho) + \begin{bmatrix} u_\ell(Q) \\ w_\ell(Q) \end{bmatrix} H_\ell^{(2)}(Qr/\rho) \right\} \exp(-j\ell\phi) \tag{6.27b}$$

for $r > \rho$, where g, o, s, u, v, and w are amplitude coefficients that are functions of the continuous variable Q, and are yet to be determined, and ℓ is an integer that assumes both positive and negative values.

The form of the unbound-mode solutions can be motivated by considering the fiber to be a perturbation of an otherwise homogeneous space; the modes of the fiber will then be a perturbation of the free space modes, which are plane waves propagating at all possible angles. We can thus think of the unbound modes as arising from the scattering of plane waves that are incident upon the fiber at all possible angles. The solution to that scattering problem is then viewed as comprising a free space portion (solution in the absence of the fiber) and an outward-traveling scattered portion, given by the Bessel function of first kind and the Hankel function of second kind, respectively; recall that $H_\ell^{(2)}(Qr/\rho)$ has an outgoing wave behavior for positive Q and the $\exp(j\omega t)$ time dependence. Solutions (6.27) reflect the requirement that the fields be periodic in ϕ with periodic 2π and that the fields within the cylinder be finite. Again, the azimuthal variation could have been written using $\sin(\ell\phi)$ and $\cos(\ell\phi)$.

It is instructive to consider a plane wave traveling in a material of wave number k_2 at angles θ' to the z axis and ϕ' to the x axis, as shown in Fig.6.7. The plane wave's propagation factor may be expressed in terms of cylinder functions as [Ref.6.100, p.136]

$$\exp[-jk_2 r \sin\theta'\cos(\phi' - \phi) - jk_2 z \cos\phi']$$

$$= \sum_{n=-\infty}^{+\infty} j^{-\ell} J_\ell(k_2 r \sin\phi')\exp[j\ell(\phi - \phi') - jk_2 z \cos\theta'] . \tag{6.28}$$

Following SNYDER [6.79], we compare the radial and z-directed variations of the free space portion of our solution (6.27b) and (6.12) with the respective variations of (6.28) and identify

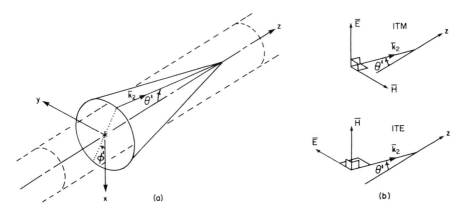

Fig.6.7a,b. Incoming plane waves. (a) Propagation direction of a wave traveling at angles θ' to the z axis and φ' to the x axis; k_2 is a vector of magnitude k_2 in the propagation direction. Each value of θ' defines a cone about
the z axis. (b) Illustration of directions for ITE and ITM plane waves, drawn
for φ' = 0 (k_2 lying in the x-z plane). In this figure, overbars denote vector quantities

$$\beta = k_2\cos\theta' \tag{6.29}$$

and

$$Q = k_2\rho \sin\theta' \ . \tag{6.30}$$

We see that each value of Q specifies the half-angle θ' of a cone on which
waves may be incident over azimuthal angles φ' covering all 2π radians.
Values of Q in the range $0 < Q \leq \rho k_2$ ($k_2 \geq \beta \geq 0$) correspond to real angles
θ' and represent propagating unbound modes, whereas values of $Q > \rho k_2$
($\beta = -j\beta''$, $0 < \beta'' \leq Q/\rho$) yield imaginary angles θ' and correspond to evanescent unbound modes. The unbound modes may thus be thought of as being selectively excitable by an infinitely extended source, located outside the guide
at infinity, which is capable of producing incident waves at both real and
imaginary angles. A plane wave decomposition of the bound modes leads to the
identification of local plane waves inside the guide whose propagation directions lie on a cone whose angle is related to $U_{\ell m}$; for these modes the source
at infinity lies within the guide.

Application of the boundary conditions at the cylinder-surround interface
again provides four equations, but, now, in the six unknowns g, o, s, u, v,
and w, yielding an underdetermined system. Unlike the bound-mode case, there
is no determinantal condition to provide an eigenvalue equation which specifies a discrete set of eigenvalues. In the present case a continuum of
values $Q \geq 0$ is permissible; $Q > 0$ is the domain over which the Hankel func-

tion of second kind describes outward-traveling waves at large r, a necessary condition for representing scattering that occurs in the vicinity of the z axis.

Unique determination of the amplitudes g, o, s, u, v, and w, or of the ratios of five of them to the sixth, is not possible owing to the underdetermined nature of the problem, and there is, thus, a degree of arbitrariness in their specification. Two approaches have been employed in order to determine those amplitudes. MARCUSE [6.111] chose a relationship between coefficients, and SNYDER [6.79] and KAPANY and BURKE [6.100] separated the modes according to whether they appear in response to incident waves of the transverse electric (ITE) or transverse magnetic (ITM) variety (Fig.6.7). The latter approach permits determination of the amplitudes by comparison with known solutions for the scattering of plane waves by dielectric cylinders.

Propagating unbound modes carry power away from the guide into the radiation field, and are important in calculations involving radiation losses. The evanescent modes describe details of the field close to the waveguide and the illumination aperture, and do not carry power away from the guide.

Expressions for unbound modes and solutions for their field amplitudes are given by SNYDER [6.79], KAPANY and BURKE [6.100] and SAMMUT et al. [6.112]; sets of approximate unbound-mode solutions are presented by SNYDER [6.79] and MARCUSE [6.102]. In Sect.6.3.4b we discuss leaky-mode representations as a means of approximating the contribution of the continuous spectrum. We now review modal expansions, mode orthogonality, and power flow; in Sect.6.2.2 we specialize these topics to the approximate bound modes.

c) Modal Expansion of Field

In general, the field on a fiber is more involved than that of a single mode. However, because the modes form a complete set, any arbitrary field $\underline{E}^{tot}$ and $\underline{H}^{tot}$ satisfying Maxwell's equations on the guide may be represented in terms of the natural modes. That representation takes the form of a generalized Fourier expansion in the transverse plane as a weighted sum of modes consisting of a summation over the finite number N of bound modes that are above cutoff, and an integration over the unbound modes. For forward-traveling (i.e., +z) waves, the fields $\underline{E}^{tot}$ and $\underline{H}^{tot}$ may be expressed as

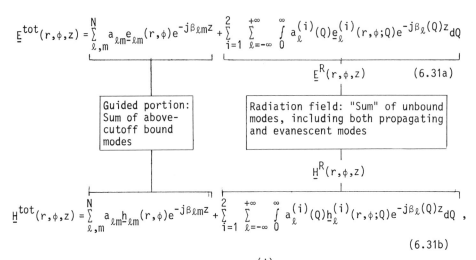

$$\underline{E}^{tot}(r,\phi,z) = \sum_{\ell,m}^{N} a_{\ell m}\underline{e}_{\ell m}(r,\phi)e^{-j\beta_{\ell m}z} + \sum_{i=1}^{2} \sum_{\ell=-\infty}^{+\infty} \int_{0}^{\infty} a_{\ell}^{(i)}(Q)\underline{e}_{\ell}^{(i)}(r,\phi;Q)e^{-j\beta_{\ell}(Q)z}dQ$$

$$\underline{E}^{R}(r,\phi,z) \qquad (6.31a)$$

| Guided portion: Sum of above-cutoff bound modes | Radiation field: "Sum" of unbound modes, including both propagating and evanescent modes |

$$\underline{H}^{R}(r,\phi,z)$$

$$\underline{H}^{tot}(r,\phi,z) = \sum_{\ell,m}^{N} a_{\ell m}\underline{h}_{\ell m}(r,\phi)e^{-j\beta_{\ell m}z} + \sum_{i=1}^{2} \sum_{\ell=-\infty}^{+\infty} \int_{0}^{\infty} a_{\ell}^{(i)}(Q)\underline{h}_{\ell}^{(i)}(r,\phi;Q)e^{-j\beta_{\ell}(Q)z}dQ ,$$

$$(6.31b)$$

where the amplitude coefficients $a_{\ell m}$ and $a_{\ell m}^{(i)}(Q)$ specify the strengths with which the modes are excited [the $a_{\ell m}^{(i)}(Q)$ actually represent the amplitude density spectrum, in Q space, of the unbound modes], and i is an index used to distinguish between the two classes of modes, e.g., ITE and ITM, or orthogonalized combinations of those modes [6.112]. The summation over the above-cutoff bound modes is written in symbolic form; because ℓm specifies a mode set rather than a single mode, care must be taken to include all launched above-cutoff modes within a set. With this in mind, we shall continue to employ that symbolism and thus avoid still another renumbering of modes. The contribution of the continuous spectrum, shown in the second set of terms for each field, is called the *radiation field* and is designated $\underline{E}^{R}$ and $\underline{H}^{R'}$ for the electric and magnetic fields, respectively. Representations of the propagating portions of the radiation field can be obtained by truncating the integrals to upper limits $Q = \rho k_2$. Calculation of the excitation amplitudes is addressed in Sect.6.3.

d) Mode Orthogonality

The modes of the dielectric waveguide are orthogonal over the infinite cross section; as a consequence, they do not exchange power with one another on the ideal waveguide and their excitations can be calculated independently. It can be shown [6.101] that with proper choice of field amplitudes, the modes of the ideal waveguide satisfy the *power orthogonality* relation

$$\int_{A_{\infty}} \underline{E}_{\nu}\times\underline{H}_{\xi}^{*} \cdot \underline{z}_0 da = \int_{A_{\infty}} \underline{e}_{\nu}\times\underline{h}_{\xi}^{*} \cdot \underline{z}_0 da = \begin{cases} \delta_{\nu\xi}, & \nu \text{ and } \xi \text{ both bound modes,} \\ \delta(\nu-\xi), & \nu \text{ and } \xi \text{ both unbound modes,} \\ 0, & \text{one bound and other unbound,} \end{cases}$$

$$(6.32)$$

where the indices ν and ξ are employed as identifying labels for individual modes, da is a differential area in the transverse plane, $\delta_{\nu\xi}$ is the Kronecker delta, which assumes the value unity when $\nu = \xi$ and is zero otherwise, and $\delta(\nu-\xi)$ is the Dirac delta function, which is infinite when $\nu = \xi$ (such that its integral over any interval containing the point $\nu = \xi$ is unity) and zero otherwise. Our definition for the Dirac delta function is a colloquial one, but one that serves us well, nonetheless.

The essence of orthogonality is that the integral in (6.32) is nonzero only when the $\underline{E}$ and $\underline{H}$ fields used in that integral belong to the same mode; viewing the modes as basis vectors, this is equivalent to orthogonal vectors whose projections upon one another are zero. Thus, the bound modes are orthogonal to one another and to the unbound modes. The question arises as to which unbound modes to use in (6.32) so that those modes will also be orthogonal to one another; it turns out that except for the axially symmetric case ($\ell = 0$), unbound modes of the ITE variety are not orthogonal to ITM unbound modes, for modes having identical values of Q. However, it is possible to obtain two orthogonal sets of unbound modes from linear combinations of the ITE and ITM modes [6.112].

We have chosen a particular form of mode normalization by requiring that the coefficients of the delta symbols in (6.32) be unity. If we recall that the equations resulting from applying the boundary conditons only permitted the specification of the field amplitudes to within a multiplicative constant, we will find that specifying this constant fixes the power carried by modes, which are excited with unit coefficients in (6.31).

Relation (6.32) is an expression of a particular form of orthogonality known as power orthogonality because, as will be seen in the following section, it has a physical interpretation in terms of mode power. This form of orthogonality holds only for the lossless waveguide [6.101]. For the lossy guide, another form of orthogonality holds, but does not have a physical interpretation although it can be used for calculating the modal excitation coefficients [6.106].

e) Power Considerations

Power flow considerations are of major importance in the study of retinal photoreceptors, and provide the basis for calculating quantities such as light collection and absorption. Power flow can be calculated via the complex power-density or *Poynting vector*; for a field specified by $\underline{E}$ and $\underline{H}$ [6.106],

$$\underline{S} = \frac{1}{2}\underline{E} \times \underline{H}^* \quad . \tag{6.33}$$

The *intensity pattern* $p(r,\phi)$ in the guide cross section is the real part of the axial component S_z of the Poynting vector

$$p(r,\phi) = \text{Re}\{S_z\} = \frac{1}{2}\,\text{Re}\{\underline{E} \times \underline{H}^*\}_z \quad , \tag{6.34}$$

and corresponds to the time averaged power density in W/m^2 flowing along the fiber. The subscript z on the braces stands for the z component of the enclosed quantity. The total time-averaged power P^A flowing through an arbitrary portion A of the infinite cross section is obtained by integrating the power density over that area; thus

$$P^A = \int_A p(r,\phi)da = \frac{1}{2}\,\text{Re}\left\{\int_A (\underline{E} \times \underline{H}^*)_z da\right\} \quad . \tag{6.35}$$

If A is identified with the infinite cross section A_∞, then P^A becomes the total power P^{tot} flowing along the waveguide axis.

A given bound mode, within the set identified by indices ℓm, that is launched with amplitude $a_{\ell m}$, has an intensity pattern

$$p_{\ell m}(r,\phi) = |a_{\ell m}|^2\,\text{Re}\{\underline{E}_{\ell m} \times \underline{H}_{\ell m}^*\}_z/2 = |a_{\ell m}|^2\,\text{Re}\{\underline{e}_{\ell m} \times \underline{h}_{\ell m}^*\}_z/2 \tag{6.36}$$

and a total mode power $P_{\ell m}$ within the infinite cross section (for the case of an individual mode we drop the "tot" superscript)

$$P_{\ell m} = \int_{A_\infty} p_{\ell m}(r,\phi)da = \frac{|a_{\ell m}|^2}{2}\,\text{Re}\left\{\int_{A_\infty} [\underline{e}_{\ell m} \times \underline{h}_{\ell m}^*]_z da\right\} = \frac{|a_{\ell m}|^2}{2} \quad , \tag{6.37}$$

where (6.32) was employed to evaluate the integral. The name "power orthogonality" for (6.32) is now transparent, and we see that for the particular choice of mode normalization in (6.32) bound modes excited with unit amplitude are normalized to carry a power of 1/2 watt. Furthermore, we see from (6.32) that individual unbound modes cannot be normalized to have finite power in the infinite cross section (recall that individual plane waves have infinite total power). This does not present a problem because unbound modes are not excited individually, but rather as a continuum; their combined effect, i.e., the radiation field, has a finite amount of power. In fact, a treatment analogous to (6.37) for the unbound modes would reveal that the quantity $|a_\ell^{(i)}(Q)|^2/2$ is the power spectral density in Q space.

To find the total power flowing along the guide through the infinite cross section for the fields $\underline{E}^{tot}$ and $\underline{H}^{tot}$ in (6.31), we substitute (6.31) into

(6.35) with $A = A_\infty$ and apply the orthogonality relation, from which [6.112] [5]

$$P^{tot} = \underbrace{\frac{1}{2} \sum_{\ell,m}^{N} |a_{\ell m}|^2}_{P_B^{tot}} + \underbrace{\frac{1}{2} \sum_{i=1}^{2} \sum_{\ell=-\infty}^{+\infty} \int_0^\infty |a_\ell^{(i)}(Q)|^2 dQ}_{P_R^{tot}} \quad . \qquad (6.38)$$

$$\quad\quad\quad\quad\quad \text{Power of} \quad\quad\quad\quad \text{Axially directed power of}$$
$$\quad\quad\quad\quad\quad \text{bound modes} \quad\quad\quad \text{radiation field (unbound modes)}$$

Owing to orthogonality, the total power in the infinite cross section is simply a sum of the powers of the individual modes. Identifying P_B^{tot} with the total bound-mode power and P_R^{tot} with the total axially directed power of the radiation field,

$$P^{tot} = P_B^{tot} + P_R^{tot} \quad . \qquad (6.39)$$

Again, the integral in (6.38) may be truncated to the upper limit $Q = \rho k_2$ to include only the propagating unbound modes, as the evanescent variety do not propagate power.

For cross sections other than A_∞, orthogonality does not apply; the power P^F flowing through the fiber cross section A_F may be written as

$$P^F = P_B^F + P_R^F + P_{cross}^F \quad , \qquad (6.40)$$

where, for purposes of performing our power accounting and of facilitating later approximations, we may mathematically separate out the contributions P_B^F due only to bound modes, P_R^F due only to radiation modes, and P_{cross}^F which is an interference contribution that results from the cross products between bound and unbound modes when (6.31) is used in (6.35). P_B^F and P_R^F have their own cross terms, too, but they contain only single mode species. Cross-term contributions are z dependent, may be either positive or negative owing to a net constructive or destructive interference at a given z plane, and vanish when the integration area becomes identical to A_∞. Needless to say, cross terms are not present when only a single mode propagates on the guide. We will do well to reemphasize the point made by SAMMUT et al. [6.112] that the separation of contributions is a mathematical one and that the observable physical quantity is the sum.

---

5 These authors call the bound-mode power by the name trapped-mode power and denote it by P_T^{tot}.

We shall have much use for a measure of the *fraction of the power of mode*
ℓm that resides within the fiber; such a measure finds extensive use in
fiber optics [6.93,104] and is defined by

$$\eta_{\ell m} = \frac{P^F_{\ell m}}{P_{\ell m}} = \frac{\text{Power of mode } \ell m \text{ within fiber cross section}}{\text{Total power of mode } \ell m \text{ within infinite cross section}} \cdot \quad (6.41)$$

$\eta_{\ell m}$ is dependent on the normalized frequency V. In a like manner, we also
define a more general η as the *fraction of total power residing within the*
fiber;

$$\eta = \frac{P^F}{P^{tot}} = \frac{\text{Total power within fiber cross section}}{\text{Total power within infinite cross section}} \cdot \quad (6.42)$$

As well as being dependent on V, η is also z dependent because it involves
the cross-term contributions.

Prior to further discussion of mode properties, we review approximate
solutions for the modes of fibers having a small difference of refractive
index from the surrounding medium, a case that is especially suited to the
study of retinal photoreceptors.

6.2.2 Approximate Bound-Mode Solutions for Small Difference in Refractive Index

Retinal photoreceptors are a special type of optical fiber; namely, they
are fibers with a small difference in refractive index between core and
cladding. The refractive index data [6.15,17,20,113] have extremes in relative refractive index (photoreceptor relative to IPM) which are less than
1.06 ($\delta < 0.11$), with the average values being well below this figure.

SNITZER pointed out some of the simplifications that occur for small
differences in refractive index [6.103,114]. That case was investigated
rigorously by SNYDER, who derived asymptotic forms for both the mode fields
and the eigenvalue $U_{\ell m}$ [6.93]. Work along these lines was continued by
GLOGE [6.115] who introduced the term *weakly guiding fiber* for the low-index-
difference case, and who derived a set of approximate linearly polarized,
almost transverse modes. GLOGE also presented a variety of design curves
for fiber optics communications, and an expression for the eigenvalue $U_{\ell m}$
that is exact at cutoff.

a) The Approximate Bound Modes

Considerable simplification in both the field expressions and the eigenvalue equation occurs for small differences in refractive index between the fiber and its surround in the limit as $n_1 - n_2 \to 0$. Because the condition $|e_{T_{\ell m}}| \gg |e_{z_{\ell m}}|$ is satisfied for small refractive-index difference, the fields of the weakly guiding fiber are nearly transverse [6.80,93,115]. Therefore, for an approximate solution, we need specify only the transverse components of the electric field as they are much larger than the longitudinal components. The transverse magnetic field can then be obtained from the transverse electric field in a straightforward manner. Thus, we can construct approximate HE and EH modes.

An alternate scheme is available for representing the modes of the weakly guiding fiber. A set of *linearly polarized approximate modes* results from a linear combination of the $HE_{\ell m}$ and $EH_{\ell - 2, m}$ modes. These linearly polarized modes are the $LP_{\ell - 1, m}$ modes derived by GLOGE. Both sets of approximate modes are orthogonal in the infinite cross section. Expressions for both sets of approximate modes, as well as other relations pertaining to those modes are given in Table 6.4.

For $\ell = 1$, every choice of m yields a set of two orthogonally polarized modes, designated either HE_{1m} or LP_{0m}. Every other choice of $\ell > 1$, yields four modes for each choice of m, designated either $HE_{\ell m}$ and $EH_{\ell - 2, m}$ or $LP_{\ell - 1, m}$. The four modes arise from even and odd azimuthal dependencies and two orthogonal polarizations. A glance at the cutoff conditions in Table 6.3 shows some near equalities (degeneracies) in $U_{\ell m}$ at cutoff for small δ; they become degenerate for $\delta = 0$. This degeneracy is present in the approximate eigenvalue equation (6.48) but is not present in the exact eigenvalue equation (6.20), which is really two equations, one for the $HE_{\ell m}$ modes and another for the $EH_{\ell - 2, m}$ modes.

The groups of modes evidencing degeneracies at cutoff in the limit $\delta \to 0$ can be listed as: (HE_{11}), $(TE_{0m}, TM_{0m}, HE_{2m})$, $(HE_{1, m+1}, EH_{1m}, HE_{3m})$ and $(HE_{\ell m}, EH_{\ell - 2, m}$; for $\ell > 3)$. In fact, the values of $U_{\ell m}$ for the modes within a group are close to one another throughout the operating range, with differences between the values of $U_{\ell m}$ vanishing for the groupings $(TE_{0m}, TM_{0m}, HE_{2m})$ and $(HE_{\ell m}, EH_{\ell - 2, m}$; $\ell \geq 3)$, for vanishing refractive-index difference. The values of $U_{\ell m}$ of the $HE_{1, m+1}$ modes remain close to, but distinct from those of the EH_{1m} and HE_{3m} modes. Thus, these groups will have identical or nearly equal values of phase velocity; they propagate along the structure in near synchronism. Hence, these modes give rise to new

Table 6.4. Approximations for bound modes on weakly guiding fibers $(n_1-n_2 \to 0)^a$

Transverse electric field[b,c] HE$_{\ell m}$ (upper sign; $\ell = 1,2,3, ...$) and EH$_{\ell-2,m}$ (lower sign; $\ell = 2,3, ...$) modes	(even) $\underline{e}_{\ell m}(r,\phi) = \psi_{\ell m}^{-1/2}[\pm\underline{x}_0\cos(\ell-1)\phi - \underline{y}_0\sin(\ell-1)\phi]f_{\ell m}(r)$ (6.43) (odd) $\underline{e}_{\ell m}(r,\phi) = \psi_{\ell m}^{-1/2}[\pm\underline{x}_0\sin(\ell-1)\phi + \underline{y}_0\cos(\ell-1)\phi]f_{\ell m}(r)$
Radial variation	$f_{\ell m}(r) = \begin{cases} J_{\ell-1}(U_{\ell m}r/\rho)/J_{\ell-1}(U_{\ell m}), & r \le \rho \\ K_{\ell-1}(W_{\ell m}r/\rho)/K_{\ell-1}(W_m), & r \ge \rho \end{cases}$ (6.44)
Normalization factor	$\psi_{\ell m} \approx (\epsilon_1/\mu_0)^{1/2}\frac{1}{\pi\rho}^2(V/U_{\ell m})^2[K_\ell(W_{\ell m})K_{\ell-2}(W_{\ell m})/K_{\ell-1}^2(W_{\ell m})]$ (6.45)
Transverse electric field Linearly polarized LP$_{\ell-1,m}$ modes[c,d] $(\ell,m = 1,2,3, ...)$	(LP,even) $\underline{e}_{\ell m}(r,\phi) = \underline{x}_0\gamma_\ell\psi^{-1/2}f_{\ell m}(r)\cos(\ell-1)\phi$ (6.46) (LP,odd) $\underline{e}_{\ell m}(r,\phi) = \underline{x}_0\gamma_\ell\psi^{-1/2}f_{\ell m}(r)\sin(\ell-1)\phi$ $\psi_\ell = \begin{cases} 1, & \ell = 1 \\ \sqrt{2}, & \ell > 1 \end{cases}$
Orthonormality	$\int_{A_\infty}[\underline{E}_{\ell m}\times\underline{H}_{\ell'm'}^*]_z da \approx (\epsilon_1/\mu_0)^{1/2}\int_{A_\infty}|\underline{e}_{\ell m}\cdot\underline{e}_{\ell'm'}|da = \begin{cases} 1; & \ell',m'=\ell,m \\ 0; & \ell',m'\ne\ell,m \end{cases}$ (6.47)
Eigenvalue equation	$U_{\ell m}J_\ell(U_{\ell m})/J_{\ell-1}(U_{\ell m}) \approx W_{\ell m}K_\ell(W_{\ell m})/K_{\ell-1}(W_{\ell m})$ (6.48)
Transverse magnetic field	$\underline{H}_{\ell m}(r,\phi,z) \approx (\epsilon_1/\mu_0)^{1/2}[\underline{z}_0\times\underline{e}_{T_{\ell m}}(r,\phi)]\exp(-j\beta_{\ell m}z)$ (6.49)
Fraction of mode power within fiber	$\eta_{\ell m} \approx (U_{\ell m}/V)^2\{(W_{\ell m}/U_{\ell m})^2 + K_{\ell-1}^2(W_{\ell m})/[K_\ell(W_{\ell m})K_{\ell-2}(W_{\ell m})]\}$ (6.50)

aThis material is from [6.80,93,115]. The forms that are closest are found in [6.80]. $\underline{x}_0$ and $\underline{y}_0$ are unit vectors.
bTM$_{0m}$ and TE$_{0m}$ modes are even and odd EH$_{0m}$ modes, respectively.
cFields are orthogonal and are normalized according to (6.32).
dFor polarization in y direction, replace $\underline{x}_0$ by $\underline{y}_0$.

patterns which are a result of superposition of the individual fields. The $LP_{\ell-1,m}$ modes, given by (6.46), are the new approximate modes which are produced as a result of that combination.

Modes having different phase velocities will beat with one another as they travel along the guide, producing an interference pattern that repeats itself in an axial distance

$$\lambda_B = 2\pi/(\beta_1 - \beta_2) \tag{6.51}$$

called the *beat wavelength*, where β_1 and β_2 are the phase constants of the two interacting modes. One consequence is that the linearly polarized modes will decompose and form a beat pattern along the guide. We thus see that whereas the transverse fields of (6.43) are approximations to the true $HE_{\ell m}$ and $EH_{\ell-2,m}$ modes of a cylinder having a finite refractive-index difference, the linearly polarized fields of (6.46) are true modes only for a cylinder of vanishing refractive-index difference and are, therefore, valid only for short fibers. A rough bound for guide lengths over which beating between the nearly degenerate $HE_{\ell m}$ and $EH_{\ell-2,m}$ modes may be ignored is given by SNYDER [6.80] as

$$(L/\rho) << \pi/\delta^{3/2} \quad , \tag{6.52}$$

where L is the fiber length and δ is the refractive-index parameter defined in (6.19). To appreciate this restriction, we examine the extremes of refractive index shown in Table 6.1. Using a value of 1.347 as the refractive index of the surround, the lowest and highest values of receptor refractive index in that table have values of $\delta = 0.0205$ and $\delta = 0.0887$, respectively; and from (6.52) these yield conditions $L/\rho << 1070$ and $L/\rho << 119$, respectively. For visual photoreceptors, the condition is easily satisfied on the lower-refractive-index myoid, but it may become restrictive on the higher-refractive-index outer segments.

b) Some Mode Properties

An expression for the mode intensity pattern that involves only the electric field can be obtained by substituting (6.49) into (6.36), whence for a mode excited with amplitude $a_{\ell m}$,

$$P_{\ell m}(r,\phi) \simeq \sqrt{\epsilon_1/\mu_0} \; |a_{\ell m}e_{\ell m}|^2/2 \quad , \tag{6.53}$$

where ϵ_1 is the permittivity of the fiber, and μ_0 its permeability, which for a dielectric is equal to that of free space. The essential radial and azimuthal intensity variations are shown in Table 6.5. The approximations to the

Table 6.5. Intensity dependence of approximate modes in guide cross section

Mode	Azimuthal dependence	Radial dependence
$HE_{\ell m}$, $EH_{\ell-2,m}$	1	$[J_{\ell-1}(U_{\ell m}r/\rho)/J_{\ell-1}(U_{\ell m})]^2$; $\quad r \le \rho$
$LP_{\ell-1,m}$	$\cos^2(\ell-1)\phi$ or $\sin^2(\ell-1)\phi$	$[K_{\ell-1}(W_{\ell m}r/\rho)/K_{\ell-1}(W_{\ell m})]^2$; $\quad r \ge \rho$

$HE_{\ell m}$ and $EH_{\ell-2,m}$ modes will have an azimuthally invariant intensity pattern, whereas combinations forming the linearly polarized $LP_{\ell-1,m}$ modes will exhibit an azimuthal variation, an exception being the LP_{0m} modes which also have azimuthally invariant intensities.

Sketches of mode intensity patterns, and values of $V^c_{\ell m}$ and $U^\infty_{\ell m}$, are shown in Fig.6.8 for the six lowest order mode sets. Those modes are the most com-

ℓm	Cutoff: $U^c_{\ell m} = V^c_{\ell m}$			Far from Cutoff (V→∞) $U^\infty_{\ell m}$	Intensity patterns $HE_{\ell m}$, $EH_{\ell-2,m}$	$LP_{\ell-1,m}$
	$HE_{\ell m}$	$EH_{\ell-2,m}$	$LP_{\ell-1,m}$			
11	0	—	0	2.405	HE_{11}	LP_{01}
21	2.405^+	2.405	2.405	3.832	HE_{21} $TM_{01}TE_{01}$	LP_{11} / $HE_{21}\cdot(TE_{01}$ or $TM_{01})$
31	3.832^+	3.832	3.832	5.136	HE_{31} EH_{11}	LP_{21}
12	3.832	—	3.832	5.520	HE_{12}	LP_{02}
41	5.136^+	5.136	5.136	6.370	HE_{41} EH_{21}	LP_{31}
22	5.520^+	5.520	5.520	7.016	$HE_{22}TM_{02}TE_{02}$	LP_{12} / $HE_{22}\cdot(TE_{02}$ or $TM_{02})$

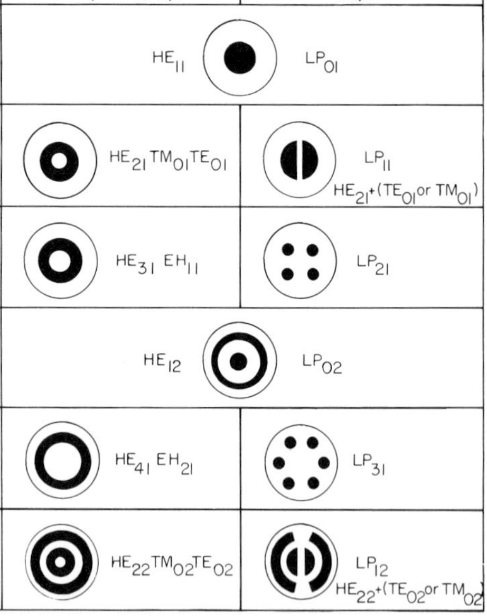

Fig.6.8. Intensity patterns, cutoff, and far-from-cutoff parameters for the six lowest order mode sets. (+) indicates that slightly higher values will occur for $\delta \ne 0$; shaded areas represent regions of highest intensity, patterns are not drawn to scale

monly observed in mammalian receptors. The form of the intensity patterns is related to the mode orders ℓ and m. For $\ell > 1$, $\ell - 1$ gives half the number of azimuthal antinodes in the pattern of the LP modes; for $\ell = 1$, the pattern is azimuthally symmetric and the number of antinodes is one. The number of radial antinodes (including that at the origin, if present) is given by m for the patterns of the LP, HE, and EH modes.

Figures 6.9a-c show the radial variation of the power density $p_{\ell m}(r,\phi)$ for the azimuthally invariant HE and EH modes, or the azimuthally averaged power density for the LP modes, for the three lowest order mode sets. For low (above-cutoff) values of V, considerable power is carried outside the fiber. In the limit as $V \to \infty$, the mode's total power is carried within the fiber. Note that whereas the first two mode sets have zero power within the guide at cutoff, the $\ell m = 31$ (LP_{21}: EH_{11}, HE_{31}) mode set has a nonzero fraction of its power within the guide at cutoff. That fraction is obtained by evaluating η_{31} at V = 3.832. An approximate expression for $\eta_{\ell m}$ is given by (6.50) in Table 6.4, and is plotted as a function of V in Fig.6.10 for the six lowest order bound-mode sets. Referring to that figure, we can confirm that whereas the $\ell m = 11$ and 21 mode sets have zero power within the guide at cutoff, the 31 set has half of its modal power within the guide at cutoff. That power will be radiated as V is decreased below cutoff.

The normalized frequency V for rod and cone inner and outer segments is shown in Fig.6.11 as a function of normalized receptor radius ρ/λ_0, for the range of receptor refractive indices shown in Table 6.1. The refractive index of the interstitial matrix was taken as 1.347, corresponding to BARER's most likely estimate [6.17]. So that one may obtain at a glance an appreciation for the range of parameters involved, approximate representive values of ρ/λ_0 are indicated for rods from various species, for $\lambda_0 = 500$ nm. Thus, whereas one would expect to observe only the lowest order modes in rods of species such as cattle and rat and in human foveal cones, one would expect to observe a myriad of mode patterns in frog rods. This is indeed in concert with experimental observations [6.10-12,27,71,117].

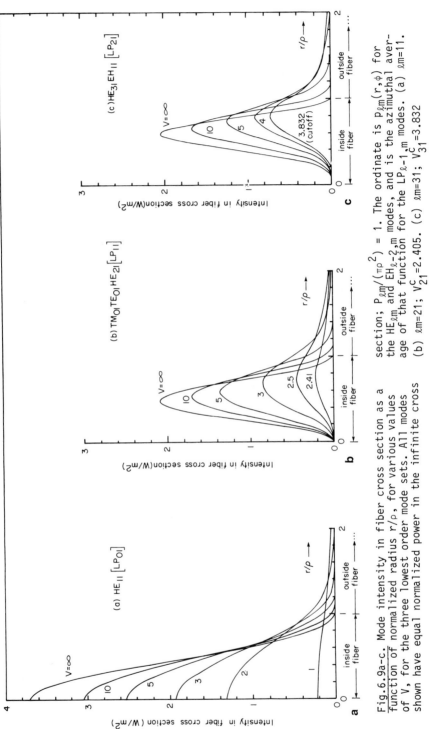

Fig.6.9a-c. Mode intensity in fiber cross section as a function of normalized radius r/ρ, for various values of V, for the three lowest order mode sets. All modes shown have equal normalized power in the infinite cross section; $P_{\ell m}/(\pi\rho^2) = 1$. The ordinate is $p_{\ell m}(r,\phi)$ for the $HE_{\ell m}$ and $EH_{\ell-2,m}$ modes, and is the azimuthal average of that function for the $LP_{\ell-1,m}$ modes. (a) $\ell m=11$. (b) $\ell m=21$; $V_{21}^c=2.405$. (c) $\ell m=31$; $V_{31}^c=3.832$

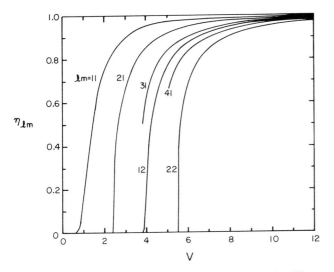

Fig.6.10. Fraction of bound-mode power within fiber cross section for the six lowest order mode sets. The curves are labeled with the values of ℓm:11 (HE_{11} or LP_{01}); 21 (TM_{01}, TE_{01}, HE_{21}, or LP_{11}); 31 (HE_{31}, EH_{11}, or LP_{21}); 12 (HE_{12} or LP_{02}); 41 (HE_{41}, EH_{21}, or LP_{31}); 22 (TM_{02}, TE_{02}, HE_{22}, or LP_{12}). These curves were calculated from (6.50) [6.93]; the continuation of the $n_{\ell 1}$ curves to demonstrate the leakage of power of the $\ell 1$ modes below $V_{\ell 1}^{c}$ is shown in [6.80,116]

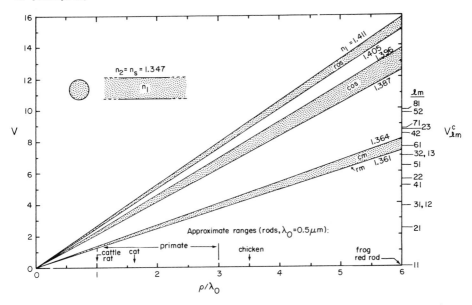

Fig.6.11. Normalized frequency V for measured ranges of receptor refractive index (see Table 6.1), with IPM index $n_s=n_2=1.347$. The species labels along the abscissa are approximate ranges for rod outer segments with $\lambda_0=0.5$ μm (500 nm), and are included to facilitate rough estimates. Values of $V_{\ell m}^{c}$ for the first 15 mode sets are indicated. cm: cone myoid; cos: cone outer segment; rm: rod myoid; ros: rod outer segment

c) Mode Combinations

We have already seen that $HE_{\ell m}$ and $EH_{\ell-2,m}$ modes combine to form the $LP_{\ell-1,m}$ modes, and that the stability of the latter modes depends upon guide length. Additional patterns, not necessarily linearly polarized, can be formed from nearly degenerate mode sets, for example, the $\ell m = 31$ and 12 mode sets. Such mode combinations, including those shown in Fig.6.8 have been observed in optical fibers [6.91,100,114,118] and in vertebrate retinal photoreceptors [6.9-12,21,27,28,71,117]. In addition, leaky modes have been observed in glass fibers [6.114] and in frog rods [6.119].

In a laboratory situation, one is often faced with an observed mode pattern which must be interpreted in terms of the modes actually present. The interpretation is hampered because many different modes and mode combinations have patterns that appear similar. Mode identification may be accomplished via two paths. 1) If one is at liberty to vary the wavelength, then cutoff data might be able to be ascertained; beyond that there is the question of which mode(s) of a given set are present. That question may be addressed by 2) appealing to the polarization properties of the modes. Thus, SNITZER and OSTERBERG [6.118] used a polarizer and analyzer interposed between the illumination source and observation optics, respectively, to obtain that information. More recently, KAPANY and BURKE [6.100] described a Mach-Zehnder interference technique to quantitatively analyze the modal power in a multimode waveguide.

In addition to the intensity patterns that are produced by combinations of modes having degenerate or nearly degenerate phase velocities, there are interference effects produced by modes with distinctly different phase velocities. These modes glide past one another at their difference velocity, thus producing a beat pattern along the axial direction. That pattern is periodic with characteristic distance given by (6.51). Such axially varying interference patterns have been observed in retinal photoreceptors by ENOCH [6.11]. As an example of that interaction, consider V in the range $2.405 < V < 3.832$ where only the HE_{11} mode and the TE_{01}, TM_{01}, HE_{21}, or LP_{11} mode set propagate. The most striking interaction occurs between the HE_{11} mode and the LP_{11} mode. The beat pattern is schematized in Fig.6.12. A number of simulated mode interference patterns produced by different illumination conditions are presented by WIJNGAARD [6.120].

It is important to note that although the modes of an ideal lossless guide interfere so as to produce an intensity pattern which varies periodically along the receptor axis, the total power carried by each mode remains constant.

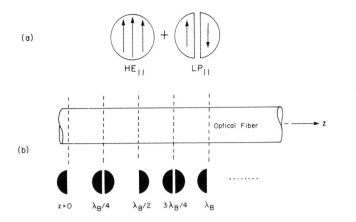

(a)

HE_{11} + LP_{11}

(b)

Optical Fiber — z

$z=0$ $\lambda_B/4$ $\lambda_B/2$ $3\lambda_B/4$ λ_B

Fig.6.12a,b. Schematization of mode beating between the HE_{11} (LP_{01}) and LP_{11} modes. (a) Intensity patterns of individual modes at $z = 0$. The arrows represent the directions of the electric field. (b) Beat pattern produced along the fiber; shaded areas are regions of greatest intensity

Thus, the modes interfere to produce new intensity patterns and also a z-dependent power within the fiber, but this does not represent an exchange of power, or coupling, between modes. Coupling requires deviations from the conditions of the ideal fiber, e.g., absorption, bends, inhomogeneity, taper, anisotropy, or deviations from circular cross sections.

d) Power Carried by Bound Modes

We have seen from (6.38) and (6.39) that, owing to mode orthogonality over the infinite cross section, the total mode power flowing along the fiber through that cross section is given exactly by the superposition of the individual mode powers. For the case of the finite fiber cross section A_F, the modes are not orthogonal over A_F, a simple superposition of power does not hold, and cross terms are present in P^F. However, the bound modes may often be regarded as approximately orthogonal over the fiber cross section and superposition may then be used to approximate the bound-mode power P_B^F within the fiber as the sum of the portions $P_{\ell m}^F = \eta_{\ell m} P_{\ell m}$ of the bound-mode powers that are contained within the fiber. Thus

$$P_B^F \approx \sum_{\ell,m}^N P_{\ell m}^F = \sum_{\ell,m}^N \eta_{\ell m} P_{\ell m} \quad , \tag{6.54}$$

where the approximation neglects the interference-produced cross terms that add a periodic, axially varying component to P_B^F; P_B^{tot} remains constant, of course. The contributions of the neglected terms are zero for modes that are orthogonally polarized, have different (odd or even) symmetry, or have different azimuthal variations (different ℓ values). This approximation is discussed in detail for the bound modes by PASK and SNYDER [6.121] who showed that the omitted terms are strongest in the neighborhood of mode cutoffs, and that the approximation becomes exact for V < 3.832. Moreover, the approximation improves at higher values of V, for which greater portions of individual-mode power lie within the guide. A statement similar to (6.54) cannot be made when unbound modes are to be included because those modes are not even approximately orthogonal over the fiber cross section [6.112].

e) Mode Coupling

Mode coupling involves the exchange of power among modes. Owing to mode orthogonality over the infinite plane, the modes of the ideal dielectric guide do not couple to one another. Any departures from the ideal waveguide, which disrupt orthogonality, will cause mode coupling either among bound modes, between bound modes and unbound modes, or among unbound modes. Such departures from the ideal waveguide include noncircular cross sections, varying cross sections including tapers, smooth or local inhomogeneities, anisotropy, absorption, and the presence of neighboring receptors. These departures from the assumed ideal conditions effectively create a new waveguide with its own boundary conditions, and the modes of the new waveguide are not those of the ideal guide. If the departures from the assumed ideal conditions are small, then the unknown modes of the new waveguide may be expressed, via perturbation, as a linear combination of the known modes of the ideal waveguide. The exact modes of the perturbed, now nonideal waveguide are orthogonal, but the modes of the ideal waveguide are not orthogonal on the new waveguide. Thus, the use of the ideal-waveguide modes on the perturbed system will give rise to mode coupling or exchange of power between the ideal modes. This application of perturbation theory is also called coupled-mode theory, and is applied in Chap.7 to study the optical interactions on arrays of photoreceptors.

Coupled-mode theory has been employed to investigate two important cases wherein the single photoreceptor differs from the idealized model; namely, tapered and absorbing (lossy) dielectric waveguides. Restrictions governing the applicability of this approach are that departures from the ideal be small or slowly varying with distance. Thus, for the cases of taper and absorption, we require that the taper be slow (small taper angle) and that the percent absorption be small per wavelength distance [6.70,102].

Many vertebrate retinal receptors have a tapered ellipsoid. From the optical point of view, we may consider the ellipsoid as a transition section between the myoid and the outer segment. In addition, the outer segments of many species are tapered. The degree of taper depends upon species, whether the receptor is a rod or a cone, and on retinal location (for an example, see [6.11]).

SNYDER studied the linear dielectric taper, using coupled-mode theory and the asymptotic forms for the modes [6.70]. That approach is valid for $\delta \ll 1$ and taper angle (Fig.6.1) $\Omega \ll 1$ (57.3°). He points out that only modes having like azimuthal symmetry will couple, and that for $V < 3.832$ (the cutoff of HE_{12} mode) there is no coupling among the bound modes, only coupling between bound modes and the radiation field. On a homogeneous taper, V decreases with axial distance and modes that become cutoff along the taper couple their power into the radiation field. Above $V = 3.832$, coupling among the bound modes begins with coupling between the HE_{11} and HE_{12} modes; and SNYDER concluded that only coupling between the HE_{11} and HE_{12} modes needs be considered for slight taper, taper angle $\Omega \ll 1$, slow taper variations, $d\Omega/dz \ll 1$, and small refractive-index difference, $\delta < 0.2$.

Calculations on model photoreceptors have usually employed the approximation wherein coupling among bound modes is neglected; furthermore, coupling between bound modes and the radiation field has also been neglected, except in the vicinity of cutoff, where coupling must be considered [6.29, 70,122,123]. Only a small amount of systematic histological data is presently available concerning ellipsoid taper; as more data become available, the question of mode coupling may need to be reassessed.

6.2.3 Absorbing Fibers

Absorption of light by the photolabile pigment contained within the lamellae of the outer segment is the beginning of the transduction process, which is the first of a series of events leading to visual perception. That absorption is central to the visual process and is in sharp contrast to the unwanted absorption, termed loss, of optical communication fibers. Retinal receptor outer segments have an absorption per unit wavelength distance that is at most a few percent [6.38]. Whereas this degree of absorption would be prohibitive in the case of communication fibers, it is nevertheless small enough to allow for useful approximations wherein the absorption can be considered a perturbation upon the lossless system.

Absorption effects within the fiber may be represented by replacing the fiber's refractive index n_1 by

$$n_1 = n_1' - jn_1'' , \qquad n_1', n_1'' > 0 , \qquad (6.55)$$

where the imaginary part of n_1 represents the loss associated with the bulk material which constitutes the receptor [6.41,105]. The loss can also be related to the bulk naperian *absorption coefficient*

$$\alpha = 2k_0 n_1'' . \qquad (6.56)$$

A complex value of refractive index n_1, and hence k_1, results in complex values of $U_{\ell m}$ and $\beta_{\ell m}$; thus, the modes of the absorbing guide are different from those of the nonabsorbing guide. Owing to the small absorption of retinal photoreceptors, we expect the new values of $U_{\ell m}$ and $\beta_{\ell m}$ to be small perturbations about the old values. For the purpose of calculating the absorbed power, we might neglect the changes in both $U_{\ell m}$ and in the real part of $\beta_{\ell m}$, and take account of only the absorption-produced attentuation introduced by the imaginary part of $\beta_{\ell m}$. This approach may be placed on a firmer footing by use of coupled-mode theory. The same considerations apply as in the case of tapered fibers; i.e., only modes of like azimuthal symmetry may couple and there is no coupling among bound modes for $V < 3.832$. Furthermore, the combination of low absorption per unit wavelength and small refractive-index difference between receptor and surround facilitates the approximation wherein coupling between modes is neglected, except near cutoff where coupling to unbound modes must be accounted for in any case [6.124]. This approach leads to approximation of the modes of the absorbing fiber by the modes of the lossless fiber, with the important exception that there is an absorption-produced attentuation in the axial direction. Validity is dependent on both small absorption per unit wavelength distance and small refractive-index difference. SNYDER gives $\delta < 0.3$ as the condition for validity of the approximation; that condition is certainly satisfied for all known vertebrate retinal receptors [6.124]. For other fibers whose $\delta > 0.3$, the approach is more complicated.

In the following discussion, we pause to consider absorption from the viewpoint of geometric optics and then treat absorption in the commonly used approximations wherein only bound modes are accounted for. Discussions relating to contributions of unbound modes were presented by SAMMUT and co-workers [6.112,125].

a) Geometric Optics Result

For a homogeneous absorbing medium having a power $P_G(0)$ at a given plane, the power $P_G(L)$ remaining after a distance L is

$$P_G(L) = P_G(0)\exp(-\alpha L) \quad , \tag{6.57}$$

and the power absorbed within that distance is

$$P_G^{abs}(L) = P_G(0)[1 - \exp(-\alpha L)] \quad . \tag{6.58}$$

In applying these results to the retinal receptor, $P_G(0)$ is interpreted as the power that is intercepted and trapped by the receptor.

b) Single Bound Mode

Consider a single bound mode on an absorbing section of guide and let $P_{\ell m}(0)$ be the total power of that mode at the beginning of the absorbing section. The total mode power remaining after a distance L is given by

$$P_{\ell m}(L) = P_{\ell m}(0)\exp(-\eta_{\ell m}\alpha L) \quad , \tag{6.59}$$

and the power absorbed within the receptor is given by

$$P_{\ell m}^{abs}(L) = P_{\ell m}(0)[1 - \exp(-\eta_{\ell m}\alpha L)] \quad , \tag{6.60}$$

thus following a modified Beer's Law [6.124,126,127]. Recall that $\eta_{\ell m}$ is that fraction of a mode's total power which is contained within the receptor cross section. $P_{\ell m}(0)$ is related to $a_{\ell m}$, the amplitude with which the mode is excited on the absorbing section, by $P_{\ell m}(0) = |a_{\ell m}|^2/2$; and $a_{\ell m}$ is determined from illumination considerations.

In the single mode case the *effective absorption coefficient* is $\eta_{\ell m}\alpha$, which reflects the fact that absorption occurs only within the receptor cross section. Furthermore, because $\eta_{\ell m}$ is wavelength dependent, the waveguide's effective absorption spectrum will be different from that given by α for the absorbing material itself, as in (6.57) and (6.58). We recall from Fig.6.10 that as V increases, $\eta_{\ell m} \to 1$; as a result $\eta_{\ell m}\alpha \to \alpha$ in (6.59) and (6.60), but $P_{\ell m}(0)$ may still be different from $P_G(0)$.

An interesting consequence of absorption is a net flow of power into the receptor from the surround [6.127,128], and is consistent with results obtained for waves at planar interfaces [6.106]. To see that this is reasonable, recall that power is absorbed only within the fiber. The absorption of power within the fiber can be thought of as causing a spatial imbalance in the intensity profile, and thus in the fields that must be matched at the boundary in a given transverse cross section. The radially inward power flow then represents a redistribution of power necessary to maintain the modes. Such inward power flow is not accounted for in the geometric optics result. In the modal approach, all power that is bound to the receptor is, eventually, available for absorption; in the geometric optics approach, only the power that is intercepted by the receptor cross section may be absorbed.

c) Multiple Bound Modes

For multiple bound modes, the absorbed power is often approximated by summing the absorption contributions of the modes as if they existed individually on the fiber [6.30,80]; the total bound-mode power absorbed in length L, $P_B^{abs}(L)$, is then given by

$$P_B^{abs}(L) \simeq \sum_{\ell,m}^{N} P_{\ell m}^{abs}(L) = \sum_{\ell,m}^{N} P_{\ell m}(0)[1 - \exp(-\eta_{\ell m}\alpha L)] \quad . \tag{6.61}$$

This further approximation is based upon the approximate orthogonality of the modes over the fiber cross section and thus neglects the interference effect of the cross terms, and was discussed earlier in connection with (6.54). For V large enough that $\eta_{\ell m} \to 1$ for all modes carrying significant amounts of power, (6.61) becomes

$$P_B^{abs}(L) \simeq P_B^{tot}(0)[1 - \exp(-\alpha L)] \quad , \tag{6.62}$$

where $P_B^{tot}(0)$ is the total bound-mode power at the entrance plane to the absorbing section. This result is reminiscent of the geometric optics result (6.58); however, depending upon the particular illumination, $P_G(0)$ may not even be a good approximation to $P_B^{tot}(0)$.

d) Tapered Outer Segments

A number of vertebrate species, notably fish and amphibia, have sharply tapered outer segments [6.3-5]. For these receptors, V varies, and in fact decreases, with axial distance. Thus, $\eta_{\ell m}$ will no longer be constant, but will be axially dependent, and a length-averaged η_{av} may be employed in (6.60) and (6.62). SNYDER and PASK have presented curves of η_{av} for the HE_{11} and TE_{01} modes [6.127,129].

6.3 Optical Excitation of Modes

In the previous section we discussed properties of the modes of the photoreceptor waveguide without considering how those modes were established on
the guide. We have specified the form of both bound and unbound modes, and
can solve the eigenvalue equation to determine which bound modes may propagate at a given value of V. In this section we discuss the determination
of the amplitude coefficients which tell the degree to which modes are actually excited for given illumination conditions. We wish to emphasize that
our use of the term excitation herein refers to the transfer of light energy
to the receptor waveguide structure and does not refer exclusively to transduction.

Our aim is to express the field produced upon the photoreceptor, by an
arbitrary illumination source, in the form of a modal expansion. We know
that it is possible to express any solution of Maxwell's equations on the
guide in the form of a superposition of modes because the modes form a complete set.

6.3.1 The Excitation Problem

Illumination is incident upon the input end of the receptor at the truncation aperture in the z = 0 plane (see Fig.6.1). For the purpose of investigating the field in the region z > 0, we can apply the equivalence principle
[Ref.6.106, Sect.1.7] and replace the incident illumination and the region
z < 0 by an equivalent source, i.e., the resultant field over the z = 0
plane. Were the field in the z = 0 plane known, the excitation coefficients
could be found by well known procedures ([6.100] and [Ref.6.106, Sect.11.8])
that make use of modal field expansions and orthogonality. The process of
calculating the excitation coefficients is a mathematical way of asking to
what degree a given mode combination looks like the field in the input
plane — which modes and how much of each are required to match the field
over the z = 0 plane?

The general problem is complicated because the field at z = 0 is unknown;
it is a superposition of the incident field, which is known, and the unknown
reflected- or scattered-field contribution over the z = 0 plane. The latter
field is itself found from the solution of the entire problem; the exact
field over the z = 0 plane is not explicitly known, but is implicit in the
solution. Moreover, there is no known exact solution to the given backscattering problem, and an approximation is necessary.

6.3.2 Simplification and Solution of the Excitation Problem

Tremendous simplification results from the use of either of two related scalar approximations, one in which backscatter from the fiber is accounted for by a field contribution that would be produced by plane-wave reflection at a plane dielectric interface representing the fiber end face, and one in which backscatter is ignored. In the former case, the field in the z = 0 plane is approximated by the sum of the incident field plus a reflected field. In the latter approximation, which is the one most commonly employed in works treating the weakly guiding fiber, the field in the z = 0 plane is approximated by the incident field alone [6.80,100,130-134]. Such a treatment is similar to the Kirchhoff, or first Born, approximations of physical optics ([Ref.6.135, p.1073] and [6.41,131]). The approximation yields good agreement with experiment, and is valid for large values of $2\pi\rho/\lambda_1$, paraxial illumination, and small differences in refractive index between the two half spaces z $\gtreqless$ 0 [6.80,131,136].

In using the above approximation in the presence of a mismatch in media, albeit small, such as occurs in the case of the retinal receptors, the results obtained are slightly different depending upon whether one chooses to match the tangential components of the electric or the magnetic field. Whereas the exact solution would maintain continuity of both fields as required by the boundary conditions, the Kirchhoff approximation prescribes both tangential components in a manner that leads to mathematical difficulties [6.41,105]. SNYDER used continuity of the electric field to calculate the modes excited by normal and oblique incidence of truncated and infinitely extended plane waves [6.130,131]. MARCUSE, on the other hand, alternately matched the electric and magnetic fields and then used the geometric mean of the coefficients resulting from the two matching conditions as the excitation coefficient of the dominant mode excited by an incident Gaussian illumination [6.132]. In a recent publication, CARDAMA and KORNHAUSER employed a rigorous modal technique to investigate the accuracy of those approximaions, and demonstrated that the results previously obtained by SNYDER and MARCUSE are in excellent agreement with their solution [6.131,132,136]. Furthermore, they showed by means of a numerical example that the excitation coefficient obtained as the geometric mean comes extremely close to the solution obtained by their rigorous modal solution.

With the total forward-traveling fields on the fiber expressed in the form of (6.31), the *modal amplitude coefficient* for excitation by a wave E_{inc} incident upon the fiber end face at an angle θ to the axis, is given by

$$a_q \simeq \int_{A_{ap}} [\underline{E}_{inc} \times \underline{h}_q^*]_z \, da \simeq \sqrt{\varepsilon_1/\mu_0} \int_{A_{ap}} \underline{E}_{inc} \cdot \underline{e}_q \, da \quad , \qquad (6.63)$$

where backscattering is ignored, q is a label used to identify a particular mode, the integration is carried out over the truncation aperture A_{ap}, and the second integral is valid to orders $\lambda_1/2\pi\rho$, δ, and θ^2 for plane-wave incidence [6.131]. Excitation coefficients of both bound and unbound modes are given by (6.63). Explicit expressions for the mode power $P_{\ell m} = |a_{\ell m}|^2/2$ of the approximate bound modes excited by truncated, obliquely incident, uniform plane waves were given by SNYDER [6.131]; amplitude coefficients for the unbound modes were presented by SAMMUT et al. [6.112].

Multiplication of (6.63) by the plane-wave transmissivity [6.41] will result in a slightly lower value for $P_{\ell m}$, and will convert (6.63) into the form wherein backscatter is accounted for by plane-wave reflection at the fiber end face. For refractive indices, in Fig.6.5, of $n_1 = n_{myoid} = 1.361$, $n_2 = n_{IPM} = 1.347$, and $1.347 \leq n_3 \leq 1.361$, that correction amounts to a difference in mode power of less than 0.004% over incidence angles ranging from normal incidence, $\theta = 0$, to incidence at $\theta_L \simeq 8.3°$.

6.3.3 Bound Mode Excitation by Specific Illumination Sources

In this section, we examine fiber excitation by coherent linearly polarized plane waves and Gaussian beams. We again employ the geometry of Fig.6.1, but now with the understanding that θ represents the central direction of the illumination incident upon the fiber input plane.

a) Truncated Uniform Plane-Wave Illumination

Although truncated plane waves are not found in practice, they can sometimes be employed as approximations to illumination sources. In addition, familiarity with the results for this type of illumination can serve as an aid in elucidating the optical excitation properties of retinal receptors and other fiber optics elements.

We use the results based on (6.63) presented by SNYDER [6.131] to calculate the variation of individual mode power within the fiber ($P_{\ell m}^F = \eta_{\ell m} P_{\ell m}$) with normalized incidence angle ($\sin\theta/\sin\theta_L \simeq \theta/\theta_L$), for illumination by a uniform plane wave having unit illumination confined to the fiber face, for $V = 5$. This value of V would, for example, correspond to illuminating the myoid of a receptor whose $\rho_m = 2.25$ µm with light of wavelength $\lambda_0 = 555$ nm.

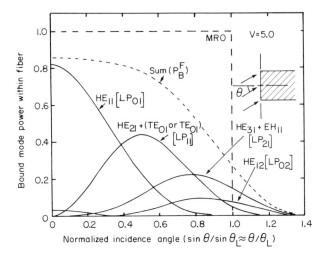

Fig.6.13. Angular variation of bound-mode power within fiber for uniform illumination with unit incident power confined to the fiber face, for V = 5. The curve labeled MRO is the result predicted using meridional ray optics. Reflection at the fiber face is neglected. These curves were calculated from the results in [6.131]

Results for the individual modes and the approximate bound-mode power within the fiber P_B^F, calculated via (6.54) are shown in Fig.6.13.

We note that meridional ray optics predicts that the power coupled into the receptor will be unity for $\theta/\theta_L < 1$, and zero for $\theta/\theta_L > 1$; this is clearly not the result pictured in Fig.6.13. SNYDER et al. have shown that the bound-mode power approaches that calculated from meridional ray optics in the limit as $V \to \infty$ $(\lambda_0 \to 0)$ [6.77]. In the present example, the bound-mode power residing within the receptor is less than that incident upon the receptor end face; thus the effective receptor aperture is smaller than its physical cross-sectional area. In Sect.6.4.1 we will pursue this property further and examine cases for which the effective receptor aperture is larger than the receptor cross section.

The variation of P_{11}, the total power in the infinite cross section coupled into the HE_{11} mode, as a function of V is shown in Fig.6.14 for two different angles of incidence. We will return to a discussion of the shape of the above curves.

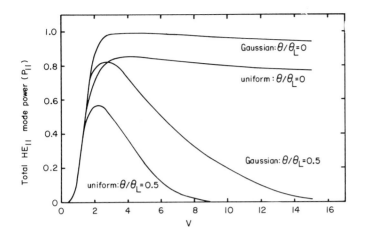

Fig.6.14. Comparison of total HE$_{11}$ mode power excited by Gaussian and uniform illumination, for two angles of incidence. Both beams have unit incident power, and the uniform illumination is confined to the fiber face. Curves for the uniform illumination were calculated using the results in [6.131], and the curves for the Gaussian illumination are taken from [Ref.6.133, Fig.2, solid curve for tilt angle = 0.10]. The intensity of the Gaussian illumination is proportional to $\exp(-2r^2/\rho_0^2)$, $\rho_0/\rho = 0.849$; these results were calculated for $n_1/n_2 = 1.0198$

b) Gaussian Beam Illumination

Another illumination source that is frequently studied is a beam having a Gaussian variation in its cross section. The dominant mode of a laser is well described by such a beam. Studies of fiber excitation by Gaussian beams were carried out by MARCUSE who considered HE$_{11}$ mode excitation for tilted and offset beams having plane wavefronts [6.132], and by IMAI and HARA who considered the general case and presented numerical results for the three lowest-order mode sets [6.133,134]. IMAI and HARA treated both tilted and offset beams and demonstrated the effect that wavefront curvature has upon mode excitation. Their results for tilted beams show that differences due to wavefront curvature are generally greatest for normal incidence and that those differences generally increase both with increases in illumination spot size and with decreases in fiber refractive index.

In Figs.6.14 and 15 we compare the effects of Gaussian and truncated uniform illumination. Figure 6.14 displays the variation of the total power in the infinite cross section that is coupled into the HE$_{11}$ mode as a function of V, for beams that are incident normally and at $\theta_L/2$; Fig.6.15 shows the angular variation of HE$_{11}$ mode power for V = 5. Figures 6.14 and 15 display a striking qualitative similarity for HE$_{11}$ mode excitation produced by

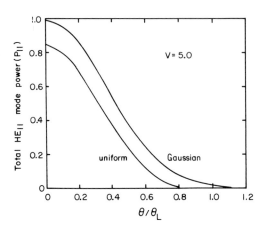

Fig.6.15. Comparison of angular variation of HE_{11} mode excitation by Gaussian illumination intensity proportional to $\exp(-2r^2/\rho_0^2)$, $\rho_0/\rho = 0.793$ and uniform illumination confined to the fiber face, for V = 5. Both beams have unit incident power. The curve for the uniform illumination was calculated using results of [6.131], and that for Gaussian illumination was taken from [Ref.6.132, Fig.8, curve for kd = 35]

a uniform plane wave whose illumination is confined to the fiber cross section, and that produced by two tightly coupled centered Gaussian beams, whose power density at the fiber boundary is less than 7% of its value at the fiber center. It is not at all surprising that the Gaussian beam is more efficient at launching the HE_{11} mode, than is the truncated plane wave, as the Gaussian variation $\exp[-(r\cos\theta/\rho_0)^2]$ matches the $HE_{11}(LP_{01})$ mode's variation $J_0(U_{11}r/\rho)$ better than does the uniform intensity profile of the truncated plane wave.

c) Qualitative Aspects of Photoreceptor Excitation

As an aid in picturing a match between the required mode or combinations of modes and the incident field, we can think of the incident illumination in the receptor input plane as composed of symmetric and asymmetric parts with respect to the y axis (Fig.6.1). Consider a linearly polarized source of illumination, which is symmetric about its central ray, incident at an angle θ in the x-z plane. For normal incidence, the fiber face is uniformly and symmetrically illuminated. The excited modes must also have electric fields that are symmetrical, and thus have nonzero intensity on the guide axis. The only modes satisfying that requirement are the $HE_{1m}(LP_{0m})$ modes. These modes are available in two orthogonal polarizations (see Table 6.4); the mode that is polarized in the direction of the incident illumination will be launched.

Oblique incidence will then produce an illumination on the fiber face with advanced phase on one side of the y axis and retarded phase on the other. That asymmetric illumination will give rise to higher order modes and will favor those having a null line along the y axis. The LP_{11} mode (Fig.6.8) is the lowest order mode that meets the latter requirement; it is composed of a combination of the HE_{21} and TE_{01} or TM_{01} modes. If the incident field is polarized with the electric vector in the plane of incidence, then the electric vector of the LP_{1m} modes will be in the x direction, or perpendicular to the null line. Thus the HE_{2m}, TM_{0m} mode combination will be excited. If the incident field is polarized perpendicular to the plane of incidence, then the electric vector of the LP_{1m} modes will be in the y direction, or parallel to the null line, and the HE_{2m}, TE_{0m} combination will be excited. Although all modes will have the central directions of their electric vectors aligned with that of the incident field, the TE_{0m} and TM_{0m} modes are the only ones whose presence is sensitive to polarization. Modes belonging to the sets $LP_{\ell-1,m}$ ($HE_{\ell m}$, $EH_{\ell-2,m}$) for $\ell \geq 3$, will each be launched with equal power for (truncated) uniform plane-wave illumination.

Some general statements can be made regarding mode launching by axially centered illumination having symmetric intensity distributions; these include infinite or truncated plane waves, Gaussian beams, and the Airy disks produced by focused spots. For on-axis illumination, the HE_{1m} modes are the only modes excited. As the angle made by the central illuminating ray and the receptor axis increases, the excitation of higher order modes increases; and the maximum value of the excitation coefficients tends to decrease for successively higher order modes. The angle at which a particular mode is maximally excited is in the neighborhood of the characteristic propagation angle of the ray bundle that represents that mode [6.100,131]. The net result is that the total power coupled into the bound modes, and thus trapped within the receptor, tends to decrease with increasing angles of illumination, even for angles smaller than the geometrically defined θ_L. This is a manifestation of the receptor's intrinsic angular sensitivity and is distinctly different from that predicted by geometric optics. We will return to the important topic of receptor directional effects in Sect.6.4.2.

In a laboratory situation, the modes excited on an optical fiber can be weighted to favor a given mode by varying the angle of incidence, spot size, spot placement, numerical aperture, wavelength, and to some degree the polarization of the incident illumination. These techniques were employed in early studies of optical fibers by SNITZER and OSTERBERG [6.118] and KAPANY [6.91], and in studies of vertebrate retinal photoreceptors by ENOCH [6.10-12,21, 117]. Although these techniques tend to favor particular modes, they also serve to launch other, possibly undesired, modes as well. In order to be as selective as possible in mode excitation, the illumination must resemble the desired mode as closely as possible. To that end, KAPANY and BURKE [6.100], have described and tested a spatial filtering technique for mode launching.

When studying the optics of the retinal photoreceptors, two objectives arise, which place different demands upon the illumination conditions: 1) to study the receptor's properties by exciting particular modes, and 2) to study a model of the phenomena that occur during life. In the former case, the optical excitation considerations outlined above will apply; in the latter, the illumination conditions must match those occurring in the eye itself [6.11]. Because the photoreceptors in the normal eye tend to align with the pupillary center, the illumination for this case will tend to be axially directed [6.2,7,8,23].

6.3.4 Excitation of Unbound Modes

Illumination, incident at the guide's input plane, that does not launch bound modes excites unbound modes which carry power away from the guide into the radiation field. Recall that unbound modes do not exist individually, but as a continuum represented by $\underline{E}^R$ and $\underline{H}^R$ in (6.31) which forms a field that spreads out in space and thus attenuates in the axial direction. Therefore, the contribution of the unbound modes to the power within the guide is strongest in the vicinity of the input end and decreases with distance along the axis. As the wave travels along the guide, it is less available for absorption. Furthermore, we expect unbound-mode contributions to be most important for small V, at which few bound modes may propagate.

Figure 6.16 illustrates the variation with normalized frequency of the total unbound-mode power P_R^{tot} for uniform, normally incident plane illumination with unit power confined to the fiber end face. Only a fraction of P_R^{tot} is contained within the fiber; it radiates from the fiber more rapidly at low values of V than at higher values of V.

SAMMUT et al. [6.112] reported that, for illumination by a normally incident truncated uniform plane wave, the power within the fiber contributed by the unbound modes has a maximum that does not always occur at the input plane. Owing to a beat phenomenon, that maximum tends to occur at an axial distance along the fiber that increases with increases in illumination spot size. For V = 4, they show the maximum unbound-mode power occurring at the aperture for a spot confined to the fiber face and at a distance of $z/\rho \simeq 8$ for a spot diameter twice that of the fiber. This power then decreases with increased distance.

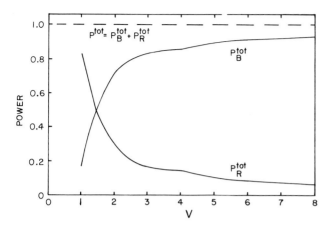

Fig.6.16. Total unbound-mode power excited on a fiber illuminated at normal incidence by plane illumination with unit power confined to the fiber face

Illumination incident upon the receptor end face at angles other than along the normal will tend to produce increasingly greater amounts of un- bound-mode power, for the receptor is presented with rays that it cannot contain. Therefore, illumination sources having numerical apertures signi- ficantly greater than that of the receptor [see (6.4)] will tend to excite strongly unbound modes.

Several examples of unbound-mode contributions to absorption in fibers are given by SAMMUT and SNYDER for fibers illuminated at the beginning of their absorbing sections by normally incident uniform plane waves confined to the fiber cross section [6.125]. Their results show that unbound modes contribute a fraction of total absorbed power that decreases with increased absorption length and decreased absorption coefficient, a result that is consistent with the axial decrease of unbound mode power residing within nonabsorbing fibers. Based upon their specific findings, they conclude that unbound modes cannot be neglected for $V \lesssim 1.5$, but for higher values of V the receptors can be successfully treated by accounting for only the bound modes, except perhaps for the shortest receptors. Suitable limits on V for larger illumination apertures and off-axis illumination remain to be specified.

Closed form solutions are not available for the continuous spectrum of the optical fiber, even for seemingly simple sources. Direct approaches for eva- luating the contribution of the unbound modes are sometimes apt to run into difficulties owing to the convergence of numerical routines employed for integration of the continuous spectrum [6.112]. An alternate approach in- volves approximating that contribution by means of leaky modes [6.80,123].

In his review, SNYDER discussed the properties of leaky modes and presented a taxonomy of leaky-mode behavior [6.80].

The advantage of the leaky-mode approach is that only a finite number of those waves need to be considered in the approximation; and furthermore, the waves describe an intuitively pleasing physical process that accounts for power transfer between the fiber and the radiation field. A brief review of the properties of leaky modes on circular optical fibers is presented in Sect.6.3.4b.

a) The Retinal Receptor Input Plane and the Significance of Unbound Modes

As discussed in Sect.6.1.1, the location of the effective input plane to the retinal fiber optics bundle is not known. The location of that plane plays a role in optical excitation of the photoreceptors. For a single model receptor that receives the bulk of its illumination at the ELM, the unbound mode power contained within the receptor may decay markedly before reaching the absorbing outer segment. In addition, owing to the different V values of the myoid and ellipsoid for a given wavelength, different amounts of bound- and unbound-mode power would be excited for input plane location at the ELM or ellipsoid level.

SAMMUT and SNYDER calculated the contribution of unbound-mode power to the absorption within a photoreceptor which was excited at the beginning of its absorbing portion by uniform illumination confined to the cross section [6.125]. The same type of calculation, but performed on a model whose absorbing section were some distance removed from the input plane, would show a smaller percent absorption arising from the unbound modes. Their results thus represent, for given V, absorption length, and absorption coefficient, estimates of maximum percent absorption contribution from the unbound modes.

For input plane location at the ELM, a portion of the unbound light will leak away before the absorbing outer segments are reached. However, the power that leaks away from one receptor may become available to its neighbors. As a result, an accounting of only the unbound light that a receptor receives through its equivalent truncation aperture is likely to result in a low estimate. An approach to the problem was provided by WIJNGAARD et al. [6.78] and WIJNGAARD and VAN KRUYSBERGEN [6.137], who modeled all the unbound light within the retina as a bulk wave, and then accounted for its absorption by geometric optics means. A possible refinement might involve a combination of two approaches, i.e., use of waveguide theory and the isolated receptor model to account for the unbound light that a receptor receives from its own unit

cell, and use of the bulk wave to model that portion of the unguided light
that radiates from the receptors. However, we must mention that such refine-
ments may be unwarranted in the face of power reckoning on the basis of the
isolated receptor.

It is interesting to note that the light scattering produced by conical
outer segments of peripheral cones has been proposed by MILLER and SNYDER
to function as a sensitivity enhancement mechanism for their neighboring
rods [6.34].

b) Leaky Modes

Leaky modes are not true modes in that they do not belong to the proper
spectrum. They exhibit an exponentially growing behavior with distance in
the transverse plane and, therefore, do not satisfy the radiation condition.
Nonetheless, a discrete sum of leaky waves can provide a convenient means for
approximating the contribution of the continuous spectrum within restricted
regions of space. These field contributions are expecially useful for representing the portion of the radiation field that exists within or very close
to the fiber and far from the excitation source [6.80,138,139].

Leaky-mode behavior is governed by complex values of $U_{\ell m}$, $Q_{\ell m}$, and $\beta_{\ell m}$
that are obtained from solutions of the characteristic equation (6.20) or
(6.48) below the cutoff frequency of mode ℓm. For the $\ell \geq 3$ modes, the solutions below cutoff represent an analytic continuation (smooth transition)
of those above cutoff [6.80]. We can see that one might expect such behavior
by examining the $\ell m = 31$ mode. Figure 6.10 shows that half of the 31 mode's
power resides within the guide at cutoff; the radial dependence of the corresponding intensity pattern is shown in Fig.6.9c. We should be extremely
surprised were that power to suddenly vanish from the guide as V were decreased below cutoff. Analytic continuation of the guide parameters below
cutoff verifies that the mode will begin to leak power out of the guide and
thereby produce an axial attenuation [6.80,138,139].

Above cutoff, the bound-mode fields decay exponentially outside the
cylinder, $r > \rho$, and have a power flow parallel to the cylinder. As V decreases below cutoff, the $\ell \geq 2$ modes exhibit a *tunneling region* (Fig.6.17)
between the guide and a radius $r_{\ell m}^{tp} = \rho(\ell - 1)/\text{Re}\{Q_{\ell m}\}$, called the *turning
point* or *caustic*, in which the field is evanescent and power flows parallel
to the cylinder. Outside that region ($r > r_{\ell m}^{tp}$) the field increases exponentially with r, and power flow is oblique to the cylinder at an angle
$\Psi_{\ell m} = \sin^{-1}[\text{Re}\{Q_{\ell m}\}/\rho k_2]$ to the cylinder axis [6.80,138,140]. As V decreases

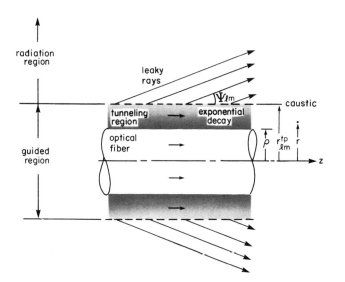

Fig.6.17. Power flow, shown by arrows, for leaky mode ℓm. Power flow in the radiation region is at an angle $\Psi_{\ell m}$ to the cylinder axis, and is parallel to that axis within the guided region. The leaky rays responsible for the radiation are skew to the cylinder axis, as they are tangential to the caustic. The darker shading corresponds to regions of greater intensity. (Adapted from [6.80], with permission)

further below cutoff $r_{\ell m}^{tp}$ moves in from infinity, and the tunneling region shrinks until a value $V_{\ell m}^{ce}$ is reached at which $\text{Re}\{Q_{\ell m}\} = \ell - 1$; the turning point is then at the cylinder surface, and the tunneling region has shrunken to zero. Further decreases in V leave the turning point at the cylinder surface. This region of V, for which the leakage arises at the cylinder surface, is termed the *refracting region*. The nomenclature is due to the fact that leakage is produced by rays that undergo electromagnetic tunneling in the first case, and refraction in the second case [6.80,140,141].

Transition between the evanescent tunneling region and the radiation region is sharp only for the $\ell \gg 1$ modes. Otherwise, a transition region exists in the neighborhood of $r_{\ell m}^{tp}$ in which the propagation vector in the surround turns from a direction parallel to the cylinder to an oblique angle $\Psi_{\ell m}$ in the far field [6.116].

For $\ell = 1$ and $m \geq 2$, no solutions exist just below cutoff, and the HE_{1m} leaky modes are refracting with $r_{1m}^{tp} = \rho$ [6.80,138]. Thus, only the HE_{1m} modes can be said to exhibit a true cutoff. Because the tunneling leaky modes leak slowly and can extend to large distances along the guide without suffering appreciable attenuation, it has been suggested that $V_{\ell m}^{ce}$ be called the *effective cutoff* of those modes [6.138]. The $\ell m = 12$ modes have large attenuation coefficients and are relatively unimportant far from the source, but are considered to play an important role for short optical fibers such as visual photoreceptors [6.123]. Leaky-mode behavior is summarized in Table 6.6.

The exponential increase of the field in the radiation region can be inferred from Fig.6.17. For fixed z and increasing r, such that $r > r_{\ell m}^{tp}$, one encounters rays that originated at points closer to the illumination source where the field is stronger, thus producing a stronger radiated field. If such a wave were to exist in unbounded space, a nonphysical condition and violation of the radiation condition would result. Although a steepest descent analysis has not been carried out for the circular fiber, such an analysis would define regions of space in which leaky-wave representations for the field are appropriate. Such computations have been performed for planar geometries, with the results that leaky waves are asymptotic representations that are defined within wedge-shaped regions as shown in Fig.6.18 [6.106,142]. The nonphysical nature of the wave is thus reconciled for the planar geometry; at any fixed value of z, increases in x produce exponential increases in the field until the boundary of the wedge is reached. For points outside that region, the leaky-wave representation is no longer valid.

Leaky modes do not obey the power orthogonality relation (6.32); thus, their excitation coefficients cannot be found via (6.63). To circumvent that, two methods have been developed for evaluating those coefficients [6.139]. The first is by defining truncated leaky modes as the guided portion of the weakly leaky modes, and applying the power orthogonality relation. In this approximation the weakly leaky modes are treated as bound, thus assuming that they behave like proper modes. The second is via an exact orthogonality relation, in which the excitation coefficients for bound and leaky modes are found by an identical process. In this case the convenience of interpretation in terms of power orthogonality is lost. It was emphasized that field expansions using such approximations are practical only when leaky modes dominate the radiation field, and that it yet remains to determine the conditions under which leaky modes account for a significant portion of the power within a fiber.

Table 6.6. Summary of leaky-mode behavior[a,b]

V	$Q_{\ell m}$	$r_{\ell m}^{tp}$	Classification	Produced by (Quasi-optics viewpoint)	Behavior
$V>V_{\ell m}^C$ above cutoff	Imaginary	∞	Bound modes	Rays totally reflected at cylinder-surround interface	Power flow along cylinder. Evanescent for $r>\rho$
$V_{\ell m}^{Ce}<V<V_{\ell m}^C$ above effective cutoff, below cutoff	Complex $0<\mathrm{Re}\{Q_{\ell m}\}<\ell-1$ $(\ell\geqq2)$	$\rho(\ell-1)/\mathrm{Re}\{Q_{\ell m}\}$ $\infty > r_{\ell m}^{tp} > \rho$	Tunneling leaky modes (weakly leaky)	Rays that cannot be contained due to curvature of cylinder	Power flow along cylinder for $0<r<r_{\ell m}^{tp}$ and oblique to cylinder at angle $0<\psi_{\ell m}<\sin^{-1}[(\ell-1)/\rho k_2]$ for $r>r_{\ell m}^{tp}$. Slow attenuation along cylinder. Evanescent for $\rho<r<r_{\ell m}^{tp}$. Leakage vanishes as $\lambda/\rho \to 0$ $(\rho k_2 \to \infty)$.
$V<V_{\ell m}^{Ce}$ $(\ell=1:\ V_{1m}^{Ce}=V_{1m}^C)$ below effective cutoff $(\ell=1,\ m\geqq2;\ \ell\geqq2)$	Complex $\mathrm{Re}\{Q_{\ell m}\} > \ell-1$	ρ	Refracting leaky modes (strongly leaky)	Rays that are refracted at cylinder-surround interface	Power flow along cylinder for $r<\rho$, and oblique to cylinder at angle $\psi_{\ell m}>\sin^{-1}[(\ell-1)/\rho k_2]$ for $r>\rho$. Rapid attenuation along cylinder. Increases exponentially for $r>\rho$. Leakage does not vanish as $\lambda/\rho \to 0$ $(\rho k_2 \to \infty)$.

[a] Table constructed from material from [6.80,138-140].

[b] $V_{\ell m}^{Ce}$ is effective cutoff, defined such that $\mathrm{Re}\{Q_{\ell m}(V^{Ce})\} = \ell-1$; $\psi_{\ell m} = \sin^{-1}[\mathrm{Re}\{Q_{\ell m}\}/\rho k_2]$.

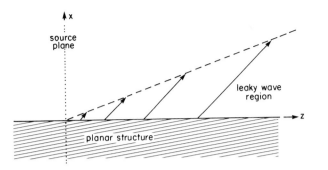

<u>Fig.6.18.</u> Region of validity of +z-traveling leaky wave along the interface to a planar structure. The source is assumed to be located in the z = 0 plane

6.4 Application to the Model Receptor

In the previous sections we treated the modeling of the retinal photorecep-tors as well as many of the approximate techniques employed in the wave optics analysis of those models. In this section we discuss the application of those methods to the calculation of receptor properties such as effective aperture, angular sensitivity, and spectral sensitivity.

For the examples presented in this section, the models employed are either the simple dielectric cylinder or the three-segment rod and cone mo-dels having untapered ellipsoids; the analysis techniques are the approximate ones outlined in previous sections, and only bound modes have been accounted for. With reference to Figs.6.1 and 6.2, the mode amplitudes for the weakly guiding fiber were calculated at the aperture plane by using the results of SNYDER [6.131], based upon the approximation (6.63). Reflection at the fiber end face, as well as that between segments, is neglected. Modes launched on the myoid are then guided by the ellipsoid to the outer segments[6]. For re-sults quoted from tapered models that have appeared in the literature, mode coupling along the taper is neglected and modes that become cutoff prior to absorption are not considered further. Absorption by the outer segment is

---

6 The myoid-ellipsoid interface produces a jump in V (see Fig.6.11) by a factor that is close to 2. Thus, our simple model will result in the nonphysical requirement of the sudden influx of power into the ellipsoid. A more refined model would account for that interface by matching the wave incident upon the ellipsoid to that segment's mode spectrum, by the procedure used in Sect.6.3.2.

calculated by using perturbation and assuming approximate orthogonality over the finite fiber cross section, i.e., via (6.61).

6.4.1 Effective Photoreceptor Aperture

The *effective receptor aperture* is the equivalent cross sectional area of a fiber that would intercept, or collect according to geometric optics, the same power as that collected by the receptor. In Sect.6.3.3a we considered an example in which the power collected within a model of a photoreceptor myoid was less than that predicted by geometric optics. In fact, for a fiber that is uniformly illuminated over its end face, the bound-mode power collected only approaches the meridional ray optics prediction for $V \to \infty$ [6.77]. Thus, for illumination that is confined to the fiber face, the effective receptor aperture never exceeds the area of the receptor end face.

For illumination that extends beyond the physical limits of the fiber, the effective aperture for light collection can be considerably greater than the fiber cross section. From the wave optics viewpoint, the important feature is the overlap and matching between the incident illumination and the modes of the fiber, and not necessarily the overlap between that illumination and the fiber cross section. Needless to say, incident illumination lying totally outside the fiber cross section would be a poor choice for launching bound modes, although it would be successful in launching a radiation field as well as comparatively weak bound modes.

SNYDER and HAMER calculated the bound-mode power absorbed by an optical fiber that received uniform on-axis illumination extending to relative radii ρ_{ap}/ρ = 1.5 and 2.0 beyond the fiber cross section [6.76]. The results obtained may also be interpreted in terms of bound-mode power collected within the receptor, owing to their use of an approximation that is independent of absorption coefficient. Their results show that for $V > 1.65$ both illumination spot sizes yield collected powers that are greter than that intercepted by the receptor end face. In fact, collection increases over geometric optics are as high as 96% for ρ_{ap}/ρ = 2.0 and $V = 4.35$, and 50% for ρ_{ap}/ρ = 1.5 and $V = 4.7$. Moreover, inclusion of the effect of the absorption coefficient can be expected to yield somewhat higher increases in absorption over geometric optics, owing to the additional influx of power along the absorbing segment (Sect.6.2.3b). Detailed curves of light acceptance properties, as functions of V and illumination aperture, for various values of incidence angle, are presented by PASK and SNYDER [6.143].

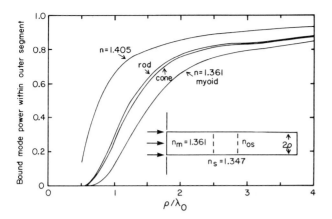

Fig.6.19. Bound-mode power collection for three-stage model with equal inner-
and outer-segment radii. Illumination is on axis, with unit incident power
confined to the receptor end face. By definition, the ordinate is also equal
to the effective receptor aperture per cross-sectional area of inner segment.
The curves marked rod and cone give light collection in the outer segments
of three-stage rod and cone models; those marked n = 1.361 and n = 1.405
give light collection in homogeneous cylinders whose refractive indices are
those of the myoids and rod outer segment, respectively. Calculated for:
n_m = 1.361; n_{os} = 1.387 and 1.405 for cones and rods, respectively; and
n_s = 1.347

The above results apply to light collection by a homogeneous cylinder.
It is of interest to calculate the power collected within the outer segment
of the three-stage receptor model. Figure 6.19 displays the light collection, as represented by bound-mode power, within the outer segments of a
rod and cone model having equal inner- and outer-segment radii, and compares these results with collection by two homogeneous cylinders, one representing the myoid, and the other having refractive index equal to that
of the rod outer segment. Illumination is by an on-axis uniform plane wave
that is truncated to the receptor cross section. The results for the rod
and cone models are virtually indistinguishable, with the rod exhibiting
slightly greater collection, and both exhibiting greater collection than
that of the myoid. The results can be explained as follows. The bound-mode
power available to the outer segment is that excited in the infinite cross
section on the myoid; the curve for the myoid displays the portion of that
power which resides within the physical confines of the myoid. When the bound
modes are transferred to the higher refractive index outer segment, a greater
portion of the total bound-mode power will then reside within the outer segment. For equal radii, on transition from the myoid to outer segment, V increases by a factor of 2.05 for rods and 1.70 for cones. For low values of

V (or ρ/λ_0) it is this increase in V which produces the largest change in tightness of coupling (see Fig.6.10). Once in the neighborhood of the higher V of the outer segment, further increases in V have a decreased effect. In fact as ρ/λ_0, hence V, becomes large, the different results merge.

Figure 6.19 also illustrates, for an untapered rod model, the effect of receptor input plane location upon light collection. The curve for a homogeneous cylinder of n = 1.405 displays the bound-mode power that would be collected by a rod outer segment, for location of the effective input plane at the level of the outer segment. Collection for this case is greater than for location of that plane at the ELM; as in the previous case, differences between curves diminish for increasing V.

In comparison to untapered receptors, receptors having identical outer segments and tapered ellipsoids can be expected to exhibit significantly increased light collection in their outer segments for on-axis illumination. The increased collection results from funneling by the higher refractive-index ellipsoid, as well as from greater light collection by the larger diameter myoid. Such an explanation is in agreement with experiment, and has been offered by TANSLEY and JOHNSON [6.144] to explain their observations of light concentration in the outer segments of the markedly tapered cones of the grass snake retina, and by ENOCH [6.11] to explain his observations that in retinal preparations outer segments of cones appear brighter than do those of rods.

6.4.2 Photoreceptor Directionality

As used herein, *directionality* refers to the variation with incidence angle of light collected and absorbed by a photoreceptor. Although often referred to as intrinsic directionality, the photoreceptor's excitation by illumination at different incidence angles depends upon the profile and spatial extent of the incident illumination [6.132-134,143,145]. Antenna engineers have circumvented this potential ambiguity by defining an antenna's receiving pattern, which is identical to its transmitting pattern, as the response to infinite plane waves incident at different angles. This convention is usually not followed in the fiber and photoreceptor optics literature, and directionality is often specified in terms of the angular acceptance of a plane wave, truncated to the fiber cross section [6.30,80]. An important exception is the measured intrinsic directionality, the Stiles-Crawford function of the first kind [6.83] (Chap.3), in which case the incident illumination of a central photoreceptor can be considered to be a plane wave;

and the effective truncation aperture would then depend upon details of the retinal mosaic. Moreover, that psychophysically determined function contains the effects of many other factors, in addition to the direction effects of a single photoreceptor (Chaps.3,4).

a) Directionality of Photoreceptor Inner Segments

In Sect.6.3.3a, we considered the case of uniform illumination confined to the fiber face and saw, in Fig.6.13, an example of how the angular dependence of individual mode excitation produces directional selectivity of the light collected by a fiber. Figure 6.20, which is calculated from results given by SNYDER [6.131], demonstrates the effect that different values of normalized frequency V have upon the directional selectivity. By identifying the fiber with a photoreceptor and the surround with the IPM, the curves represent the directionality of bound-mode power collected within the myoid. Different values of V result both in different amounts of light collection for fixed angle of incidence, and in different shapes, hence, in different widths, of the respective selectivity curves.

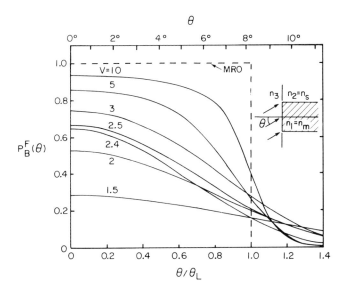

Fig.6.20. Angular variation of bound-mode power within fiber for uniform illumination with unit incident power confined to the fiber face, for various values of V. Reflection at the fiber face is neglected, and the curves were calculated from the results in [6.131]. The curve labeled MRO is the meridional ray optics result. Values of θ shown are illustrative and are for a receptor myoid with $n_3 = n_1 = n_m = 1.361$ and $n_2 = n_s = 1.347$, for which $\theta_L = 8.23°$

286

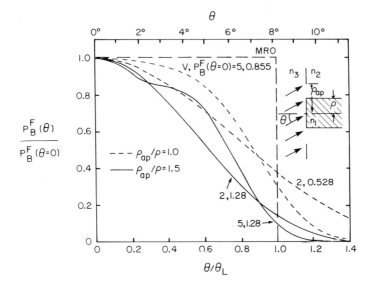

Fig.6.21. Normalized bound-mode power within fiber excited by truncated
uniform illumination with unit power incident over the fiber end face, for
two different truncation apertures, and two values of V. The curves may be
unnormalized by multiplying the ordinate by $P_B^F(\theta = 0)$, the bound-mode power
collected within the fiber for normal incidence. The curve labeled MRO gives
the normalized geometric optics collected power predicted by meridional ray
optics; the unnormalized values are 1.0 and 0.444 for $\rho_{ap}/\rho = 1.0$ and 1.5,
respectively. The values of θ shown are illustrative and are calculated for
a receptor myoid with $n_3 = n_1 = n_m = 1.361$ and $n_2 = n_s = 1.347$, for which
$\theta_L = 8.23°$

In order to demonstrate the dependence of directional selectivity upon
the spatial extent of the incident illumination, we again use results given
by SNYDER [6.131] to calculate the angular variation of the bound-mode power
residing within the myoid for uniform plane-wave illumination through truncating apertures with normalized radii $\rho_{ap}/\rho = 1.0$ and 1.5; results for
V = 2 and 5 are displayed in Fig.6.21. The curves have been normalized with
respect to their values for normal incidence, so that their shapes may be
compared. Unbound modes have not been accounted for and it is expected that
their inclusion would modify the undulation in the curve for V = 5 and
$\rho_{ap}/\rho = 1.5$ because the cutoff for the $\ell m = 41$ leaky mode occurs at V = 5.136.
In addition, as the illumination spot size increases beyond the fiber boundary, unbound modes are excited with increasing amplitude. Furthermore, unbound modes can be expected to play an increasingly important role in regions
where the bound-mode power is small; this includes the region $\theta \gtrsim \theta_L$ [6.78,
123].

Increasing the aperture from $\rho_{ap}/\rho = 1.0$ to $\rho_{ap}/\rho = 1.5$ results in an increase of incident power from unity to 2.25; as seen from the normalization factors in Fig.6.21, this results in power increases within the receptors by factors of 2.42 and 1.73 for V equal to 2 and 5, respectively. For V = 2 the power in the guide increases by a larger factor than does the incident power, and for V = 5 by a smaller factor. This occurs because the modes for V = 5 are fairly tightly bound to the cylinder (Fig.6.10) and are better matched by incident illumination that is more concentrated in the neighborhood of the receptor, than is the more loosely bound mode for V = 2, for which the increased aperture provides a better match. Moreover, for $\rho_{ap}/\rho = 1.5$ and both V = 2 and 5, the normalization factors are greater than unity, the power intercepted by the receptor end face; hence, for these values there is a range of angles, containing normal incidence, for which the effective receptor aperture is greater than the area of the fiber face.

b) Directionality of Receptor Models and Comparison with the Stiles-Crawford Effects

In this discussion, we touch on some theoretical treatments of photoreceptor direction effects and their comparison to the Stiles-Crawford effects. The interested reader will find an extensive discussion of these effects in Chap.3.

There are two psychophysically defined Stiles-Crawford effects, that of the first kind (SC-I) and that of the second kind (SC-II). SC-I describes the relative sensitivity of the visual system to illumination, in the form of parallel beams of light, that is incident at different points in the eye's entrance pupil. In the absence of aberration, these beams will strike the retina at the same point, but at different angles. In the Gullstrand schematic eye, a 1-mm displacement in the entrance pupil corresponds to a 2.5° change in incidence at the retina, for foveal points; the equivalence may change somewhat for eccentric points. Typically, the retinal areas involved in the measurement contain large numbers of photoreceptors. Defining the sensitivity s(d) as the ratio of the amount of light entering the pupil center to elicit a given response to that entering distance d from the pupil center to produce the same response, and for displacement along the horizontal meridian, STILES fit empirically a function

$$s(d) = s_{max}\ 10^{-\zeta(\lambda)(d-d_m)^2} \tag{6.64}$$

to describe the known data [6.84][7]. Here, d is the displacement in millimeters from the pupil center and is positive for temporal displacements
and negative for nasal displacements, d_m is that displacement at which the
maximum sensitivity s_{max} occurs, and $\zeta(\lambda)$ is a wavelength-dependent parameter that specifies the width of the sensitivity curve; larger values of
$\zeta(\lambda)$ correspond to narrower sensitivity curves. Using the retinal incidence angle equivalent of pupillary displacement,

$$s(\theta) = s_{max} \; 10^{-\zeta(\lambda)(\theta/2.5)^2} \; , \qquad (6.65)$$

where $\theta = 0$ is referenced to $d = d_m$ and is assumed to coincide with the average direction of the local receptor axis.

The particular description given corresponds to measurement along the
horizontal meridian; a traverse along the vertical meridian will produce a
similar function, with somewhat different values for ζ and d_m. In addition
to the dependence of ζ on wavelength, ζ is strongly dependent upon retinal
location, hence visual angle. From measurements performed within 35° of the
central fovea, it appears that ζ increases from its value at the fovea to a
maximum value at approximately 3.75° and then decreases for further increases
in visual angle[8]. ζ also exhibits differences between subjects; and mean
values, averaged over subjects for both the horizontal and vertical meridians, lie in the approximate range 0.04 to 0.092. Individual differences
are evident in d_m, and eccentricities of 1-2 mm are common. Moreover, d_m
may vary a couple of millimeters, depending on retinal location.

In addition to the directional sensitivity and its variation with wavelength that is described by SC-I, there is an apparent hue change associated
with illumination that is obliquely incident upon the retina, a phenomenon
described by SC-II [6.84].

Retinally based phenomena appear to account for a considerable portion
of the Stiles-Crawford effects; measured results are reported either as
measured at the cornea, or as the function that is obtained by correcting
for preretinal effects which arise in the optical media. Such corrections
may account for wavelength-dependent absorption through varying thicknesses

---

7 In order to avoid confusion with previously defined quantities, we use
 s and ζ for quantities that STILES called η and p, respectively. This
 designation differs from that in Chap.3. SAFIR and HYAMS made an argument
 for fitting log s(d) by a Gaussian function [6.146].

8 From measurements at 0°, 2°, 3.75°, 5°, 10°, and 35°, reported in [6.147].

of lens material [6.148-150], absorption by the macular pigment, and fore-shortening of the entrance pupil, a strong effect for measurement at eccentric retinal locations. Retinal effects comprise photoreceptor optics effects and a portion of neural summation effects; physical parameters such as photoreceptor dimensions, refractive index, mosaic pattern, interphotoreceptor spacing, and photoreceptor and group orientation will influence directionality. The SC functions for foveal and near-parafoveal cones correlate well with various theoretical calculations performed on single photoreceptor models. Rod directionality is seen as an enigma, however [6.151].

A number of different analytical models provide directional sensitivity curves that resemble SC-I. A typical foveal value of $\zeta = 0.05$ specifies a directional sensitivity curve that falls to one-half of its peak value at $\theta = 6.1°$ and one-tenth peak value at $\theta = 11.2°$. For a very rough comparison, we can compare those angles with angles describing light collection by photoreceptor myoids; recall that the geometric optics limiting incidence angle, defined via (6.3), is approximately 8.2°. Further comparison with myoid values can be made from Figs.6.20 and 21, in which the angular dependence of bound-mode power collected in a receptor myoid is shown. From such comparisons, we can take heart that we are on the right track in looking for a receptor optics description of SC-I. However, we must realize that we must look at the light that is absorbed in the outer segments, and not merely that collected by the myoids. Moreover, any description of SC-I must match not only the angular dependence of the directional sensitivity, but also its variation with wavelength; to be complete, the model must also be consistent with SC-II and the photopic luminous efficiency function.

SNYDER and PASK constructed an averaged three-stage tapered model for the foveal cones, in which they accounted for the bound modes, to explain SC-I [6.29]. Their model is capable of supporting the three lowest order mode sets, whose behavior at cutoff provides the structure for $\zeta(\lambda)$. Outer segment refractive indices used in this model were based upon estimates made from electron micrographs, and fall outside the range of values reported by SIDMAN [6.15] and BARER [6.17]; this point may be worthy of further study as the calculated results exhibit strong effects for small changes in both refractive index and dimensions. Mode launching was calculated based upon truncation of the incident light to the myoid cross section. Comparison of calculated results with STILES's data for SC-I [6.84], results in reasonable agreement for the choice of model parameters. In addition, SNYDER and PASK showed that for increased inner-segment diameter, with the other parameters held constant, the angular sensitivity becomes narrower, which is in agreement with the psychophysical measurements of the photopic directional sensitivity of the foveal and near-parafoveal retina [6.147,152,153]. More recently, SAMMUT included leaky modes in Snyder and Pask's model and showed that the inclusion of those modes removed a wavelength-dependent discontinuity in the calculated values of $\zeta(\lambda)$ [6.123]. Snyder and Pask's results, together with Sammut's modification, provide a theoretical description of directionality at the retinal level. In order to compare those results with the psychophysical function, as measured at the cornea, the effect of the ocular media needs to be accounted for. As pointed out by VOS and VAN OS [6.150], lens absorption provides a strong contribution in the blue end of the spectrum, and will affect results obtained for $\zeta(\lambda)$. In addition, the lens may also affect the symmetry of the angular sensitivity owing to the asymmetry of the lens with respect to the SC-I peak, which is usually somewhat eccentrically located in the pupil. In a later paper, PASK and SNYDER again employed the three-segment model, together with a particular choice of parameters that used different values of V for both inner and outer segments of red and green cones, and demonstrated that color data calculated for that model provide a match for SC-II for wavelengths longer than 500 nm [6.122].

A somewhat different approach was taken by WIJNGAARD, BOUMAN, and BUDDING, who studied light absorption by two trichromatic models of the foveal cone [6.78]. Both were tapered two-stage model receptors; one was based upon geometric optics and the other upon single (HE_{11})-mode propagation. In those models, the directional sensitivity of SC-I was attributed to leakage from the inner segments, and the wavelength dependence of SC-I and the color change (SC-II) were explained by self-screening (i.e., the dependence of the effective absorption spectrum upon the absorption length) of the guided light. Unguided light within the retina was modeled as a bulk wave and geometric optics was employed to calculate the absorption of that wave for both models. In the geometric optics model, a function having a Gaussian variation with incidence angle was chosen to represent light collection by the inner segments. For the purpose of calculating HE_{11} mode excitation in the second model, it was assumed that receptor inner segments were touching in a hexagonal lattice and the truncation aperture area was chosen equal to that of a unit cell in the mosaic; i.e., 1.1 times the myoid's cross-sectional area. Absorption by the lens and macular pigment was accounted for. These authors obtained values for many of the parameters of their models by performing an approximate fit of their calculated results to STILES's data [6.84]. The fit for the HE_{11} model requires equal refractive indices for the inner and outer segments; this is a good approximation for the outer segment and the ellipsoid of the inner segment, but, if extended to the myoid, yields refractive index values for that section that are higher than the measured values. Visual inspection of their curves suggests a better fit to $\zeta(\lambda)$ for the geometric optics model than for the single-mode model. Results for the geometric optics model provide a reasonable fit to SC-II over segments of the spectrum; similar results are reported for the HE_{11} model. WIJNGAARD and VAN KRUYSBERGEN [6.137] discussed these models further and compared the results with those obtained from a modification of SNYDER and PASK's model [6.29], in order to account for absorption of the unguided light. In addition, WIJNGAARD and VAN KRUYSBERGEN measured the SC effects of a subject whose peak sensitivity resides at a markedly eccentric location in the entrance pupil. They were, therefore, able to perform measurements using relatively large incidence angles at which the unguided light can be expected to predominate over the guided light. That angular region is characterized by a fairly flat, low limiting value of directional sensitivity.

SNYDER and PASK also studied the sensitivity of the results that they calculated for $\zeta(\lambda)$ to variations in the physical parameters of the photoreceptors [6.29]. As they pointed out, such investigations are valuable because they can demonstrate the types of visual anomalies that can arise from small changes in photoreceptor dimensions and refractive index. Those sensitivities, together with the sensitivity to truncation aperture diameter, shown in Fig.6.21, underscore the need for more definitive characterization of the physical parameters; this will then provide a firmer basis for theoretical calculations.

6.4.3 Spectral Sensitivity

Light that is absorbed by the photolabile pigment contained within the photoreceptor outer segment will have passed through a wavelength-selective filter — the physical structure of the photoreceptor itself. That filtering has been observed by ENOCH in bleached retinal preparations [6.10-12]. The wavelength selectivity of the structure can be appreciated by noting the wavelength dependence of mode excitation, as displayed in Figs.6.14 and 19, as well as the wavelength dependence of the fraction of excited mode power that resides within the guide, Fig.6.10. Thus, both $P_{\ell m}(0)$ and $\eta_{\ell m}$ in (6.61) are wavelength dependent, and $P_B^{abs}(L)$ will have a wavelength dependence that is not governed solely by that of the absorption coefficient α. Thus, the spectral sensitivity of an individual receptor may be biased by that filtering.

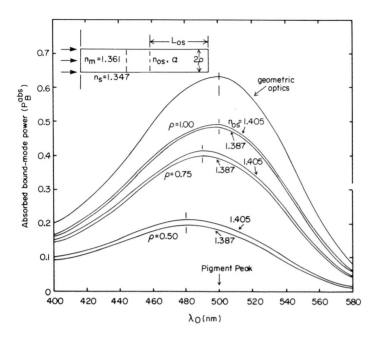

Fig.6.22. Absorption by outer segment of three-stage receptor model for equal inner- and outer-segment radii, for two different outer-segment indices of refraction and three different radii. The pigment model is bovine rhodopsin, $\alpha_0 L_{os}$ is chosen as unity, and absorption by the photoproduct is neglected. Illumination is on-axis, uniform, of unit power, and confined to the fiber face. The short vertical bars mark the receptors' absorption peaks

This dependence can be demonstrated as follows. Let the pigment absorption coefficient α be expressed as

$$\alpha = \alpha_0 \alpha_R(\lambda_0) \tag{6.66}$$

where $\alpha_R(\lambda_0)$ will account for the wavelength dependence of α and is normalized to unity at the absorption peak; α_0 is then independent of wavelength and accounts for pigment concentration. For illustrative purposes we choose bovine rhodopsin [Ref.6.154, Fig.3], with absorption peak at $\lambda_0 = 500$ nm, as the model pigment. Maximum absorption, α_0, has been reported to be a few percent per wavelength distance [6.46]. In the calculational model, a value of maximum absorption times outer-segment length $\alpha_0 L_{os} = 1$ is used.

Absorption in the outer segment is calculated from (6.61) using the approximations outlined in the previous sections. Illumination is on-axis, with unit total power confined to the receptor face.

The results are shown in Fig.6.22 for receptors having equal inner- and outer-segment radii of 0.5, 0.75, and 1.00 μm. The myoid refractive index is chosen as n_m = 1.361; two outer-segment refractive indices, n_{os} = 1.387 and 1.405 are employed; and the refractive index of the surround is chosen as n_s = 1.347. The geometric optics result (6.58), employing Beer's Law, is shown for comparison.

The curves for the different outer-segment refractive indices show little difference, owing to the dominance of the inner segment in the light collection process, as discussed in Sect.6.4.1. Curves for photoreceptors of different radii differ in three respects. 1) Absorption is greater for the larger diameter receptors which, having higher V values, are more efficient at collecting light (see Fig.6.19). 2) The absorption peak occurs at decreasing wavelengths for decreasing receptor radius. The peak for ρ = 1.00 μm is virtually indistinguishable from the geometric optics absorption peak, which coincides with the pigment peak. Peaks for ρ = 0.75 and 0.50 μm are displaced from the pigment peak by 10 and 20 nm, respectively, toward shorter wavelengths. 3) The absorption curves broaden for decreasing radius. While the half-absorption width is 120 nm for the geometric optics result, and virtually indistinguishable from that for ρ = 1.00 and 0.75, the halfwidth for ρ = 0.5 is 130 nm. Similar results are shown by SNYDER and MILLER [6.155] and SNYDER and PASK [6.156] for the invertebrate dipteran retinula cells. In addition, TANNENBAUM has demonstrated spectral shaping for the modes of a model receptor having an assumed circular parabolic refractive index profile [6.40]. The foregoing results would have bearing upon responses and measurements of individual photoreceptors.

6.5 Concluding Remarks

Our excursion through the waveguiding properties of retinal photoreceptors has exhibited many of the optical effects associated with a single photoreceptor. Whereas the modes themselves play no known role in vision, they do provide a means for accurately assessing the optical properties of the photoreceptors. The waveguide approach can therefore aid both in the interpretation and planning of experiments wherein calculation of photoreceptor optical properties are involved. These properties influence the transduction of energy within the receptor. Because this is an important step in the visual process, it behooves us to understand the underlying optical processes.

References

6.1 A.W. Snyder, R. Menzel (eds.): *Photoreceptor Optics* (Springer, Berlin, Heidelberg, New York 1975)

6.2 J.M. Enoch, B.R. Horowitz: "The Vertebrate Retinal Receptor as a Waveguide", in *Optical and Acoustical Micro-Electronics*, Proc. Symposium, New York, April 16-18, 1974, ed. by J. Fox (Polytechnic, New York 1975) pp.133-159

6.3 S.L. Polyak: *The Retina* (Univ. Chicago Press, Chicago 1941)

6.4 S.L. Polyak: *The Vertebrate Visual System* (Univ. Chicago Press, Chicago 1957)

6.5 G.L. Walls: *The Vertebrate Eye* (Cranbrook Inst. Sci., Bloomfield Hills, MI 1942)

6.6 A. Laties, P.A. Liebman, C.E.M. Campbell: Photoreceptor orientation in the primate eye. Nature *218*, 172-173 (1968)

6.7 A. Laties: Histochemical techniques for the study of photoreceptor orientation. Tissue Cell *1*, 63-81 (1969)

6.8 A. Laties, J.M. Enoch: An analysis of retinal receptor orientation. I. Angular relationship of neighboring photoreceptors. Invest. Ophthalmol. *10*, 69-77 (1971)

6.9 J.M. Enoch: Waveguide modes in retinal receptors. Science *133*, 1353-1354 (1961)

6.10 J.M. Enoch: Nature of the transmission of energy in the retinal receptors. J. Opt. Soc. Am. *51*, 1122-1126 (1961)

6.11 J.M. Enoch: Optical properties of the retinal receptors. J. Opt. Soc. Am. *53*, 71-85 (1963)

6.12 J.M. Enoch: Some waveguide characteristics of retinal receptors. G. Fis. *IV*, pp. cover, 242, opposite 248 (1963)

6.13 J.E. Dowling: Foveal receptors of the monkey retina: fine structure. Science *147*, 57-59 (1965)

6.14 A.I. Cohen: "Rods and Cones", in *Physiology of Photoreceptor Organs*, ed. by M.G.F. Fuortes, Handbook of Sensory Physiology, Vol.7/2 (Springer, Berlin, Heidelberg, New York 1972) Chap.2, pp.63-110

6.15 R.L. Sidman: The structure and concentration of solids in photoreceptor cells studied by refractometry and interference microscopy. J. Biophys. Biochem. Cytol. *3*, 15-30 (1957)

6.16 R.L. Sidman: Histochemical studies on photoreceptor cells. Ann. N.Y. Acad. Sci. *74*, 182-195 (1958)

6.17 R. Barer: Refractometry and interferometry of living cells. J. Opt. Soc. Am. *47*, 545-556 (1957)

6.18 B.S. Fine, L.E. Zimmerman: Observations on the rod and cone layer of the human retina. Invest. Ophthalmol. *2*, 446-459 (1963)

6.19 P. Röhlich: The interphotoreceptor matrix: electron microscopic and histochemical observations on the vertebrate retina. Exp. Eye Res. *10*, 80-96 (1970)

6.20 J.M. Enoch, F.L. Tobey, Jr.: Use of the waveguide parameter V to determine the difference in the index of refraction between the rat rod outer segment and the interstitial matrix. J. Opt. Soc. Am. *68*, 1130-1134 (1978)

6.21 J.M. Enoch: "The Retina as a Fiber Optics Bundle", in *Fiber Optics*, by N.S. Kapany (Academic, New York 1967) App. B, pp.372-396

6.22 J.M. Enoch: "Comments on Excitation of waveguide modes in retinal receptors". J. Opt. Soc. Am. *57*, 548-549 (1967)

6.23 J.M. Enoch: Retinal receptor orientation and the role of fiber optics in vision. Am. J. Optom. Arch. Am. Acad. Optom. *49*, 455-471 (1972)

6.24 J.M. Enoch, L.E. Glismann: Physical and optical changes in excised retinal tissue. Invest. Ophthalmol. *5*, 208-221 (1966)

6.25 H. Ohzu, J.M. Enoch, J. O'Hair: Optical modulation by the isolated retina and retinal receptors. Vision Res. *12*, 231-224 (1972)

6.26 H. Ohzu, J.M. Enoch: Optical modulation by the isolated human fovea. Vision Res. *12*, 245-251 (1972)

6.27 J.M. Enoch, F.L. Tobey, Jr.: A special microscope-microspectrophotometer: optical design and application to the determination of waveguide properties of frog rods. J. Opt. Soc. Am. *63*, 1345-1356 (1973)

6.28 F.L. Tobey, Jr., J.M. Enoch, J.H. Scandrett: Experimentally determined optical properties of goldfish cones and rods. Invest. Opthalmol. *14*, 7-23 (1975)

6.29 A.W. Snyder, C. Pask: The Stiles-Crawford effect — explanation and consequences. Vision Res. *13*, 1115-1137 (1973)

6.30 A.W. Snyder: "Photoreceptor Optics — Theoretical Principles", in *Photoreceptor Optics*, ed. by A.W. Snyder, R. Menzel (Springer, Berlin, Heidelberg, New York 1975) pp.38-55

6.31 W. Wijngaard: Guided normal modes of two parallel circular dielectric rods. J. Opt. Soc. Am. *63*, 944-950 (1973)

6.32 W. Wijngaard: Some normal modes of an infinite hexagonal array of identical circular dielectric rods. J. Opt. Soc. Am. *64*, 1136-1144 (1974)

6.33 A.W. Snyder, R.L. Kyhl: Surface mode propagation along an array of dielectric rods with all elements excited identically. IEEE Trans. AP-*14*, 510-511 (1966)

6.34 W.H. Miller, A.W. Synder: Optical function of human peripheral cones. Vision Res. *13*, 2185-2194 (1973)

6.35 W.J. Schmidt: Doppelbrechung, Dichroismus und Feinbau des Aussengliedes der Sehzellen von Frosch. Z. Zellforsch. *22*, 485-522 (1935)

6.36 J.J. Wolken: *Vision: Biophysics and Biochemistry of the Vertebrate Eye* (Thomas, Springfield, IL 1966)

6.37 F. Crescitelli: "The Visual Cells and Visual Pigments of the Vertebrate Eye", in *Photochemistry of Vision*, ed. by H.J.A. Dartnall, Handbook of Sensory Physiology, Vol.7/1 (Springer, Berlin, Heidelberg, New York 1972) Chap.8, pp.245-263

6.38 P.A. Liebman: "Birefringence, Dichroism and Rod Outer Segment Structure", in *Photoreceptor Optics*, ed. by A.W. Synder, R. Menzel (Springer, Berlin, Heidelberg, New York 1975) pp.199-214

6.39 B.S. Fine, M. Yanoff: *Ocular Histology* (Harper and Row, New York 1972)

6.40 P.M. Tannenbaum: Spectral shaping and waveguide modes in retinal cones. Vision Res. *15*, 591-593 (1975)

6.41 M. Born, E. Wolf: *Principles of Optics*, 3rd ed. (Pergamon, Oxford 1964)

6.42 R.A. Weale: Optical properties of photoreceptors. Br. med. Bull. *26*, 134-137 (1970)

6.43 R.A. Weale: On the birefringence of rods and cones. Pfluegers Arch. Gesamte Physiol. Menschen Tiere *329*, 244-257 (1971)

6.44 R.A. Weale: Rod birefringence and light. Vision Res. *11*, 1387-1393 (1971)

6.45 E.J. Denton: A method of easily observing the dichroism of the visual rods. J. Physiol. *124*, 16-17 (1954)

6.46 P.A. Liebman: "Microspectrophotometry of Photoreceptors", in *Photochemistry of Vision*, ed. by H.J.A. Dartnall, Handbook of Sensory Physiology, Vol.7/1 (Springer, Berlin, Heidelberg, New York 1972) Chap.12, pp.481-528

6.47 R.A. Weale: On the linear dichroism of frog rods. Vision Res. *11*, 1373-1385 (1971)

6.48 T.I. Shaw: "The Circular Dichroism and Optical Rotary Dispersion of Visual Pigments", in *Photochemistry of Vision*, ed. by H.J.A. Dartnall, Handbook of Sensory Physiology, Vol.7/1 (Springer, Berlin, Heidelberg, New York 1972) Chap.6, pp.180-199

6.49   F.I. Hárosi: Absorption spectra and linear dichroism of some amphibian photoreceptors. J. Gen. Physiol. *66*, 357-382 (1975)

6.50   F.I. Hárosi: "Linear Dichroism of Rods and Cones", in *Vision in Fishes*, ed. by M.A. Ali (Plenum, New York 1975)

6.51   F.I. Hárosi, E.F. MacNichol, Jr.: Dichroic microspectrophotometer: a computer-assisted, rapid, wavelength-scanning photometer for measuring linear dichroism in single cells. J. Opt. Soc. Am. *64*, 903-918 (1974)

6.52   F.I. Hárosi, E.F. MacNichol, Jr.: Visual pigments of goldfish cones: spectral properties and dichroism. J. Gen. Physiol. *63*, 279-304 (1974)

6.53   W.A.H. Rushton: "Visual Pigments in Man", in *Photochemistry of Vision*, ed. by H.J.A. Dartnall, Handbook of Sensory Physiology, Vol.7/1 (Springer, Berlin, Heidelberg, New York 1972) Chap.9, pp.364-394

6.54   R.A. Weale: "Human Vision", in *An Introduction to Photobiology*, ed. by C.P. Swanson (Prentice-Hall, Englewood Cliffs, NJ 1969) pp.23-51

6.55   W.B. Marks, W.H. Dobelle, E.F. MacNichol, Jr.: Visual pigments of single primate cones. Science *143*, 1181-1183 (1964)

6.56   P.K. Brown, G. Wald: Visual pigments in human and monkey retinas. Nature *200*, 37-43 (1963)

6.57   P.K. Brown, G. Wald: Visual pigments in single rods and cones of the human retina. Science *144*, 45-52 (1964)

6.58   W.B. Marks: Visual pigments of single goldfish cones. J. Physiol. *178*, 14-32 (1965)

6.59   P.A. Liebman, G. Entine: Visual pigments of frog and tadpole. Vision Res. *8*, 761-775 (1968)

6.60   C.R. Hackenbrock: Ultrastructural bases for metabolically linked mechanical activity in mitochondria: I. reversible ultrastructural changes with change in metabolic steady state in isolated liver mitochondria. J. Cell Biol. *30*, 269-297 (1966)

6.61   L.B. Arey: The movement in the visual cells and the retinal pigment of the lower vertebrates. J. Comp. Neurol. *26*, 121-201 (1916)

6.62   M.A. Ali: Les responses retinomotrices: caractères et mécanismes. Vision Res. *11*, 1225-1288 (1971)

6.63   W.H. Miller, A.W. Snyder: Optical function of myoids. Vision Res. *12*, 1841-1848 (1972)

6.64   A.W. Snyder, P. Richmond: Anomalous dispersion in visual photoreceptors. Vision Res. *13*, 511-515 (1972)

6.65   A.W. Snyder, P. Richmond: Effect of anomalous dispersion on visual photoreceptors. J. Opt. Soc. Am. *62*, 1278-1283 (1972)

6.66   D.G. Stavenga, H.H. Van Barneveld: On dispersion in visual photoreceptors. Vision Res. *15*, 1091-1095 (1975)

6.67   D. Miller, G. Benedek: *Intraocular Light Scattering* (Thomas, Springfield, IL 1973) This book contains an extensive list of references

6.68   R.H. Steinberg, I. Wood: Pigment epithelial cell ensheathment of cone outer segments in the retina of the domestic cat. Proc. R. Soc. London B*187*, 461-478 (1974)

6.69   N. Bülow: Light scattering by pigment epithelium granules in the human retina. Acta Ophthalmol. *46*, 1048-1053 (1968)

6.70   A.W. Snyder: Coupling of modes on a tapered dielectric cylinder. IEEE Trans. MTT-*18*, 383-392 (1970)

6.71   F.L. Tobey, Jr., J.M. Enoch: Directionality and waveguide properties of optically isolated rat rods. Invest. Ophthalmol. *12*, 873-880 (1973)

6.72   W. Fischer, R. Röhler: The absorption of light in an idealized photoreceptor on the basis of waveguide theory - II: the semiinfinite cylinder. Vision Res. *14*, 1115-1125 (1974)

6.73   G. Österberg: Topography of the layer of rods and cones in the human retina. Acta. Ophthalmol. Suppl. *6*, 1-102 (1935)

6.74 G. Falk, P. Fatt: Distinctive properties of the lamellar and disk-edge structures of the rod outer segment. J. Ultrastruct. Res. *28*, 41-60 (1969)

6.75 V.L. Borovjagin, T.A. Ivanina, D.A. Moshkov: The ultrastructural organization of the photoreceptor membranes and the intradisc spaces of the vertebrate retina as revealed by various experimental treatments. Vision Res. *13*, 745-752 (1973)

6.76 A.W. Snyder, M. Hamer: The light-capture area of a photoreceptor. Vision Res. *12*, 1749-1753 (1972)

6.77 A.W. Synder, C. Pask, D.J. Mitchell: Light-acceptance property of an optical fiber. J. Opt. Soc. Am. *63*, 59-64 (1973)

6.78 W. Wijngaard, M.A. Bouman, F. Budding: The Stiles-Crawford colour change. Vision Res. 951-957 (1974)

6.79 A.W. Snyder: Continuous mode spectrum of a circular dielectric rod. IEEE Trans. MTT-*19*, 720-727 (1971)

6.80 A.W. Snyder: Leaky-ray theory of optical waveguides of circular cross section. Appl. Phys. *4*, 273-298 (1974)

6.81 A.W. Snyder, C. Pask: Optical fibre: spatial transient and steady state. Opt. Commun. *15*, 314-316 (1975)

6.82 E. Brücke: cited in H. von Helmholtz: *Physiological Optics*, Opt. Soc. Am., Washington, D.C. 1924, p.229

6.83 W.S. Stiles, B.H. Crawford: The luminous efficiency of rays entering the eye pupil at different points. Proc. R. Soc. London B*112*, 428-450 (1933)

6.84 W.S. Stiles: The luminous efficiency of monochromatic rays entering the eye pupil at different points and a new colour effect. Proc. R. Soc. London B*123*, 90-118 (1937)

6.85 W.S. Stiles: The directional sensitivity of the retina. Ann. R. Coll. Surg. Engl. *30*, 73-101 (1962)

6.86 B.H. Crawford: "The Stiles-Crawford Effects and Their Significance in Vision", in *Visual Psychophysics*, ed. by D. Jameson, L.M. Hurvich, Handbook of Sensory Physiology, Vol.7/4 (Springer, Berlin, Heidelberg, New York 1972) Chap.18, pp.470-483

6.87 W.D. Wright, J.H. Nelson: The relation between the apparent intensity of a beam of light and the angle at which the beam strikes the retina. Proc. Phys. Soc. London *48*, 401-405 (1936)

6.88 B. O'Brien: A theory of the Stiles-Crawford effect. J. Opt. Soc. Am. *36*, 506-509 (1946)

6.89 B. O'Brien: Vision and resolution in the central retina. J. Opt. Soc. Am. *41*, 882-894 (1951)

6.90 R. Winston, J.M. Enoch: Retinal cone receptor as an ideal light collector. J. Opt. Soc. Am. *61*, 1120-1121 (1971)

6.91 N.S. Kapany: *Fiber Optics* (Academic, New York 1967)

6.92 R.J. Potter, E. Donath, R. Tynan: Light-collecting properties of a perfect circular optical fiber. J. Opt. Soc. Am. *53*, 256-260 (1963)

6.93 A.W. Snyder: Asymptotic expressions for eigenfunctions and eigenvalues of a dielectric or optical waveguide. IEEE Trans. MTT-*17*, 1130-1138 (1969)

6.94 D. Marcuse: Cutoff condition in optical fibers. J. Opt. Soc. Am. *63*, 1369-1371 (1973)

6.95 E.A.J. Marcatili: Extrapolation of Snell's law to optical fibers. J. Opt. Soc. Am. *63*, 1372-1373 (1973)

6.96 A.W. Snyder, J.D. Love: Reflection at a curved dielectric interface - electromagnetic tunnelling. IEEE Trans. MTT-*23*, 134-141 (1975)

6.97 G. Toraldo Di Francia: Per una teoria dell'effetto Stiles-Crawford (Letter). Nuovo cimento *5*, 589-590 (1948)

298

6.98 G. Toraldo Di Francia: Retinal cones as dielectric antennas. J. Opt. Soc. Am. *39*, 324 (1949)
6.99 L.B. Felsen: Rays and modes in optical fibres. Electron. Lett. *10*, 95-96 (1974)
6.100 N.S. Kapany, J.J. Burke: *Optical Waveguides* (Academic, New York 1972)
6.101 D. Marcuse: *Light Transmission Optics* (Van Nostrand Reinhold, New York 1972)
6.102 D. Marcuse: *Theory of Dielectric Optical Waveguides* (Academic, New York 1974)
6.103 E. Snitzer: Cylindrical dielectric waveguide modes. J. Opt. Soc. Am. *51*, 491-498 (1961)
6.104 G. Biernson, D.G. Kinsley: Generalized plots of mode patterns in a cylindrical dielectric waveguide applied to retinal cones. IEEE Trans. MTT-*13*, 345-356 (1965)
6.105 J.A. Stratton: *Electromagnetic Theory* (McGraw-Hill, New York 1941)
6.106 R.E. Collin: *Field Theory of Guided Waves* (McGraw-Hill, New York 1960)
6.107 M. Javid, P.M. Brown: *Field Analysis and Electromagnetics* (McGraw-Hill, New York 1963) p.330
6.108 F.W.J. Olver: "Bessel Functions of Integer Order", in *Handbook of Mathematical Functions*, ed. by M. Abramowitz, I.A. Stegun (National Bureau of Standards, Washington, D.C. 1965) Chap.9, pp.355-433
6.109 D.G. Stavenga: Refractive index of fly rhabdomeres. J. Comp. Physiol. *91*, 417-426 (1974)
6.110 D.G. Stavenga: Waveguide modes and refractive index in photoreceptors of invertebrates. Vision Res. *15*, 323-330 (1975)
6.111 D. Marcuse: Radiation losses of the dominant mode in round dielectric waveguides. Bell Syst. Tech. J. *49*, 1665-1693 (1970)
6.112 R. Sammut, C. Pask, A.W. Snyder: Excitation and power of the unbound modes within a circular dielectric waveguide. Proc. IEE *122*, 25-33 (1975)
6.113 J.M. Enoch, J. Scandrett, F.L. Tobey, Jr.: A study of the effects of bleaching on the width and index of refraction of frog rod outer segments. Vision Res. *13*, 171-183 (1973)
6.114 E. Snitzer: "Optical Dielectric Waveguides", in *Advances in Quantum Electronics*, ed. by J.R. Singer (Columbia Univ. Press, New York 1961) pp.348-369
6.115 D. Gloge: Weakly guiding fibers. Appl. Opt. *10*, 2252-2258 (1971)
6.116 R. Sammut, A.W. Snyder: Leaky modes on circular optical waveguides. Appl. Opt. *15*, 477-482 (1976)
6.117 J.M. Enoch: Visualization of waveguide modes in retinal receptors. Am. J. Ophthalmol. *51*, 1107-1118 (1961)
6.118 E. Snitzer, H. Osterberg: Observed dielectric waveguide modes in the visible spectrum. J. Opt. Soc. Am. *51*, 499-505 (1961)
6.119 W. Wijngaard: Depolarization of plane-polarized light by light-guiding frog rods. J. Opt. Soc. Am. *61*, 1187-1189 (1971)
6.120 W. Wijngaard: Mode interference patterns in retinal receptor outer segments. Vision Res. *14*, 889-893 (1974)
6.121 C. Pask, A.W. Snyder: Power of modes propagating inside a dielectric rod. J. Opt. Soc. Am. *64*, 393-395 (1974)
6.122 C. Pask, A.W. Snyder: "Theory of the Stiles-Crawford Effect of the Second Kind", in *Photoreceptor Optics*, ed. by A.W. Snyder, R. Menzel (Springer, Berlin, Heidelberg, New York 1975) pp.145-148
6.123 R. Sammut: Effect of leaky modes on the response of short optical fibers. J. Opt. Soc. Am. *67*, 1284-1285 (1977)
6.124 A.W. Snyder: Coupled mode theory for optical fibers. J. Opt. Soc. Am. *62*, 1267-1277 (1972)

6.125 R. Sammut, A.W. Snyder: Contribution of unbound modes to light absorption in visual photoreceptors. J. Opt. Soc. Am. *64*, 1711-1714 (1974)

6.126 A.W. Snyder: Power loss on optical fibers. Proc. IEEE *60*, 757-758 (1972)

6.127 A.W. Snyder, C. Pask: Waveguide modes and light absorption in photoreceptors. Vision Res. *13*, 2605-2608 (1973)

6.128 R. Röhler, W. Fischer: Influence of waveguide modes on the light absorption in photoreceptors. Vision Res. *11*, 97-101 (1971)

6.129 A.W. Snyder, C. Pask: Absorption in conical optical fibers. J. Opt. Soc. Am. *63*, 761-762 (1973)

6.130 A.W. Snyder: Surface waveguide modes along a semi-infinite dielectric fiber excited by a plane wave. J. Opt. Soc. Am. *56*, 601-606 (1966)

6.131 A.W. Snyder: Excitation and scattering of modes on a dielectric or optical fiber. IEEE Trans. MTT-*17*, 1138-1144 (1969)

6.132 D. Marcuse: Excitation of the dominant mode of a round fiber by a Gaussian beam. Bell Syst. Tech. J. *49*, 1695-1703 (1970)

6.133 M. Imai, E.H. Hara: Excitation of fundamental and low-order modes of optical fiber waveguides by Gaussian beams. 1: tilted beams. Appl. Opt. *13*, 1893-1899 (1974)

6.134 M. Imai, E.H. Hara: Excitation of fundamental and low-order modes of optical fiber waveguides with Gaussian beams. 2: offset beams. Appl. Opt. *14*, 169-173 (1975)

6.135 P.M. Morse, H. Feshbach: *Methods of Theoretical Physics*, Pt. II (McGraw-Hill, New York 1953) p.1073

6.136 A. Cardama, E.T. Kornhauser: Modal analysis of coupling problems in optical fibers. IEEE Trans. MTT-*23*, 162-169 (1975)

6.137 W. Wijngaard, J. Van Kruysbergen: "The Function of the Nonguided Light in Some Explanations of the Stiles-Crawford Effect", in *Photoreceptor Optics*, ed. by A.W. Snyder, R. Menzel (Springer, Berlin, Heidelberg, New York 1975) pp.175-183

6.138 A.W. Snyder, D.J. Mitchell: Leaky mode analysis of circular optical waveguides. Opto-electronics *6*, 287-296 (1974)

6.139 R. Sammut, A.W. Snyder: Leaky modes on a dielectric waveguide: orthogonality and excitation. Appl. Opt. *15*, 1040-1044 (1976)

6.140 A.W. Snyder, J.D. Love: Tunnelling leaky modes on optical waveguides. Opt. Commun. *12*, 326-328 (1974)

6.141 A.W. Snyder, D.J. Mitchell: Leaky rays on circular optical fibers. J. Opt. Soc. Am. *64*, 599-607 (1974)

6.142 T. Tamir, A.A. Oliner: Guided complex waves, part 1. fields at an interface. Proc. IEE *110*, 310-324 (1963)

6.143 C. Pask, A.W. Snyder: Light acceptance property of an optical fiber. Appl. Opt. *13*, 1889-1892 (1974)

6.144 K. Tansley, B.K. Johnson: The cones of the grass snake's eye. Nature *178*, 1285-1286 (1956)

6.145 C. Pask, A.W. Snyder: "Angular Sensitivity of Lens-Photoreceptor Systems", in *Photoreceptor Optics*, ed. by A.W. Snyder, R. Menzel (Springer, Berlin, Heidelberg, New York 1975) pp.159-166

6.146 A. Safir, L. Hyams: Distribution of cone orientations as an explanation of the Stiles-Crawford effect. J. Opt. Soc. Am. *59*, 757-765 (1969)

6.147 H.E. Bedell, J.M. Enoch: A study of the Stiles-Crawford (S-C) function at 35° in the temporal field and the stability of the foveal S-C function peak over time. J. Opt. Soc. Am. *69*, 435-442 (1979)

6.148 R.A. Weale: Notes on the photometric significance of the human crystalline lens. Vision Res. *1*, 183-191 (1961)

6.149 J. Mellerio: Light absorption and scatter in the human lens. Vision Res. *11*, 129-141 (1971)

6.150 J.J. Vos, F.L. Van Os: The effect of lens density on the Stiles-Crawford effect. Vision Res. *15*, 749-751 (1975)

6.151 J.M. Enoch: "Vertebrate Rod Receptors are Directionally Sensitive", in *Photoreceptor Optics*, ed. by A.W. Snyder, R. Menzel (Springer, Berlin, Heidelberg, New York 1975) pp.17-37

6.152 G. Westheimer: Dependence of the magnitude of the Stiles-Crawford effect on retinal location. J. Physiol. London *192*, 309-315 (1967)

6.153 J.M. Enoch, G.M. Hope: Directional sensitivity of the foveal and parafoveal retina. Invest. Ophthalmol. *12*, 497-503 (1973)

6.154 H.J.A. Dartnall: "Photosensitivity", in *Photochemistry of Vision*, ed. by H.J.A. Dartnall, Handbook of Sensory Physiology, Vol.7/1 (Springer, Berlin, Heidelberg, New York 1972) Chap.4, pp.122-145

6.155 A.W. Snyder, W.H. Miller: Fly colour vision. Vision Res. *12*, 1389-1396 (1972)

6.156 A.W. Snyder, C. Pask: Spectral sensitivity of dipteran retinula cells. J. Comp. Physiol. *84*, 59-76 (1973)

7. Theoretical Consideration of Optical Interactions in an Array of Retinal Receptors

W. Wijngaard

With 14 Figures

The absorption of light in a retinal receptor depends on the optical properties of its surroundings and therefore also on the optical properties of neighboring receptors. For example, light may travel through one receptor before being absorbed by another. In this chapter optical interaction is discussed. By optical interaction between retinal receptors is understood the influence on the absorption in one receptor exerted by the presence of the other receptors.

7.1 Overview

Initially, it is of interest to discuss the problem by use of geometrical optics. In Fig.7.1 two receptors, a rod and a cone, are depicted schematically. When the rod is not present some rays (ray B, for example) are guided through the outer segment of the cone; others like A and C are not guided by the cone. The photopigment inside the cone outer segment will absorb light from the guided and possibly from the unguided rays. When the rod is placed near the cone in Fig.7.1 we see that ray B remains a guided ray; ray C, however, is intercepted by the rod; its course through the rod is depicted by the dashed lines.

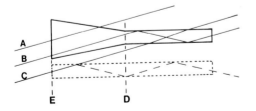

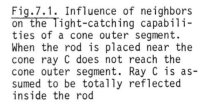

Fig.7.1. Influence of neighbors on the light-catching capabilities of a cone outer segment. When the rod is placed near the cone ray C does not reach the cone outer segment. Ray C is assumed to be totally reflected inside the rod

Many other conditions may be found where geometrical optics predict a difference in optical properties of a receptor with and without neighbors. However, interaction effects become more interesting when wave optical effects are included. The use of wave optics suggests the possibility of wavelength-dependent effects which may be of interest for theories of colour vision. In the following discussion we shall concentrate on wave optical arguments.

A theoretical wave optical treatment of optical interactions in an array of retinal receptors can only be of a qualitative nature at this moment. Even to estimate the absorption of light by a single receptor is difficult owing to its complex form. A complicating factor is the uncertainty which exists about refractive indices, distances between the receptors, etc. The aim of our theoretical consideration is to gain some insight in the kind of optical influence which a receptor may exert on its neighbors.

To show the difference between geometrical optics and wave optical arguments we shall describe the situation of Fig.7.1 in wave optical terms. The argument will be as follows: a plane wave is incident on the receptors, its wave normal directed along the line depicted by A. When the rod is not present this wave is transformed by the cone inner segment and causes guided waveguide modes (see Appendix A for symbols and definitions) to be excited on the outer segment. Apart from guided modes an unguided (radiation) field is excited.

When the rod is placed near the cone, assuming parallel outer segments, the plane wave is transformed by the part of the receptors to the left of plane D. The part to the right of D is a pair of dielectric cylinders; on this structure guided and unguided modes are possible [7.1]. The wave incident on plane D excites the guided and unguided modes of the pair of dielectric cylinders.

In the following discussion we shall concentrate mainly on two effects. The first effect is a competition between the receptors for the available power at the level of their apertures (plane E in Fig.7.1). This effect is already evident from the geometrical optics discussion of Fig.7.1; it shows that the cone, when placed alone, will absorb more light than when the rod is present. This effect will be called an aperture effect and will be further analyzed in wave optical terms in the next section. The second effect is related to the phenomenon of frustrated total reflection. For two receptors, the light may beat back and forth between the receptors. This effect is also similar to the effect which occurs in a single receptor, in that the light may beat back and forth in the transverse direction between one side of the receptor and the other.

We shall often refer to the modes of the circular dielectric cylinder because these modes are well known and because slightly noncircular or even rectangular guides have similar modes (compare Fig.7.4). In particular the $HE_{1,1}$ mode of the circular dielectric cylinder will often be considered as an example. The reason for this is convenience in the discussion, and the fact that the $HE_{1,1}$ mode is excited with a large amplitude on a dielectric rod when the incident wave is directed along the axis. Besides, the $HE_{1,1}$ mode does not have a cut-off; it exists independent of wavelength.

For retinal receptors the approximations of SNYDER for small differences in refractive index apply [7.2]. Absorption is essential, but for retinal receptors the fraction of light absorbed per wavelength is ≪ 1 and, therefore, a perturbation calculation may be used to calculate the absorption. In this case it is assumed that the absorption does not change the field pattern of the modes except in amplitude.

7.2 The Aperture Effect

When a plane wave is incident on a single receptor the power guided along the receptor may be larger than the power incident on the geometrical area of the receptor inner segment [7.3,4]. This is due to the fact that the mode field extends into the outside medium. When the receptors are situated in an array, the power available from the incident plane wave will be distributed over the receptors. In such a situation there is a form of competition between the receptors for the available power.

As an example let us consider a model retina consisting of circular cylindrical dielectric rods. First, the excitation of modes on the single isolated dielectric rod will be discussed. Afterwards, the effects of placing the rods in an array will be discussed.

The amplitude with which a mode of the single rod will be excited by a plane wave depends on the fit between the incident field and the mode field. The $HE_{1,1}$ mode, for example, does not have phase changes in its E field over a transverse plane. Therefore the $HE_{1,1}$ mode will be excited with a large amplitude by a plane wave incident along the axis [7.4]. When the plane wave is incident obliquely so that two points with distance ρ (ρ is the radius

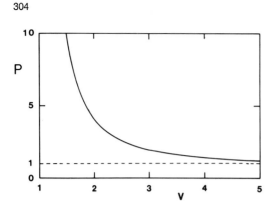

Fig.7.2. Excitation of the $HE_{1,1}$ mode of a semi-infinite dielectric cylinder by an axially incident plane wave. P is the power guided by the $HE_{1,1}$ mode normalized to the power incident on the geometric cross section of the cylinder. The effective aperture of the cylinder can be much larger than its geometric cross section. P is calculated following [7.4] for a small difference in refractive index of the inner and the outer medium

of the rod) are in antiphase, the $HE_{2,1}$ mode will be excited with a large amplitude when this mode exists.

The field of a mode extends outside the rod. Therefore, when a plane wave is incident axially on a single rod, the power in the $HE_{1,1}$ mode, which is excited with a large amplitude in this case, may be larger than the power incident on the geometrical area of the rod. Figure 7.2 gives the power in the $HE_{1,1}$ mode excited by a plane wave incident axially with unity incident power on the area of the rod. In Fig.7.2 the power P is given as a function of the cutoff parameter v. Here $v = (2\pi\rho/\lambda_0)(n_1^2 - n_2^2)^{1/2}$, n_1 and n_2 are the refractive indices of the inside and the outside medium, respectively, ρ is the radius of the rod, λ_0 is the vacuum wavelength of the incident light. Figure 7.2 is valid for the single rod. When the rod is placed in an array of identical rods it is evident that not as much power is available to the rod as in the isolated situation. A difficulty is that, in principle, one should not speak of the $HE_{1,1}$ mode of a rod in the array. The $HE_{1,1}$ mode of the single dielectric rod is not a normal mode of the array of dielectric rods. To circumvent this difficulty we shall at first make no assumptions about the mode fields; the rods are only assumed to be identical. For the single isolated rod we assume that 50% of the power in the $HE_{1,1}$ mode is absorped; in this case, for v = 2.0 the power absorbed from the $HE_{1,1}$ mode is twice as large as the power incident on the geometric area of the rod. The totally absorbed power will be even more due to absorption of unguided light.

When the rod is placed in an array of identical rods every rod will absorb the same power from the incident plane wave. When the rods are touching,
the power available to each rod is only 1.1 times the power incident on the
geometric area of one rod. Therefore, for touching identical rods, the power
absorbed by each rod is less than 1.1 times the power on the geometric area
of one rod. Therefore, an aperture effect exists, even in the case when the
receptors do not quite touch. There may be significant competition between
receptors for the available power, especially when the receptors are near
each other.

7.3 The Modes of a Pair of Dielectric Rods

The normal modes of the isolated dielectric rod do not fulfill the boundary
conditions for a pair of parallel rods. However, some results for the two
rod problem may be obtained from the field of the normal modes of the single
rod. Assume a single circular dielectric rod is guiding the $HE_{1,1}$ mode. The
field of this mode extends into the outside medium. The field components decay approximately exponentially as a function of radial distance from the
rod. In Fig.7.3 the transverse E field is given along a line through the axis
of the rod for $v = 3.0$ and for $v = 1.4$. Another rod approaching the first
from infinity will disturb the field. This kind of approach leads to a coupled
mode approximation [7.5,6].

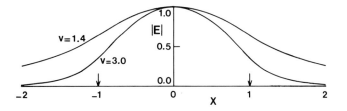

Fig.7.3. The transverse E field for the $HE_{1,1}$ mode along a line through the
center of the circular cross section of the cylinder. The coordinate along
that line is called x. The E field is normalized ot unity at the center. The
arrows indicate the boundary of the rod. The field was calculated following
[7.2] for $\delta \to 0$. Here $\delta = 1-(n_2^2/n_1^2)$ with n_1 and n_2 the refractive indices of
the inner and the outer medium, respectively. The parameter v in this figure
is defined by $v = (2\pi\rho/\lambda_0)(n_1^2-n_2^2)^{1/2}$, where ρ is the radius of the cylinder
and λ_0 the vacuum wavelength of the incident light

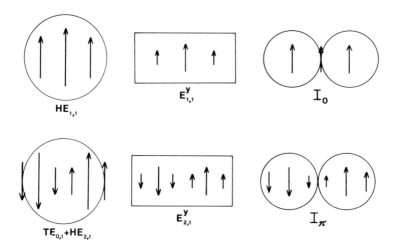

Fig.7.4. Schematic E field of some modes of a single circular dielectric cylinder, a rectangular dielectric cylinder, and a pair of touching identical circular dielectric cylinders

Another way of looking at the problem, which is particularly useful when the rods are near to each other, is to guess the field of the modes by analogy with the modes of similar dielectric structures [7.7]. As a simplifying step, let us assume the rods to be identical and touching. Such a structure exhibits certain similarities to a dielectric rod of rectangular cross-section. The modes of the rectangular dielectric rod have already been investigated by GOELL [7.8].

The lowest order guided modes of this structure can also easily be compared with the modes of the circular dielectric rod. The $HE_{1,1}$ mode of the circular rod may be compared with the $E_{1,1}^y$ mode of the rectangular rod. The combination of modes $TE_{0,1} + HE_{2,1}$ of the circular rod may be compared with the $E_{2,1}^y$ mode of the rectangular rod (Fig.7.4). By way of analogy these modes and the modes of the pair of touching circular rods are depicted in Fig.7.4. The $HE_{1,1}$-like mode is indicated by I_0 and the $TE_{0,1} + HE_{2,1}$-like mode (bilobe) is depicted by I_π.

7.3.1 Derivation of the Mode Fields for a Pair of Dielectric Rods

The propagation constants and the fields of the guided modes of a pair of circular dielectric rods have been derived in several ways [7.1]. Here these modes will be discussed by use of the coupled-mode theory [7.6]. The geometry of the structure is given in Fig.7.5.

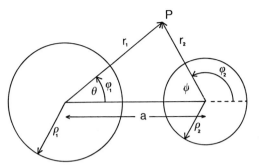

Fig.7.5. Geometry of the two-fibre problem; the z axes are directed toward the reader [7.1]

In coupled-mode theory the normal modes of the pair of dielectric rods are approximated by superpositions of the normal modes of each rod without its neighbor. The fields of modes I_0 and I_π (Fig.7.4) may be described approximately as superpositions of the $HE_{1,1}$ mode of each rod. I_0 is approximately an inphase superposition and I_π an antiphase superposition [7.1]. Stated otherwise, the $HE_{1,1}$ mode of the left rod is coupled to the $HE_{1,1}$ mode of the right rod (Fig.7.5).

To derive the mode fields we shall assume that only one mode of the left rod is coupled to one mode of the right rod. Later some remarks about this assumption will be made. The E field of the mode of the left rod will be given by $a_1(z)\underline{e}_1(r_1,\phi_1)\exp(j\omega t)$; the E field of the mode of the right rod will be given by $a_2(z)\underline{e}_2(r_2,\phi_2)\exp(j\omega t)$. When there is no coupling, $a_1(z) = \bar{a}_1 \exp(-jh_1 z)$ and $a_2(z) = \bar{a}_2 \exp(-jh_1 z)$ with $\bar{a}_1$ and $\bar{a}_2$ constant. With coupling, $|a_1(z)|$ and $|a_2(z)|$ are functions of the axial direction z in general. $a_1(z)$ and $a_2(z)$ obey approximately a set of two differential equations [7.6,9].

$$\frac{da_1}{dz} + jh_1 a_1 = -jc_{12}a_2 \quad , \tag{7.1}$$

$$\frac{da_2}{dz} + jh_2 a_2 = -jc_{21}a_1 \quad , \tag{7.2}$$

where c_{12} and c_{21} are mutual coupling coefficients; h_1 and h_2 are the propagation constants of modes 1 and 2, respectively.

$$c_{12} = \frac{1}{2}\omega(\varepsilon_{1,2} - \varepsilon_2) \iint \underline{e}_1(r_1,\phi_1) \cdot \underline{e}_2(r_2,\phi_2) \, dS \quad . \tag{7.3}$$

right rod

$$c_{21} = \frac{1}{2}\omega(\varepsilon_{1,1} - \varepsilon_2) \iint \underline{e}_1(r_1,\phi_1) \cdot \underline{e}_2(r_2,\phi_2) \, dS \quad . \tag{7.4}$$

left rod

It is assumed that the fields are normalized so as to give

$$\sqrt{\frac{\varepsilon_2}{\mu}} \iint \underline{e}_1(r_1,\phi_1) \cdot \underline{e}_1(r_1,\phi_1) \; dS = 1$$

and a similar expression for the mode of the right rod. The normalization integral should be extended throughout all space. $\varepsilon_{1,1}$ and $\varepsilon_{1,2}$ are the dielectric constants of the left and the right rod, respectively; ε_2 is the dielectric constant of the matrix medium; μ is the permeability, assumed to be constant throughout all space. The expressions for the coupling coefficients are overlap integrals; therefore, the coupling coefficients will increase in absolute value when the distance between the rods decreases, if they are not identically zero. The system of two differential equations is solved by inserting formally

$$a_1(z) = a_{1,0} \, \exp(-j\alpha z) \quad , \tag{7.5}$$

$$a_2(z) = a_{2,0} \, \exp(-j\alpha z) \quad . \tag{7.6}$$

$a_{1,0}$ and $a_{2,0}$ are assumed to be independent of z; this implies that when a solution is found in this way, it is approximately a normal mode of the pair of dielectric rods because $a_1(z)/a_2(z) = a_{1,0}/a_{2,0}$ is constant.

Insertion of (7.5) and (7.6) in (7.1) and (7.2) gives

$$(h_1 - \alpha)a_{1,0} + c_{12}a_{2,0} = 0 \quad ,$$

$$c_{21} \cdot a_{1,0} + (h_2 - \alpha)a_{2,0} = 0 \quad .$$

For the solution to be nontrivial we require that

$$\begin{vmatrix} h_1-\alpha & c_{12} \\ c_{21} & h_2-\alpha \end{vmatrix} = 0 \quad .$$

For each of the two solutions of this equation, to be called α_1 and α_2, there is a normal mode. The E field of a normal mode is given by

$$[a_{1,0}\underline{e}_1(r_1,\phi_1) + a_{2,0}\underline{e}_2(r_2,\phi_2)] \cdot \exp[j(\omega t - \alpha z)] \quad ,$$

where α is either α_1 or α_2.

It should be remembered that only coupling between two modes has been included here; modes I_0 and I_π can be described in this way as superpositions of the $HE_{1,1}$ modes of the left and the right rod. The modes II_0, II_π, III_0, and III_π of two identical rods, for which intensity patterns are given in Fig.7.10, are approximately superpositions of the $TE_{0,1}$ and the $HE_{2,1}$ modes of the left and the right rod. We shall comment on this in the next section.

7.3.2 Beating of Light Between the Rods

For two coupled waveguides it is well known that the power may beat back and forth between the guides [7.10]. This effect is relevant for the analysis of optical effects in double cones. Our aim in this section is to find the relevant parameters. The discussion parallels that given by WIJNGAARD and STAVENGA [7.11], which is concerned with the power transfer between fly rhabdomeres. For fly rhabdomeres total transfer of light from one rhabdomere to its neighbor may occur. However, this kind of analysis generally suffers from the inaccurate knowledge of several parameters.

The phenomenon of beating is described here as an interference effect of normal modes with slightly different phase velocities. For example, when the modes I_0 and I_π (Fig.7.4) occur together they will interfere; when they are in phase most light is in one of the rods, when they are in antiphase most light is in the other rod. Intensity patterns are given in Fig.7.6. The intensity patterns of Fig.7.6 are superpositions of modes I_0 and I_π with equal amplitudes inphase or antiphase, respectively. The choice as to which pattern belongs to the inphase condition is arbitrary. When modes I_0 and I_π are excited with equal amplitudes, the situations depicted in Fig.7.6 alternate along the axial direction, owing to the difference in phase velocity between these modes. Therefore, the light beats between the rods. A beat length λ_B is defined as the distance between two successive maxima inside one rod. The effect corresponds to a similar interference effect for the single rod, between the $HE_{1,1}$ mode and the mode combination $TE_{0,1} + HE_{2,1}$ depicted in Fig.7.7 [Ref.7.7, Fig.1]. For the single rod the light beats back and forth between the left and the right part of the rod.

Using the simplified coupled-mode theory discussed above, it is possible to obtain analytical expressions for the power transfer between the rods. When we require that for $z = 0$ the power in mode 2 (or rod 2) is zero $[a_2(z) = 0]$ and the power in mode 1 (proportional to $|a_1(z)|^2$) is unity, then the two (mode) solutions with propagation constants α_1 and α_2, respectively, should be superposed. For the power P_2 in the second mode belonging to the second rod we obtain [7.9].

$$P_2 = F_{21} \sin^2\left(\frac{\pi}{\lambda_B} z\right) \quad , \tag{7.7}$$

$$F_{21} = \frac{1}{\left|\dfrac{C_{12}}{C_{21}}\right| + \dfrac{(h_1-h_2)^2}{4|C_{21}|^2}} \quad , \tag{7.8}$$

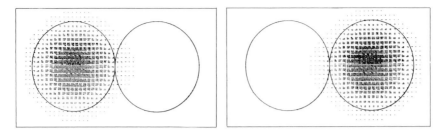

Fig.7.6. Interference pattern of modes I_0 and I_π of two identical touching dielectric rods, superposed with equal amplitudes inphase and antiphase, respectively. The choice as to which pattern belongs to the inphase condition is arbitrary. For each rod v = 3.5 and δ = 0.1 (from [7.7]). These patterns were obtained by calculating the flux density P = Re(1/2 EXH) for the points of a grid inside a rectangle enclosing the rods. The flux density was visualized on an oscilloscope screen by displaying spots around each point of the grid. The number of spots was proportional to the flux density. For the maximum flux density obtained, the number of spots was normalized to 25. The rectangle and the boundaries of the rods were also displayed [7.1]

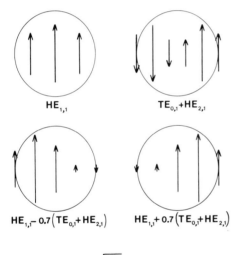

Fig.7.7. Schematic representation of the transverse E field of the $HE_{1,1}$ mode and some mode superpositions for the single circular dielectric rod. Inphase and antiphase superposition of the $HE_{1,1}$ mode and the mode combination $TE_{0,1} + HE_{2,1}$ leads to situations in which the power is largely confined to the right or to the left side of the rod, respectively. v = 3.5 and $\delta \rightarrow 0$ [7.7]

$$\frac{\pi}{\lambda_B} = |c_{21}| \sqrt{\frac{1}{F_{21}}} \quad , \tag{7.9}$$

where λ_B is the beat length, the distance between two successive maxima of mode 2 on rod 2.

In (7.7) it is seen that the maximum power transfer to the nonexcited mode is F_{21}. The maximum value of F_{21} is obtained when the modes have equal phase velocity ($h_1 = h_2$). In this case $c_{12} = c_{21} = c$ and $F_{21} = 1$, whereas $\pi/\lambda_B = |c|$. This means that all power will be transferred from the excited to the nonexcited mode over a length equal to half the beat length. From (7.7,8) it follows that two modes are nearly independent when

$(h_1 - h_2)^2/(4|c_{21}|^2) \gg 1$; in this case the maximum power transfer F_{21} is small compared to the total power.

To further elucidate the coupling between the modes on two parallel circular dielectric cylinders it is necessary to include a discussion of the symmetry of the modes. For example, the $TE_{0,1}$ and the $TM_{0,1}$ mode do not couple at all because $c_{12} = c_{21} = 0$ in this case. To characterize the symmetry of a mode on the pair of circular dielectric cylinders it is enough to know the symmetry of E_z, the E field in the axial direction. WIJNGAARD [7.1] has shown that the pair of dielectric cylinders has two classes of modes: modes with $E_z = 0$ on the line through the centers of the rods to be called E_z-sine modes and the other modes, to be called E_z-cosine modes. In a similar way the single-rod modes of each rod can be classified as modes with $E_z = f(r_i) \sin(n\phi_i)$ to be called E_z-sine modes and modes with $E_z = f(r_i) \cos(n\phi_i)$ to be called E_z-cosine modes (i is 1 or 2 for the left and the right rod of Fig.7.5, respectively). The E_z-sine modes do not couple to the E_z-cosine modes [7.1].

For the $HE_{1,1}$ mode this means that an $HE_{1,1}$ mode polarized along the line though the centers of the rods, which is classified as an E_z-cosine mode, does not couple to the perpendicularly polarized $HE_{1,1}$ mode which is classified as an E_z-sine mode. The $TE_{0,1}$ mode is an E_z-sine mode; this mode does not couple to the $TM_{0,1}$ mode which is an E_z-cosine mode. Figure 7.4 was drawn with this discussion in mind; the E_z-sine modes are depicted there.

Let us consider the special case of the coupling between the $HE_{1,1}$ E_z-sine modes of two identical rods; in the limit for a small difference between the refractive index of the inside and the outside medium the results for the coupling between the $HE_{1,1}$ E_z-cosine modes are identical with the results for the coupling between the $HE_{1,1}$ E_z-sine modes. The beat length will depend on v and δ ($\delta = 1 - n_2^2/n_1^2$) because these parameters determine the form of the field of the mode; evidently, the beat length also depends on the normalized distance a/ρ between the centers of the rods. Therefore, in general, the beat length is a function of three variables. However, for small index differences ($\delta \ll 1$) $\lambda_B \sqrt{\delta}/\rho = \pi\sqrt{\delta}/(|c|\rho)$ is only a function of v and a/ρ.

In Figure 7.8, $\lambda_B \cdot \sqrt{\delta}/\rho$ is given as a function of v with parameter a/ρ. How to obtain the *power transfer* for the case of retinal receptors $L \sqrt{\delta}/\rho$ is indicated in Fig.7.8, where L is the length of the receptor. For example, if $L = 25$ µm, $\rho = 0.5$ µm, $n_1 = 1.39$, $n_2 = 1.34$, and $\lambda_0 = 0.5$ µm then $L \sqrt{\delta}/\rho = 13.3$ and $v = 2.32$; this situation is indicated by point D in Fig.7.8. From this figure the ratio L/λ_B can be obtained for a particular normalized

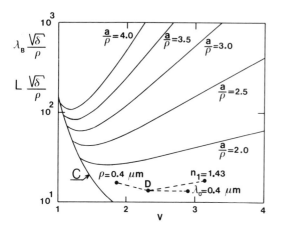

<u>Fig.7.8.</u> Continuous curves. The normalized beatlength $\lambda_B\sqrt{\delta}/\rho$ for the $HE_{1,1}$ modes of two identical circular cylindrical dielectric rods as a function of the cutoff parameter v for $\delta \ll 1$. ρ is the radius of the rods and a is the distance between their axes. Curve C is a cutoff curve below which mode I_π would have a larger phase velocity than the phase velocity of plane waves in the outside medium. Point D represents the normalized length $L \cdot \sqrt{\delta}/\rho$ of a hypothetical receptor pair consisting of two parallel cylinders with circular cross section. Point D was calculated with $L = 25$ µm, $\rho = 0.5$ µm, $n_1 = 1.39$, $n_2 = 1.34$, and $\lambda_0 = 0.5$ µm. The dots at the ends of the dashed lines were obtained in the same way, but by sequentially changing each of the parameters ρ, λ_0 and n_1. The value of the parameter which is changed is indicated near the dot in question

distance a/ρ. The power transfer to the nonexcited mode is given by $P_2 = \sin^2(\pi L/\lambda_B)$ because $F_{21} = 1$ for identical rods. The power transfer is therefore also known. For $a/\rho = 2$ (touching rods) the power at the end of the nonexcited rod is 96.5% of the total power; for $a/\rho = 3$ only 4.6% of the total power is transferred to the other rod.

Further insight can be gained from Fig.7.8 by sequentially changing each of the parameters in this example. The resulting values of $L\sqrt{\delta}/\rho$ are indicated by a dot at the end of each dashed line radiating from point D. The respective values of λ_0, ρ, and n_1 are inserted near the dots.

When $h_1 \neq h_2$, it may initially be assumed that $c_{12} = c_{21} = c$ for small differences in propagation constant between the modes. Then from (7.7,8) it follows that the power transfer is negligible when $|h_1 - h_2|/(2|c|) \gg 1$. This implies that two modes are independent when $|h_1 - h_2|/(2|c|) \gg 1$. Let us consider the case where both rods have the same dielectric constant $(\varepsilon_{1,1} = \varepsilon_{1,2})$ with, for example, $1 - \varepsilon_2/\varepsilon_{1,1} = 0.1$ and the parameter v for the left rod is 3.5. The modes of the right rod (Fig.7.5) will be distinguished

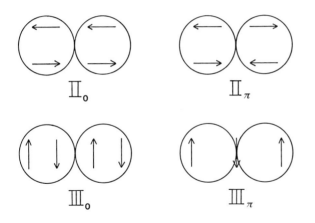

Fig.7.9. Schematic representation of the transverse E field of some E_z-sine modes of two identical touching dielectric cylinders for $\delta \ll 1$ and $v = 3.5$ [7.1]

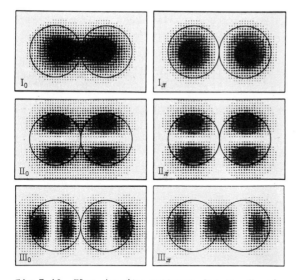

Fig.7.10. Flux density patterns for the E_z-sine modes of two identical touching dielectric cylinders with $v = 3.5$ and $\delta = 0.1$ [7.1]; δ is small enough to make a comparison with Fig.7.9 possible

from the modes of the other rod by a bar, for example $\overline{HE}_{1,1}$. E_z-sine modes do not couple to E_z-cosine modes. For identical rods we have three modes of E_z-sine symmetry in each rod, $HE_{1,1}$; $TE_{0,1}$ and $HE_{2,1}$; and $\overline{HE}_{1,1}$, $\overline{TE}_{0,1}$ and $\overline{HE}_{2,1}$. $HE_{1,1}$ and $\overline{HE}_{1,1}$ couple strongly owing to the fact that $h_1 = h_2$ for these modes. $HE_{1,1}$ does not couple to $\overline{TE}_{0,1}$ or $\overline{HE}_{2,1}$ because for these combinations $|h_1 - h_2|/(2|c|) \gg 1$. However, for $TE_{0,1}$, $HE_{2,1}$, $\overline{TE}_{0,1}$, and $\overline{HE}_{2,1}$ the situation is more difficult. $TE_{0,1}$ may couple to $\overline{HE}_{2,1}$ because $|h_1 - h_2|$ may be less than $2|c|$ when the distance between the rods is small enough. So in this case four modes are generally interdependent. Schematic field patterns and intensity patterns for the two-fiber modes in the situation discussed above are given in Figs.7.9,10, respectively.

For large distances between the rods the coupling coefficient c is small and only a small difference in, for example, the rod radius is necessary to make the coupling negligible.

If we now diminish the radius of the right rod in Fig.7.5, the single-rod modes will become independent when the change in radius is large enough and the distance between the rods is not too small. This is caused by the fact that the modes generally do not have equal phase velocity for nonidentical rods ($h_1 \neq h_2$). However, when ρ_1/ρ_2 is 2.0 we again obtain a *resonance condition*. In this case, $TE_{0,1}$ and $HE_{2,1}$ are approximately in resonance with $\overline{HE}_{1,1}$, whereas $\overline{TE}_{0,1}$ and $\overline{HE}_{2,1}$ no longer exist because for the right rod the cutoff parameter v is 1.75 in this case. The $HE_{1,1}$ mode of the left rod remains relatively undisturbed because it does not couple to other modes. Several resonance conditions for unequal rods have been investigated by McINTYRE and SNYDER [7.9].

For identical rods, resonance is caused by symmetry and is, therefore, independent of exciting wavelength λ_0. For nonidentical rods, however, there is only resonance between two modes for one particular exciting wavelength $\bar{\lambda}_0$. When one of the modes is excited, the transfer of power to the other mode is most significant near that wavelength $\bar{\lambda}_0$.

7.3.3 Possible Interaction Effects for Double Cones

In this section we shall discuss two effects which may be of some physiological interest for *double cones*.

For two guides, with one guide excited in the $HE_{1,1}$ mode, the leak to the other guide is faster for red than for blue light, owing to the fact that the field of the $HE_{1,1}$ mode extends farther into the outside medium for red than for blue light. A possible consequence for a double-cone system may be that one of the cone outer segments will receive relatively too much red light, with the result that the effective absorption spectrum of the receptor will shift to the red relative to the absorption spectrum of the photopigment. There are two conditions which should be fulfilled, namely, the incident light should only excite the $HE_{1,1}$ mode of one outer segment, for example, owing to an asymmetry of the inner segments; furthermore, the length of the outer segments should be less than half the beat length.

The shift to the red mentioned above is the complement of the shift of the effective absorption spectrum of a receptor to the blue discussed by SNYDER and MILLER in relation to the rhabdomeres of flies [7.12]. This shift to the blue is also caused by the fact that the part of the power of a mode travelling outside the rod increases when the wavelength of the light increases. This results in too much blue light in the rod when, for example, white light is incident, and too much red light possibly going to a neighboring rod. A more complete discussion has been given by McINTYRE [7.13].

An even more interesting effect due to optical coupling may occur when the rods are not identical. When two modes travelling along these rods have equal phase velocity for a particular wavelength λ_0, they do not have the same phase velocity for other wavelengths. This may lead to a *resonance effect*; light incident at one of the rods will only go to the other rod when the wavelength is near λ_0. For this effect to occur with double cones the length of the outer segments should not be much less than the beat length.

7.4 Guided and Radiated Light in Infinite Regular Arrays of Dielectric Rods

We consider an *array if dielectric rods* with light incident on one rod (Fig.7.11). When the distance between the rods is not too small, the concept of guided and radiated light remains approximately valid. The incident field will give a spreading radiation field as well as light guided along the excited rod. However, owing to *frustrated total reflection* the guided light leaks to neighboring rods. This leak of guided light may be compared with the beating of light between two dielectric rods discussed above. As an example the characteristics of an infinite *linear array* of identical dielectric rods are discussed [7.5]. Coupled-mode theory is sufficient for the qualitative effects to be considered, and only $HE_{1,1}$ modes (either E_z-sine or E_z-cosine modes) will be treated for simplicity. In Fig.7.12 the geometry of the array is indicated; each rod is numbered. The E field of the $HE_{1,1}$

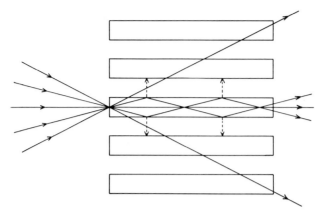

Fig.7.11. Schematic picture of the spreading of light in the retina when the light is focussed at the center of one receptor. The dashed arrows indicate the power transfer due to frustrated total reflection

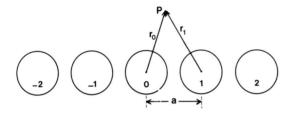

Fig.7.12. Geometry of the infinite linear array of circular dielectric rods. Only five rods are depicted

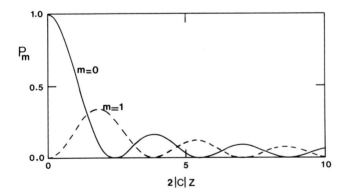

Fig.7.13. The power P_m in the $HE_{1,1}$ mode of rod m of an infinite linear array of dielectric rods as a function of $2|c|z$ for $m = 0$ and $m = 1$. The $HE_{1,1}$ mode of rod 0 is excited with unity power at $z = 0$

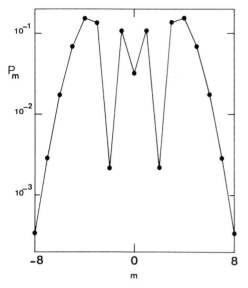

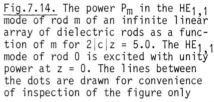

Fig.7.14. The power P_m in the $HE_{1,1}$ mode of rod m of an infinite linear array of dielectric rods as a function of m for $2|c|z = 5.0$. The $HE_{1,1}$ mode of rod 0 is excited with unity power at $z = 0$. The lines between the dots are drawn for convenience of inspection of the figure only

modes of rod m is $a_m(z)\underline{e}_{1,1}(\underline{r}_m) \exp(j\omega t)$, here $a_m(z)$ can be called the amplitude of the mode. The power in the $HE_{1,1}$ mode of rod m is proportional to $|a_m(z)|^2$.

The $HE_{1,1}$ mode of rod m will be coupled directly to the $HE_{1,1}$ modes of all other rods; however, the influence of the two nearest neighbors is most important. Therefore, only nearest-neighbor coupling is taken into account. Let us assume that the $HE_{1,1}$ mode of the rod numbered 0 is excited with unity power at $z = 0$. In this case the power P_m of the $HE_{1,1}$ mode of rod m is given as a function of z by $P_m = |J_m(2|c|z)|^2$ (Appendix B).

The power of the $HE_{1,1}$ mode is given in Fig.7.13 for the excited rod (P_0) and for its two nearest neighbors ($P_1 = P_{-1}$). Here it will be recalled that $|c| = \pi/\lambda_B$, implying that $2|c|z = 2\pi z/\lambda_B$. Therefore, $2|c|z$ is the axial distance normalized to the beat length λ_B for the beating of power between the $HE_{1,1}$ modes of two identical dielectric rods.

In Fig.7.14 the spread of power over the rods is given for a value of $2|c|z = 5.0$. As would be expected, the power diffuses through the array. A kind of beating effect is present; when the excited rod has lost all its power, some power is transferred back to the excited rod (Fig.7.13). The excited rod has lost all its power after a distance $z_0 = 2.405/(2|c|)$. For the pair of identical rods the length after which the power in the excited rod is zero for the first time is given by $z_0 = \pi/(2|c|)$. These values for z_0 do not differ much, therefore, the results for the pair of dielectric rods give an order-of-magnitude estimate for the coupling in more general arrays.

A distinction between guided and radiated light in an array is only useful when the spread of the radiated light is much faster than the spread of the guided light (owing to frustrated total reflection). As an example we again choose the linear array of identical dielectric rods.

The $HE_{1,1}$ mode of one of the rods will be excited with a large amplitude when a point source of light is imaged on the axis of the rod. The numerical aperture of the exciting objective will be chosen in this example to give an Airy disk for which the first dark ring is identical with the rim of the rod for a particular wavelength λ_0. In this case the numerical aperture (N.A.) of the objective is given by N.A. $= (3.832/v)(\lambda_0/\lambda_0)\sqrt{n_1^2 - n_2^2}$. For a homogeneous medium ($n_1 = n_2$) the light would spread again inside a cone with half angle $\bar{\theta}$ = N.A./n_1, ($\bar{\theta} \ll 1$). When $n_1 > n_2$ it is estimated that the unguided light spreads approximately inside a cone with the same half angle. The spread due to frustrated total reflection is estimated as follows. The power in the first neighbors of the excited rod is at first at its maximum for $z_1 = 1.84/(2|c|)$. For this value of z the power has travelled in the transverse direction along the array over the distance a (Fig.7.12). A distinction between guided and unguided light is only useful when $z_1 \cdot \bar{\theta} \gg a$. This leads to $\lambda_B\sqrt{\delta/\rho} \gg 0.89(a/\rho)(\lambda_0/\lambda_0)v$. Even for touching rods (a/ρ = 2), this condition is fulfilled as can be seen from Fig.7.8. (It is assumed that λ_0 is approximately equal to λ_0).

On the analogy of the description of the beating of light between two receptors, frustrated total reflection in an array can be described as a *superposition of normal modes* travelling with different phase velocities in the axial direction. The modes of the linear array of Fig.7.12 are infinitely extended along the array (Appendix B). For two-dimensional arrays a similar situation exists. Some normal modes of the infinitely extended *hexagonal array* of identical dielectric rods have been studied by WIJNGAARD [7.14]; the whole spectrum of modes cannot be determined using this method. VANCLOOSTER and PHARISEAU have discussed the related scattering problem in which a plane wave is incident on an infinite hexagonal array of dielectric rods [7.15]. Some insight may be gained from the comparison of the retina to a thick holographic grating [7.16,17].

7.5 Optical Interactions in the Retina

The results for a pair of dielectric rods are not only of interest for the study of double cones, but are also useful for the power transfer in an array (finite or infinite) of receptors. In general, an order-of-magnitude estimate of *power transfer* in an array may be obtained from a consideration of the pair of dielectric rods.

In this way WIJNGAARD and STAVENGA have discussed the power transfer between the rhabdomeres of the fly retinula [7.11].

Estimates of power transfer between receptors are generally uncertain owing to insufficient knowledge of refractive indices, distances, etc. in the retina. Besides, waveguide theory has only been developed for a few very idealized structures. Microwave models may be of some use to circumvent difficult computations (see [7.18,19]). At this moment it is only possible to derive from waveguide theory some effects which may be of some physiological relevance. In some cases it seems as if receptors are built so as to avoid power transfer to their neighbors. In the fly retinula, for example, the rhabdomeres bulge outward, avoiding a large influence of the power transfer in this way [7.11]. In the human fovea some power transfer between $HE_{1,1}$ modes of neighboring cones may occur [7.14], although it does not seem to be of much importance. Other modes and even unguided light may be of more importance in this connection. It will be useful to compare the experimentally determined modulation transfer functions (MTFs) of excised retinas [7.20,21] with the theoretical notions discussed above. At the moment a quantitative comparison is lacking. The observations of TOBEY et al. on the modal patterns in the goldfish retina are very interesting [7.22]. Both double cones and an array of rods have been investigated by these authors. However, in this case, too, a quantitative comparison with theory is lacking.

Appendix A. Symbols and Definitions

Mode	An electromagnetic field pattern which will travel undisturbed along a cylindrical structure.
Guided modes	Modes of which the fields decay approximately exponentially in going to infinity in the transverse direction.
$HE_{1,1}$, $HE_{2,1}$, $TE_{0,1}$ etc.	Guided modes of the circular cylindrical dielectric rod.
E_1^y, $E_{2,1}^y$ etc.	Guided modes of the rectangular dielectric rod in the notation in [7.8]
I_0, I_π, II_0 etc.	Guided modes of the pair of circular dielectric cylinders in the notation in [7.1]
Array modes	Modes of an array of dielectric cylinders.
E_z-sine (cosine) modes of a pair of dielectric rods	Modes for which E_z, the component of the E field in the axial direction, is identically zero along the line through the centers of the rods are called E_z-sine modes, the other modes are E_z-cosine modes.
Beat length (λ_B)	When two dielectric cylinders are placed near each other, the power (light) may leak from one cylinder to its neighbor and back again. The beat length is the distance between two successive points of maximum power in one of these cylinders.
λ_0	Vacuum wavelength of the incident light.
ρ	Radius of a circular dielectric cylinder.
n_1	Refractive index of the inside medium of a dielectric rod.
n_2	Refractive index of the matrix medium.
Cutoff parameter	$v = (2\pi\rho/\lambda_0)(n_1^2 - n_2^2)^{1/2}$ $\delta = 1 - n_2^2/n_1^2$

Appendix B. The Linear Array

Here we shall derive some results for the *transfer of power* between the $HE_{1,1}$ modes of the rods of a *linear array* of identical dielectric rods (see [7.5]). The geometry of the array is indicated in Fig.7.12; each rod is numbered. The amplitude of the $HE_{1,1}$ mode of rod m, including the z dependence, is indicated by $a_m(z)$. The $HE_{1,1}$ mode of rod m will be coupled directly to the $HE_{1,1}$ modes of all other rods; however, the influence of the nearest neighbors is most important. Therefore, only nearest-neighbor coupling is taken into account. The coupling coefficient between the modes of two nearest neighbors is equal to the coupling coefficient for mode coupling between the pair of rods. Therefore, the *coupled-mode equations* are

$$\frac{da_m(z)}{dz} + jha_m(z) = -jc[a_{m+1}(z) + a_{m-1}(z)] \quad , \quad (-\infty < m < \infty) \quad . \quad (7.10)$$

First we shall circumvent a solution for the normal modes. We define $\bar{a}_m = a_m \exp(jhz)$. Substitution in (7.10) leads to

$$\frac{d\bar{a}_m}{dz} = -jc(\bar{a}_{m+1} + \bar{a}_{m-1}) \quad . \quad (7.11)$$

Transformation to the variable $\bar{z} = -2jcz$ delivers

$$2\frac{d\bar{a}_m}{d\bar{z}} = \bar{a}_{m+1} + \bar{a}_{m-1} \quad . \quad (7.12)$$

From a comparison with the recurrence relations for $I_m(x)$ we obtain that $I_m(\bar{z})$ is a solution of (7.12). Here I_m is the modified Bessel function of order m (see [7.23]). Therefore we have

$$\bar{a}_m(\bar{z}) = I_m(\bar{z}) = (-j)^m J_m(2cz) \quad .$$

The power in the $HE_{1,1}$ mode of rod m is proportional to $|a_m|^2$. For the solution given above, the power in the $HE_{1,1}$ mode of rod m is proportional to $|J_m(2|c|z)|^2$. For z = 0 this function is equal to 1 for the rod numbered 0, and zero for the other rods. Therefore, we found a solution for the case where an $HE_{1,1}$ mode is excited in one rod at z = 0. When the power with which the $HE_{1,1}$ mode is excited is unity, the power P_m of the $HE_{1,1}$ mode of rod m is given by

$$P_m = |J_m(2|c|z)|^2 \quad .$$

A *normal mode* solution of (7.10) will now be obtained. We substitute in (7.10) $a_m = g(z)\exp(jkma)$, where k is a variable which is characteristic for the mode. The result is

$$\frac{dg(z)}{dz} = -jg(z)[h + 2c \cdot \cos(ka)] \quad .$$

The solution of this equation is

$$g(z) = g_0 \exp\{-j[h + 2c \cdot \cos(ka)]z\} \quad .$$

The E field of the normal mode may now be written as follows:

$$\underline{E}_{t,k}(x,y,z) = \sum_{m=-\infty}^{+\infty} \underline{e}_{,1,1}(\underline{r}_m)\exp(jkma)\exp\left(j\{\omega t-[h + 2c \cdot \cos(ka)]z\}\right) \quad .$$

Here, $\underline{e}_{1,1}(\underline{r}_m)\exp[j(\omega t-hz)]$ is the transverse E field of the $HE_{1,1}$ mode of the single guide. The same mode is obtained when ka is increased by $q \cdot 2\pi$ where q is an integer. Therefore ka may be restricted to the range $-\pi \leq ka \leq \pi$.

Superposition of all modes with equal amplitude delivers the solution given above. This can be shown in the following way:

$$\frac{1}{2\pi} \int_{-\pi}^{+\pi} \underline{E}_{t,k}(x,y,z)d(ka) = \sum_{m=-\infty}^{+\infty} \underline{e}_{1,1}(\underline{r}_m)\exp[j(\omega t-hz)]$$

$$\times \frac{1}{2\pi} \int_{-\pi}^{+\pi} \exp\{j[mka - 2cz \cdot \cos(ka)]\}d(ka)$$

$$= \sum_{m=-\infty}^{+\infty} \underline{e}_{1,1}(\underline{r}_m)\exp[j(\omega t-hz)](-j)^m J_m(2cz) \quad .$$

In this superposition the power in the $HE_{1,1}$ mode of rod m is proportional to $|J_m(2|c|z)|^2$.

References

7.1 W. Wijngaard: Guided normal modes of two parallel circular dielectric rods. J. Opt. Soc. Am. *63*, 944-950 (1973)

7.2 A.W. Snyder: Asymptotic expressions for eigenfunctions and eigenvalues of a dielectric or optical waveguide. IEEE Trans. MTT *17*, 1130-1138 (1969)

7.3 A.W. Snyder: Surface waveguide modes along a semi-infinite dielectric fiber excited by a plane wave. J. Opt. Soc. Am. *56*, 601-606 (1966)

7.4 A.W. Snyder: Excitation and scattering of modes on a dielectric or optical fiber. IEEE Trans. MTT *17*, 1138-1144 (1969)

7.5 A.L. Jones: Coupling of optical fibers and scattering in fibers. J. Opt. Soc. Am. *55*, 261-271 (1965)
7.6 A.W. Snyder: Coupled-mode theory for optical fibers. J. Opt. Soc. Am. *62*, 1267-1277 (1972)
7.7 W. Wijngaard, H. Heyker: *Photoreceptor Optics*, ed. by A.W. Snyder, R. Menzel (Springer, Berlin, Heidelberg, New York 1975) pp.167-174
7.8 J.E. Goell: A circular-harmonic computer analysis of rectangular-dielectric waveguides. Bell Syst. Tech. J. *48*, 2133-2160 (1969)
7.9 P.D. McIntyre, A.W. Snyder: Power transfer between optical fibers. J. Opt. Soc. Am. *63*, 1518-1527 (1973)
7.10 N.S. Kapany, J.J. Burke: *Optical Waveguides* (Academic, New York 1972)
7.11 W. Wijngaard, D.G. Stavenga: On optical crosstalk between fly rhabdomeres. Biol. Cybern. *18*, 61-67 (1975)
7.12 A.W. Snyder, W.H. Miller: Fly colour vision. Vision Res. *12*, 1389-1396
7.13 P.D. McIntyre: Cross talk in absorbing optical fibers. J. Opt. Soc. Am. *65*, 810-813 (1975)
7.14 W. Wijngaard: Some normal modes of an infinite hexagonal array of identical circular dielectric rods. J. Opt. Soc. Am. *64*, 1136-1144 (1974)
7.15 R. Vanclooster, P. Phariseau: Diffraction of an electromagnetic wave by a fiber bundle. Physica Utrecht *50*, 308-316 (1970)
7.16 H. Kogelnik: Coupled wave theory for thick hologram gratings. Bell Syst. Tech. J. *48*, 2909-2947 (1969)
7.17 F.G. Kaspar: Diffraction by thick, periodically stratified gratings with complex dielectric constant. J. Opt. Soc. Am. *63*, 37-45 (1973)
7.18 B. O'Brien: Vision and resolution in the central retina. J. Opt. Soc. Am. *41*, 882-894 (1951)
7.19 J.M. Enoch, G.A. Fry: Characteristics of a model retinal receptor studied at microwave frequencies. J. Opt. Soc. Am. *48*, 899-911 (1958)
7.20 H. Ohzu, J.M. Enoch, J.C. O'Hair: Optical modulation by the isolated retina and retinal receptors. Vision Res. *12*, 231-244 (1972)
7.21 H. Ohzu, J.M. Enoch: Optical modulation by the isolated human fovea. Vision Res. *12*, 245-251 (1972)
7.22 F.L. Tobey, J.M. Enoch, J.H. Scandrett: Experimentally determined optical properties of goldfish cones and rods. Invest. Ophthalmol. *14*, 7-23 (1975)
7.23 M. Abramovitz, I.A. Stegun: *Handbook of Mathematical Functions* (Dover, New York 1965)

8. The Visual Receptor as a Light Collector

R. Winston

With 11 Figures

In their external shape and optical properties, certain visual receptors strikingly resemble ideal light collectors. Ideal light collectors are nonimaging optical devices which optimize the concentration of rays with a specified angular divergence onto the smallest possible exit aperture. This chapter reviews the principles of light collection and introduces the concept of an ideal light collector. Several examples of light-concentrating visual receptors are discussed on the basis of an ideal light collector model. This geometrical optics description is useful for comprehending the principal features of light collection and angular selectivity in retinal cone receptors.

8.1 Principles of Light Collection

The problem of light collection for axially symmetric systems may be stated as follows: Given a set of light rays with a specified angular divergence θ_{max} distributed over an entrance aperture, how can we direct these rays efficiently onto the smallest possible exit aperture?

In discussing solutions to this problem, it is essential to distinguish between light collection and imaging. In the usual theory of image formation, the well-known Abbe sine law provides the relation between image size, object size, and θ_{max}. For the purposes of light collection, however, imaging is not required; we can optimize light collection by constructing nonimaging systems. Although the theory of image formation is as old as the science of optics itself, the recognition that optimized light collectors are intrinsically nonimaging is relatively recent; this recognition has evolved through applications to such diverse fields as radiation detectors and solar energy collectors, as well as the optics of visual receptors [8.1].

8.2 Theoretical Maximum Concentration

One of the fundamental concepts we use in the theory of light collection is the concept of a beam of light of a certain diameter and angular extent. These two can be combined giving a quantity known as the étendue. This quantity is, in fact, an invariant through the optical system, provided that there are no obstructions in the light beam and provided we ignore certain losses due to properties of the materials such as absorption and scattering. We shall use this invariant étendue to derive the theoretical maximum-concentration ratio.

Let the system be bounded by homogeneous media of refractive indices n_1 and n_2 as in Fig.8.1 and suppose we have a ray traced exactly between the points P_1 and P_2 in the respective input and output media. We wish to consider the effect of small displacements of P_1 and of small changes in direction of the ray segment through P_1 on the emergent ray, so that these changes define a beam of rays of a certain cross section and angular extent. In order to do this we set up a cartesian coordinate system $0_1x_1y_1z_1$ in the input medium and another $0_2x_2y_2z_2$ in the output medium. We specify the input ray segment by the coordinates of $P_1(x_1,y_1,z_1)$ and the direction cosines of the ray (L_1,M_1,N_1), and similarly for the output segment. We can now represent small displacements of P_1 by increments dx_1 and dy_1 of its x_1 and y_1 coordinates and we can represent small changes in the direction of the ray by increments dL_1 and dM_1 to the direction cosines for the x_1 and y_1 axes. Thus we have generated a beam of area dx_1dy_1 and the angular extent defined by dL_1 and dM_1; this is indicated in Fig.8.2 for the y_1 section. It is necessary to note tha the increments dL_1 and dM_1 are in direction cosines, not angles; thus, in Fig.8.2 the notation on the figures should not be taken to mean that dM_1 is the angle indicated, merely that it is a measure of this angle. Corresponding increments dx_2, dy_2 dL_2 and dM_2 will occur in the output ray position and direction.

Then the invariant étendue turns out to be n^2 dxdydLdM, i.e., we have

$$n_2^2 dx_2 dy_2 dL_2 dM_2 = n_1^2 dx_1 dy_1 dL_1 dM_1 \quad . \tag{8.1}$$

We can now use the étendue invariant to calculate the theoretical maximum concentration ratios of light collectors. Consider first a two-dimensional light collector as in Fig.8.3. From (8.1) we have for any ray bundle which traverses the system

$$n_1 dy_1 dM_1 = n_2 dy_2 dM_2 \quad , \tag{8.2}$$

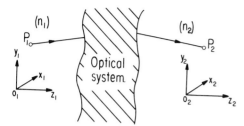

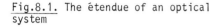

Fig.8.1. The étendue of an optical system

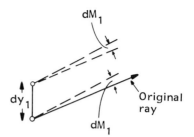

Fig.8.2. The étendue in the y_1 section

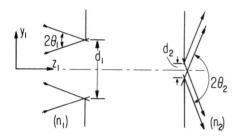

Fig.8.3. The theoretical maximum concentration ratio for a two-dimensional optical system

and integrating over y and M we obtain

$$2n_1 d_1 \sin\theta_1 = 2n_2 d_2 \sin\theta_2 ,$$ (8.3)

so that the concentration ratio is

$$C = d_1/d_2 = \frac{n_2 \sin\theta_2}{n_1 \sin\theta_1} .$$ (8.4)

In this result d_2 is a dimension of the exit aperture large enough to permit any ray that reaches it to pass and θ_2 is the largest angle of all the emergent rays. Clearly, θ_2 cannot exceed $\pi/2$ so the theoretical maximum concentration ratio is

$$C_{max} = \frac{n_2}{n_1 \sin\theta_{max}}$$ (8.5)

where θ_{max} is the maximum value attained by θ_1, i.e., the input semiangle.

Similarly for the three-dimensional case we can show that for an axisymmetric concentrator the theoretical maximum is

$$C_{max} = \left(\frac{d_1}{d_2}\right)^2 = \left(\frac{n_2}{n_1 \sin\theta_{max}}\right)^2 \qquad (8.6)$$

where, again, θ_{max} is the input semiangle [8.2].

8.3 Ideal Light Collectors

We call an optical system that actually attains the theoretical maximum concentration ratio an ideal light collector. Imaging systems fall significantly short of ideal-collector performance. For example, for a system corrected for spherical aberration and coma, this would correspond to an f number of 0.5, a physically unrealizable lower limit. However, nonimaging ideal light collectors have been developed in the United States [8.3] and independently in the Soviet Union and Western Germany [8.4,5];[1] these are optically homogeneous light guides that derive their characteristic properties from the specific shape of the exterior surface, which is made specularly reflecting.

Figure 8.4 illustrates the design of an ideal light collector for a constant index of refraction $(n_1 = n_2)$. All light rays incident upon the entrance aperture d_1 at an angle $\theta < \theta_{max}$ are channeled through the exit aperture d_2 after undergoing perhaps one or more reflections. The rays emerge with a diffuse angular spread, approaching 90° to the optic axis.[2] The flux concentration $C = (d_1/d_2)^2$ is related to the angular acceptance θ_{max} by

$$C = 1/\sin^2\theta_{max} \quad , \qquad (8.7)$$

in accordance with the sine limit (8.6). The overall collector length L is given by

$$L = (1/2)(d_1 + d_2)\cot\theta_{max} \quad . \qquad (8.8)$$

A complementary property of the ideal light collector is the exclusion of any stray light incident at angles greater than θ_{max}.

---

1 The concept arose in the context of optimizing the collection of the extremely faint radiation produced in a transparent medium by faster-than-light charged particles (Cherenkov light). Detection of such radiation is used to signal the passage of fast particles in high-energy physics experiments. See [8.3].

2 For meridional rays, the angular cutoff is discontinuous at $\theta = \theta_{max}$. The cutoff averaged over all rays occurs over an extremely narrow interval $\Delta\theta$ where $\Delta\theta \le 0.05\ \theta_{max}$ for the angles of interest. See [8.1].

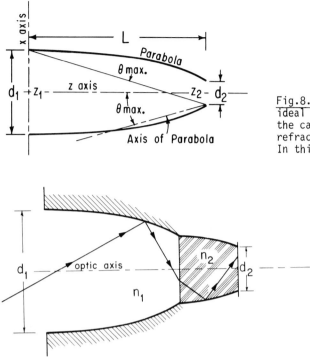

Fig.8.4. Construction of an ideal light collector for the case of constant index of refraction; taken from [8.2]. In this example, θ_{max} = 16°

Fig.8.5. Construction of an ideal light collector for the case $n_2 > n_1$; taken from [8.2]. A typical light ray traversing the entire system is shown. In this example, n_2/n_1 = 1.5 and θ_{max} = 35° at the entrance aperture

The design of an ideal light collector for distinct indices of refraction ($n_2 > n_1$), illustrated in Fig.8.5, consists of two collectors, of the type discussed above, positioned in tandem. The first of these, immersed in a medium of index n_1, is designed for a value of θ_{max} appropriate to the angular acceptance at the entrance aperture d_1, whereas the second, immersed in a medium of index n_2, is designed for a value of θ_{max} = arc sin(n_1/n_2), the critical angle. The overal flux concentration is, as required by (8.6),

$$C = (n_2/n_1)^2(1/\sin^2\theta_{max}) \quad . \tag{8.9}$$

8.4 The Visual Receptor as an Ideal Light Collector

8.4.1 The Retinal Cone Receptor

A striking similarity between the design of an ideal light collector and the structure of certain visual photoreceptors was first noticed by ENOCH for the case of the ellipsoid portion of highly tapered cone receptors in the human retina [8.6]. Detailed analysis revealed an excellent fit to the external shape of the ellipsoid portion by assuming a value θ_{max} = 13° for this parametric angle (Fig.8.6). In Fig.8.6 the diameters d_1, d_2 of the collector correspond to the diameters of the inner and outer segments of the cone cell, whereas the length of the ellipsoid corresponds to the collector length L. Corresponding to the specularly reflecting exterior surface of the artificial collector, total internal reflection of light rays takes place at the interface between the more refractive interior medium of the cone (index = n_i) and the surround (index = n_0). This total internal reflection mirror is efficient only for angles of incidence to the boundary surface greater than θ_c, the critical angle; therefore, some rays will leak out of the cone.

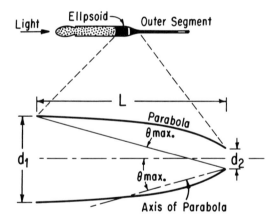

Fig.8.6. Top: schematic diagram of a cone receptor in the human retina; taken from [8.7]; original in [8.8]. The ellipsoid region is highly tapered, with a ratio of inner-segment to outer-segment diameter $\approx$ 4.5. Bottom: construction of an ideal light collector for the case θ_{max} = 13°

Quantatatively, one may expect the collection mechanism to be effective for acceptance angles θ_{max} considerably less than $(\pi/2 - \theta_c)$. From observational data quoted in the literature [8.9], one may take for the ratio of indices a value of $(n_i/n_0) \approx 1.04$, corresponding to $\theta_c \approx 74^{\circ}$. Using this value, the calculated angular collection efficiency of the ellipsoid is shown as curve B of Fig.8.7. The same figure (curve C) shows how the effective efficiency is modified by taking into account the probable orientation of the absorption axis of the photolabile pigmpnt perpendicular to the optic axis. The effect

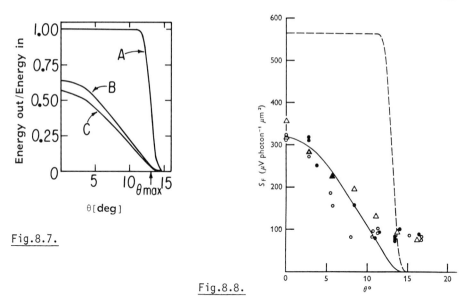

Fig.8.7.

Fig.8.8.

Fig.8.7. Angular acceptance of a θ_{max} = 13° light collector. Curve A shows the acceptance of an ideal collector having walls with perfect reflectance at all angles of incidence. Curve B includes the effect of critical reflection at the walls, assuming a relative index of refraction n_i/n_0 = 1.04. Curve C shows how B is modified if the absorption axis of the photolabile pigment is perpendicular to the optic axis

Fig.8.8. Directional selectivity of the red-sensitive cones; taken from [8.11] compared to the theoretical curve in [8.6] for a cone with diameters of inner and outer segments in the ratio of 4.5. The different symbols (open circles, closed circles, and triangles) represent results collected from individual cells. The curves were scaled vertically so that the solid curve (C in the original paper) fit the experimental points. Experimental angles have been corrected to allow for refraction at the surface of the retina. The dashed line shows the performance of an ideal collector

of pigment orientation is to reduce the absorption probability of unpolarized light by $(1 + \cos^2\theta)/2$, where θ is the angle that the light ray in the outer segment makes with the optic axis. In curves B and C (Fig.8.7) only energy delivered to the outer segment is calculated. Path length within the outer segment or the probability of propagation of energy to the cell terminations are not considered. Nevertheless, this simple model is capable of correlating a number of known characteristics of a cone receptor: 1) the value of the parametric angle θ_{max} = 13° required to fit the ellipsoid shape is compatible with the numerical aperture of the retinal receptor [8.10], 2) the same angle is consistent with the maximum angle of incidence allowed by the limit of the human exit pupil [8.10], and 3) the internal medium of the ellipsoid is sufficiently refractive to permit the light-collecting mechanism to operate.

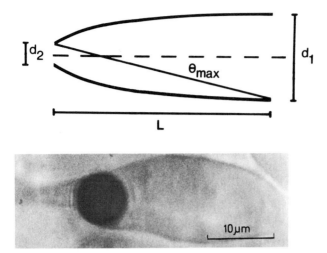

Fig.8.9. Shape of the inner segment of a red-sensitive cone compared to that of an ideal collector; taken from [8.11]. The diameters of entrance and exit apertures are in the ratio of 4.5. The dark object at the left of the inner segment is a red oil droplet; the base of the outer segment is also present

An opportunity to quantitatively test the predictions of the light collector model has recently become available through the work of BAYLOR and FETTIPLACE on turtle photoreceptors [8.11]. In an impressive series of experiments, these authors determined the directional sensitivity of individual receptors by directly measuring, in vivo, the linear response to light stimuli as a function of angle of incidence to the cone axis. Figure 8.8 shows the remarkably good quantitative agreement between their experimental results and the predictions of the model. The flat region of response beyond $\theta \approx 11°$ is attributed to a weak secondary collection arising from the leakage of light from the inner segment of an adjacent cone. This mechanism would be expected to cause cross talk between adjacent cones. Figure 8.9 compares the shape of a typical red-sensitive cone with that of an ideal collector of $\theta_{max} = 13°$. The authors [8.11] observed: "The similarities between the behavior of the real cone and that of the model suggest that collection is determined by the shape of the inner segment and its index of refraction relative to that of the surround. The success of this geometrical treatment implies that diffraction and interference effects are of secondary importance, and this is consistent with the relatively large dimensions of the inner segment." As discussed in Sect.8.5, this surmise is further supported by experiments on artificial ideal collectors.

Figure 8.9 shows an additional optical structure which may have a light-collecting function. The red oil droplet at the base of the inner segment is known to have a considerably higher index of refraction that the surrounding ellipsoid.[3] The similarity of this arrangement with the design of an ideal collector with distinct indices of refraction (Fig.8.5) was already noted in [8.11]. It is suggestive that the highly refractive droplet, through which the light must pass before reaching the visual pigment, provides additional concentration in accordance with the principles discussed in Sect.8.3.

8.4.2 Another Example of Light Collectors in Nature

Although outside the strict scope of this monograph, the crystalline cones of the ommatidia of *Limulus polyphemus* provide a sufficiently striking example of optimized light collector design to deserve mention (this qualitative similarity was first noticed by R. LEVI-SETTI et al. [8.12]).

The external shape is well represented by the profile curve of an ideal light collector with the parametric angle $\theta_{max} = 19°$ (Fig.8.10). From the observations of LEVI-SETTI et al. [8.12] the interior of the cone is substantially optically homogeneous of refractive index $n_i \approx 1.54$ surrounded by a fluid medium of much lower refractive index $n_0 \approx 1.34$. The ratio $n_i/n_0 \approx 1.15$ is sufficiently large to produce a highly efficient collector. Figure 8.11a shows the individual cones as they appear on the interior surface of the cornea. The gross light-concentrating properties of the crystalline cones are dramatically illustrated in Fig.8.11b. The remarkable optical properties of the ommatidia of *Limulus* have been a subject of long-standing interest [8.13]. The simple geometrical optics model has been shown to account for the observed optical characteristics. For a more detailed comparison, measurements of the directional sensitivity, analogous to the experiments on turtle photoreceptors discussed earlier, would be valuable.

8.4.3 Optimized Dielectric Concentrators

Recently it was found that nonimaging collectors operating solely by total internal reflection can be designed to eliminate any leakage of radiation at the external walls [8.14]. In as much as these design principles may have relevance for visual structures we shall summarize their characteristics here.

---

3 Measured by P.A. Liebman and communicated to the author by J.M. Enoch.

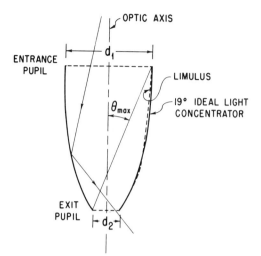

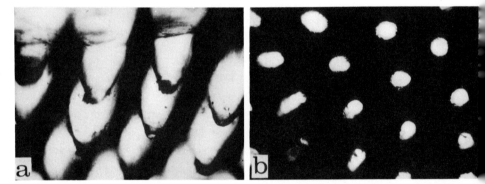

Fig.8.10. Profile of ideal light concentrator for maximum angle of acceptance $\theta_{max} = 19°$. On the right branch of the profile, the shape of a crystalline cone of *Limulus* is also shown. Taken from [8.12]

Fig.8.11. (a) Interior surface of the central area of the cornea in the lateral (compound) eye of *Limulus polypyemus*. Fresh specimen, immersed in tissue fluid. (b) Interior of the visual surface, as in (a), when illuminated by a diffuse source from the exterior side of the cornea. The light entering the crystalline cones is channelled to their tips. Taken from [8.12]

The profile curve of the new design is determined by the condition that the slope assume its maximum value consistent with totally internally reflecting the rays incident at θ_{max} onto the exit aperture. For a certain range of refractive indices and angular acceptance, this curve coincides with the ideal light collector shape already described. Thus, in case

$$(n_i/n_0) > \sqrt{2} \tag{8.10}$$

and

$$\sin\theta_{max} \leq [1 - 2(n_0/n_i)^2] \quad , \tag{8.11}$$

we simply follow the previous design and recover ideal concentration. However, when the parameters fall outside this range, as is actually the case for the visual structures considered here, the maximum slope curve for that portion of the collector adjacent to the exit aperture becomes a straight section of slope angle $\pi/2 - (\theta_c + \theta_{max})$ which joins smoothly to the parabolic shape of the ideal-collector design. The resulting flux concentration is

$$C = (\sin\theta_2/\sin\theta_{max})^2 \tag{8.12}$$

where

$$\theta_2 = \pi - 2\theta_c - \theta_{max} \quad . \tag{8.13}$$

The physical interpretation of θ_2 is that it is the maximum angle of incidence of the light rays on the plane of the exit aperture. Thus, a complementary property of the optimized dielectric concentrator is that the angular spread of the light rays at the exit aperture is limited to a specific half angle $\theta_2 < \pi/2$.

8.5 Limitations of a Geometrical Optics Description

Implicit in the foregoing discussion has been the assumption that diffraction effects were of secondary importance. Although this is a manifestly justifiable procedure for the relatively large crystalline cones of *Limulus*, this approach can be only approximately valid for the retinal cones. In fact, excellent treatments of the waveguide properties of retinal cones exist in the literature [8.7,15]. However, one may expect the geometric optics model to provide a useful description, even down to the scale of dimensions encountered in retinal receptors. This expectation is based on measurements performed on artificial ideal collectors at wavelengths comparable to the exit aperture d_2. These measurements were performed in a totally different context, for the purpose of optimizing detection of far-infrared radiation [8.16]. The results show that diffraction effects are secondary for values of λ/d_2 as large as 1.7, a considerably more severe condition than that encountered in retinal receptors. Thus, while a completely rigorous treatment would require inclusion of waveguide effects, the geometric optics description is useful for comprehending the principal features of light collection and angular selectivity in retinal cone receptors.

Acknowledgement. I am grateful to J.M. Enoch for guidance throughout the course of these investigations and to D. Baylor for making available his work on turtle photoreceptors prior to publication. The author's research on light collection has been supported by a number of agencies including The Sloan Foundation, The U.S. Atomic Energy Commission, The National Science Foundation, and The U.S. Energy Research and Development Administration.

References

8.1 W.T. Welford, R. Winston: *Optics of Nonimaging Concentrators* (Academic, New York 1978)
8.2 R. Winston: J. Opt. Soc. Am. *60*, 235 (1970)
8.3 H. Hinterberger, R. Winston: Rev. Sci. Instrum. *37*, 1094 (1966)
8.4 M. Ploke: Optik *25*, 31 (1967)
8.5 V.K. Baranov, G.K. Melnikov: Sov. J. Opt. Technol. *33*, 408 (1966)
8.6 R. Winston, J.M. Enoch: J. Opt. Soc. Am. *61*, 1120 (1971)
8.7 J.M. Enoch: in *Fiber Optics*, ed. by N.S. Kapany (Academic, New York 1967)
8.8 G.L. Walls: *The Vertebrate Eye* (Cranbook Inst. Sci., Bloomfield Hills, MI 1942)
8.9 R. Barer: J. Opt. Soc. Am. *47*, 545 (1957)
8.10 J.M. Enoch: J. Opt. Soc. Am. *53*, 71 (1963)
8.11 D.A. Baylor, R. Fettiplace: J. Physiol. *248*, 433 (1975)
8.12 R. Levi-Setti, D.A. Park, R. Winston: Nature *253*, 116 (1975)
8.13 S. Exner: *Die Physiologie der facettierten Augen von Krebsen und Insekten* (Fr. Duetecke, Leipzig, Wien 1891)
8.14 R. Winston: Appl. Opt. *15*, 291 (1976)
8.15 A.W. Snyder: IEEE Trans. MTT *18*, 383 (1970)
8.16 D.A. Harper, R.H. Hildebrand, R. Stiening, R. Winston: Appl. Opt. *15*, 53 (1976)

9. Microspectrophotometry and Optical Phenomena: Birefringence, Dichroism, and Anomalous Dispersion

F. I. Hárosi

With 8 Figures

This chapter deals with optical phenomena that are pertinent to the study of photoreception. Optical theory is reviewed in Sect.9.1. Although incomplete, it is intended as an elementary introduction, yet sufficiently detailed for the understanding of subsequent discussions. Sect.9.2 deals with some contributions that microspectrophotometric investigations have made toward the knowledge of visual processes. Again, it is an incomplete review. In selecting the highlights, emphasis was given to those results that were obtained from transversely oriented single cells. In sect.9.3 the optical properties of the retinal elements in situ are discussed. Topics were selected to illustrate our incomplete understanding of several phenomena in the hope of stimulating research interest.

Microspectrophotometry (MSP) is a term commonly used to mean a technique by which spectrophotometry is performed through the light microscope. The light used for measurements, of course, need not be visible, and indeed, the earliest successes of MSP were attained with ultraviolet light. CASPERSSON demonstrated, among other things, that nucleic acids were abundant in the nucleus and that proteins containing cyclic amino acids were present in the cytoplasm of isolated cells [9.1,2]. The principles of MSP and the technical details of instrumentation have been treated not only by its originator, CASPERSSON [9.1], but also by several other investigators in a number of more recent publications (see, for instance [9.3-6]).

9.1 Interactions of Light with Matter

It is common knowledge that light is a form of transverse electromagnetic radiation. It propagates at its maximum speed in vacuum ($c = 2.998 \times 10^{10}$ cm/s) in the form of advancing oscillations in electric and magnetic field intensities. Due to the electromagnetic character of matter, their interaction is electromagnetic. For most purposes it is sufficient to consider the electric field variations in the light wave and the impressionable outer electrons in the atoms and molecules. In their encounter, they modify each other in a number of possible ways.

When light falls upon matter, their interaction results in either reflection or penetration (refraction) which, in turn, may lead to absorption, scattering, or transmission. The nature of interactions between light and biological material are discussed below for two fundamental processes: spectral absorption and refraction. With the inclusion of linear and circular polarization of light, related phenomena are explored such as birefringence and dichroism. Dispersion (including anomalous dispersion) is also discussed. As more specialized topics are reached, the scope of treatment is narrowed to the biological materials of interest: vertebrate photoreceptors and their visual pigments.

9.1.1 Absorption of Light

For the present purpose, absorption of light is defined as a complete conversion of energy carried by the wave train of light (photon) into other forms, such as internal potential (electronic) energy or vibrational (heat) energy. When there is no immediate concern for the absorbed energy, absorption may be viewed as loss of light. For homogeneous media, this loss is describable on the basis of the principle, that equal thicknesses of the same material absorb equal fractions of the incident light flux (Φ). Thus, $d\Phi/\Phi = -a\,dx$ which, after integrating over a finite thickness, yields the exponential law of absorption

$$\Phi = \Phi_0 \exp(-ax) \quad . \tag{9.1}$$

(For more details, see [Ref.9.7, Chap.11]). The flux (intensity) at any level through the medium (Φ) is therefore proportional to the incident flux (Φ_0) times an exponential term that includes thickness (x). The factor a is called absorption coefficient; it is a characteristic constant of the medium, signifying a distance over which Φ falls to $1/e$ of Φ_0.

The above absorption coefficient is a function of both the pigment concentration and the wavelength. It can be expressed as a product of concentration and a new wavelength-dependent absorption coefficient. Thus, the law of absorption (the Bouguer-Lambert-Beer law) in chemical spectroscopy is written as

$$\Phi = \Phi_0 \exp(-\alpha_\lambda cd) \quad , \tag{9.2}$$

where the exponent is the product of the absorption coefficient α_λ (determined at wavelength λ), concentration c, and pathlength d. When the natural logarithm is taken of (9.2), we obtain one of the two expressions for optical density

$$D = \ln(\Phi_0/\Phi) = \alpha_\lambda cd \quad . \tag{9.3}$$

The second and perhaps more common definition for optical density is expressed as the decadic logarithm of the ratio of the incident to transmitted fluxes as

$$D = \log(\Phi_0/\Phi) = \epsilon_\lambda cd \quad , \tag{9.4}$$

where ϵ_λ is the molar extinction coefficient, c the molar concentration, and d the pathlength of light in the absorbing medium. The relationship between the two densities is apparent after taking the decadic logarithm of (9.2) and then comparing the result with (9.4). As long as c and d are in the same system of units, the comparison yields $\epsilon_\lambda = 0.4343\alpha_\lambda$. However, the absorption coefficient can be expressed in different units. For instance, if α'_λ is defined as the molecular (chromophoric) extinction coefficient, measured in cm^2/molecule, and ϵ_λ as the molar extinction coefficient, measured in l/mol cm, then their relation is

$$\alpha'_\lambda = 3.82 \times 10^{-21} \epsilon_\lambda \quad . \tag{9.5}$$

Equation (9.5) involves the conversion $1 = 10^3 \ cm^3$, and the application of Avogardro's Law, according to which $N = 6.0225 \times 10^{23}$ molecules/mole.

Example 9.1: Provided that ϵ_{max} = 42,000 l/mole cm is the molar extinction coefficient of a typical vertebrate rhodopsin, what is the average absorption cross section (extinction of a chromophore at λ_{max}) for each rhodopsin molecule? From (9.5), $\alpha'_{max} = 1.6 \times 10^{-16} \ cm^2$/chromophore (cf. [9.8]).

It appears logical at this point to introduce a concept, needed subsequently for describing lossy dielectrics (i.e., nonconductors), of the complex refractive index $(\bar{n})$ defined as

$$\bar{n} = n - jk \quad , \tag{9.6}$$

where n is the refractive index, k the absorption index, and $j = (-1)^{1/2}$.

Although the absorption index has other definitions (e.g., $k = \kappa_0 = n\kappa = \alpha\lambda/4\pi$), its relation to the absorption coefficient used in (9.1) will be used throughout this chapter,

$$k = a\lambda/4\pi \quad . \tag{9.7}$$

If we express a from (9.7), substitute it in (9.1), take the decadic logarithm, and compare the optical density so obtained with that of (9.4), we may write that $0.4343(4\pi k/\lambda)x = \epsilon_\lambda cd$. From this, if x and d are distances in the same unit, we obtain the relationship

$$k = 2.303\epsilon_\lambda c\lambda/4\pi = a\lambda/4\pi \quad . \tag{9.8}$$

Example 9.2: A pigment with ϵ_λ = 42,000 l/mole cm, in a concentration of 3.5 mmole/l, at λ = 500 nm, would yield by (9.8) a = 339/cm for the absorption coefficient and $k = 1.35 \times 10^{-3}$ for the absorption index. The numerical value of a, according to (9.1), means that the light intensity in such a medium falls to 37% of its initial value in a distance of 30 μm, to 14% in 60 μm, to 3% in ca. 0.1 mm. Note that $k \ll 1$. This property will be used subsequently.

Although (9.4) is the most important expression in absorption spectroscopy, others are also used. If the radiant flux absorbed by a sample is designated as Φ_a, the incident flux as Φ_0, and the transmitted flux as Φ, then

$$A(\lambda) = \Phi_a(\lambda)/\Phi_0 = [\Phi_0 - \Phi(\lambda)]/\Phi_0 = 1 - T(\lambda) \quad , \tag{9.9}$$

where $A(\lambda)$ is the absorptance and $T(\lambda) = \Phi(\lambda)/\Phi_0$ is the transmittance as functions of wavelength. The decadic logarithm of the reciprocal of transmittance is the optical density (also called absorbance, extinction, or simply density)

$$D(\lambda) = -\log[T(\lambda)] = \epsilon(\lambda)cd \quad . \tag{9.10}$$

The definition expressed in (9.10) provides the link between absorptance and density, because the combination of (9.9) and (9.10) yields

$$D(\lambda) = -\log[1 - A(\lambda)]$$

and

$$A(\lambda) = 1 - 10^{-D(\lambda)} \quad . \tag{9.11}$$

Therefore, $A(\lambda)$ and $D(\lambda)$ are completely equivalent and interconvertible quantities. However, whereas densities are additive, which is apparent in (9.4), absorptances are not, as may be readily seen in (9.11).

The phenomenon of self-screening also follows from the second relation of (9.11). It usually refers to the fact that the relative absorptance spectrum of a pigment is dependent for its shape on the absolute optical

density at its peak. In case of very low densities (dilute solutions), absorptance and density spectra are closely similar in shape. For higher densities, however, they deviate more and more with the relative absorptance
spectrum becoming broader and flatter. This broadening of the shape is known
as self-screening (for more details, see [9.9,10]).

Absorption spectrophotometry, including MSP, operates on the basic pre-
mise that light missing from the test beam was absorbed. The issue here is
as follows: All devices based on the photoelectric effect (practically all
modern spectrophotometers) detect photons; their output is proportional at
each wavelength to the light flux incident upon the detector per unit time.
Therefore, such instruments measure transmitted fluxes (Φ_t). The quantity
of interest, however, is the absorbed flux (Φ_a), which is inaccessible to
measurement. It is determined indirectly according to the equation
$\Phi_a = \Phi_i - \Phi_t - \Phi_r - \Phi_s$. If Φ_i, the incident flux, is known from a blank
measurement, and reflection (Φ_r) and scattering (Φ_s) are negligible, the
transmitted flux is sufficient to characterize the absorption properties
of the sample. Thus, absorption spectroscopy is based on the measurement
of transmitted fluxes. It should be remembered that the determination of
absorptance (or extinction) is only as good as the assumption concerning
the negligibility of reflection and scattering losses. Scattering of light
requires particular attention in MSP, where low light intensities are
often used to illuminate small areas in relatively thick samples. For details of the technique of MSP, see the references cited at the beginning
of the chapter.

9.1.2 Refraction and Reflection

Maxwell's equations demand that the propagation constant of free space for
a transverse electromagnetic radiation, c, bear the relationship
$c = 1/(\varepsilon_0\mu_0)^{1/2}$, where ε_0 is the electric permittivity and μ_0 the magnetic
permeability of free space. The fact that the experimentally determined
velocity of light and that computed from the measured electric and magnetic
constants agree within experimental error shows that light is indeed a form
of electromagnetic radiation. In a transparent, nonmagnetic material medium
the velocity is reduced from c to $v = 1/(\varepsilon\mu_0)^{1/2}$, where $\varepsilon = \varepsilon_r\varepsilon_0$ is the absolute, and ε_r the relative, dielectric constant of the medium. Thus, the
electric and magnetic properties of the medium determine the propagation
velocity, and consequently, discontinuities in media cause velocity changes
for light propagation. When light enters from one dielectric medium (air)
to another (glass), there is a change in velocity; the refractive index, n,

expresses the ratio of the two bulk velocities as

$$n = c/v = (\varepsilon_r\varepsilon_0\mu_0)^{1/2}/(\varepsilon_0\mu_0)^{1/2} = (\varepsilon_r)^{1/2} \tag{9.12}$$

(since air and free space have nearly equal electromagnetic properties).

In general, light is both reflected and refracted at boundaries of different media. The nature of reflected and refracted light depends not only on the media, as just discussed, but also on the wavelength, polarization, and angle of incidence. Although their derivations may be involved, they are obtainable as solutions to boundary-value problems in electromagnetic theory. For the laws of refraction and reflection, as well as for their derivations, however, the reader is referred to textbooks, such as BORN and WOLF [Ref.9.11, Chap.1], DITCHBURN [Ref.9.12, Chap.14], or JENKINS and WHITE [Ref.9.7, Chap.25].

9.1.3 Forms of Polarization

Light, as all electromagnetic radiation, may be polarized. The transverse electromagnetic waves that light consists of are characterized by the amplitude and the orientation of their component electric (E) and magnetic (H) fields. The electric ($\underline{E}$) and magnetic ($\underline{H}$) field vectors are inseparable; they oscillate together while remaining perpendicular to each other and to the direction of propagation at every instant of time and location in space. Polarization refers to a spatial description of waves. For instance, when the oscillations of $\underline{E}$ are confined to a single plane, the wave is said to be plane polarized or linearly polarized. Figure 9.1 shows such a wave.

Linearly polarized light may be regarded as a resultant wave of two orthogonal linearly polarized light waves of the same frequency, moving in phase with each other in the same direction. This is illustrated in Fig.9.2. Note that the resultant $\underline{E}$ is produced by the vectorial addition (i.e., superposition) of the two orthogonal components $\underline{E}_x$ and $\underline{E}_y$. The significance of this representation is that it facilitates the description of circular polarization. The electric field variations are illustrated in Fig.9.3 for a right-circularly polarized wave. It is produced as a resultant of two linearly polarized and mutually orthogonal waves of equal amplitude and frequency, one of which is phase-shifted with respect to the other by $\lambda/4$. Although the resultant electric field vector, $\underline{E}$, has a constant magnitude, it is not confined to any one plane, but rotates about the direction of propagation, making one revolution per cycle. When viewed toward the source, $\underline{E}$ may rotate

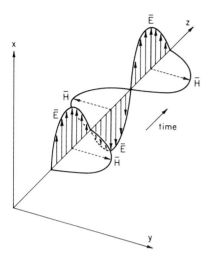

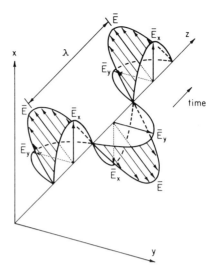

Fig.9.1. Electric and magnetic field intensities in a linearly polarized monochromatic wave. The polarization is in the x-z plane

Fig.9.2. Production of linearly polarized light from two orthogonal linearly polarized components. For simplicity, only the electric field intensities are shown

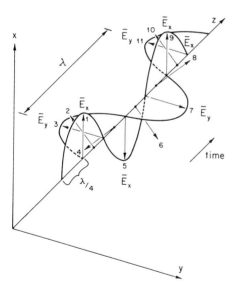

Fig.9.3. Right-circularly polarized light obtained from the superposition of two orthogonal linear waves of equal amplitude. E_y lags E_x by $\lambda/4$. The resultant electric vectors (numbered at their tips) follow a helical path, tuning clockwise when viewed against the direction of propagation

clockwise or counterclockwise (when projected into the x-y plane). By convention, the former is termed right-circular, the latter, left-circular polarization.

In reference to the illustration in Fig.9.3, it can be readily imagined that elliptically polarized waves result if either the linear components are of unequal amplitude or their phase difference is not exactly λ/4 (or its integer multiples), or if both of these conditions prevail simultaneously. Indeed, the most general polarization is elliptical, and linear and circular polarizations may be regarded as its special cases. Linearly polarized waves are obtained at phase differences of λ/2 and its integer multiples. Circularly polarized waves, as shown above, require equal amplitudes and phase shifts of λ/4 or its odd multiples. All other conditions yield elliptical polarization. Visible demonstration of the addition of two harmonic waves at right angles is possible with a cathode-ray oscilloscope. Equipped with two pairs of orthogonal deflection plates, a sinusoidal voltage may be displayed to produce Lissajous's patterns, easily demonstrating the cited conditions. Analytical treatment of the vectorial addition of two orthogonal sinusoidal time functions is also available, which supports the foregoing discussion [Ref.9.7, Sect.12.9].

An equivalent representation of circularly and elliptically polarized light is possible by the realization that two circularly polarized waves, one left-handed and one right-handed, may be summed to yield, in general, an elliptical wave. If the two amplitudes are equal, the result is linear, if unequal, they produce an elliptically polarized wave. These conditions are illustrated in Fig.9.4. Such representation is often found advantageous in the analysis of certain optical phenomena.

Light is nearly always produced by excited atoms or molecules in transition between two energy states as emitted electromagnetic radiation. A random collection of emitters usually radiates waves with a random set of vibrational directions. The resulting light is unpolarized, meaning that it is made up of waves with no specific direction of polarization. However, various optical devices may be used to separate from a beam of unpolarized light the desired forms of polarization. For example, unpolarized light may be rendered linearly polarized by 1) the repeated use of reflection and refraction at dielectric surfaces (such as a pile of glass plates), 2) the use of birefringent crystal (e.g., Nicol prism), and 3) the utilization of dichroic absorption (e.g., Polaroid filter). Circularly polarized light may be produced from a beam of linearly polarized light by the interposition of a quarter-wave plate (see Sect.9.1.4).

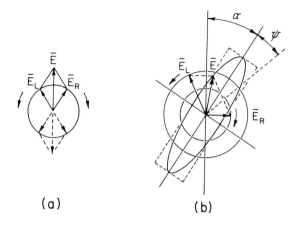

Fig.9.4a,b. Superposition of two circularly polarized waves. (a) Equal-amplitude right-handed and left-handed waves add up to a linearly polarized wave. (b) Circular waves with unequal amplitudes result in elliptical polarization

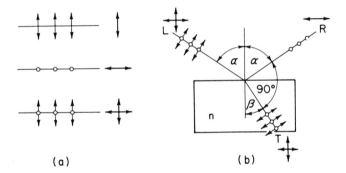

Fig.9.5a,b, Conventional graphic representations of light rays. (a) Double-pointed arrows at right angle to the direction of propagation designate direction and magnitude of the oscillating electric field vectors in the plane of the paper. Linear polarization perpendicular to the plane of the paper is shown by dots along the direction of propagation. Unpolarized light is indicated with both sets of symbols, meaning a lack of preference for either direction. (b) Light reflection and refraction at a dielectric interface from air to glass, having refractive indices of 1 and n. Rays are drawn for the illustration of Brewster's law

In concluding this section, two basic laws of optics are reviewed briefly. Referring to Fig.9.5b, it is supposed that an unpolarized beam (L) traveling in air falls upon a polished surface of a denser medium (such as glass) with refractive index n, at an angle α. Generally, the incident ray gives rise to two rays, one reflected (R) and one refracted (T). The deviation of the refracted ray is related to the incident ray through Snell's law: $\sin\alpha/\sin\beta = n$.

Moreover, R and T are found to be partially plane polarized for most values of α. At a special angle of incidence when $\alpha + \beta = 90°$, the two secondary rays R and T are exactly $90°$ apart and the reflected ray is completely plane polarized. This special angle of incidence is related to the refractive index as: $\tan\alpha = n$, which is Brewster's law. As shown in Fig.9.5b, the reflected ray contains no oscillations in the plane of incidence. This is a consequence of the fact that the field vectors in light are always perpendicular to the direction of propagation [Ref.9.7, Sect.24.3].

9.1.4 Linear Birefringence

Many crystalline substances are optically anisotropic. Such solids, whose atoms are arranged in a repetitive array, show optical properties that are not the same in all directions within any given sample. In contrast, solids whose physical properties at each point are independent of direction are isotropic. Birefringence or double refraction is a manifestation of optical anisotropy. Linear birefringence refers to the phenomenon in which a sample exhibits two refractive indices in linearly polarized light.

Calcite ($CaCO_3$) and quartz (SiO_2) form anisotropic (uniaxial) crystals. These, for each unpolarized incident ray, generally give rise to two refracted rays: one, for which Snell's law holds, termed ordinary or o ray and one, for which Snell's law does not hold, termed extraordinary or e ray. Both the o and e rays are linearly polarized at right angles to one another, as illustrated in Fig.9.6. Such crystals are used for the fabrication of practical devices. For instance, linear polarizers can be produced by the isolation and removal of one of the two linearly polarized beams. The Nicol prism is made from a special cut of natural calcite, split diagonally, then recemented again with Canada balsam such that the e ray is transmitted undeviated through the device, whereas the o ray is removed by total reflection at the cemented interface [Ref.9.7, Chap.24]. The Glan-Thompson prism is similar to the Nicol except that the crystal is cut rectangularly so that light may enter and leave normal to the end faces. In order to improve ultraviolet transmission, the latter type of prism is made with a film of air between the two halves instead of cement, and called the Glan-Foucault polarizer.

All uniaxial crystals possess a single direction called the optic axis in which the o and e rays do not separate because they propagate with equal velocity. The perpendicular direction to the optic axis is also special because in that direction the e ray velocity (v_e) is extreme; in calcite (a "negative" crystal) v_e reaches its greatest value, whereas in quartz (a "posi-

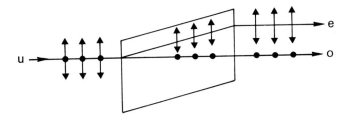

Fig.9.6. Double refraction in a calcite crystal. The unpolarized beam of light (u) is split into two linearly polarized beams (e and o), with mutually perpendicular directions of polarization

tive" crystal), v_e is at a minimum. Since velocity and refractive index are inversely related, as expressed in (9.12), the refractive indices in uniaxial crystals are related as $n_e < n_o$ in calcite, but $n_e > n_o$ in quartz. What is important here is that negatively birefringent samples have greatest retardation parallel to the optic axis, whereas positively birefringent ones have the least retardation in the direction of the optic axis. These properties make possible the fabrication of retarders or wave plates.

For example, a thin plate of calcite cut with faces parallel to the optic axis is maximally birefringent for normally incident light. If the plate thickness is d, the optical paths for the e and o rays (polarized parallel and perpendicular to the optic axis) are $n_e d$ and $n_o d$, respectively. Their difference is $d(n_e - n_o)$, and the corresponding phase retardation

$$\Delta\phi = d(n_e - n_o)2\pi/\lambda \quad . \tag{9.13}$$

Since $n_o > n_e$, the e wave advances ahead of the o wave within the plate; these, upon leaving the crystal, will interfere to form an elliptically polarized wave. As a special case, circularly polarized light is obtained from such a plate if the incident beam is linearly polarized with its plane of oscillation (perpendicular to the surface of the plate) inclined at 45° with respect to the optic axis, and the phase difference, $\Delta\phi$, determined by (9.13) is either $\pi/2$ (left-handed) or $3\pi/2$ (right-handed). By appropriate choice of thickness, d, computed from (9.13) for a specific wavelength, quarter-wave or half-wave plates may be fabricated. Whereas the former is the simplest device for producing and detecting circularly polarized light, the latter is useful for changing the state of polarization of a beam of linearly polarized light.

The optical anisotropy responsible for linear birefringence may be natural or induced; in either case it is due to some ordered arrangement of nonabsorbing molecules whose retardation of light depends on the direction of polarization. Although isotropic substances, such as glass, can be induced to become doubly refractive in strong electric fields or under mechanical stress, our discussion is restricted to natural anisotropies. We further subdivide linear birefringence into 1) intrinsic and 2) form birefringence.

a) Intrinsic Birefringence

As the name implies, intrinsic birefringence is due to the molecular makeup of the sample, namely, the array of anisotropic atomic groups or the regular arrangement of electrically anisotropic molecules. As a result of an empirical classification of the optical properties of natural crystals, it has been found (e.g., [9.12]) that strong double refraction tends to occur when 1) atoms or molecules are arranged in parallel layers (e.g., calcite); 2) ionic groups are strongly planar (CO_3, NO_3, etc., form planar ions, as in calcite); 3) the system forms a chain lattice (e.g., NaN_3).

   In experimental investigations of birefringence the first task is to find the optic axis of the sample. To accomplish this, the sample is placed in linearly polarized light and rotated until a direction is found for which the double refraction disappears. For uniaxial materials there are one optic axis and two principal indices of refraction. The ordinary index can be determined along the optic axis; in such an o ray the $\underline{E}$ vibrations are perpendicular to the optic axis, moving at a speed of $v_\perp$. Therefore, $n_0 = c/v_\perp$. The extraordinary index is determined in planes normal to the optic axis; then, the $\underline{E}$ vibrations of the e ray will be parallel to the optic axis, moving with $v_\shortparallel$. Thus, $n_e = c/v_\shortparallel$. In this manner, therefore, uniaxial samples can be fully characterized for linear birefringence. However, such a determination cannot by itself distinguish between intrinsic and form birefringence; that question must be pursued separately.

b) Form Birefringence

WIENER made the surprising discovery that parallel platelets can yield negative birefringence, and parallel rodlets can result in positive birefringence even though the constituting materials are themselves isotropic (i.e., possess no intrinsic birefringence [9.13]. Since this type of anisotropy depends on the texture or shape of the substratum, it is termed form birefringence. Although WIENER and others (e.g., [9.11,14,15]) have analyzed various assemblies of isotropic particles in several idealized forms, it is instructive to derive the expected birefringence for two simple models. The

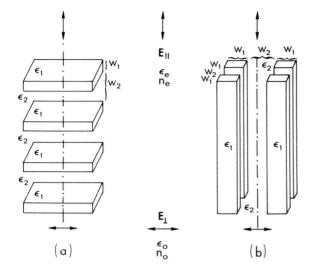

Fig.9.7a,b. Two assemblies of optically isotropic, dielectric particles.
(a) A regular array of rectangular platelets with w_1, ε_1, separated by a second medium of w_2, ε_2. The dimensions w_1 and w_2 are small, whereas the planar
extent of the platelets are large compared with the wavelength of light.
(b) A regular collection of rectangular rodlets with w_1, ε_1, separated from
each other by a second medium of w_2, ε_2. The dimensions w_1 and w_2 are small
but the length of the rodlets are large compared with the wavelength of
light. Illumination is normal to the axis of both models; double-ended arrows indicate the direction of electric vectors of the incident light

model assemblies are depicted in Fig.9.7; they were intended to resemble
photoreceptor structures of vertebrate disks and invertebrate microvilli.

Figure 9.7a shows a regular assembly of submicroscopic particles consisting of rectangular platelets of thickness w_1, dielectric constant ε_1,
separated by a second medium of thickness w_2 and dielectric constant ε_2.
The irradiation is assumed to be a collimated beam of light incident normal
to the length ("axis") of the stack. It is further assumed that w_1 and w_2
are small, whereas the planar extent of the plates are large compared to
the wavelength, so that the $\underline{E}$ field of monochromatic plane waves incident
upon the assembly will be uniform across the boundaries of the plates.

For the ordinary ray, the electric field vectors, $\underline{E}$, are parallel to the
plates (normal to the axis). Since the tangential component of the electric field is continuous at the boundary, the electric displacements in the
two regions are $D_1 = \varepsilon_1 E$ and $D_2 = \varepsilon_2 E$. The mean electric displacement,
$D = (w_1\varepsilon_1 E + w_2\varepsilon_2 E)/(w_1 + w_2)$, yields an effective dielectric constant of

$$\varepsilon_0 = D/E = f_1\varepsilon_1 + f_2\varepsilon_2 \quad , \tag{9.14}$$

where $f_1 = w_1/(w_1 + w_2)$, $f_2 = w_2/(w_1 + w_2) = 1 - f_1$ are thickness fractions (corresponding to the volume fractions).

For the extraordinary ray, the electric field vectors, $\underline{E}$, are normal to the plates (parallel to the axis). In this case the normal component of D must be continuous across the boundaries, so that $E_1 = D/\varepsilon_1$ and $E_2 = D/\varepsilon_2$. The mean electric field

$$E = (w_1 D/\varepsilon_1 + w_2 D/\varepsilon_2)/(w_1 + w_2) \quad ,$$

yields an effective dielectric constant of

$$\varepsilon_e = D/E = \varepsilon_1\varepsilon_2/(f_1\varepsilon_2 + f_2\varepsilon_1) \quad , \tag{9.15}$$

where f_1 and f_2 are again the volume fractions. After combining (9.14) and (9.15), and then replacing the dielectric constants with the appropriate squared refractive indices, one obtains

$$n_e^2 - n_o^2 = f_1 f_2 (n_1^2 - n_2^2)^2/(f_1 n_2^2 + f_2 n_1^2) \quad . \tag{9.16}$$

It is apparent in (9.16) that the system under consideration is optically anisotropic. Moreover, since $(n_e^2 - n_o^2)$ is negative, $n_e < n_o$ (i.e., $n_{\shortparallel} < n_{\perp}$), so that the birefringence $(n_e - n_o)$ due to form is also negative. Thus, the system behaves much like a negative uniaxial crystal (calcite) in which the optic axis is normal to the platelets.

Example 9.3: Calculate the magnitude of form birefringence for a lamellar array resembling the structure and the appropriate parameters of a vertebrate rod outer segment. Assume for membrane refractive index $n_1 = 1.50$ (for references and discussion, see Sect.9.2.2d) for cytoplasmic refractive index $n_2 = 1.34$ (e.g., [9.16]), equal volume fractions for the two media, $f_1 = f_2 = 1/2$ [9.17], and small dielectric losses so that $\varepsilon = n^2$ causes negligible error. With these assumptions, (9.14) simplifies to

$$n_o^2 = (n_1^2 + n_2^2)/2 \tag{9.17}$$

and (9.15) converts to

$$n_e^2 = 2n_1^2 n_2^2/(n_1^2 + n_2^2) \quad . \tag{9.18}$$

For small birefringence one may also assume that $(n_o + n_e) \simeq (n_1 + n_2)$, with which (9.16) reduces to

$$n_e - n_o = -(n_1 - n_2)^2(n_1 + n_2)/2(n_1^2 + n_2^2) \quad . \tag{9.19}$$

The numerical substitution in (9.19) shows that the expected form bire-fringence is $n_e - n_o = -0.009$. For comparison with measured values, see Sect.9.2.2c.

The second model system, depicted in Fig.9.7b, has only an approximate morphological resemblance to invertebrate microvilli. It is assumed that the assembly is made up of long, parallel, rectangular rods of dielectric constant ε_1 and cross section w_1^2 so that each volume element with base area $(w_1 + w_2)^2$ is filled with one rod in the corner and an L-shaped medium of dielectric constant ε_2. Again, w_1 and w_2 are assumed to be small but the length of the rods to be great compared with the wavelength of light.

For the extraordinary ray, the electric field vectors, $\underline{E}$, are parallel to the long dimension of the rods, and the tangential component of $\underline{E}$ will be continuous at the boundaries of the two media. Thus, $D_1 = \varepsilon_1 E$ and $D_2 = \varepsilon_2 E$, and the mean electric displacement is

$$D = \{\varepsilon_1 E w_1^2 + \varepsilon_2 E [w_1 w_2 + w_2(w_1 + w_2)]\}/(w_1 + 2_w)^2 \ .$$

From this the effective dielectric constant is obtained as

$$\varepsilon_e = D/E = f_1 \varepsilon_1 + (f_2 + f_3)\varepsilon_2 \ , \tag{9.20}$$

where $f_1 = w_1^2/(w_1 + w_2)^2$, $f_2 = w_2^2/(w_1 + w_2)^2$, $f_3 = 2w_1 w_2/(w_1 + w_2)^2$ $= 2(f_1 f_2)^{1/2} = 1 - f_1 - f_2$ are area fractions which, in the model, corres-pond to volume fractions.

For the ordinary ray, the electric field vectors, $\underline{E}$, are transverse to the rods; the o rays will encounter a composite body of two layers. In the first layer there are two slabs side-by-side for which the normal electric dis-placement is continuous. Thus, $\overset{\bullet}{E} = D/\varepsilon_1$, $E_2 = D/\varepsilon_2$. The average electric field in the first layer then, $E_{\ell 1} = (w_1^2 D/\varepsilon_1 + w_1 w_2 D/\varepsilon_2)/w_1(w_1 + w_2)$, yield-ing an average dielectric constant of

$$\varepsilon_{\ell 1} = D/E_{\ell 1} = \varepsilon_1 \varepsilon_2/[(f_1)^{1/2}\varepsilon_2 + (f_2)^{1/2}\varepsilon_1] \ . \tag{9.21}$$

The second layer is a continuous slab with cross section $w_2(w_1 + w_2)$ and dielectric constant ε_2, so that

$$\varepsilon_{\ell 2} = \varepsilon_2 \ . \tag{9.22}$$

Since the boundary between the two layers is parallel with $\underline{E}$, the tangen-tial electric field will be continuous, so that $D_{\ell 1} = \varepsilon_{\ell 1} E$, $D_{\ell 2} = \varepsilon_{\ell 2} E$. Thus, the mean electric displacement is

$$D = [\varepsilon_{\ell 1} E w_1 (w_1 + w_2) + \varepsilon_{\ell 2} E w_2 (w_1 + w_2)]/(w_1 + w_2)^2 \quad .$$

The substitution of (9.21) and (9.22) into this equation yields the mean dielectric constant for the ordinary ray as

$$\varepsilon_o = D/E = \varepsilon_2 \{f_2 (f_1)^{1/2} \varepsilon_2 + [(f_1 f_2)^{1/2} + f_2 (f_2)^{1/2}] \varepsilon_1\}/[(f_1 f_2)^{1/2} \varepsilon_2 + f_2 \varepsilon_1]$$

$$(9.23)$$

After subtracting (9.23) from (9.20), rearranging, and then replacing the dielectric constants with the squares of the refractive indices, one obtains

$$n_e^2 - n_o^2 = f_1 f_2 (n_1^2 - n_2^2)^2 / [(f_1 f_2)^{1/2} n_2^2 + f_2 n_1^2] \quad . \tag{9.24}$$

Equation (9.24) shows that the system of rodlets is optically anisotropic. Because $(n_e^2 - n_o^2)$ is positive, $n_e > n_o$ (i.e., $n_\shortparallel > n_\perp$), and the birefringence $(n_e - n_o)$ due to form must also be positive. The system thus behaves as a positive uniaxial crystal (quartz) whose optic axis follows the direction of the rods.

Formulae like (9.16) and (9.24) have been used in biological microscopy for many years in both qualitative and quantitative studies (e.g., [9.11,15]). Note in these relationships that form birefringence disappears if either one of the volume fractions vanishes or if n_2 is made equal to n_1. Provided that n_2 is due to a cytoplasmic medium accessible to external manipulation, form birefringence may be affected. This is the basis of "imbibition", which is a technique aimed at experimentally distinguishing form birefringence from the intrinsic. The relevance of the above to vertebrate photoreceptors is further discussed in Sect.9.2.2c.

9.1.5 Linear Dichroism

The term linear dichroism (LD) refers to the differential absorption of the two orthogonal components of a linearly polarized incident beam. It is the manifestation of anisotropy in absorption. A dichroic material may be regarded as a "uniaxial absorbing crystal." As such, it has two absorption coefficients for two mutually orthogonal directions of polarization and an optic axis. By way of analogy to LB, the optic axis of a dichroic sample is the direction of propagation in which no LD exists. The ratio of the largest to the smallest absorption coefficient is termed dichroic ratio. This parameter is often used in characterizing the LD property of biological materials. Based on LD measurements, the major orientation of the light-absorbing molecules (chromophores) can be deduced which, in turn, provides information on structure and, possibly, on biological function. For an excellent review of these topics, see [9.18].

Linear dichroism may be comprised of two components: intrinsic and form dichroism (just as in the case of LB). Although these two components may be difficult, or impossible, to separate, the contribution of form to the total LD must be assessed separately. The analysis of form dichroism is discussed in Sect.9.1.5b).

a) Intrinsic Dichroism

The majority of molecules absorb light anisotropically. Intrinsic dichroism is due to some regular arrangement of anisotropically absorbing molecules. The direction of the sample in which absorption is most effective may be represented by an "absorption vector"; this shows the direction of linear polarization most effective for absorption and the strength (probability) at which it occurs.

The term absorption vector is often used synonymously with the quantum-mechanical expression of electric dipole transition moment. By definition, the transition moment (for short) represents a spatial direction and a probability amplitude with which an electronic transition is inducible. In this sense linear dichroism of a sample results from a spatial anisotropy of its transition moments. Consequently, the knowledge of LD may become informative in one of two ways. 1) When the molecular structure of a sample is known (i.e., from X-ray diffraction), the direction of transition moments may be determined with respect to molecular coordinates. 2) When a transition moment is already known (i.e., from theoretical calculation) with respect to a set of molecular coordinates, the orientation of the molecule may be inferred (within the framework of molecular distribution models). It is mostly this second way that LD measurements are being utilized in vision research [9.18-21]. The LD of photoreceptors is discussed in Sect.9.2.2d.

b) Form Dichroism

WIENER discovered that a composite body consisting of light-adsorbing particles, isotropic themselves, embedded in a nonabsorbing medium in the form of a regular array, can exhibit anisotropic absorption [9.22]. This is referred to as form or textural dichroism, and is strictly analogous to form birefringence (Sect.9.1.4b).

In the subsequent analysis, absorption is regarded as a dielectric loss (Sect.9.1.1), represented by the complex refractive index, as defined in (9.6). Following Wiener's formalism, the relationship will be derived for form dichroism of a lamellar system resembling vertebrate rod outer segments.

Assume an array of absorbing lamellae interspersed with a pure dielectric substance. The arrangment is just like the stack of lamellae shown in Fig. 9.7a, except that now ε_1 is complex, leading to a complex refractive index

$$\bar{n}_1^2 = (n_1^2 - k_1^2) - j2n_1k_1 \quad . \tag{9.25}$$

The second medium is lossless, hence

$$\bar{n}_2^2 = \varepsilon_2 = n_2^2 \quad . \tag{9.26}$$

By utilizing the former analysis of the model developed in Sect.9.1.4b, equations (9.14) and (9.15) may be rewritten in terms of (9.25) and (9.26). After substitution and separation of the real and imaginary parts, one obtains the following equations

$$n_o^2 - k_o^2 \equiv d_o = f_1(n_1^2 - k_1^2) + f_2 n_2^2 \equiv a \quad ,$$

$$2n_ok_o \equiv p_o = 2n_1k_1f_1 \equiv b \quad , \tag{9.27}$$

and

$$n_e^2 - k_e^2 \equiv d_e = n_2^2(cg + dh)/(g^2 + h^2) \quad ,$$

$$2n_ek_e \equiv p_e = n_2^2(dg - ch)/(g^2 + h^2) \quad , \tag{9.28}$$

where $c \equiv (n_1^2 - k_1^2)$, $d \equiv 2n_1k_1$, $g \equiv f_1n_2^2 + f_2(n_1^2 - k_1^2)$, $h \equiv 2n_1k_1f_2$. The general solutions of these quadratic equations are of the same form. For (9.27) they can be written as

$$2n_o^2 = d_o[1 + (1 + p_o^2/d_o^2)^{1/2}] \quad ,$$

$$2k_o^2 = d_o[-1 + (1 + p_o^2/d_o^2)^{1/2}] \quad . \tag{9.29}$$

At this point, it appears necessary to restrict the scope of the analysis by making certain assumptions about the system. It will be assumed that the lamellar stack has 1) small birefringence and 2) weak absorption. Specifically, that $|n_e - n_o| < 10^{-2}$ and $k_1 < 10^{-2}$. The first assumption is justified by the result of Example 9.3, the second, because of Example 9.2. Note that assumption 2) would be inadmissible in the case of metals, which are strong absorbers; e.g., $k = 4$ for silver [9.22].

Because of assumption 2), p_0^2/d_0^2 is small. Thus $(1 + p_0^2/d_0^2)^{1/2}$ may be approximated with $(1 + p_0^2/2d_0^2)$ by use of the Binomial Theorem (with third- and higher-order terms neglected), so that (9.29) becomes

$$2n_0^2 = 2d_0 + p_0^2/2d_0 = 2a + b^2/2a \quad ,$$

$$2k_0^2 = p_0^2/2d_0 = b^2/2a \quad . \tag{9.30}$$

After substitution of the parameters in (9.30) one obtains

$$k_0^2 = n_1^2 k_1^2 f_1^2 / [f_1(n_1^2 - k_1^2)^2 + f_2 n_2^2] \quad ,$$

$$n_0^2 = \{[f_1(n_1^2 - k_1^2) + f_2 n_2^2]^2 + n_1^2 k_1^2 f_1^2\} / [f_1(n_1^2 - k_1^2) + f_2 n_2^2] \quad . \tag{9.31}$$

Note in (9.31) that for $k_1 = 0$ and $f_1 = f_2 = 1/2$, n_0^2 yields the expression obtained in (9.17) for the purely birefringent case. Also, that $k_0 = 0$ not onyl for $k_1 = 0$, but also for $f_1 = 0$.

The solutions for (9.28) are found in an analogous manner. Again, the form will be those of (9.29) and, although d_e and p_e differ from d_0 and p_0, their ratio will be small (i.e., $p_e^2/d_e^2 \ll 1$) because of assumption 2). Therefore,

$$2n_e^2 = 2d_e + p_e^2/2d_e = n_2^2[n^4 b^2 + 4(cg + dh)^2]/2(cg + dh)(g^2 + h^2) \quad ,$$

$$2k^2 = p_e^2/2d_e = n_2^6 b^2/2(cg + dh)(g^2 + h^2) \quad . \tag{9.32}$$

Following substitution of the parameter values and the simplification that $(cg + dh) = n_1^2(f_1 n_2^2 + f_2 n_1^2)$ and that $(g^2 + h^2) = (f_1 n_2^2 + f_2 n_1^2)^2$, based on assumption 2), (9.32) may be reduced to

$$k_e^2 = f_1^2 k_1^2 n_2^6 / (f_1 n_2^2 + f_2 n_1^2)^3 \quad ,$$

$$n_e^2 = n_2^2[f_1^2 k_1^2 n_2^4 + n_1^2(f_1 n_2^2 + f_2 n_1^2)^2]/(f_1 n_2^2 + f_2 n_1^2)^3 \quad . \tag{9.33}$$

Observe in (9.33) that for $k_1 = 0$ and $f_1 = f_2 = 1/2$, n_e^2 reverts to the expression obtained on (9.18) for the purely birefringent case. Also, that $k_e = 0$ for either $k_1 = 0$, or for $f_1 = 0$. This is, of course, consistent with f_1 being the volume fraction occupied by the lossy dielectric medium.

Further simplifications may be implemented in (9.31) and (9.33) by regarding $k_1^2 \ll n_1^2$ and, (9.14-16) to have continued validity. With these supplements to the initial assumptions, the ordinary absorption index becomes

$$k_o = f_1 k_1 n_1 / n_o \quad , \tag{9.34}$$

and the extraordinary absorption index

$$k_e = f_1 k_1 n_e^3 / n_1^3 \quad . \tag{9.35}$$

The ratio of (9.34) and (9.35), defined now as the dichroic ratio due to form, R_f, of a laminated system, consisting of isotropic, weakly absorbing lamellae alternating with layers of a purely dielectric material is

$$R_f \equiv k_o / k_e = n_1^4 / n_o n_e^3 \simeq n_1^4 / n_{av}^4 \quad . \tag{9.36}$$

The latter approximation stems from assumption 1); in weakly birefringent systems, n_o and n_e differ little from the average refractive index n_{av}, of the array (cf. Example 9.3).

Example 9.4: Consider a lamellar system resembling a rod outer segment from the frog retina. Assuming $n_1 = 1.50$ (for references, see Example 9.3) and $n_{av} = 1.41$ [9.23], what is its form dichroism? By (9.36), the expected form contribution to the dichroic ratio is $R_f = 1.28$. Thus, the rod's absorption along the lamellae (normal to the axis) is expected to be 28% higher than in the perpendicular direction (axially) even in the absence of intrinsic dichroism. For additional discussion and references, see Sect.9.2.2d.

Since intrinsic dichroism often coexists with form dichroism, the observable ratio (R) will be a composite of the intrinsic (R_i) and the form (R_f) components. The presence of intrinsic dichroism in a lamellar system considered above would have necessitated the inclusion of the absorption anisotropy from the start of analysis. However, the modification of the foregoing results appears possible by the modification of the absorption index such that k_1 has two distinct values: $k_\perp$ for the ordinary and $k_{\shortparallel}$ for the extraordinary direction of polarization. With these, (9.34) may be written as

$$k_o^\perp = k_\perp f_1 n_1 / n_o \tag{9.37}$$

and (9.35) as

$$k_e'' = k_n f_1 n_e^3/n_1^3 \ .$$ \hfill (9.38)

The division of (9.37) by (9.38) now yields the composite dichroic ratio

$$R = k_o^\perp/k_e'' = R_i R_f \ ,$$ \hfill (9.39)

where $R_i = k_\perp/k_n$ is the intrinsic dichroic ratio and R_f, defined in (9.36), the form contribution to the linear dichroism.

Example 9.5: Consider two lamellar samples with equal form ($R_f = 1.28$) but drastically different composite dichroism, one having mainly transverse absorption with $R_1 = 4.6$, and one with a strong axial absorption so that $R_2 = 0.3$. The intrinsic dichroic ratios, based on (9.39), are $R_{i1} = R_1/R_f = 3.59$ and $R_{i2} = R_2/R_f = 0.23$. Thus, the intrinsic dichroic ratio in both samples are numerically smaller than the measured composite values. However, whereas the smaller number in the first sample means a relatively lesser regularity in alignment of transverse chromophores, the same in the second sample implies a relatively greater axial alignment. Thus, considering the effect of form dichroism upon the intrinsic dichroic ratio, the conclusion is that transverse lamellae appear to "amplify" an intrinsic transversely dichroic system. However, transverse lamellae will "attenuate" an intrinsic axially dichroic system. This has relevance for the interpretation of dichroic ratio measurements obtained from dark-adapted and bleached vertebrate photoreceptors, a topic not pursued in this chapter.

9.1.6 Optical Activity

When a linearly polarized light beam is directed along the optic axis of a quartz crystal, it is found that the direction of polarization turns steadily as the wave progresses. This phenomenon of the rotation of the plane of polarization is called optical activity (OA), a property possessed by not only quartz but many solids and liquids such as crystalline sugar, turpentine, or sugar in solution.

A quartz crystal may rotate the plane of vibration either to the right or to the left. Substances which rotate to the right are called dextrorotatory and those which rotate to the left are called levorotatory. In right-handed rotation, the plane of polarization turns clockwise when looking against the oncoming wave; in left-handed, it rotates counterclockwise. In the illustration of Fig.9.4a, clockwise rotation of $\bar{E}$ would result if $\bar{E}_R$ rotated faster than $\bar{E}_L$. The reason for a differential retardation to circular waves is matter itself in which bound electrical charges are structurally constrained to move along helical paths [9.24]. Accordingly, charge displacements due to external fields would have circular components in optically active substances. Optical rotation in liquids is attributable to the presence of molecules with asymmetric centers. Since the magnitude of rotation is closely proportional to the concentration of the optically active substance, polarimetry has become an analytic method in measuring the amounts of sugar, for instance, in the presence of nonactive impurities.

a) *Circular Birefringence*

Circular birefringence (CB) is a measure of optical activity, meaning that right- (R-) and left- (L-) circularly polarized beams are propagated with different velocities. The phase difference ϕ_λ between L and R through an active sample of thickness d at wavelength λ is related to the refractive indices as

$$\phi_\lambda = d(n_L - n_R)2\pi/\lambda \quad . \tag{9.40}$$

The specific rotation or rotatory power $[\phi]_\lambda$ is usually defined by the observed angle of rotation at some λ, ϕ_λ, produced by a 1-mm thick crystal, or a 10-cm column of liquid containing 1 gr/cm^3 concentration C of active substance. Thus,

$$[\phi]_\lambda = 10\phi_\lambda/C.d \quad , \tag{9.41}$$

where d is the pathlength of light through the sample, in centimeters.

b) *Rotatory Dispersion*

A significant aspect of optical activity is that the rotation of the plane of polarization is wavelength dependent; the phenomenon is called optical rotatory dispersion (ORD). Although a discussion on dispersion is deferred to Sect.9.1.7, it is important to note that ORD is the dispersion of CB, i.e., the dependence of circular birefringence upon wavelength. ORD spectra, therefore, present profiles of optical rotation over a range of wavelengths, and thus ORD may contain more information about a transparent medium than does CB. Optical rotatory dispersion is related to another phenomenon known as circular dichroism, arising from differential absorption of right- and left-circularly polarized light (see below).

c) Circular Dichroism

The type of optical activity in which circularly polarized light is absorbed anisotropically is called circular dichroism (CD). A quantitative expression of CD is obtained when the difference is formed between the molar extinction coefficients of the substance for left- and right-circularly polarized light. This difference, $(\varepsilon_L - \varepsilon_R)_\lambda = \Delta\varepsilon_\lambda$, is called the differential circular dichroic extinction. While $\Delta\varepsilon_\lambda$ can be either negative or positive, for enantiomorphs (crystals of identical composition but with nonsuperimposable mirror image forms) the CD extinction are equal but opposite in sign at all wavelengths. When optical density measurements are available for L- and R-circular polarization, $(D_L - D_R)_\lambda = \Delta D_\lambda$, the differential CD extinction may be obtained as $\Delta\varepsilon_\lambda = \Delta D_\lambda / c.d$, where c is the concentration in moles per liter and d is the pathlength of light in centimeters. The molar extinction for unpolarized light is taken as the average between ε_L and ε_R, thus, $\varepsilon_\lambda = (\varepsilon_L + \varepsilon_R)/2$.

An alternative approach for the quantitative expression of CD is possible by the measure of ellipticity of the emerging light produced by the sample from linearly polarized light (e.g., [9.24]). This is done by characterizing the ellipse by an angle (Ψ) whose tangent is equal to the ratio of the minor and major axes (see Fig.9.4b). The ellipticity, ϕ_λ, in degrees, is related to ΔD_λ by the equation $\Delta D_\lambda = \phi_\lambda/33$, when ΔD_λ and ϕ_λ are expressed in equivalent units [9.25]. The molar ellipticity, (θ) is commonly related to the differential CD extinction as

$$(\theta) = 3300(\varepsilon_L - \varepsilon_R) \quad , \tag{9.42}$$

where the dimensions are 1/cm mole for $(\varepsilon_L - \varepsilon_R)$ and deg cm^2/decimole for (θ). Equation (9.42) has practical importance. For further details, see VELLUZ et al. [9.26], BEYCHOK [9.27], or CHIGNELL and CHIGNELL [9.28].

Since CD is observed only in the wavelength region where the optically active chromophore absorbs light, the CD band often resembles the absorption band in shape and location. The interpretation of CD spectra, however, is largely empirical. It has been found, for example, that the presence of α helix radically alters the CD and ORD spectra; thus, they are used to characterize the helical content of polypeptide and protein solutions.

In strict analogy with the linear case (i.e., LB and LD), the spectra of CD and ORD are not independent. The Kronig-Kramers relations establish their interdependence. For these, however, the reader is referred to specialized texts (e.g., [9.25,28,29]). The phenomenon called Cotton effect, which is the anomalous behavior of optical rotatory dispersion in a CD band, is the circular-polarization analogue of the anomalous behavior of the ordinary dispersion in an absorption band (Sect.9.1.7).

d) Circular Dichroism of Visual Pigments

Extracted rhodopsins from two species of frog, cattle, as well as porphyropsin from the carp retina [9.30] revealed that visual pigments possess several relatively strong CD bands (for a review, see [9.25]). On the basis of studies performed mainly on cattle rhodopsin [9.31-33] there appear to be at least 5 CD bands, 3 positive and 2 negative. The 2 positive CD bands associated with the α and β bands (at 495 and 340 nm) disappear upon bleaching and are thought to be due to a dissymmetric environment of the retinylidine chromophore (extrinsic Cotton effect). The two strong negative CD bands in the UV (207 and 221 nm) are attributable to the α-helical structure. In contrast to solubilized rhodopsin, there is only one CD band at 225 nm in rod particles, and it does not decrease on bleaching. The far-UV band (185-200 nm) is positive and remains to be studied. All three of the latter CD bands are related to protein conformation (intrinsic Cotton effect). From them it was derived that the apparent α-helix content of cattle rhodopsin is 50-60%.

Circular dichroism has not been detected in single photoreceptors, although an attempt was reported by WEALE [9.34]. On the basis of the CD spectrum obtained by SHICHI [9.32] from sonicated bovine rod outer segment particles, one can estimate the differential CD extinction to be ca. 2×10^{-4} at the peak of the α band for a 6-μm-diameter rod outer segment (measured transversely). This optical density appears measurable under optimal conditions. The reason for wanting to extend the technique of MSP to include circular dichroism is the desire to learn about in situ conformations of visual pigments, and their changes on bleaching. For example, if each rhodopsin contains α-helical polypeptides and they are arranged normal to the plane of disk membranes (i.e., axial in the cell), then this should manifest itself as differences in strength of certain CD bands when measured at various orientations. Experiments of this kind are yet to be undertaken.

9.1.7 Optical Dispersion

Newton's discovery that a glass prism breaks up white light into its constituent colors, the spectrum, is a consequence of the phenomenon of dispersion. The fact that blue light is deviated to a greater extent than red by a prism shows that the refractive index between air and glass is wavelength dependent.

The refractive index (n) is a number expressing the relative speed of light in a material medium (v) compared to that in vacuum (c): $n = c/v$. Since c is a true constant, not only n but also v must be wavelength dependent. Indeed, it has been found that blue light in water, for instance, travels more slowly than red, even though the speed of light in vacuum is the same for all wavelengths [9.12].

The variation of n with λ for commonly used glasses is similar in that the refractive index diminishes from blue to red. The decreasing n with increasing λ is called normal dispersion. For the visible region, this can be represented quite well by Cauchy's equation

$$n = A + B/\lambda^2 + C/\lambda^4 \quad , \tag{9.43}$$

where A, B, and C are characteristic constants of the medium (e.g., [Ref. 9.7, Chap.23]).

a) Anomalous Dispersion

When the variation of n is tested for various glasses beyond the visible spectrum, the dispersion curves do not follow (9.43), but in regions of selective absorption n first falls more rapidly than expected, then, on the long-wavelength side of the band, increases to attain much higher values before falling to a somewhat higher plateau. The phenomenon of increasing n (instead of decreasing) with increasing λ that occurs in bands of selective absorption is called anomalous dispersion.

An accounting for dispersion including the anomalous regions was achieved by Sellmeier, who derived a formula based on the supposition that there were oscillators acted upon by elastic forces in the transmission media which then affected the velocity of the light wave. He obtained the equation

$$n^2 = 1 + \sum_i A_i \lambda^2/(\lambda^2 - \lambda_i^2) \quad , \tag{9.44}$$

where A_i and λ_i are constants. The index i allows for the existence of several independent oscillators (A_i), with natural frequencies (ν_i) such that the corresponding wavelengths in vacuum (λ_i) satisfy the equation $\nu_i \lambda_i = c$. Though successfully representing dispersion in most regions, (9.44) still fails at wavelengths near the absorption peaks, because it takes no account of the dissipation of energy by absorption. This was first pointed out by Helmholtz; indeed, the shortcoming of (9.44) can be overcome by the introduction of a damping (dissipative) term in the oscillator model [Ref.9.7, Sect.23.6].

The classical theory of dispersion is based on the assumption that every dielectric medium is made up of dipole oscillators which may absorb, emit, or scatter light. Each of these oscillators is represented by an electron acted upon by an elastic restoring force, damping and inertial forces, on the one hand, and the electric field of the light wave and polarization field of the dielectric medium, on the other. The equation of motion may thus be set up and solved for various conditions. Obtained in this manner, the fundamental equation of classical dispersion theory is as follows [9.12]

$$\bar{n}^2 - 1 = 4\pi \frac{Ne^2}{m} \sum_s \frac{f_s}{\omega_s^2 - \omega^2 + j\gamma_s\omega} \quad , \tag{9.45}$$

where $\bar{n}$ is the complex refractive index of the medium; N is the number of dispersing molecules per unit volume, e the electronic charge, m its mass, ω the circular frequency of the incident light; f_s is the fraction or strength of oscillator s whose resonance frequency is ω_s and damping γ_s; and $j = (-1)^{1/2}$.

In a region near to one of the natural frequencies of the oscillators (9.45) may be written as

$$\bar{n}^2 = n_0^2 + 4\pi \frac{Ne^2}{m} \frac{f_s}{\omega_s^2 - \omega^2 + j\gamma_s\omega} \quad . \tag{9.46}$$

The assumption in writing (9.46) is that absorption is due only to the sth oscillator; the other oscillators contribute a constant amount to the real part of the complex index, which is included in n_0.

Because of the formalism introduced in (9.6), the real and imaginary parts of (9.46) may be separated and compared with (9.25). If $(n - n_0)$ and k are both small compared with n_0, the approximation that $(n^2 - k^2 - n_0^2) = 2n_0(n - n_0)$ is justified and thus yields

$$n = n_0 + \frac{4\pi Ne^2 f_s}{2n_0 m} \frac{(\omega_s^2 - \omega^2)}{(\omega_s^2 - \omega^2)^2 + (\gamma_s\omega)^2} \tag{9.47}$$

and

$$k = \frac{4\pi Ne^2 f_s}{2nm} \frac{(\gamma_s\omega)}{(\omega_s^2 - \omega^2)^2 + (\gamma_s\omega)^2} \quad . \tag{9.48}$$

It may also be shown that γ_s is the half-width of the extinction spectrum (i.e., the total width at 50% of the maximum). In quantum theory the dispersion formula is identical in form with the classical result. There, however, γ_s is made equal to the Einstein absorption coefficient. Also, f_s is no longer a small integer corresponding to the number of dipole oscillators per molecule. Quantum theory associates an oscillator with each possible electronic transition, and assigns to each an oscillator strength or f value. This quantity is defined to be the number of classical oscillators per molecule which would absorb the same amount of radiation from a parallel beam of light. The fact that f_s is usually less than unity, and is often a small fraction, is reassuring (cf. [9.12]).

By differentiation of (9.47) with respect to ω, the extreme values of n can be determined. It is found that n has a minimum when $\omega^2 = \omega_s^2(1 + \gamma_s/\omega_s)$, and a maximum at $\omega^2 = \omega_s^2(1 - \gamma_s/\omega_s)$.

Further simplifications may be applied to (9.47) and (9.48) in the case of weak absorption. In an isolated, weak, absorption, band the refractive index does not undergo large variations so that $n \simeq n_0 \simeq n_{ave}$, and the circular frequencies and wavelengths are related as $\omega = 2\pi c/n_0\lambda$, $\omega_s = 2\pi c/n_0\lambda_s$, and $\gamma_s = 2\pi c\Delta\nu/n_0$. In the latter relations c is the velocity of light in vacuum and $\Delta\nu$ is the bandwidth of the absorption band at half maximum. Thus (9.47) and (9.48) are expressible as

$$n = n_0\left(1 + KNf_s \frac{a}{a^2 + b^2}\right) \tag{9.49}$$

and

$$k = n_0 KNf_s \frac{b}{a^2 + b^2} , \tag{9.50}$$

where $K = e^2/2\pi c^2 m$, $a = \left(\frac{1}{\lambda_s^2} - \frac{1}{\lambda^2}\right)$, $b = \frac{\Delta\nu}{\lambda}$.

Example 9.6: The magnitude of anomalous dispersion is to be estimated in a rod outer segment filled with rhodopsin. The values used for the various parameters are chosen realistically to be:
λ_s = 500 nm, $\Delta\nu$ = 4,000/cm, n_0 = 1.400; rhodopsin concentration of 3.5 mmole/l, f_s = 0.5 (for the α band of rhodopsin). By the use of common physical constants one may obtain that $e^2/m = 2.53 \times 10^8$ cm^3/s^2, $K = 4.48 \times 10^{-14}$ cm, $N = 2.11 \times 10^{18}$ molecules/cm^2, $KNf_s = 4.73 \times 10^4$/cm^2.

The maximum and minimum in n occur, respectively, at 559 and 456 nm. The incremental change in n at these wavelengths are $+ 4.60 \times 10^{-4}$ and $- 3.76 \times 10^{-4}$, so that total excursion in n is 8.4×10^{-4}. Note that if f_s = 1.0 had been assumed, the total excursion in n would be twice as large, 1.7×10^{-3}. The variation of n and k as functions of wavelength are plotted in Fig.9.8.

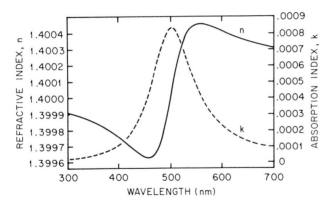

<u>Fig.9.8.</u> Wavelength dependence of n, the refractive index described by (9.49), and of k, the absorption index described by (9.50) as obtained in Example 9.6. The parameter values were chosen to correspond to the main absorption band of rhodopsin, λ_s = 500 nm, $\Delta\nu$ = 4,000/cm, f_s = 0.5, rhodopsin concentration 3.5 mmole/l, n_0 = 1.4000

b) Anomalous Dispersion in Photoreceptors

The possible roles of anomalous dispersion in photoreceptors have been investigated by SNYDER and RICHMOND [9.35,36], STAVENGA and VAN BARNEVELD [9.37], and JAGGER and LIEBMAN [9.38]. In the theoretical study of SNYDER and RICHMOND, it was shown that dispersion of the visual pigment in a cylindrical receptor would cause a refractive index change [9.35]. This, in turn, would alter the internal reflection property of the receptor and hence affect its spectral absorption. They suggested, on the basis of certain assumptions, that the in situ axial absorption spectrum of photoreceptors should be appreciably depressed on the shortwave side of the peak. That conclusion, however, was challenged by STAVENGA and VAN BARNEVELD, who found a vanishing effect of dispersion upon the spectral sensitivity of receptor cells. The latter authors estimated the peak-to-peak excursion of refractive index throughout the main absorption band of the rhodopsin-containing cells to be 4.5×10^{-4}.

The investigation of JAGGER and LIEBMAN also showed a rather small index variation due to dispersion. They measured linear retardation as a function of wavelength in frog rod outer segments by MSP. The analysis of their empirical results yielded a peak-to-peak variation of refractive index in the visible spectrum of about 1.7×10^{-3}. The agreement between the computed estimate obtained in Example 9.6 of 8.4×10^{-4} and the results of the above two publications appear to be good. Thus, it may be concluded that the index

variation in vertebrate photoreceptors due to anomalous dispersion of rhodopsin is about 0.1 -0.05%. The effect of such small variations, however,
would probably be undetectable in either the spectral absorption or the action
spectrum of photoreceptors.

9.2 Contributions of Microspectrophotometry in Vision Research:
Historical Notes

The efforts of CASPERSSON to combine light microscopy and chemical spectroscopy resulted in the development of microspectrophotometry (MSP) [9.1,2].
Even earlier, however, SCHMIDT conducted pioneering optical studies on vertebrate photoreceptors in search of structure-function relationships [9.39,40].
With the polarizing microscope, he was the first to investigate visual cells
for linear birefringence and linear dichroism of their outer segments. Moreover, on the basis of those measurements, he was able to infer the essential
features of the subcellular structure with amazing accuracy.

Using a microspectrophotometer developed by JOHNSON [9.41], DOBROWOLSKI
et al. were among the first to determine visual pigment absorption spectra
through the side of single rods [9.42]. Their method involved taking a series
of photomicrographs of isolated cells at different wavelengths of monochromatic light and then measuring the relative density of the developed
images on photographic plates. A similar photographic method was also employed initially by DENTON [9.43], and DENTON and WYLLIE [9.44], who used
a photoelectric densitometer to compare film images. DENTON confirmed and
extended SCHMIDT's observations on rod dichroism, and introduced the "edge-
fold" preparation of the retina, which was to be used later by so many other
investigators [9.40,45,46].

Although not MSP, the application of phase contrast and interference
microscopy to retinal rods and cones was also important in providing us
with additional empirical data. BARER, in developing the method of immersion
refractometry, enabled SIDMAN to determine the average refractive index
of photoreceptor outer segments and other parts [9.23,47]. From the refractive indices, it was possible to calculate the concentration of solids for
each subcellular compartment.

Further development of MSP now equipped with a photoelectric detector,
led HANAOKA and FUJIMOTO to attempt the detection of the light-sensitive
visual pigments in single cones [9.48]. Their objective was to ascertain
whether the various pigments inferred by other means are all contained in

each cell or segregated in various types. As is well known, the answer to their question was not established unequivocally until some years later. Although not with MSP, HAGINS and JENNINGS were able to obtain evidence for a very important principle [9.49]. As they were unable to induce retinal dichroism by polarized bleaches, they concluded that rhodopsin probably enjoys complete rotational freedom in the transverse disk membranes, a conclusion that was later corroborated by others. They also deduced from their analysis that form dichroism alone could not explain the large observable transverse dichroism of vertebrate photoreceptors which then must be due to intrinsic dichroism (i.e., chromophore orientation).

In the meantime, automatically recording spectrophotometers with great sensitivity were being designed and, after the appearance of the dual-beam ratio-recording MSP of CHANCE et al. [9.50], the field started an explosive development. Several instruments were being built almost simultaneously (e.g., [9.3,51-54]) and various ways to use them were being comtemplated (e.g., [9.55]).

Several other instruments were built and applied to the study of various aspects of vision during the current decade as well. For example, MACNICHOL and HÅROSI designed a digital computer-controlled single-beam MSP to study linear dichroism in single receptors [9.56,57]; ENOCH and TOBEY developed a special microscope microspectrophotometer for the determination of waveguide properties of photoreceptors [9.58]; and MURRAY initiated the development of a rapid MSP in which spectral scanning is accomplished without the use of moving mechanical parts [9.59]. To the knowledge of this writer, the most sophisticated instrument of its kind was developed by MACNICHOL [9.60]. This MSP is also digital computer controlled and uses a single light beam; moreover, it utilizes photon-counting light detection, infrared television monitoring for viewing the specimen, and a large memory to facilitate gathering and processing of data. Some of the salient results that were obtained by the use of MSP are dealt with in the following sections.

9.2.1 A Classification of the Technique as Applied to Photoreceptors

As discussed previously, photoreceptor MSP may be classified according to various criteria [9.6]: by cell type. (A) vertebrate or (B) invertebrate; by method of sample preparation, (A_1) single-cell, (A_2) multicell (i.e., retinal patch or edge-fold retina), or (A_3) cell-fragment suspension; and by orientation of cell with respect to the measuring light. (A_{11}) transverse or (A_{12}) axial.

Although any such division tends to be arbitrary (and that it is, can be concluded from the fact that one and the same instrument may serve in any or all the above classes), this writer finds it convenient for the purpose of discussion. Aside from personal choices that each investigator makes, such as choosing between (A) and (B), there are good reasons for the selection of one method over the other. For example, if one intends to study linear dichroism or linear birefringence in vertebrate rods, the choice should be (A_{11}), whereas for the study of modal propagation in the "visual fibers" (A_{12}) must be used (cf. [9.55]). Of course, dichroism and birefringence of rod cells are also observable in a transversely oriented multicell preparation $(A_{2,11})$, the edge-folded retina, as demonstrated by DENTON [9.45], and later adopted by LIEBMAN et al. [9.61] and by ERNST and KEMP [9.62]. In measuring many pigment-containing outer segments simultaneously, the advantage is a possible increase in signal-to-noise ratio because of an increased area and depth of pigment accessible to the measuring light beam. However, when compared with a transversely oriented single-cell preparation $(A_{1,11})$, scattering between cells and through a thicker bathing medium may be so troublesome so as to outweigh the advantage. A multicell sample can be measured axially as well $(A_{2,12})$, as DENTON and WYLLIE [9.44] showed. In this case, again, the signal should increase because of the increased path length along the axis of the long cylindrical rods, but so would scattering increase because of the extra retinal tissue present. Retinal patches are often measured axially; BROWN [9.51] used it on frog, BROWN and WALD [9.63] applied it to human and monkey retinae, MUNTZ [9.64] to fishes, and ENOCH and TOBEY [9.58] used it routinely on all retinae to demonstrate modal propagation in photoreceptors.

The axial measurment may also be attempted on single cells $(A_{1,12})$. Although this has been a relatively infrequently used method, probably because of the technical difficulties involved, it nevertheless yielded valuable information on the parafoveal cone and rod pigments of human and monkey retinae in [9.65, 66].

MURRAY applied the technique of MSP to fragmented photoreceptor suspensions (A_3) [9.67]. His objective was to determine whether color perception is mediated in the primate retina by multiple cone pigments, or by wavelength-selective waveguide or interference filter properties of the foveal cones. For this he punched out the central foveal portions of monkey retinae, fragmented them by sonication, and then measured the spectral absorptance of the fragments in suspension. He found, by the method of partial bleaches, that the suspensions contained at least two spectrally distinct pigments even though the cone outer segments were broken into fragments of 2 μm or smaller. Thus, he concluded that color discrimination in the primate fovea is primarily encoded in a multiplicity of visual pigment absorption spectra.

By far the most generally used and, perhaps, the most advantageous method is the transverse single-cell MSP ($A_{1,11}$). It has in its favor that 1) the light path is well defined, 2) there is no intervening structure, and 3) the preparation is thin, easy to make, and easy to align with the light beam. The method has been applied to many preparations, from frog rods (e.g., [9.19, 51,56]) through goldfish cones (e.g., [9.52-54,68]) to primate rods and cones (e.g., [9.69,70]).

9.2.2 Salient Results

The technique of single visual cell MSP has provided us with two celebrated results in recent memory. The first was the demonstration by MARKS and MAC-NICHOL [9.71] and MARKS [9.53,54], confirmed by LIEBMAN and ENTINE [9.52], that a higher vertebrate, the goldfish, that was behaviorally known to perceive colors independently of brightness, and that required three colors to match a given spectral light (trichromatic vision) actually has three types of spectroscopically distinguishable visual pigments segregated into separate cells. These results not only established the principles sought earlier by HANAOKA and FUJIMOTO [9.48], but provided evidence for the physical basis of the Young-Helmholtz theory of color vision which presupposed that there must be a small number of, probably three, basic sensitivities in the eye, each being active in different parts of the spectrum, that enable us to discriminate between the wavelengths (colors or hues) of light stimuli (cf. [9.3, 72]). The second historically prominent result obtained by MARKS et al. [9.65] and BROWN and WALD [9.66] was concerned with primate cone pigments. They showed that what is true for goldfish is also true for man, namely that there are, in addition to a rod pigment, three spectrally distinct cone pigments in the primate retina. The absorption maxima (λ_{max}) of the corresponding pigments in man and in goldfish were of course different, but that came as no surprise for the diversity in visual pigment absorption spectra has already been well documented (for reviews see [9.73,74]).

Thus emerged the principle that one photoreceptor cell contains one visual pigment. A word of caution is advisable on this point, however, for there are many exceptions to this among the rods and cones of amphibian and fish retinae that use "mixed" visual pigments. It has been established that all

known visual pigments utilize either one of two closely related molecules as
the chromophore (light-absorbing molecule): the aldehyde of vitamin A_1 (retinal) or the aldehyde of vitamin A_2 (3,4-dehydroretinal); these may combine
with various "opsins" (species-specific proteins) to form the many "rhodopsins"
and "porphyropsins" and corresponding cone pigments (cf. [9.75,76]). The
same opsin may utilize either one or the other and, in doing so, it produces
a homologous pair of pigments with modified spectroscopic properties (e.g.,
[9.77,78]). Animals that contain both vitamins A_1 and A_2 in their eyes make
both types of pigments and their visual cells contain an admixture of the
A_1- and A_2-based pigments in varying proportions. While BRIDGES and YOSHIKAMI
[9.79] and REUTER et al. [9.80] showed that the pigment epithelium determines
the chromophore pool and that the retina regenerates either one or both types
of visual pigment in accordance with the availability of the chromophores, it
was MSP that established that the various rods and cones accept both chromophores equally well, so that all outer segments in the same retina (or adjacent cells in a retinal patch) contain about equal proportions of the homologous pigments [9.21,81,82]. MSP measurements also established that vertebrate rods and cones are uniformly filled with one or a certain mixture of
the homologous pair of pigments throughout their outer segment. The only possible exception to this so far is the recent finding of LOEW and DARTNALL
[9.82], according to which special situations such as changing light conditions may cause a shift in the A_1:A_2 ratio in some animals; the new proportion
may then manifest itself in the newly formed pigment at the basal portions
of the rod outer segments, in accordance with the renewal process for rod
outer segment membranes found by YOUNG [9.83].

Another fact concerning spectra from vertebrate visual pigments of numerous species is that while specimens have been recorded from hundreds of them
using various techniques including MSP, not every possible λ_{max} value has
been found. In other words, their spectral maxima appear to cluster instead
of forming a continuous distribution [9.84,85]. It was believed until recently that the λ_{max} of all vertebrate pigments fall between 433 and 620 nm
with clusters at about 6-10-nm intervals. As far as it is known, the clustering of λ_{max} values still holds even though the total range appears to expand
on the short-wavelength side to close to 400 nm (cf. [9.86-88]).

a) Human and Monkey Visual Pigments

In a review by MACNICHOL et al. this topic was discussed in detail [9.4]. The current discourse is an updated summary of that review. The first spectroscopic measurement of a digitonin-extracted human retina was apparently done by CRESCITELLI and DARTNALL [9.89]. The alkaline bleaching difference spectrum of that extract (i.e., no hydroxylamine present), presumably due to rhodopsin, peaked at 497 ± 2 nm. By the use of a fundus reflection photometer RUSHTON was able to obtain a bleaching difference spectrum for rhodopsin in the normal, living, human eye [9.90]. His difference spectrum was similar to that of CRESCITELLI and DARTNALL but with λ_{max} at 5-10-nm-longer wavelength (i.e., 502-507 nm). Although the spectroscopic accuracy of Rushton's determination is low in part because of the very likely presence of blue- and green-absorbing photoproducts in the living retina (cf. [9.91]), it provided strong support to the idea that the extracted pigment was indeed rhodopsin.

WALD and BROWN also determined the absorption spectrum of human rhodopsin in the extracted form as well as in suspended rods [9.92]. Their digitonin extract had a λ_{max} if 493 nm, whereas that of suspended rods was 500 nm. Later they again measured human rhodopsin; this time they used axial MSP on single rods and obtained a λ_{max} at 505 nm [9.66]. In the later determination, as contrasted with the previous two, hydroxylamine was not used and thus some displacement of the peak toward the red would be expected. In a study of rhodopsin bleaching kinetics in the isolated human retina BAUMANN and BENDER used axial MSP of large (several millimeters in diameter) retinal patches [9.94]. The bleaching difference spectrum in their determination peaked at about 500 nm; this was obtained without the use of hydroxylamine. Digitonin extraction of human retinae was once again performed by BRIDGES and QUILLIAM [9.95]. In their determination, both the alkaline and hydroxylamine difference spectra peaked at 493-494 nm.

Rhesus monkey rhodopsin was measured in digitonin extract by BRIDGES [9.96] with λ_{max} at 497 ± 1 nm and recently by BOWMAKER et al. [9.70]. The latter investigators determined its λ_{max} in digitonin extracts at 499 ± 1 nm (in the presence of hydroxylamine) as well as with transverse single-cell MSP at 502 ± 2.5 nm.

Human cone pigment absorptions were first measured by RUSHTON with a fundus reflection densitometer [9.97]. The bleaching difference spectra obtained from the normal fovea indicated the presence of two pigments, one with λ_{max} at 540 nm, the other at 590 nm. There was no evidence for the presence of a blue-absorbing pigment. BROWN and WALD used axial MSP in measuring small retinal

patches of human and rhesus monkey retinae [9.63]. Their bleaching difference spectra indicated a red-sensitive pigment at 565 nm in both human and
monkey foveas, whereas the green-sensitive pigments were slightly different,
at 535 nm for human and at 527 nm for monkey.

Subsequent studies were carried out with single-cell MSP, the earlier
ones in axial orientation, the later ones transversely. MARKS et al. [9.65]
reported absorption difference spectra obtained from ten parafoveal cones
of human and monkey (two species) retinae at λ_{max} values of 445, 535, and
570 nm. The slightly revised figures published by MACNICHOL placed the λ_{max}
at 447, 540, and 577 nm, respectively [9.3]. BROWN and WALD also measured
the parafoveal cones in the human retina with axial MSP [9.66]. The bleaching difference spectra they obtained for four cones had λ_{max} at 450, 525,
and 555 nm. WALD and BROWN presented five additional single human cone difference spectra with peaks at 440, 529, and 565 nm [9.93]. By the application
of chromatic bleaches in sonicated foveal suspensions, MURRAY was able to
identify two pigments in the rhesus monkey [9.67]; their λ_{max} values were
placed at 526 and 573 nm. He found no evidence for the presence of a blue-
absorbing type. Strangely enough, in the most comprehensive study of the
rhesus monkey retina to date by BOWMAKER et al., blue-sensitive cones were
not found, either [9.70]. The 82 cones measured with transverse single-cell
MSP indicated 2 cone types with λ_{max} at 536 ± 3.5 nm and at 565 ± 2.5 nm. The
results of the foregoing studies are summarized in Table 9.1. Although, as
years pass, there is a modest increase in data accumulation, ambiguities
still remain even in the exact λ_{max} values.

Not included in Table 9.1 are the transverse single-cell MSP measurements
of DOBELLE et al. because of the lack of λ_{max} information [9.69]. Also excluded are the numbers given by LIEBMAN for the λ_{max} of human and monkey cone
pigments, because neither a single spectrum nor any experimental detail concerning the circumstances in which they were obtained, were ever published
[9.5]. The brief report of BOWMAKER et al. was also excluded [9.88]. In it
the claim is made for the finding of blue-absorbing pigments in human (at
λ_{max} of 420 nm) and in the cynomolgus monkey (at λ_{max} of 415 nm), but none
in the rhesus monkey. Although these findings would be significant if confirmed, more details are needed for being able to assess them properly.

Table 9.1. Wavelength of peak absorption of human and rhesus monkey visual pigments: λ_{max} [nm]

Reference	Method	Rod (rhodopsin)	Blue-abs.	Cones green-abs.	Red-abs.	Note
CRESCITELLI and DARTNALL [9.89]	digitonin extract	497 ± 2	-	-	-	human
RUSHTON [9,90,97]	fundus reflection photometry	502-507	-	540	590	human
WALD and BROWN [9.92]	digitonin extract } rod suspension }	493 / 500	-	-	-	human / human
BRIDGES [9.96]	digitonin extract }	497 ± 1	-	-	-	monkey
BROWN and WALD [9.63]	axial MSP (patch)	- / -	- / -	535 / 527	565 / 565	human / monkey
MARKS, DOBELLE and MACNICHOL [9.65]	axial MSP (single)	-	445	535	570	human and monkey
MACNICHOL [9.3]	axial MSP (single)	-	447	540	577	average human and monkey
BROWN and WALD [9.66]	axial MSP (single)	505	450	525	555	human
WALD and BROWN [9.93]	axial MSP (single)	-	440	529	565	human
MURRAY [9.67]	sonicated foveal suspension	-	-	526	573	monkey
BAUMANN and BENDER [9.94]	axial MSP (patch) }	500	-	-	-	human
BRIDGES and QUILLIAM [9.95]	digitonin extract }	493-494	-	-	-	human
BOWMAKER et al. [9.70]	transverse MSP / digitonin extract }	502 ± 2.5 / 499 ± 1	- / -	536 ± 3.5 / -	565 ± 2.5 / -	monkey / monkey

b) Rotational and Translational Diffusion of Visual Pigments in situ

Following the earlier conclusion concerning the rotational freedom of rhodopsin in the rod disk membrane by HAGINS and JENNINGS [9.49], BROWN [9.98] and
CONE [9.99] convincingly demonstrated the phenomenon. BROWN found that when
frog rods are appropriately fixed in glutaraldehyde, axial dichroism can be
photoinduced in them, whereas formaldehyde is ineffective under similar conditions. Thus, it appears that glutaraldehyde with two reactive aldehyde
groups may form crosslinks between rhodopsin molecules and thus prevent their
motion which normally makes their angular orientation random. The time course
of randomization was measured by CONE. He found that the transient dichroism
induced by a brief polarized flash in a fresh frog retina decays with a half-
time of 3.0 ± 1.5 µs at 20°C. Such a decay, provided that it is due to rotational diffusion, which is very likely in view of the glutaraldehyde effect,
corresponds to a relaxation time of about 20 µs. From that value he calculated
the viscosity of the membrane site for rhodopsin to be in the range 0.7-6
poise, comparable with the viscosity of light oils, such as olive oil.

In addition to rotational diffusion, POO and CONE also showed that rhodopsin undergoes rapid lateral diffusion in the disk membranes of single
bullfrog and mudpuppy rods [9.100]. Measuring at single wavelengths, they
found that longitudinally bleaching half an outer segment does not immediately
affect the other half, but that the unbleached pigment (as measured by its
optical density at its λ_{max}) then rapidly mixes with that of the bleached
portion. The half-time of this process to a uniform distribution at 20°C was
found to be 35 ± 7 s for frog and 23 ± 5 s for mudpuppy. The redistribution of
bleached and unbleached visual pigment molecules was stopped by glutaraldehyde fixation and, also, when the bleaching boundaries were transverse to the
long axis of symmetry. By solving the two-dimensional diffusion equation they
proceeded to calculate the diffusion constants and the membrane viscosity.
Again, for membrane viscosity, POO and CONE obtained 1-4 poise. Moreover, they
estimated the collision frequency between rhodopsin molecules to be in the
range of 10^5 - 10^6 collisions per second.

LIEBMAN and ENTINE also studied the lateral diffusion of visual pigments
in the photoreceptors of leopard frog and mudpuppy [9.101]. The half-time of
the redistribution process in their measurements was 4.9 ± 0.5 s for frog and
15.4 ± 3 s for mudpuppy. Although, these results are numerically different,
they confirmed the essential features in the findings of CONE and of POO and
CONE [9.99,100]. Lateral diffusion of visual pigment was also found in toad
rods by WILLIAMS et al. [9.102], who used rapid and repeating spectral scans

(and not merely single wavelengths) so that the products of bleaching could also be monitored (cf. [9.60]). In contrast to vertebrate photoreceptor membranes which are found highly fluid, the visual pigment molecules in the rhabdomeric membranes of crayfish appear to be much more restricted, according to [9.103]. Should high viscosity prove to be a general property of invertebrate photoreceptor membranes, this difference with respect to vertebrates will surely be significant in the elucidation of the intervening processes between light absorption and cellular excitation.

There are at least two new concepts that follow from these experiments. First, rhodopsin is not inserted in a fixed matrix, but enjoys two translational (in addition to one rotational) degrees of freedom to move about and collide with its neighbors. Thus, a disk membrane in a rod is not like a solid semiconducting crystal with relatively constant distances and forces between its centers. For this reason, electron or hole conduction over long distances in the plane of the disk membrane is a highly unlikely mechanism for signaling to the plasma membrane about an event of absorption. Second, disk membrane viscosity is only about one-hundred-fold greater than that of water. In such a fluid environment the collision frequency is high (a visual pigment molecule may collide with its neighbors perhaps as often as a million times a second) so that rapid reactions may take place between membrane proteins. As pointed out by POO and CONE [9.100], by partitioning into phospholipid membranes, protein molecules may achieve high effective concentrations as well as favorable orientations for high reaction rates.

c) Linear Birefringence in Rod and Cone Outer Segments

The early observations on vertebrate rod birefringence have been reviewed by MOODY [9.104]. Even though not the discoverer of the phenomena, SCHMIDT carried out the most comprehensive study of the linear birefringence and dichroism of the frog rod outer segment [9.39,40,105]. He found that the observed birefringence, $n_e - n_o = 0.0017$, was due to the combination of two components, a positive intrinsic and a negative form birefringence. He suggested that the negative form birefringence was caused by a pile of parallel platelets, whereas the positive intrinsic LB results from the lipid hydrocarbon chains arranged parallel to the rod axis. His model for the fine structure of the rod outer segment turned out to be remarkably accurate (cf. [9.40]).

Extensive studies on the LB and LD of photoreceptors were also carried out by WEALE [9.34,106-108]. In addition to frog rods, he also tested goldfish cones (as well as rods and cones from several amphibian species). He found that bleached photoreceptors, when fixed in formaldehyde or glutaraldehyde, change their LB from positive to negative, and that rods switch sooner than cones. There were also species variations, but cones were always more birefringent (by a factor of 2-4) than rods [9.106]. Interesting though these observations may be, the underlying reasons remain obscure.

The static birefringence of single frog rod outer segments was also measured by LIEBMAN et al. [9.61]. By the use of glycerol imbibition, they found the total birefringence for dark-adapted rods to be +0.001, being made up by a form component of -0.004 and an intrinsic component of +0.005. In addition to the static birefringence of dark-adapted photoreceptors, there are transient processes associated with bleaching [9.108,109]. The latter investigators found a rapid loss of intrinsic birefringence, following a flash exposure, followed by a slower secondary loss of net birefringence (also intrinsic). While the rapid intrinsic birefringence loss appears to be correlated with the formation of metarhodopsin II, the slower component remains unidentified. On a much slower time scale birefringence recovers to the prebleach level or beyond; in this process the plasma membrane seems to play some role.

Perhaps the most interesting observation concerning birefringence is that, in the leopard frog rod, it varies along the length of the outer segment [9.110,111]. Birefringence appears to be highest at the basal end of the outer segment near the ellipsoid region and diminishes toward the distal end. This birefringence gradient along the length exists whether the rod was dark adapted, bleached, osmotically active or inactive, fixed in glutaraldehyde, or impregnated with glycerol. Although the source of the gradient was found to be due mainly to a gradient in the intrinsic birefringence and not the textural component, its molecular basis has not been identified [9.111].

Although the polarization microscope in the measurement of birefringence provides us with perhaps the most sensitive technique for studying an optical property of biological materials, its application has contributed relatively little to the understanding of visual processes. One possible reason for this is that cells, though commonly transparent, are enormously complex optical systems, even if they are seemingly as "simple" as vertebrate photoreceptors. It is hoped that, in the continued application of the technique, it will be possible to resolve birefringence so that the underlying processes become separable and thus accessible for investigation.

d) *Linear Dichroism of Rod and Cone Outer Segments*

It is a well-known fact that SCHMIDT discovered the linear dichroism of frog rod outer segment [9.39,40]. He found that light absorption is maximal through the side of the rod when the illuminating (blue-green) light is polarized transversely to the length and minimal when the polarization is parallel to the long axis of the cell. The original observation of LD in frog rods was subsequently confirmed first by DENTON [9.43,45,46], who found it also in the retinal rods of a few other vertebrate species, and then by other investigators (e.g., [9.107]). Transverse MSP measurements on isolated cells proved that the LD of the outer segment of vertebrate rods and cones is a general phenomenon. It was shown to exist, for example, in frog rods [9.5,49,56,57, 112], in goldfish cones [9.52-54,68], in frog cones [9.81] and in nine species of fish cones [9.72]. With the demonstration by DOBELLE et al. that primate rod and cone outer segments also preferentially absorb light with transverse polarization, there should be no doubt that dichroism is a common feature in vertebrate photoreceptors [9.69].

As was discussed in Sect.9.1.5, LD may be intrinsic or form in origin. The former is due to a regular alignment of the electric dipole transition moments due to visual pigment molecules, while the latter is a result of the lamellar structure. When we measure the transverse linear dichroism of a photoreceptor, the absorptances or optical densities of the cell are determined in the principal directions, namely, along the long axis of symmetry (z axis) and perpendicular to that direction (x or y axis). The ratio of these two densities at λ_{max} is commonly used to characterize the LD of the cell; this is the dichroic ratio, R. It is usually defined such that the larger is divided by the smaller so that $R \geq 1$. Accordingly, for vertebrate rods and cones the dichroic ratio is $R = D_\perp/D_\parallel$, where $\perp$ and $\parallel$ refer to the direction of polarization with respect to the cell's axis of symmetry. Measured values of R for vertebrate photoreceptors are usually in the range of 4-5 (e.g., [9.21,49,112]), although lower values, especially for cones, are also found (e.g., [9.68,113]). Claims for higher values occasionally appear (e.g., [9.5]), but the published records fail to bear them out. In an extensive study of the leopard frog rhodopsin-containing rods, R was found to be $4.2^{+0.7}_{-0.3}$. This was obtained from multiscan recordings of the dichroic absorptance spectra from 38 isolated rods. Their absolute averages at λ_{max} plus and minus one standard deviation were converted to optical densities and then the appropriate ratios formed [9.57].

The interpretation of dichroic ratio measurements is of obvious importance. It was recognized early on that if we are to go beyond qualitative statements concerning in situ visual pigments, molecular models are needed for aid in quantitative evaluation. However, before proceeding to the possible interpretations of LD measurements, the effect of form dichroism upon the total LD should be evaluated.

HAGINS and JENNINGS were the first to analyze form dichroism in a lamellar system modeled after the vertebrate rod [9.49]. They concluded that it could not be large enough to account for the total observed dichroism. As pointed out by MOODY [9.104], their published formula was erroneous. Also, the analytical description of form dichroism by SNYDER and LAUGHLIN [9.114] appears to be in error, as discussed by HAROSI [9.115]. The same criticism also applies to the derivation of ISRAELACHVILI et al. [9.116]. [There is an indication of an inconsistency in their equation (12) which predicts strong form dichroism for f = 0. However, when the disk volume fraction is zero, the system becomes spatially isotropic in which form dichroism must also vanish. See Sect.9.1.5b for more details]. The relationship published by MOODY, which is identical with (9.36), is believed to be correct. For the estimation of the magnitude of form dichroism, therefore, we would need to know the refractive indices of the disk membrane and that of the cell as a whole.

The refractive index of the disk membrane is unknown. Estimates for it, based on glycerol imbibition experiments dealing with frog rods, range from 1.475 [9.61] to 1.492-1.508 [9.111]. Thus a value $n_1 = 1.50$ is on the high side but, perhaps, not too unreasonable to assume. The average refractive index for frog rod outer segments was first determined by SIDMAN at 1.410 (by the use of white light) [9.23]. BLAUROCK and WILKINS redetermined it with green light at $4°C$ and found it between 1.40 and 1.41 [9.117]. In the latest and probably most careful determination by ENOCH et al., infrared light was used and the average index of refraction for frog rod outer segments was found to be 1.400 (for both dark-adapted and bleached cells) [9.16]. Using the last figure for the average index and $n_1 = 1.50$, therefore, the expected form factor in frog rod LD is about $(1.50/1.40)^4 = 1.32$. Thus, according to (9.36), the intrinsic dichroism of such a cell is about 30% smaller than the total, due to the lamellar stacking of the disks.

The quantitative evaluation of photoreceptor LD was initiated by LIEBMAN [9.19] and WALD et al. [9.112]. The simplest possible molecular model for a rhodopsin-filled cylindrical cell may be summarized as follows (cf. [9.118]). Assume that rhodopsin molecules are independent of each other, that each can be represented with an absorption vector, $\underline{M}$, inclined at a fixed angle to the

plane of the disk membrane (x-y), and that each enjoys three degrees of freedom within the plane (two translational and one rotational). Thus, on the
average, the in-plane extinction will be equal in the x and y directions
($E_x = E_y$). If the total extinction of the cylinder is $E = E_x + E_y + E_z$ and
each molecule has an equal contribution in the z direction as $M_z = E_z/E$, then,
from the side (say x direction), $R = E_y/E_z$, and the fraction pointing in the
z direction is

$$M_z = \frac{E_z}{E} = \frac{E_z}{2E_y + E_z} = \frac{1}{1 + 2R} \quad . \tag{9.51}$$

Thus, by (9.51) the dichroic ratio, R, permits us to compute how effective
the system is for the absorption of light propagating in the physiological
direction (along the z axis). In this way we can estimate that about 90% of
the total rhodopsin extinction is effective in the transverse plane (cf.
[9.112]).

Another possibility for making use of the dichroic ratio measurements is
by computing the angle of inclination of the absorption vector in a specific
molecular distribution model. In the simplest model described above, LIEBMAN
[9.19] obtained $\theta_1 = 16°$, whereas HÁROSI and MALERBA [9.20] calculated
$\theta_1 = 18.5°$.

Perhaps the most useful information from R is the estimation of in situ
visual pigment concentration. This can be done by evaluating the following
equation [9.118]:

$$(c\epsilon_{max}) = (D_\perp/\ell)(2/3 + 1/3R) \quad , \tag{9.52}$$

where ϵ_{max} is the molar extinction coefficient, c is the molar concentration,
$(D_\perp/\ell)$ is the transverse specific optical density of the cell, and R is the
dichroic ratio as defined above. In deriving (9.52), the use of collimated
light was assumed; high-numerical aperture condensors require additional correction [9.20]. Also, the R in (9.52) should correspond to the intrinsic
dichroic ratio and not the total, as discussed above.

The LD properties of the photoproducts have been investigated as well.
Single-cell transverse MSP measurements on frog rods have established that
the metarhodopsins II and III possess essentially the same transverse dichroism as the original rhodopsin [9.19,56,57]. Similar results were obtained
from folded retinae [9.62,119,120]. Concerning retinol in situ, DENTON was
the first to show that its dichroism is axial, i.e., that it is significantly
changed from the strong transverse dichroism of rhodopsin [9.46]. Although
this result was disputed by WALD et al. [9.112], all subsequent determinations

confirmed DENTON's finding (e.g., [9.56,57,62,119,121]). The artificial product retinal oxime, which is formed when bleaching proceeds in the presence
of hydroxylamine, also shows axial dichroism, i.e., similar to that of retinol
but in reversed direction with respect to that of rhodopsin (e.g., [9.56,57,
112]). In a study of the LD properties of amphibian photoreceptors, some of
which contained rhodopsins, some porphyropsins, while some contained their
admixtures, it was found that rhodopsin-containing cells were of stronger
dichroism than those of the porphyropsin-containing counterparts [9.21].
This is consistent with the relatively low dichroic ratios obtained from goldfish cones [9.68,113], and with the recent findings by ERNST and KEMP [9.62]
concerning the LD of the red rods of the axolotl, which contains variable
rhodopsin-porphyropsin mixtures. It appears, therefore, that the photoproducts through at least the meta II but probably through the meta III stage
show no major change in linear dichroism with respect to the parent pigment.
The late products of retinal and retinal oxime, exhibit axial dichroism,
implying a more or less axial chromophore. The interesting studies of WRIGHT
et al. support the above conclusion [9.122,123]. First, they showed that
cattle rhodopsin solubilized in aqueous digitonin and dried into a film of
gelatin behaves just as if it were in a disk membrane; it orients in the plane
of the film and, when bleached in the presence of hydroxylamine, it turns
perpendicular to the film. Second, they established that shear-oriented rho-
dopsin-digitonin pastes orient in the direction of the shear by becoming
highly dichroic in that direction; moreover, the dichroism remains essentially unchanged through the early photoproducts of bathorhodopsin, lumirhodopsin, meta I, and meta II. Again, dichroism of retinal oxime is found
mainly across the direction of shear. The low-temperature measurements of
TOKUNAGA et al. [9.124] and KAWAMURA et al. [9.125-127] resulted in a somewhat different conclusion from the above. According to these authors, the
chromophore orientation in bathorhodopsin is parallel with the disk plane,
whereas those of rhodopsin, lumirhodopsin, meta I, and isorhodopsin all deviate from that plane by about 20°. The apparent unique behavior of bathorhodopsin has neither been explained, nor has the observation been confirmed.
In conclusion, the question of chromophore orientation in visual pigments
and in their photoproducts is somewhat unsettled. Moreover, its significance
in the molecular mechanism of photic excitation remains obscure.

e) Concentration of Visual Pigments in situ

Among the first to measure the photosensitive pigment densities in individual receptors of the frog were DENTON and WYLLIE [9.44]. They found that the axial density change on bleaching of individual rhodopsin-containing rods, on the average, was 0.76 optical density unit at the wavelength of 500 nm. A density of 0.76 means that 83% of the light entering the rods axially would be absorbed, which, when compared with contemporary determinations based mostly on total amounts of pigment extracted from a retina, appeared rather high. The early measurements by LIEBMAN performed on flattened frog rods also yielded relatively high estimates for transverse densities as well as for dichroic ratios [9.19]. His calculated rhodopsin concentration was given as 2.5 ± 0.5 mmole/l, a value that was still maintained by LIEBMAN and ENTINE [9.81]. Meanwhile, others were obtaining lower densities. WALD et al. reported side-on measurements of single frog rods as well as axial measurements of retinal patches (0.2 mm in diameter) in the frog retina [9.112]. The optical density (OD) for the former was 0.072 at λ_{max} for transversely polarized light, whereas the latter had OD = 0.546 at the peak of the difference spectrum (in hydroxylamine). If we take the outer segment of a frog rod to be cylindrical with 6-μm diameter and 50-μm length [9.112], then these densities translate into 0.012/μm and 0.011/μm specific densities (for transversely polarized light), respectively. MARKS measured not only goldfish cone but also frog rod outer segments in transverse orientation [9.54]. He reported changes in absorptance upon bleaching for each type of cell, for transversely polarized light, of about 1.8%/μm, corresponding to a specific density of 0.008/μm. DOBELLE et al. (1969) obtained the same numerical value for the specific density of primate rods and foveal cones [9.69].

LIEBMAN and GRANDA [9.128], while studying the visual pigments in the rods and cones of two species of turtle, estimated the specific density at 0.013/μm, and contended that turtle receptors, like those of other animals, are about 2 mmole/l in visual pigment concentration. This was further amplified by LIEBMAN [9.5]. He stated that the transverse specific density of 0.013 ± 0.002/μm holds constant over a large variety of rods and cones from both the retinal and dehydroretinal pigment families; for the concentration of visual pigments he calculated 2-2.5 mM in situ. It appears now, however, that the above generalization is wrong on two counts; the transverse specific density is neither constant among the various species (there are at least two groups of values, one for each of the two pigment families), nor does the 0.013/μm value hold any special significance (see below).

This author has consistently obtained higher values from rhodopsin-containing cells. For example, the transverse absorptance of a leopard frog rod outer segment was 25% at λ_{max} = 502 nm [9.56]. Since the estimated cell diameter was 7 µm, the transverse specific optical density would be 0.0195/µm. On the basis of numerous spectral recordings from the same species, but only visual estimations of cell diameter, we obtained the following specific densities: 0.0175 ± 0.0015/µm [9.4], 0.0182 ± 0.002/µm [9.68], 0.0170 ± 0.0019/µm [9.57]. The last figure was derived from the uncorrected average prebleach absorption spectra of 38 isolated rods. The density at λ_{max} for transversely polarized light was divided by 6 µm, which is the assumed average cell diameter (taken to be equal with true pathlength). The tolerance figure is one standard deviation, representing the absolute variability at the peak of the spectra. No correction for bleaching or for any other reason was applied to that figure. The former transverse specific density (0.0182 ± 0.002/µm) was obtained as the corrected average of 2 determinations in which the rods were suspended in hydroxylamine; moreover, their dimensions were photographically determined. This author has subsequently studied amphibian rods of other species [9.21], when the cells were routinely photographed after each spectral recording. In that work it was found that although the transverse specific density varied between 0.0123/µm and 0.0195/µm, the equivalent visual pigment concentration appeared less variable at 3.5 mmole/l ± 10%. Furthermore, recent measurements on goldfish cones [9.113] were consistent with an in situ pigment concentration of 3.75 mmole/l, even though the specific densities were found at 0.0139-0.0158/µm.

Among the more recent publications that deal with the issues of specific density and visual pigment concentration are LIEBMAN [9.110] and BOWMAKER et al. [9.70]. In the former the rhodopsin concentration in the leopard frog rod outer segment is given as 3.1 ± 0.1 mmole/l. The latter paper deals with rhesus monkey rod and cone measurements; for rods the specific density is given as 0.019 ± 0.004/µm, for cones as 0.015 ± 0.004/µm. Although the lack of dichroic ratio data prevents one from calculating the in situ pigment concentration in those monkey photoreceptors, the high specific-density values imply concentrations comparable to those of the amphibian rods. It appears, therefore, that more recent determinations yield higher values in both specific density and pigment concentration. Moreover, the visual pigment concentration in vertebrate photoreceptors is maintained at a fairly constant level of about 3-4 mmole/l or, at least, 3.3 ± 0.3 mmole/l.

Example 9.7: By solving (9.52), calculate the rhodopsin concentration for a cell with the following parameters: $D_\perp/\ell$ = 0.015/µm, R = 4. Assuming R_f = 1 (no form dichroism) and ε_{max} = 42,000 1/mole cm, we obtain c = 2.68 mmole/l. If R_f = 1.33, when R_i = 4/1.33 = 3, then (9.52) yields c = 2.78 mmole/l. Thus c is not particularly sensitive to the magnitude of R (as long as R > 1).

f) Shape Variations in Visual Pigment Absorption Spectra

Spectroscopic measurements on extracted visual pigments showed that although the peak of the main absorption band (λ_{max}) may vary from species to species, the overall shape of absorption spectra remain remarkably similar. DARTNALL discovered that, if the relative extinction (optical density) of a rhodopsin pigment is plotted on a reciprocal wavelength or wavenumber (or frequency) scale, its shape is nearly invariant with respect to the extinction of other rhodopsins plotted the same way [9.129]. Based on this observation, he constructed a nomogram in which his template rhodopsin curve, peaking at 502 nm, could generate the shape of rhodopsin spectra with any other λ_{max}. Over the years this proved to be one of the most useful generalizations in the study of primary vision processes because it can be used as an aid in establishing the identity and λ_{max} of previously unknown visual pigments. With some qualifications it will probably remain very useful, even though the spectral shapes are not exactly invariant but depend somewhat on the absolute value of λ_{max}.

The first qualification to the "Dartnall principle" resulted from the realization that visual pigments that use 3,4-dehydroretinal as chromophore (porphyropsins) possess absorption spectra with slightly broader α bands and higher β bands than those using retinal (rhodopsins). For the porphyropsins a new standard curve [9.77] and a new nomogram [9.130] was produced. Meanwhile, DARTNALL improved the accuracy of his data on the rhodopsin standard [9.131] so that there were two nomograms available for the simulation of two families of curves.

In the course of new and improved measurements, data began to accumulate that indicated anomalies in spectral shape. MARKS found that the red-absorbing cones of goldfish yielded somewhat narrower, and the blue somewhat broader absorption spectra than what he expected [9.53,54]. Similarly, TOMITA et al. found in the carp narrower action spectra (electrically recorded from single cones) than either of the two standard curves [9.132]. It was then pointed out by BRIDGES [9.77] that the absorption spectrum of cyanopsin, a synthetic pigment resulting from the condensation of 11-cis, 3,4-dehydroretinal with chicken cone opsin by WALD et al. [9.133], is also unusually narrow.

LIEBMAN and ENTINE [9.81] obtained evidence from the rods and cones of frogs and tadpoles, while LIEBMAN and GRANDA [9.128] from those of turtles, indicating that the red-absorbing cone pigments have narrower spectra, the blue-absorbing rod pigments broader, and the green-absorbing rod pigments about as wide spectra as the appropriate rhodopsin and porphyropsin standards. BAYLOR and HODGKIN, on the basis of electrical recordings from turtle cones, also found systematic differences between spectral sensitivity curves and the appropriate standard [9.134].

The narrowness of the red-absorbing goldfish cone pigment spectrum was confirmed by HÅROSI and MACNICHOL [9.68] and a systematic variation was found among the three goldfish cone pigment spectra by HÅROSI [9.113]. The blue-absorbing ("green") rod spectra were also found to be broader than the existing templates in the larval and adult tiger salamanders and in the toad Bufo marinus [9.21]. As a result of a spectral analysis of MSP records obtained from goldfish cones by HÅROSI [9.113], a spectroscopic continuity was found among the extinction spectra of the chromophore (as measured in solution) and those of the three cone pigments. The width of the main band (at 50% of the peak extinction) was found to vary approximately as the following function

$$WMB(\nu_m) = 3.3 \times 10^3 + 3 \times 10^{-23} \nu_m^6 \, , \tag{9.53}$$

where ν_m is the peak frequency of the main band in cm^{-1}. A hypothesis was also proposed for the conservation of oscillator strengths (as measured by the area under the first three Gaussian components) and, as a consequence of it, that as the main peak narrows toward the red end of the spectrum, the peak extinction increases, whereas toward the blue, the extinction curves are lower and wider. No confirmation of this exists as yet. Although (9.53) should describe width variation for visual pigments belonging to the porphyropsin family, a qualitatively similar scheme has been found for the rhodopsins. The constants for that set, however, are yet to be determined accurately. New wavelength-dependent nomogram have been devised by EBREY and HONIG [9.135] that are essentially in agreement with the predictions of (9.53). As of late, it has been suggested by METZLER and HARRIS [9.136] that the use of lognormal distribution curves permits the description of the shape of extinction curves to a greater accuracy than if Gaussian curves are used. This is not surprising in view of the fact that an extra parameter for skewness is needed for each lognormal component.

Additional evidence for wavelength-dependent shape variation has been found in the cones of chicken [9.137], pigeon [9.138], and monkey [9.70].

Thus, it may be concluded that systematic variations in the spectral shape of visual pigments do, in fact, exist and that the standard curves of DARTNALL [9.129] and BRIDGES [9.77] should be used only near the center of the visible spectrum (500-530 nm). Shape deviations should be expected from the standards in each pigment family for red-absorbing and blue-absorbing visual pigments. For such deviations, however, corrective measures are now at hand.

9.3 Special Topics Aimed at Spurring Further Research

Several issues are brought up in the course of this concluding section. These include transverse membrane specialization in rods and cones, presumably to increase detection efficiency, foveal specialization for increasing spatial resolution, the problems of self-screening and waveguiding, as well as other issues. Emphasis is placed either on the lack of understanding of functional aspects, or on the lack of evidence for mechanisms that are thought to be understood. In all cases the objective is to stimulate research interest by illustrating the need for further investigations.

9.3.1 Structural Specializations in Receptors

Vertebrate photoreceptors are highly specialized cells. Morphologically and functionally they may be either rods or cones (for exhaustive reviews see [9.139,140]). Although the division between them is often tenuous, rods are believed to mediate dim light or scotopic vision and contain rhodopsins whereas cones mediate bright light or photopic vision and contain visual pigments that usually differ in spectral absorption from the rhodopsin present in the same retina. There are two types of rods (i.e., the amphibian "red" and "green" rods) and several types of cones (i.e., single cones, double cones, twin cones, etc.) some of which may be morphologically distinct or only spectroscopically distinguishable (such as in the case of the three cone types of the primate retina). Both cell types have several anatomically recognizable parts: the outer segment, containing membraneous disks bearing the visual pigment molecules; the ellipsoid, enclosing many mitochondria; the myoid, that connects the ellipsoid to the nuclear region of the cell; and an appendage that bears the synaptic apparatus (for a review, see [9.141]).

The outer segment of the rods and cones is highly variable even though the basic structure is remarkably constant. The diversity is mainly in the degree of tapering along the length of the outer segment, and in the dimensions of the diameter and length. Although most cones have tapered outer segments, the majority of rods resemble long, thin cylinders. Exceptions are known in both types. For example, the foveal cones of primates are rodlike [9.142] and the green rods of Bufo marinus are tapered [9.21]. The diameter of a rod or cone outer segment may be 1 μm or slightly smaller such as in the case of the foveal cones of primates [9.142]. Rods found in the amphibian class are commonly much thicker, reaching outer segment diameters of about 12 μm in mudpuppy [9.143]. The thickest rod outer segments, with diameters 25-30 μm, found so far, were in the retina of the African lungfish [9.144]. The outer segment length is also variable. POLYAK found the foveal cones to be 35 μm long in the human retina, but 42-67 μm long in the rhesus monkey [9.142]. While the outer segment of frog rods may be 55 μm long [9.145], that of the African lungfish is 100 μm [9.144]. The longest outer segments are to be found in the rods of some deep-sea fish where they may reach 600 μm in length [9.146].

The fine structure of the outer segments of rods and cones exhibit remarkable similarity in that they all contain hundreds of layers of transverse membranes in a tightly packed array [9.147]. Although the cone membranes may be continuous with the plasma membrane, the rod disks appear to be pinched off (for a review, see [9.141]). On the whole, both rods and cones are well structured for catching light that traverses them axially, provided that they utilize linearly absorbing pigment molecules confined to the transverse planes. In that manner the absorption efficiency is increased by 50% as compared with a random ensemble of molecules (i.e., a solution) at the same concentration and pathlength (i.e., from 1/3 to 1/2). It appears, therefore, that the purpose of the arrangement is to optimize absorption for axially incident light.

Until very recently it was believed that every rod and cone was built on a common plan that utilizes transverse membranes in a cylindrical or conical outer segment. The finding of longitudinal membranes in the cones of two species of anchovy by FINERAN and NICHOL appears contradictory [9.148]. Upon reflection, however, it occurred to the authors that these membranes in the cones of the anchovy probably specialize in detecting photons that are reflected by the tapetal stacks of guanine platelets surrounding the cone outer segments rather than those photons that arrive straight from the entrance pupil of the eye. In this assessment anchovy cones are no exceptions, and that tends to support the idea that absorption efficiency was a primary evolutionary force in the development of photoreceptor structure. Having thus acquired the transverse membranous structure, vertebrate photoreceptors could further increase absorption efficiency in one of two ways (provided that the visual pigment was the most efficient absorber available to the cell and that the pigment molecules in the membranes and the membrane themselves were already as tightly packed as possible), both involving the increase of the number of absorbing layers: 1) by stacking more membranes into longer cells or 2) by stacking the photoreceptors on top of one another. Indeed, both of these solutions were found to exist in deep-sea fish retinae. For them, we tend to think, light must be so scarce that detection efficiency is of utmost importance.

As discussed by LOCKET, the environmental adaptations found in the deep-sea fish retinae present several puzzles [9.146]. For instance, the pigment epithelium is weakly developed in many of these fishes. How then, can a practically flat pigment epithelium mediate pigment regeneration hundreds of micrometers away in long rods or in the inner banks of rods of a tiered retina? Another question is how the signals can be transmitted from the outer segment to the synaptic region (for a review, see [9.149]) in a multilayered retina over the existing long distances? In a recent investigation of the receptor mechanisms in the tiered retina of the conger eel, GORDON and SHAPLEY found that only the innermost layer of rods yields a physiological response to light whereas the other four or more outer banks of rods act merely as passive spectral filters [9.150]. Thus, under certain experimental conditions, the conger eel in the tested phase of its life cycle does not support the hypothesis that multilayered rods are a form of adaptation to an increased retinal sensitivity.

There are other optical problems to which no satisfactory solutions exist at the present time. For instance, why are spectral filters present in the ellipsoid portion of inner segments in some cones but not in others? Birds and reptiles make extensive use of such colored "oil droplets" (for a recent review see [9.151]), whereas fish rarely, and mammals never, use them. So what is the visual function served by such optical filtering that is important for birds and reptiles but unimportant for most other animals? Another problem concerns the myoid portions of rod and cone cells. MILLER and SNYDER proposed that the green-rod and single-cone myoids in the frog retina act as optical waveguides [9.152]. Indeed, these structures under some conditions are filamentous with cross-sectional diameters of 1-2 μm and thus would be expected to propagate visible light in modes. The question is, however, what advantage, if any, would be derived from waveguide properties by the selective use of thin myoids?

9.3.2 Retinal Specializations

It has been well established that while cephalopods, such as squid and octopus, evolved the direct retina, vertebrates arrived at the indirect retina by way of a different developmental process (e.g., [9.153]). Although both types of eye resemble the photographic camera and have an elaborate dioptrical apparatus, they differ in the location of the photosensitive layer. In the direct retina of the cephalopod, light falls first upon the photoreceptors, whereas in the indirect retina of the vertebrate light must pass through the neural layers of the retina before reaching the light-detecting rods and cones. As a consequence of this optical "defect" of the vertebrate eye, the retina had to evolve as a transparent tissue, on the one hand, and apparently needed additional specializations, on the other.

Although many vertebrate possess an area centralis, a specially developed central area of the retina, the highest degree of specialization was reached in primates and in birds. There the area is referred to as the central fovea or fovea centralis. In the human eye the fovea corresponds to a small depression on the vitreal surface of the retina of ca. 1.5 mm in diameter, representing $\pm 2.5°$ visual angle from the ideal fixation point [9.142]. The fovea provides us with the highest degree of visual acuity (i.e., finest spatial resolution), and in its center is the point where the real image is formed, of a small object in three-dimensional space, that the eye is fixated on (i.e., point of fixation). The central portion of the fovea, also called foveola, forms the nearly flat floor of the foveal pit. In man this pit is about 0.4 mm in diameter. This central area is completely rod free (d = 0.5-0.6 mm). The cones here are tightly packed, much longer and thinner than elsewhere. Also, the retina is reduced in thickness by the removal of the inner layers of cells; vitreal to the cone nuclei everything is swept out of the center. No blood vessels or nerve fibers are to be found here except the "cone fibers" (i.e., the connectives to the synaptic region) running radially out of the center to and beyond the foveal ridge (cf. [9.142]).

An additional specialization of the central foveal region in primates is its yellow pigmentation. The area (at least 3 mm across) is called the yellow spot or macula lutea. The pigmentation is most intensive in the slopes and in the margin of the fovea centralis [9.142]. The macular pigment was spectroscopically characterized by WALD [9.154] and in a recent study of three species of monkey retinae, by SNODDERLY et al. [9.155]; the pigment was tentatively localized in two cell types: the cones and the bipolars. In the latter study the yellow bands were found in the "fiber layer" of the fovea, probably in the cone fibers, and it was suggested that the anisotropy thus produced by the pigmented fibers could account for the phenomenon known as Haidinger's brushes (Sect.9.3.3).

As a consequence of the anatomy of the fovea centralis, the thin myoids (cone fibers) cannot serve as light pipes because of their tangential orientation to the sloping foveal surface. Light, therefore, arriving from the entrance pupil of the eye through the vitreous humor to the foveal retina must enter the cone fibers through their sides before penetrating the inner and outer segments. The better the index matching between the adjacent media, the better the efficiency of light propagation and the smaller are the reflection losses and path deviations due to refraction. For a complete optical description, therefore, we need accurate data not only on cellular refractive index of rods and cones, but also on the optical properties of all the forms of biological tissue that contact photoreceptors, such as pigment epithelial processes and glial cells as they exist and interact in the living eye.

9.3.3 Polarized Light Reception in Vertebrates

Although responses to polarized light occur in vertebrates, including man, there is very little known about the mechanism, or, for that matter, the role such a sense plays in the behavior of the organism. The brief summary to follow is based on the recent review of the field by WATERMAN [9.156].

There are a few special circumstances in which the human eye can detect polarized light. Perhaps the best-known example of these is commonly referred to as Haidinger's brushes. The phenomenon is observable with either linearly or circularly polarized light under various conditions (cf. [9.9]). For instance, if the sky is looked at through a Nicol prism, a pattern of four sectors is seen centered about the fixation point: two appear slightly yellower (in the direction of polarization) than the surrounding field, the other two slightly bluer (in the perpendicular direction). In monochromatic light the sectors appear as contrast of brightness and not of color, and can be elicited only at wavelengths that the macular pigment absorbs (i.e., blue), but not at others (such as red). Thus, all these experimental facts implicate the macular pigmentation in forming an anisotropic optical filter in front of the receptors. In agreement with the discussion of the previous section, the cone fibers running out of the foveola seem to be involved. Nevertheless, it remains to be determined whether the macular pigment molecules are preferentially aligned with their absorption vectors parallel to the fiber axes, or the pigmented fibers, while isotropic themselves, exhibit form dichroism by the filamentous arrangement.

Teleost fish appear to respond behaviorally (by orientation) and electro-physiologically (when recording from the brain) to polarized light [9.156]. Although retinal recordings failed to show polarization-dependent responses, in the optic tectum of goldfish such units can be found. Based on the ana-lyses of tectal responses, it has been concluded that large entoptic images (those that are attributable to structures within the eye) must fall on the retina in linearly polarized light. As possible mechanisms, WATERMAN sug-gested selective intraocular scattering and differential oblique absorption by rod and cone outer segments. However, neither of these has been correlated with the observed responses.

9.3.4 Self-Screening in Rods and Cones

As discussed in Sect.9.1.1, absorption and density are logarithmically re-lated. Self-screening is a manifestation of this nonlinear relationship. The larger the peak density, the more pronounced is the nonlinearity. Since the expected D_{max} of a rod or cone cell is large due to its high visual pigment concentration, and the photoreceptor is a photon counting device, its spec-tral sensitivity curve represents the relative quantal absorption (as a func-tion of wavelength) of its pigment at the prevailing optical density. Thus, at a given geometry and molar extinction, the shape of the spectral sensiti-vity spectrum should correspond to the concentration of the pigment mediat-ing the response. This principle has been utilized extensively in vision re-search; an illustrious example is discussed below.

In the course of investigating the absolute sensitivity of dim (scotopic) vision in humans, HECHT et al. needed to estimate the fraction of light caught by rhodopsin in the eye [9.157]. For this they plotted the experimentally de-termined scotopic quantum spectral sensitivity curve of human vision, correc-ted for the transmissivity of the ocular media, and compared it with the re-lative absorptance of rhodopsin spectra drawn for various presumed peak ab-sorptivities. The surprising result of the comparison was that the shape of the spectral sensitivity fell between those of the absorptance spectra with 5% and 20% peak values. Although the 10% curve gave the best fit, they as-sumed 20% as the upper limit for the absorbtion of 510-nm light by rhodopsin in the dark-adapted human retina (20° temporal from the fixation point). Using the 20% value, they estimated that at the threshold of human vision (tested with flashes of 1-ms duration) 5-14 quanta are absorbed by rhodopsin within about 500 rods. HECHT and his co-workers also determined frequency of seeing curves and found, on the basis of Poisson statistics, that 5-8 "critical events" appear to be required to produce the sensation of vision. The agreement between the values of the two determinations support the notion that the critical events are the actual quanta absorbed by the retinal receptors. Indeed, these results had a great deal to do with establishing the human eye as a quantum detector whose operation is close to the theoretical limit of light detection.

In the present context, attention is focused on a dilemma, namely, the apparent close agreement between the shapes of the rhodopsin density spectrum and the scotopic spectral sensitivity of human vision at threshold. Considering the high visual pigment concentrations found in various dark-adapted rods and cones, the agreement was not to be expected. For example, if we assume the length of a human rod outer segment to be 28 μm [9.142], and filled with rhodopsin to yield a transverse specific density of 0.015/μm (Sect.9.2.2e), the total optical density at λ_{max} for light travelling axially would be ca. 0.5, corresponding to 68% absorption. Thus, the 10% figure quoted above is significantly out of line; moreover, it seems unlikely that an error in the spectral sensitivity determination could be large enough to account for the discrepancy. It appears, therefore, that the reason for the lack of broadening in the spectral sensitivity spectrum should be sought elsewhere, namely, in the apparent low optical density of the sensory device. How could such an apparent loss of absorptivity of light come about?

The most plausible answer to the question is that the effective density of a receptor is determined by the effective pathlength of light through its pigment. This follows from the definition of optical density, provided that all other parameters are fixed, such as the molar extinction coefficient, the molar concentration, and the linear dichroism of the outer segment. Since the effective density appears to be low, the effective pathlength should be short, and, in the case of human rods, it may be much shorter than the physical length of the outer segments. There are several possibilities for this occurrence. For instance, the effective pathlength would be shortened if 1) the sensory response originated from a short initial portion of the outer segment, while the rest acted merely as a passive filter (this is a corollary to the situation found in the conger eel mentioned in Sect.9.3.1); 2) light traversed the sensory pigment in short distances due to oblique incidence; and 3) light would "leak out" along the length of the rod structure even though it were "guided" axially.

Although none of the above alternatives is attractive from the viewpoint
of detection efficiency, some are more improbable than others. For example,
possibility 1) appears quite unlikely to occur for several reasons, the most
important being that there is no evidence contrary to the expectation that
rods (and cones) are equally effective in eliciting responses to light applied
along the entire length of the outer segment. There is no evidence to support
proposition 3), either. In fact, waveguiding would be expected to confine most
of the propagating energy to within the guide instead of leaking it out. Thus,
the most probable mechanism is 2), the oblique incidence of light in those
rods that mediate the scotopic spectral sensitivity response of human vision.
Just how oblique should the rays be to account for an effective peak absorptance of 10%? If we assume 1.3 μm for the diameter of a human rod [9.141], and
that the rod in vivo would show 10% peak absorptance (D_{max} = 0.05) while performing its sensory function, the average pathlength and angle of incidence
with respect to the rod axis would be, respectively, 3 μm and 24°. (The estimation is based on a simple geometrical optical model: a dichroic cylinder,
with transverse specific density of 0.018/μm, traversed by oblique rays confined to diagonally transecting planes). These results suggest that extrafoveal rods operate under unfavorable optical conditions. There appear to be
only two processes that could produce oblique incidence of light in photoreceptors within the eye: gross misalignment of receptors and light scattering.
The former is considered improbable in view of the evidence indicating that
all rods and cones tend to point toward the center of the entrance pupil of
the eye (for a recent review see [9.158]). Light scattering within the eye,
therefore, appears to be important. No satisfactory explanation exists at
present and further investigations are necessary.

Although conclusive demonstration of any manifestation of in situ visual
pigment self-screening proved elusive, it has been invoked in the interpretations of several phenomena. STILES suggested that the apparent hue shifts
on viewing lights entering the pupil off center (Stiles-Crawford effect of
the second kind) may be due in part to the difference in shape of the spectral sensitivity curves for different directions of incidence [9.159]. BRIND-
LEY proposed a similar mechanism to account for the upset of color matches
due to adaptation to bright lights [9.160,161]. He, in fact, found the two
hue shifts to be similar and was able to account for both in a theoretical
model for which he assumed that 1) three classes of receptors are active in
foveal vision and 2) the "red" receptors have a maximum optical density of
0.5. An unattractive feature of the model was the very low optical density

ascribed to the green-absorbing receptors. That assumption, however, may not be an essential one [9.9]. WALRAVEN and BOUMAN [9.162] presented a quantitative model based on self-screening of the photolabile receptive pigments to account for the hue-shift phenomenon of STILES. In their analysis D_{max} = 0.7 (80% absorptance) was assumed for each of the three receptor pigments. Even though they used two additional and arbitrary parameters, a "leak fraction" and an "inner segment factor", their theoretical predictions agreed reasonably well with the empirical hue shifts. Unfortunately, however, there are at least two other models capable of accounting equally well for the same phenomenon, as was discussed in a recent analysis by WIJNGAARD and VAN KRUYSBERGEN [9.163]. Thus, we are still faced with the task of designing experiments to establish the place and the extent of self-screening in vertebrate vision.

Acknowledgement. I wish to express my gratitude to Drs. Frank L. Tobey and Edward F. MacNichol, Jr., for reading the manuscript and for providing me with many suggestions toward its improvement. I also thank Mr. Richard Waltz for computer programming and plotting the curves for Figure 9.8. This work was supported in part by grants from the National Eye Institute, EY 02399, and from the Rowland Foundation.

References

9.1   T. Caspersson: Methods for the determination of the absorption spectra of cell structures. J. R. Microsc. Soc. *60*, 8-25 (1940)
9.2   T.O. Caspersson: *Cell Growth and Cell Function. A Cytochemical Study.* (Norton, New York 1950)
9.3   E.F. MacNichol, Jr.: Three-pigment color vision. Sci. Am. *211*, 48-56 (1964)
9.4   E.F. MacNichol, Jr., R. Feinberg, F.I. Hárosi: "Colour Discrimination Processes in the Retina", in *Colour 73* (Adam Hilger, London 1973) pp.191-251
9.5   P.A. Liebman: "Microspectrophotometry of Photoreceptors", in *Photochemistry of Vision*, ed. by H.J.A. Dartnall, Handbook of Sensory Physiology, Vol.7/1 (Springer, Berlin, Heidelberg, New York 1972) pp.481-528
9.6   F.I. Hárosi: "Microspectrophotometry: The Technique and Some of Its Pitfalls", in *Vision in Fishes. New Approaches in Research*, ed. by M.A. Ali (Plenum, New York 1975) pp.43-54
9.7   F.A. Jenkins, H.E. White: *Fundamentals of Optics*, 3rd ed. (McGraw-Hill, New York 1957)
9.8   H.J.A. Dartnall: "Photosensitivity", in *Photochemistry of Vision*, ed. by H.J.A. Dartnall, Handbook of Sensory Physiology, Vol.7/1 (Springer, Berlin, Heidelberg, New York 1972) pp.122-145
9.9   G.S. Brindley: *Physiology of the Retina and Visual Pathway*, 2nd ed. (Williams & Wilkins, Baltimore 1970)

9.10 R.W. Rodieck: *The Vertebrate Retina* (Freeman, San Francisco 1973)
9.11 M. Born, E. Wolf: *Principles of Optics*, 5th ed. (Pergamon, Oxford 1975)
9.12 R.W. Ditchburn: *Light* (Interscience, New York 1953)
9.13 O. Wiener: Die Theorie des Mischkörpers für das Feld der stationären
      Strömung. Abh. Math. Phys. Kl. Königl. Sächs. Ges. Wiss. *32*, 507-604
      (1912)
9.14 W.L. Bragg, A.B. Pippard: The form birefringence of macromolecules.
      Acta Crystallogr. *6*, 865-867 (1953)
9.15 G. Oster: "Birefringence and Dichroism", in *Physical Techniques in Bio-
      logical Research I*. Optical Techniques, ed. by G. Oster, A.W.
      Pollister (Academic, New York 1955) pp.439-460
9.16 J.M. Enoch, J. Scandrett, F.L. Tobey, Jr.: A study of the effects of
      bleaching on the width and index of refraction of frog rod outer seg-
      ments. Vision Res. *13*, 171-183 (1973)
9.17 M. Chabre, A. Cavaggioni: X-ray diffraction studies of retinal rods.
      II. Light effects on the osmotic properties. Biochim. Biophys. Acta
      382, 336-343 (1975)
9.18 J. Hofrichter, W.A. Eaton: Linear dichroism of biological chromophores.
      Annu. Rev. Biophys. Bioeng. *5*, 511-560 (1976)
9.19 P.A. Liebman: In situ microspectrophotometric studies on the pigments
      of single retinal rods. Biophys. J. *2*, 161-178 (1962)
9.20 F.I. Hárosi, F.E. Malerba: Plane-polarized light in microspectrophoto-
      metry. Vision Res. *15*, 379-388 (1975)
9.21 F.I. Hárosi: Absorption spectra and linear dichroism of some amphibian
      photoreceptors. J. Gen. Physiol. *66*, 357-382 (1975)
9.22 O. Wiener: Formdoppelbrechung bei Absorption. Kolloidchem. Beih. *23*,
      189-198 (1926)
9.23 R.L. Sidman: The structure and concentration of solids in photoreceptor
      cells studied by refractometry and interference microscopy. J. Biophys.
      Biochem. Cytol. *3*, 15-30 (1957)
9.24 E.U. Condon: Theories of optical rotatory power. Rev. Modern Phys. *9*,
      432-457 (1937)
9.25 T.I. Shaw: "The Circular Dichroism and Optical Rotatory Dispersion of
      Visual Pigments", in *Photochemistry of Vision*, ed. by H.J.A. Dartnall,
      Handbook of Sensory Physiology, Vol.7/1 (Springer, Berlin, Heidelberg,
      New York 1972) pp.180-199
9.26 L. Velluz, M. Legrand, M. Grosjean: *Optical Circular Dichroism* (Academic,
      New York 1965)
9.27 S. Beychok: Circular dichroism of biological macromolecules. Science
      154, 1288-1299 (1966)
9.28 C.F. Chignell, D.A. Chignell: "The Application of Circular Dichroism
      and Optical Rotatory Dispersion to Problems in Pharmacology", in
      Methods in Pharmacology, Vol.2, Physical Methods, ed. by C.F. Chignell
      (Meredith, New York 1972) pp.111-156
9.29 L.D. Landau, E.M. Lifshitz: *Electrodynamics of Continuous Media*
      (Pergamon, Oxford 1960)
9.30 F. Crescitelli, W.F.H.M. Mommaerts, T.I. Shaw: Circular dichroism of
      visual pigments in the visible and ultraviolet spectral regions. Proc.
      Nat. Acad. Sci. U.S.A. *56*, 1729-1734 (1966)
9.31 H. Shichi, M.S. Lewis, F. Irreverre, A.L. Stone: Biochemistry of visual
      pigments. I. Purification and properties of bovine rhodopsin. J. Biol.
      Chem. *244*, 529-536 (1969)
9.32 H. Shichi: Spectrum and purity of bovine rhodopsin. Biochem. *9*, 1973-
      1977 (1970)
9.33 H. Shichi: Circular dichroism of bovine rhodopsin. Photochem. Photobiol.
      13, 499-502 (1971)

9.34 R.A. Weale: Optical properties of photoreceptors. Br. Med. Bull.
     26, 134-137 (1970)
9.35 A.W. Snyder, P. Richmond: Effect of anomalous dispersion on visual
     photoreceptors. J. Opt. Soc. Am. *62*, 1278-1283 (1972)
9.36 A.W. Snyder, P. Richmond: Anomalous dispersion in visual photoreceptors.
     Vision Res. *13*, 511-515 (1973)
9.37 D.G. Stavenga, H.H. Van Barneveld: On dispersion in visual photorecep-
     tors. Vision Res. *15*, 1091-1095 (1975)
9.38 W.S. Jagger, P.A. Liebman: Anomalous dispersion of rhodopsin in rod
     outer segments of the frog. J. Opt. Soc. Am. *66*, 56-59 (1976)
9.39 W.J. Schmidt: Doppelbrechung, Dichroismus und Feinbau des Aussengliedes
     der Sehzellen vom Frosch. Z. Zellforsch. u. Mikrosk. Anat. *22*, 485-522
     (1935)
9.40 W.J. Schmidt: Polarisationsoptische Analyse eines Eiweiss-Lipoid-Sys-
     tems, erläutert am Aussenglied der Sehzellen. Kolloid Z. *85*, 137-148
     (1938)
9.41 B.K. Johnson: Microspectrophotometry. J. Quekett Microsc. Club Ser. 4,
     Vol.3, 392-396 (1952)
9.42 J.A. Dobrowolski, B.K. Johnson, K. Tansley: The spectral absorption of
     the photopigment of *Xenopus laevis* measured in single rods. J. Physiol.
     130, 533-542 (1955)
9.43 E.J. Denton: On the orientation of molecules in the visual rods of
     Salamandra maculosa. J. Physiol. *124*, 17-18 (1954)
9.44 E.J. Denton, J.H. Wyllie: Study of the photosensitive pigments in the
     pink and green rods of the frog. J. Physiol. *127*, 81-89 (1955)
9.45 E.J. Denton: A method of easily observing the dichroism of the visual
     rods. J. Physiol. *124*, 16-17 (1954)
9.46 E.J. Denton: The contributions of the orientated photosensitive and
     other molecules to the absorption of whole retina. Proc. R. Soc.
     London B*150*, 78-94 (1959)
9.47 R. Barer: Refractometry and interferometry of living cells. J. Opt.
     Soc. Am. *47*, 545-556 (1957)
9.48 T. Hanaoka, K. Fujimoto: Absorption spectra of a single cone in carp
     retina. Jpn. J. Physiol. *7*, 276-285 (1957)
9.49 W.A. Hagins, W.H. Jennings: Radiationless migration of electornic ex-
     citation in retinal rods. Discuss. Faraday Soc. *27*, 180-190 (1959)
9.50 B. Chance, R. Perry, L. Akerman, B. Thorell: Highly sensitive record-
     ing microspectrophotometer. Rev. Sci. Instrum. *30*, 735-741 (1959)
9.51 P.K. Brown: A system for microspectrophotometry employing a commercial
     recording spectrophotometer. J. Opt. Soc. Am. *51*, 1000-1008 (1961)
9.52 P.A. Liebman, G. Entine: Sensitive low light level microspectrophoto-
     meter: detection of photosensitive pigments of retinal cones. J. Opt.
     Soc. Am. *54*, 1451-1459 (1964)
9.53 W.B. Marks: "Difference Spectra of the Visual Pigments in Single Gold-
     fish Cones"; Ph.D. Thesis, The Johns Hopkins University (Baltimore,
     MD 1963)
9.54 W.B. Marks: Visual pigments of single goldfish cones. J. Physiol. *178*,
     14-32 (1965)
9.55 J.M. Enoch: Retinal microspectrophotometry. J. Opt. Soc. Am. *56*,
     833-835 (1966)
9.56 F.I. Härosi: "Frog Rhodopsin in situ: Orientational and Spectral Changes
     in the Chromophores of Isolated Retinal Rod Cells", Ph.D. Thesis, The
     John Hopkins University (Baltimore, MD 1971)
9.57 F.I. Härosi, E.F. MacNichol, Jr.: Dichroic microspectrophotometer: A
     computer-assisted, rapid, wavelength-scanning photometer for measuring
     linear dichroism in single cells. J. Opt. Soc. Am. *64*, 903-918 (1974)

9.58 J.M. Enoch, F.L. Tobey, Jr.: Special microscope microspectrophotometer: Optical design and application to the determination of waveguide properties of frog rods. J. Opt. Soc. Am. *63*, 1345-1356 (1973)

9.59 J.A. Resnik, F.E. Malerba, T.E. Colburn, G.C. Murray, T.G. Smith, Jr.: A novel rapid scanning microspectrophotometer and its use in measuring rhodopsin photoproduct pathways and kinetics in frog retinas. J. Opt. Soc. Am. *68*, 937-948 (1978)

9.60 E.F. MacNichol, Jr.: " A Photon Counting Microspectrophotometer for the Study of Single Vertebrate Photoreceptor Cells", in *Frontiers in Visual Science*, ed. by S.J. Cool, E.:. Smith, III, Proc. Dedication Symp., Univ. Houston College of Optometry, Houston, TX, March, 1977; Springer Series in Optical Sciences, Vol.8 (Springer, Berlin, Heidelberg, New York 1978) pp.194-208

9.61 P.A. Liebman, M.W. Kaplan, W.S. Jagger, F.G. Bargoot: Membrane structure changes in rod outer segments associated with rhodopsin bleaching. Nature *251*, 31-36 (1974)

9.62 W. Ernst, C.M. Kemp: Studies on the effect of bleaching amphibian rod pigments in situ. III. Linear dichroism in axolotl red rods before and during bleaching. Exp. Eye Res. *27*, 101-116 (1978)

9.63 P.K. Brown, G. Wald: Visual pigments in human and monkey retinas. Nature *200*, 37-43 (1963)

9.64 W.R.A. Muntz: Yellow filters and the absorption of light by the visual pigments of some Amazonian fishes. Vision Res. *13*, 2235-2254 (1973)

9.65 W.B. Marks, W.H. Dobelle, E.F. MacNichol, Jr.: Visual pigments of single primate cones. Science *143*, 1181-1182 (1964)

9.66 P.K. Brown, G. Wald: Visual pigments in single rods and cones of the human retina. Science *144*, 45-52 (1964)

9.67 G.C. Murray: "Visual Pigment Multiplicity in Cones of the Primate Fovea"; Ph.D. Thesis, The John Hopkins University (Baltimore, MD 1968)

9.68 F.I. Hárosi, E.F.MacNichol, Jr.: Visual pigments of goldfish cones. Spectral properties and dichroism. J. Gen. Physiol. *63*, 279-304 (1974)

9.69 W.H. Dobelle, W.B. Marks, E.F. MacNichol, Jr.: Visual pigment densities in single primate foveal cones. Science *166*, 1508-1510 (1969)

9.70 J.K. Bowmaker, H.J.A. Dartnall, J.N. Lythgoe, J.D. Mollon: The visual pigments of rods and cones in the rhesus monkey, *Macaca mulatta*. J. Physiol. *274*, 329-348 (1978)

9.71 W.B. Marks, E.F. MacNichol, Jr.: Difference spectra of the visual pigments in single goldfish cones. Abstr. 6th Annu. Meeting Biophys. Soc. TEZ (1962)

9.72 G. Svaetichin, K. Negishi, R. Fatehchand: "Cellular Mechanism of a Young-Hering Visual System", *Physiology and Experimental Psychology*, Ciba Foundation Symposium on Color Vision, ed. by A.V.S. Dereuck, J. Knight (Little, Brown, Boston 1965) pp.178-207

9.73 J.N. Lythgoe: "List of Vertebrate Visual Pigments", in *Photochemistry of Vision*, ed. by H.J.A. Dartnall, Handbook of Sensory Physiology, Vol.7/1 (Springer, Berlin, Heidelberg, New York 1972) pp.604-626

9.74 M.A. Ali, H.-J. Wagner: "Visual Pigments: Phylogeny and Ecology", in *Vision in Fishes, New Approaches in Research*, ed. by M.A. Ali (Plenum, New York 1975) pp.481-516

9.75 G. Wald: Visual purple system in fresh-water fishes. Nature *139*, 1017-1018 (1937)

9.76 G. Wald: The molecular basis of visual excitation. Nature *219*, 800-807 (1968)

9.77 C.D.B. Bridges: Spectroscopic properties of porphyropsins. Vision Res. *7*, 349-369 (1967)

9.78 C.D.B. Bridges: "The Rhodopsin-Porphyropsin Visual System", in *Photochemistry of Vison*, ed. by H.J.A. Dartnall, Handbook of Sensory Physiology, Vol.7/1 (Springer, Berlin, Heidelberg, New York 1972) pp.417-480

9.79   C.D.B. Bridges, S. Yoshikami: The rhodopsin-porphyropsin system in freshwater fishes. 2. Turnover and interconversion of visual pigment prosthetic groups in light and darkness: role of the pigment epithelium. Vision Res. *10*, 1333-1345 (1970)

9.80   T.E. Reuter, R.H. White, G. Wald: Rhodopsin and porphyrospin fields in the adult bullfrog retina. J. Gen. Physiol. *58*, 351-371 (1971)

9.81   P.A. Liebman, G. Entine: Visual pigments of frog and tadpole (*Rana pipiens*). Vision Res. *8*, 761-775 (1968)

9.82   E.R. Loew, H.J.A. Dartnall: Vitamin A_1/A_2-based visual pigment mixtures in cones of the rud. Vision Res. *16*, 891-896 (1976)

9.83   R.N. Young: Visual cells and the concept of renewal. Invest. Ophthalmol. *15*, 700-725 (1976)

9.84   H.J.A. Dartnall, J.N. Lythgoe: The spectral clustering of visual pigments Vision Res. *5*, 81-100 (1965)

9.85   C.D.B. Bridges: Absorption properties, interconversions, and environmental adaptation of pigments from fish photoreceptors. Cold Spring Habor Symp. Quant. Biol. *30*, 317-334 (1965)

9.86   E.F. MacNichol, Jr., Y.W. Kunz, J.S. Levine, F.I. Härosi, B.A. Collins: Ellipsosomes: Organelles containing a cytochrome-like pigment in the retinal cones of certain fishes. Science *200*, 549-552 (1978)

9.87   J.S. Levine, E.F. MacNichol, Jr., T. Kraft, B.A. Collins: Intraretinal distribution of cone pigments in certain teleost fishes. Science *204*, 523-526 (1979)

9.88   J.K. Bowmaker, H.J.A. Dartnall, J.D. Mollon: The violet-sensitive receptors of primate retinae. J. Physiol. *292*, 31 (1979)

9.89   F. Crescitelli, H.J.A. Dartnall: Human visual purple. Nature London *172*, 195-197 (1953)

9.90   W.A.H. Rushton: The difference spectrum and the photosensitivity of rhodopsin in the living human eye. J. Physiol. *134*, 11-29 (1956)

9.91   R.A. Weale: On an early stage of rhodopsin regeneration in man. Vision Res. *7*, 819-827 (1967)

9.92   G. Wald, P.K. Brown: Human rhodopsin. Science *127*, 222-226 (1958)

9.93   G. Wald, P.K. Brown: Human color vision and color blindness. Cold Spring Harbor Symp. Quant. Biol. *30*, 345-361 (1965)

9.94   C.H. Baumann, S. Bender: Kinetics of rhodopsin bleaching in the isolated human retina. J. Physiol. *235*, 761-773 (1973)

9.95   C.D.B. Bridges, T.A. Quilliam: Visual pigments of men, moles and hedgehogs. Vision Res. *13*, 2417-2421 (1973)

9.96   C.D.B. Bridges: Visual pigments of some common laboratory mammals. Nature *184*, 1727-1728 (1959)

9.97   W.A.H. Rushton: Kinetics of cone pigments measured objectively on the living human fovea. Ann. N.Y. Acad. Sci. *74*, 291-304 (1958)

9.98   P.K. Brown: Rhodopsin rotates in the visual receptor membrane. Nature London New Biol. *236*, 35-38 (1972)

9.99   R.A. Cone: Rotational diffusion of rhodopsin in the visual receptor membrane. Nature London New Biol. *236*, 39-43 (1972)

9.100 M. Poo, R.A. Cone: Lateral diffusion of rhodopsin in the photoreceptor membrane. Nature *247*, 438-441 (1974)

9.101 P.A. Liebman, G. Entine: Lateral diffusion of visual pigment in photoreceptor disk membranes. Science *185*, 457-459 (1974)

9.102 T.P. Williams, E.F. MacNichol, Jr., H.E. Johnson: Lateral diffusion of photopigments in the outer segments of rods of *Bufo marinus*. Biol. Bull. Woods Hole, Mass. *153*, 450 (1977)

9.103 T.H. Goldsmith, R. Wehner: Restrictions on rotational and translational diffusion of pigment in the membranes of a rhabdomeric photoreceptor. J. Gen. Physiol. *70*, 453-490 (1977)

9.104 M.F. Moody: Photoreceptor organelles in animals. Biol. Rev. Cambridge Philos. Soc. *39*, 43-86 (1964)

9.105 W.J. Schmidt: Polarisationsoptische Analyse der Verknüpfung von Protein und Lipoidmolekeln, erläutert am Aussenglied der Sehzellen der Wirbeltiere. Pubbl. Stn. Zool. Napoli *23* Suppl., 158-183 (1951)

9.106 R.A. Weale: On the birefringence of rods and cones. Pflügers Arch. *329*, 244-257 (1971)

9.107 R.A. Weale: On the linear dichroism of frog rods. Vision Res. *11*, 1373-1385 (1971)

9.108 R.A. Weale: Rod birefringence and light. Vision Res. *11*, 1387-1393 (1971)

9.109 M.W. Kaplan, P.A. Liebman: Slow bleach-induced birefringence changes in rod outer segments. J. Physiol. *265*, 657-672 (1977)

9.110 P.A. Liebman: "Birefringence, Dichroism and Rod Outer Segment Structure", in *Photoreceptor Optics*, ed. by A.W. Snyder, R. Menzel (Springer, Berlin, Heidelberg, New York 1975) pp.199-214

9.111 M.W. Kaplan, M.E. Deffebach, P.A. Liebman: Birefringence measurements of structural inhomogeneities in *Rana pipiens* rod outer segments. Biophys. J. *23*, 59-70 (1978)

9.112 G. Wald, P.K. Brown, I.R. Gibbons: The problem of visual excitation. J. Opt. Soc. Am. *53*, 20-35 (1963)

9.113 F.I. Hárosi: Spectral relations of cone pigments in goldfish. J. Gen. Physiol. *68*, 65-80 (1976)

9.114 A.W. Snyder, S.B. Laughlin: Dichroism and absorption by photoreceptors. J. Comp. Physiol. *100*, 101-116 (1975)

9.115 F.I. Hárosi: Comments on form dichroism in vertebrate photoreceptors. Vision Res. *18*, 353-354 (1978)

9.116 J.N. Israelachvili, R.A. Sammut, A.W. Snyder: Birefringence and dichroism of photoreceptors. Vision Res. *16*, 47-52 (1976)

9.117 A.E. Blaurock, M.H.F. Wilkins: Structure of frog photoreceptor membranes. Nature *223*, 906-909 (1969)

9.118 F.I. Hárosi: "Linear Dichroism of Rods and Cones", in *Vision in Fishes. New Approaches in Research*, ed. by M.A. Ali (Plenum, New York 1975)

9.119 C.M. Kemp: "Dichroism in Rods During Bleaching", in *Biochemistry and Physiology of Visual Pigments*, ed. by H. Langer, Symposium, Inst. f. Tierphysiologie, Ruhr-Universität, Bochum, Germany, Aug. 27-30, 1972 (Springer, Berlin, Heidelberg, New York 1973) pp.307-312

9.120 R.A. Weale: "Metarhodopsin III", in *Biochemistry and Physiology of Visual Pigments*, ed. by H. Langer, Symposium, Inst. f. Tierphysiologie, Ruhr-Universität, Bochum, Germany, Aug. 27-30, 1972 (Springer, Berlin, Heidelberg, New York 1973(pp.101-104

9.121 P.A. Liebman: Microspectrophotometry of retinal cells. Ann. N.Y. Acad. Sci. *157*, 250-264 (1969)

9.122 W.E. Wright, P.K. Brown, G. Wald: The orientation of rhodopsin and other pigments in dry films. J. Gen. Physiol. *59*, 201-212 (1972)

9.123 W.E. Wright, P.K. Brown, G. Wald: Orientation of intermediates in the bleaching of shear-oriented rhodopsin. J. Gen. Physiol. *62*, 509-522 (1973)

9.124 F. Tokunaga, S. Kawamura, T. Yoshizawa: Analysis by spectral difference of the orientational change of the rhodopsin chromophore during bleaching. Vision Res. *16*, 633-641 (1976)

9.125 S. Kawamura, F. Tokunaga, T. Yoshizawa: Absorption spectra of rhodopsin and its intermediates and orientational change of the chromophore. Vision Res. *17*, 991-999 (1977)

9.126 S. Kawamura, S. Wakabayashi, A. Maeda, T. Yoshizawa: Isorhodopsin: conformation and orientation of its chromophore in frog disk membrane. Vision Res. *18*, 457-462 (1978)

9.127 S. Kawamura, F. Tokunaga, T. Yoshizawa, A. Sarai, T. Kakitani: Orientational changes of the transition dipole moment of retinal chromophore on the disk membrane due to the conversion of rhodopsin to bathorhodopsin and to isorhodopsin. Vision Res. *19*, 879-884 (1979)

9.128 P.A. Liebman, A.M. Granda: Microspectrophotometric measurements of visual pigments in two species of turtle, *Pseudemys scripta* and *Chelonia mydas*. Vision Res. *11*, 105-114 (1971)

9.129 H.J.A. Dartnall: The interpretation of spectral sensitivity curves. Br. Med. Bull *9*, 24-30 (1953)

9.130 F.W. Munz, S.A. Schwanzara: A nomogram for retinene$_2$-based visual pigments. Vision Res. *7*, 111-120 (1967)

9.131 G. Wyszecki, W.S. Stiles: *Color Science*. Concepts and Methods, Quantitative Data and Formulas (Wiley, New York 1957)

9.132 T. Tomita, A. Kaneko, M. Murakami, E.L. Paulter: Spectral response curves of single cones in the carp. Vision Res. *7*, 519-531 (1967)

9.133 G. Wald, P.K. Brown, P.H. Smith: Cyanopsin, a new pigment of cone vision. Science *118*, 505-508 (1953)

9.134 D.A. Baylor, A.L. Hodgkin: Detection and resolution of visual stimuli by turtle photoreceptors. J. Physiol. *234*, 163-198 (1973)

9.135 T.G. Ebrey, B. Honig: New wavelength-dependent visual pigment nomograms. Vision Res. *17*, 147-151 (1977)

9.136 D.E. Metzler, C.M. Harris: Shapes of spectral bands of visual pigments. Vision Res. *18*, 1417-1420 (1978)

9.137 J.K. Bowmaker, A. Knowles: The visual pigments and oil droplets of the chicken retina. Vision Res. *17*, 755-764 (1977)

9.138 J.K. Bowmaker: The visual pigments, oil droplets and spectral sensitivity of the pigeon. Vision Res. *17*, 1129-1138 (1977)

9.139 G.L. Walls: *The Vertebrate Eye and Its Adaptive Radiation* (Hafner, New York 1963)

9.140 F. Crescitelli: "The Visual Cells and Visual Pigments of the Vertebrate Eye", in *Photochemistry of Vision*, ed. by H.J.A. Dartnall, Handbook of Sensory Physiology, Vol.7/1 (Springer, Berlin, Heidelberg, New York 1972) pp.245-263

9.141 A.I. Cohen: "Rods and Cones", in *Physiology of Photoreceptor Organs*, ed. by M.G.F. Fuortes, Handbook of Sensory Physiology, Vol.7/2 (Springer, Berlin, Heidelberg, New York 1972) pp.63-110

9.142 S.L. Polyak: *The Retina* (Univ. Chicago Press, Chicago 1941)

9.143 P.K. Brown, I.R. Gibbons, G. Wald: The visual cells and visual pigment of the mudpuppy, *Necturus*. J. Cell Biol. *19*, 79-106 (1963)

9.144 N.A. Locket: Landolt's club in the retina of the African lungfish *Protopterus aethiopicus*, Heckel. Vision Res. *10*, 299-306 (1970)

9.145 J.J. Wolken: Structure and molecular organization of retinal photoreceptors. J. Opt. Soc. Am. *53*, 1-19 (1963)

9.146 N.A. Locket: "Some Problems of Deep-Sea Fish Eyes", in *Vision in Fishes. New Approaches in Research*, ed. by M.A. Ali (Plenum, New York 1975) pp.645-655

9.147 F.S. Sjöstrand: The ultrastructure of the outer segments of rods and cones of the eye as revealed by the electron microscope. J. Cell Comp. Physiol. *42*, 15-44 (1953)

9.148 B.A. Fineran, J.A.C. Nicol: Novel cones in the retina of the anchovy (Anchoa). J. Ultrastruc. Res. *54*, 296-303 (1976)

9.149 W.A. Hagins: The visual process: Excitatory mechanisms in the primary receptor cells. Annu. Rev. Biophys. Bioeng. *1*, 131-158 (1972)

9.150 J. Gordon, R.M. Shapley: Receptor mechanisms in the tiered retina of the conger eel (Conger conger). Invest. Ophthalmol. Vis. Sci. Supp. *18*, 7 (1979) (abstract)

9.151 W.H. Miller: "Ocular Optical Filtering", in *Comparative Physiology and Evolution of Vision in Invertebrates*. A. Invertebrate Photoreceptors, ed. by H. Autrum, Handbook of Sensory Physiology, Vol.7/6A (Springer, Berlin, Heidelberg, New York 1979) pp.69-143

9.152 W.H. Miller, A.W. Snyder: Optical function of myoids. Vision Res. *12*, 1841-1848 (1972)

9.153 H.V. Zonana: Fine structure of the squid retina. Bul. Johns Hopkins Hosp. *109*, 185-205 (1961)

9.154 G. Wald: The photochemistry of vision. Doc. Ophthalmol. *3*, 94-137 (1949)

9.155 D.M. Snodderly, J. Auran, F.C. Delori: Localization of the macular pigment. Invest. Opthalmol. Vis. Sci. Supp. *18*, 80 (1979) (abstract)

9.156 T.H. Waterman: "Natural Polarized Light and e-Vector Discrimination by Vertebrates", in *Light as an Ecological Factor II*, ed. by G.C. Evans, R. Bainbridge, O. Rackham (Blackwell Scientific, Oxford 1975) pp.305-335

9.157 S. Hecht, S. Shlaer, M.H.L. Pirenne: Energy, quanta, and vision. J. Gen. Physiol. *25*, 819-840 (1942)

9.158 J.M. Enoch: Vertebrate photoreceptor orientation. Int. J. Quantum Chem. Symp. 3, 65-88 (1976)

9.159 W.S. Stiles: The luminous efficiency of monochromatic rays entering the eye pupil at different points, and a new colour effect. Proc. R. Soc. London B*123*, 90-118 (1937)

9.160 G.S. Brindley: The effects on colour vision of adaptation to very bright lights. J. Physiol. *122*, 332-350 (1953)

9.161 G.S. Brindley: A photochemical reaction in the human retina. Proc. Phys. Soc. London B*68*, 862-870 (1955)

9.162 P.L. Walraven, M.A. Bouman: Relation between directional sensitivity and spectral response curves in human cone vision. J. Opt. Soc. Am. *50*, 780-784 (1960)

9.163 W. Wijngaard, J. van Kruysbergen: "The Function of the Nonguided Light in Some Explanations of the Stiles-Crawford Effects", in *Photoreceptor Optics*, ed. by A.W. Snyder, R. Menzel (Springer, Berlin, Heidelberg, New York 1975) pp.175-183

10. Tapeta Lucida of Vertebrates

J. A. C. Nicol

With 12 Figures

Tapeta lucida are reflecting layers that lie immediately outside the retina. They reflect light into the retina and participate in vision. Another shiny layer, the stratum argenteum, is found in the outer chorioid against the sclera. The argenteum is usually obscured by a pigment layer and it is only rarely that it reflects light into the retina. Tapeta lucida occur in invertebrates and vertebrates. This account is limited to vertebrate tapeta. Although commonly present, tapeta lucida have received less attention than other intraocular structures, possibly because they do not lie directly in the optical pathway to the retina. However, a tapetum is capable of directing a significant amount of light back into the retina, light that augments the photoreceptor response and affects the quality of the image. In addition, the tapetum sometimes acts as a screen for the photoreceptors.

10.1 Types and Appearance of Tapeta Lucida

10.1.1 Types of Tapeta Lucida and Their Distribution in Vertebrates

By way of introduction the principal kinds of tapeta lucida (according to location and structure) and their phyletic occurrence in vertebrates will be briefly described.

Tapeta lucida occur in either the pigment epithelium or the chorioid, and are termed retinal and chorioidal. Light reflection from tapeta depends upon physical mechanisms and the tissue contains minute paraplasmic or extracellular bodies that form the structural basis for these mechanisms.

a) Retinal Tapeta Lucida

Retinal tapeta lucida occupy the cell processes or cell bodies. The reflecting particles that constitute the physical basis of reflectivity are of two main types. Either they are small granules, in the form of spheres, cubical crystals and the like, which are densely and massively packed in the cell processes; or the reflecting entities are thin platelets, which are regularly arranged and stacked in tiers.

b) Chorioidal Tapeta

Chorioidal tapeta lie in the inner region of the chorioid and two types are known. One is the tapetum cellulosum composed, as the name implies, of reflecting cells stacked in depth. The cells contain numerous refractile bodies, of uniform size and shape, which have an orderly arrangement. For historical reasons the chorioidal tapeta of fishes are referred to as guanine tapeta, those of mammals, as cellular tapeta. In the following pages the term tapetum cellulosum is used to include both of these tapeta. The other type is the tapetum fibrosum, which is a layer of densely packed connective tissue fibers.

c) Distribution of Tapeta Lucida

A tapetum cellulosum is found in many fishes, including selachians, chimaeroids, sturgeons, the bichir *Polypterus* and reed-fish *Calamoichthys* (Brachiopterygii), several families of teleosts (glasseyes Priacanthidae; snake mackerels Gempylidae), the Australian lungfish or barramunda *Neoceratodus* (Dipnoi), and the coelacanth *Latimeria*. Among mammals it is known in carnivores, seals and prosiminians (lemurs, bush babies, etc.) [10.1-9].

A tapetum fibrosum is found in most ruminant ungulates (Artiodactyla); in horese, asses and zebras (Equidae, Perissodactyla); in elephants (Proboscidea); and in whales (Cetacea).

In addition, there are reports of tapeta lucida, either of the fibrosum or cellulosum type, in the Tasmanian wolf *Thylacinus*, the Australian native cat *Dasyurus*, the aardvark *Orycteropus*, the American tapir, the rock cony *Procavia*, the spotted cavy *Cuniculus*, and some others [10.3].

A retinal tapetum has been found in bony fishes (Actinopterygii), alligators and crocodiles (Crocodilia), in goatsuckers (Caprimulgidae) among birds, in the American opossum and in some Old World fruit-bats (Megachiroptera) [10.4,10]. Noteworthy are the wide distribution, common occurrence, and diverse composition of retinal tapeta among fishes, occurring in gars (Lepisosteidae) and in many families of teleosts [10.11-13].

10.1.2 Historical Note

A nocturnal phenomenon so conspicuous as eyeshine has patently long been a matter of common knowledge. A summarized historical account and literature review are to be found in FRANZ [10.3], who drew upon a treatise by MURR [10.14].

A reflector within the eye was recognized in the seventeenth century, when it was termed a tapis or tapetum. Studies of carnivore and ruminant eyes led to a differentiation of cellular and undulatory (fibrous) types of tapeta. The appearance of the fundus in freshly dissected eyes of mammals was described by SOEMMERRING [10.15], and the fundus as it appeared in the ophthalmoscope, by JOHNSON [10.16,17].

Tapeta lucida of fishes were recognized by DELLE CHIAJE [10.18] and BRÜCKE [10.19] and have been studied repeatedly since [10.1,2,20-23].

10.1.3 Eyeshine and Appearance of the Fundus

Among lower vertebrates that have tapeta lucida, eyeshine is visible in dark-adapted eyes and may appear red because of the visual pigment. That of birds can be seen at surprisingly great distances, up to 200 μm in thick-knees (Burhinidae) [10.24].

The tapetal color in mammals is varied. Among carnivores the fundus in the tapetal region is gold to green, sometimes mixed in the same eye. Among ungulates it ranges from yellow to blue. Slow lories and galagos have golden tapeta.

The tapetum generally in carnivores and ungulates is restricted to the dorsal fundus, where it appears as a triangular area, base horizontal, apex above. In cetaceans, it extends over most of the fundus [10.25]. It is restricted to the dorsal temporal fundus of poor-wills (goatsuckers) and alligators [10.10,26,27]. Generally, in coastal and freshwater fishes the tapetum fills the upper twothirds of the fundus [10.28]. But in others, including deep-sea species, it may occupy the entire fundus, or occur in many other patterns. The pigment epithelium is usually without pigment over chorioidal tapeta but, again, there are exceptions, for example the sturgeon [10.6].

Anomalous eyeshine has been reported in the human [10.4], and there is a rare inherited retinal anomaly known as Oguchi's disease wherein a reflector or tapetum is formed following adaptation to high light levels. Both fovea and periphery appear to be affected; the physical basis of the reflection is not well understood [10.29].

10.2 Structure of Tapeta

Chorioidal tapeta will be described first. They lie immediately outside the choriocapillaris and are penetrated by small blood vessels which connect with the latter.

10.2.1 Tapetum Cellulosum

The tapetum cellulosum is a thin band of reflecting cells which are stacked in depth, sometimes in regular rows or columns, and which contain an orderly arrangement of refractile bodies.

In fishes, as far as known, the tapetum is a layer of flat cells, four or five deep. As a rule, the cells are parallel to the retinal surface in the central fundus and obliquely directed towards the posterior pole near the periphery (Fig.10.1). The cells overlap one another and contain piles of thin flat crystals which lie parallel to each other and to the planar surface of the cell. Crystals tend to be spaced at fairly regular intervals; each is closely invested by a crystal membrane and lies in a crystal sac formed by an undulatory unit membrane [10.31].

Tapeta of selachians and chimaerids range in color from blue (deep-sea squaloids) to yellow-green (stingrays); generally the surface has a light metallic sheen. The cells contain about 12 crystals, and there are about 60 superposed crystals through the thickness of the tapetum. The crystals are hexagons, variable in planar dimensions, and there are a number of them along the length of the cell, overlapping in tandem. The tapetum is backed by a layer of melanophores, the processes of which extend inwards between the reflecting cells in many species; the pigment is able to migrate inwards through the processes so as to occlude the tapetum [10.3,30,32-34].

Chorioidal tapeta of other fishes are generally similar to those of se-lachians, e.g., sturgeon [10.6] and *Latimeria* [10.9]. In the latter, the thickness of the tapetum is sufficient to accommodate several hundred crystals. The oil fish *Ruvettus* contains very large hexagonal crystals in each of which the thickness varies in different sectors [10.35]. In the tapetum of glass-eyes the crystals are regularly arranged, and each cell contains one or two stacks, recalling the orderly arrangment found in the reflecting cells of the ratfish skin [10.36]. The crystals, hexagonal in shape, are thinner towards their margins. The tapetum of the glasseye is orange and very shiny. In none of these fishes is there an occlusion mechanism [10.8,37].

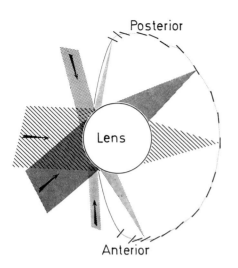

Posterior

Lens

Anterior

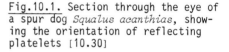

Fig.10.1. Section through the eye of a spur dog *Squalus acanthias*, showing the orientation of reflecting platelets [10.30]

The tapetum cellulosum of carnivores and seals is made up of closely spaced cells stacked above one another, forming more or less regular columns, up to 35 deep. The cells are flat hexagons or polygons, rectangular in cross section, which are parallel to the retinal surface; they contain a mass of birefringent rodlike structures [10.38-40].

In the cat's tapetum the rodlike structures observed by light microscopy are bundles of tiny rods (rodlets), arranged in groups with their long axes parallel to the retinal surface. Rodlets are 4.8 μm long and 0.23 μm in mean diameter (range 0.19 to 0.35 μm). Each rodlet consists of a dense outer zone enclosing a uniform interior of lighter density, and is surrounded externally by irregular strands, which sometimes are continuous with strands of adjacent rodlets.

Usually there are 12 to 14 layers of rodlets in a bundle. The arrangement is highly regular; rodlets within a bundle are parallel, but the direction in adjacent bundles to planar view may be quite diverse. On the other hand, the rodlets in a vertical plane are nearly all aligned in one direction through the thickness of one cell, and the majority of rodlets are aligned in the same direction in a vertical section of adjacent superposed cells. An orderly arrangment of this kind has been observed through ten cell layers. Rodlets within a bundle are layered in a regular manner and exhibit two-dimensional periodicity. Mutually parallel, in section they show an array like a crystal lattice; one set of lattice planes may lie within 15° of the chorioidal surface, and others lie at various other angles [10.41,42].

The tapetum of the dog consists of 10 to 12 layers of closely packed reflecting cells which appear as irregular polygons in surface view, and rectangles in sections. They contain arrays of rodlets arranged in parallel. Each rodlet is enclosed by a double membrane and has a microtubule in its center [10.43].

A cellular tapetum is present in many nocturnal lemurs (Stripsirhines) [10.7,44]. In the bush baby *Galago* it is brilliant gold, and extends over most of the fundus. It is very thin, about 12 μm, and the flattened cells of which it is composed form six or seven layers. Each cell has a cytoplasmic rind enclosing 18 to 25 regularly spaced lamellae separating flat crystals. Protrusions and irregularities on the surfaces of lamellae and crystals possibly indicate that the crystal stacks are keyed into the lamellae and are interlocked. In a vertical section through the tapetum, there are about 130 superposed lamellae or 260 thin films [10.45].

10.2.2 Tapetum Fibrosum

This type of tapetum is found in most artiodactyls (except pigs). The bovine eye has been most studied. The tapetum is a specialized development of chorioidal connective tissue and consists of a compact layer of regularly arranged fibers which are undulatory and lie more or less parallel to the retina. The fibers are organized in a spiral or in concentric rings about the tapetal center; the tapetum, therefore, is a layer of long fusiform fibers arranged in a whorl. The fibers are about 10 μm thick, densely packed and lying, perhaps, in some kind of a matrix; between them are thin elongated cells that are fibrocytes [10.3,5,14,19,46,47].

10.2.3 Retinal Tapeta Lucida

They fall into two types, those containing small particles (tapetal spheres and the like), and those containing thin films. The former are diffuse, the latter specular reflectors. In describing each kind, a phyletic arrangement is adopted; it is noteworthy that both kinds can occur in the same eye.

a) Fishes

Retinal tapeta among fishes were first described in the bream *Abramis* and the ruffe *Acerina*; they have since been discovered in many other species [10.3, 12,48]. The commonest type is one in which the pigment epithelium cells contain numerous spherules or tapetal spheres. It occurs in gar-pikes (Lepisosteiformes) and in many families of teleosts, including carps and characins

(Cypriniformes), cats (Siluriformes), weakfishes, grunts, spadefishes, cusk-eels, etc. (Perciformes). Long processes of the pigment epithelium cells are packed with the tapetal spheres; indeed, little else is visible except some smooth endoplasmic reticulum. There are different arrangements related to the organization of the photoreceptor layer. Sometimes the tapetum forms a layer largely sclerad (distal) to the receptors, e.g., cusk-eels; sometimes the processes, containing spherules, extend vitread between the photorecep-tor cells almost to the external limiting membrane, e.g., characins, and may envelop the outer segments of the visual cells, e.g., walleyes [10.13, 49-52].

The tapetal spheres contain refractile material, of which several kinds are known (see p.412). In lipid tapeta the spherules are surrounded by a boundary layer, possibly phospholipid [10.53]. In melanoid tapeta they are invested by a distinct membrane apparently continuous with the smooth endo-plasmic reticulum, e.g., catfishes [10.54]. Generally, the spherules are of about the same size, showing a unimodal size distribution; mean diameters are about 0.5 μm in *Malacosteus*, 0.36 μm in sea trout and catfishes, and 0.24 μm in gars. The density is about 11 spherules per μm^3 in the catfish [10.11,13,54,55].

The gizzard shad has a particularly dense tapetum which extends far vitread, almost reaching the external limiting membrane. The epithelial cells are loaded with pteridine crystals which are about 0.26×0.49 μm in cross section and randomly distributed. They are cuboidal or rectangular polyhedrons and are membrane enclosed [10.56]. The reflecting particles of the sauger and walleye are randomly distributed, and measure about 0.35×0.60 μm [10.51]. Presumably the tapetum of the pike-perch *Lucioperca* has the same structure [10.51].

Small particles of guanine occur in the retinal tapeta of many teleosts, including the anchovy, bream, mormyrids, ruffe, etc. [10.21,28,57,58]. In-formation about ultrastructure in the last three species does not seem to be available. Two reflecting systems having dissimilar roles are found in the pigment epithelium of several teleosts, for example, the anchovy (*Anchoa*). The anchovy has a diffuse reflector which occupies the basal regions of the epithelium; it is made up of a large number of minute cubical crystals and serves the rods. It also has a specular reflector which occupies the distal region of the cell processes and serves the cones. Cell processes extend inwards between the outer segments of two adjacent long cones, and the two sloping opposing faces contain thin flat crystals, which are stacked in regu-lar piles, several deep. Long cones and intervening epithelial processes are

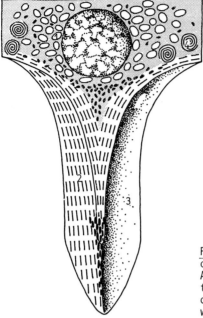

Fig.10.2. Diagram of a pigment epithelium cell in the main retina of *Scopelarchus*. A layer of pigment granules lies vitread to the nucleus and in the apical center of the cell. Most of the apical region is packed with oriented crystalline platelets [10.62]

aligned in rows running dorsoventrally around the optic cup; between these rows the cell processes contain a single layer of needle-shaped crystals [10.59-61].

Eyes of deep-water teleosts are sometimes regionally differentiated and very complex. Part of the retina of *Scopelarchus*, for example, contains closely packed rods which are backed by a diffuse reflector of randomly oriented platelets. Part is a grouped retina in which the rods are in bundles, which are inserted into pits lined with stacks of parallel, reflecting platelets facing the photoreceptors (Fig.10.2). Because of this arrangement the bundles of rods are spatially and optically isolated from one another [10.13, 62]. Other interesting arrangements of this sort, involving rods and cones of bathypelagic species, have been described by FREDERIKSEN [10.63] and MUNK [10.64].

Attempts to categorize these reflecting systems fail because of their variety. Often, but not exclusively as we have seen, dual retinal reflectors serve rods and cones, respectively, or are associated with grouped retinae. Several taxa which have grouped duplex retinae and dual retinal reflectors are Elopidae, Mormyridae, and Hiodontidae. In the latter (mooneyes), the cones are grouped and are ensheathed by a reflector of regularly arranged rod-shaped crystals; the rods are surrounded by randomly distributed disk-shaped crystals [10.49,57,65-69].

In lanternfishes (Myctophidae) the epithelium contains a row of oriented platelets and simulates a chorioidal tapetum cellulosum [10.70].

b) Crocodiles

The tapetum of alligators and crocodiles is a horizontal band lying slightly above the optic nerve [10.71]. It is white in vivo; the cell bodies and processes contain minute granules of reflecting material and melanosomes. In the central retinal region melanin is sparse; above and below that region the melanin increases substantially. Outer segments of the visual cells are embedded in the tapetal layer [10.14,26,71,73].

c) Goatsuckers

A lipid tapetum lucidum remarkably like that of teleost fishes is found in the eyes of goatsuckers (Caprimulgiformes). This is a group of crepuscular and nocturnal birds which include poor-wills, nighthawks, nightjars and oil-birds. The tapetum is white, diffusely reflecting, and occupies the posterior and dorsal fundus. Processes of the pigment epithelium cells in the tapetal region are packed with tapetal spheres, randomly arranged. The spheres are about 0.5 μm in diameter, and have a thin surface membrane [10.10].

d) Mammals

The tapetum of the American opossum (*Didelphis*) forms a semicircular area in the superior fundus. Processes of the pigment epithelium cells in this region contain lipid spheres and are without melanin [10.4,5,74,75].

A retinal tapetum occurs in Old World fruit-bats or flyingfoxes, e.g., *Pteropus*, *Hysignathus*, etc., in which the choriodoretinal interface is peculiarly organized. The chorioid is thrown into conical protruberances on which the pigment epithelium and outer retina are folded, like so many papillary caps. In *Pteropus giganteus* the tapetum is limited to the upper half of the fundus. The pigment epithelium consists of a single layer of cuboidal cells which in the tapetal region are filled with a dense accumulation of small spheres. Of variable diameter, they have a dense perimeter and a lighter center, and are embedded in closely packed endoplasmic reticulum [10.3,4,32, 76-81].

10.3 Retinomotor Activity and Pigment Migration

Retinomotor movements, as generally understood, refer to radial displacement of cones, rods and retinal pigment in lower vertebrates, and involve the tapetum lucidum, when present. In brief, rods move outwards, cones and retinal pigment inwards when the eye is illuminated, and reverse changes occur in darkness or dim light (scotopic conditions). Catfishes (*Ictalurus*) and bream (*Abramis*) are species in which retinomotor activity has been much investigated.

In dark-adapted eyes the melanin pigment is retracted into the basal region of the pigment epithelium cells and the tapetum is unmasked. The rods, being shortened, come to lie inside the tapetum. In light-adapted eyes the melanin extends into the cell processes, obscuring the tapetum, cones are shortened and rods are pushed into the pigment epithelium. In the former condition, therefore, the rods are in a position to receive light reflected from the tapetum; in the latter, they are screened both by melanin and tapetal reflecting material (Fig.10.3). Similar conditions obtain in many other fishes [10.20,28,54,51,71,82-91]. It has been estimated that migration of retinal pigment (in the frog) reduces the effective intensity of light to one-third [10.92].

This is the conventional pattern obtaining in a great many teleostomes, and there are many variations and exceptions. In gar fishes (*Lepisosteus*), for example, there is little or no movement of cones. In the walleye only the rods move; the cones lie close to the external limiting membrane; and the melanin remains concentrated at the tips of the epithelial processes at the level of the cone ellipsoids [10.51]. No retinomotor movements take place in sturgeon, cusk-eels, deep-sea teleosts, etc.

Retinomotor movements are fairly slow, occurring over 20 to 70 min; usually light adaptation is faster than dark adaptation. Based upon photomechanical changes the levels of illumination at which eyes change from the light-adapted to the dark-adapted state range from 10^{-1} to 10^{-5} lx (approximately 10^{-2} to 10^{-6} $\mu\omega$ cm^{-2}) depending on the species. For the trout the threshold for pigment retraction is 2 $\mu\omega$ sr^{-1}m^{-2} and the maximum of the action spectrum is 520 nm. In species which have tapeta lucida such as bream and carp, the changeover occurs in the same range of light levels as in other fishes [10.93-97].

Migration of pigment occurs in the chorioidal tapeta of many selachians, obscuring the tapetal surface when the eye is illuminated, and retreating in darkness. Expansion occupies 60-90 min, retraction, 30-60 min [10.30,34,98-101].

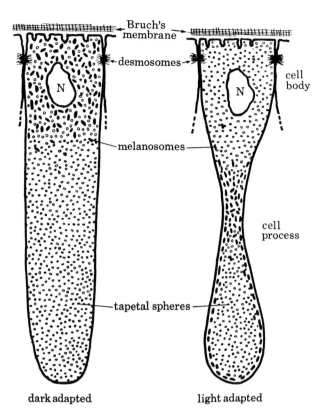

Fig.10.3. Diagrammatic representation of pigment epithelial cells in the tapetal region of the seatrout under dark- and light-adapted conditions (left and right, respectively). Note changes in position of melanosomes and tapetal spheres, as well as external shape [10.53]

10.4 Chemical Composition

Reflection by the tapetum depends upon its chemical composition and structural organization. At least twelve chemical compounds are thought to occur in tapeta of vertebrates [10.5].

Guanine, 2-amino-6-oxypurine, as a reflecting material, occurs in tapeta of fishes and crocodiles. Guanine was discovered by Unger in 1845, and KÜHNE and SEWALL [10.21] recognized guanine in the tapetum of the bream. They based their identification upon solubility, crystalline appearance, and a positive murexide reaction. Subsequently it became customary to describe the tapeta of fishes, whenever they were encountered, as guanine tapeta, sometimes on the basis of their histological appearance.

Guanine has been identified as the tapetal reflecting material in selachians and chimaerids [10.30,102,103], anchovies [10.104], bigeyes Priacanthidae [10.8,56], oil fish *Ruvettus* [10.35], ladyfish *Elops* [10.105], bream *Abramis*, mormyrid *Gnathonemus*, lizardfish *Bathysaurus* [10.68], and *Latimeria* [10.9]. Quantities present in chorioids of selachians and chimaerids range from 0.1 to 0.9 mg cm^{-2} surface area, depending on the region sampled. The tapeta of bony fishes contain 0.3 to 5 mg cm^{-2} (references cited above).

Guanine crystals occur in at least six forms in organisms, four of them in tapeta lucida [10.30,59,60]; they have been described in previous pages. The diversity of shapes is peculiar; that it may be connected with small amounts of substances other than guanine, especially hypoxanthine, has been conjectured.

The tapeta of alligators and crocodiles are believed to contain guanine. The original evidence is a positive murexide reaction [10.4,26,73].

Uric acid, 2,6,8-trioxypurine, occurs in retinal tapeta of mooneyes *Hiodon*, as the acid or urate (the chemical form has not been determined). Two kinds of crystals are present, namely rods and disks. An eye of a mature fish contains about 1 mg of uric acid [10.67-69].

A pteridine, 7,8-dihydroxanthopterin, occurs in the retinal tapetum of the gizzard shad, the walleye, and sauger [10.51,104]. The amount present is surprisingly large, about 5 mg cm^{-2} in the shad. 7,8-Dihydroxanthopterin has a minor absorption peak at ca. 400 nm, and shows rapidly falling absorbance to 450 nm, negligible at longer wavelengths. It fluoresces in alkaline solution but the tapetum is not fluorescent in vivo.

The reflecting material is a lipid in the retinal tapeta of some fishes, birds, and mammals, and occurs in tapetal spheres. Among fishes a lipid tapetum has been found in carps, weakfishes, grunts, cusk-eels, brotulids, spade fishes, and threadfins [10.12,52]. In the sand trout, pigfish, snook, and cusk-eel it consists predominantly of a triglyceride, glyceryl tridocosahexaenoate. Docosahexaenoic acid (C22:ω6) comprises 85% to 95% of the fatty acids present. The tapetum of the sand trout contains about 1.5 mg of glyceryl tridocosahexaenoate per cm^2 of surface area [10.53,55,104,106]. Absorbance in the visible range is negligible.

A lipid tapetum is also present in one order of crepuscular birds, viz., the goatsuckers [10.10]. The lipid occurs in tapetal spheres, it has not yet been identified (cholesterol is suspected). In the American opossum the reflecting material consists of cholesterol and cholesteryl esters, lies in tapetal spheres, and is said to be birefringent and crystalline [10.5,107].

The reflecting material of the fruit-bat is believed to be largely phospholipid; it is birefringent in frozen sections [10.5].

There is a carmine tapetum in the deep-sea fish *Malacosteus*. The pigment is astaxanthin [10.13,108].

Two kinds of melanoid compounds are found in the reflectors of fishes. One is present in gars where the tapetal pigment is 2,5-S,S-dicysteinyldopa. This pigment is derived from one molecule of dopa and two molecules of cysteine. It is a colorless powder and has ultraviolet absorption maxima at 303 and 316 at pH 1 and pH 6.8, respectively [10.109,110].

The other melanoid reflecting material occurs in siluroids, characoids, and gymnotoids, and consists of two components, one acidic and the other basic. The former, termed tapetal pigment, is a mixture of oligomers (mainly tetramers) derived from oxidative coupling of 5,6-dihydroxyindole-2-carboxylic acid [10.48,111]. The monomers are joined mainly by C-C linkages at positions 4 and 7. The pigment has been synthesized by slow enzymatic oxidation of 5,6-dihydroxyindole-2-carboxylic acid with tyrosinase, and in vivo it is probably derived from tyrosine. The chief basic component is an adenine nucleoside, decarboxylated S-adenosylmethionine. When a solution of tapetal pigment at neutral pH is mixed with the same molar quantity of decarboxylated S-adenosylmethionine, a precipitate forms. One of the functions of the decarboxylated S-adenosylmethionine, apparently, is that it makes the tapetal pigment less soluble as a noncovalently bound complex. Smaller amounts of other basic components are also present, including decarboxylated S-adenosylhomocysteine, thiamine, and others [10.111-113].

The tapeta cellulosa of carnivores contain zinc-complexes of amino acids, peptides and protein. The zinc content of the chorioid and of the tapetum is very high, and lies between 8.5% and 13.8% of tapetum dry weight in the fox and dog. The reflecting material of these animals is soluble in 0.5 M HCl and precipitates at neutrality. The atomic ratio Zn:S in dried preparations of tapeta is 1:1, and the reflecting material is probably zinc cysteine hydrate (1:1:1) [10.114]. The zinc functions as a three-pronged ligand, one polar bond is joined to carboxyl; one more, covalent, to sulphur; and one, coordinating, to amino nitrogen. The valency number is completed by a firmly bound water molecule. The reflecting material is amorphorous, it is water insoluble and it is almost certainly a polymer. The tapeta of the pine marten and of the seal also contain zinc cysteine hydrate, and the zinc content of the seal tapetum is 4.2% to 14.6% dry weight [10.115,116].

Tapetal cells of the cat contain crystals of a metallo-protein. The crystals are minute and of uniform size; they have the same shape as those of the dog but are several times larger. The reflecting material is an insoluble, zinc-containing protein, probably a simple polypeptide in a single chain. Sixteen amino acids have been identified, including cysteine, tryosine and phenylalanine. A minimum molecular weight of 6400 for the polypeptide chain and a content of six peptides are indicated. It is likely that zinc-complexed protein has a higher molecular weight owing to polymerization via the zinc atoms or by disulphide bridges. The zinc content in elementary analysis is 2.5%. Crystals of zinc protein isolated from the tapetal cells are not birefringent [10.115,117].

The reflecting material of the bush baby *Galago* is riboflavin, laid down as irregular flat crystals. Riboflavin absorbs at 445 nm and fluoresces at 520 nm; the riboflavin crystals of the tapetum are fluorescent [10.45,118].

No especial study appears to have been made of the collagen of fibrous tapeta lucida.

10.5 Reflection

There are inherent difficulties in measuring reflection, in part instrumental, in part because of the fragility of the tissues. The tapetum has been illuminated normal to its surface and examined in the same axis (cat [10.119]), illuminated obliquely with a beam of light (bush baby [10.45]), illuminated laterally with diffuse light (selachians, sharks, and rays [10.30]), or placed in an integrating sphere (fishes [10.37,55]). The situations are quite different in specular and diffuse reflectors. Spectra are desirable in order to estimate the contribution of reflection to retinal absorption.

Reflection from the tapeta of selachians (sharks, rays) is highly specular, spread of an incident beam being $\sim$ 4°. The spectrum forms a broad band, the maximum ranging from about 480 nm in squaloid fishes to 530 nm in stingrays (Fig.10.4). Reflectivity is high at the maximum, approaching 88% in dogfishes and sharks. The spectrum shifts to shorter wavelengths at higher angles of reflection, as expected for systems depending upon constructive interference of light reflected from the surfaces of thin films [10.30].

Chorioidal tapeta of other fishes are mostly specular reflectors, that of the glasseye, for example, shines like burnished metal. Reflection is maximal at long wavelengths, approaching 90% (Fig.10.5) [10.37]. The tapetum

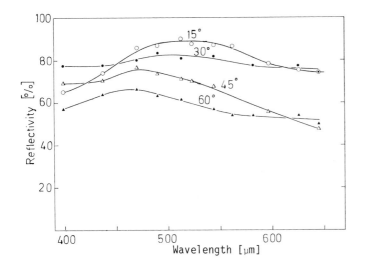

Fig.10.4. Spectral reflectivity of dogfish (*Scyliorhinus*) tapetum at different angles to the normal [10.30]

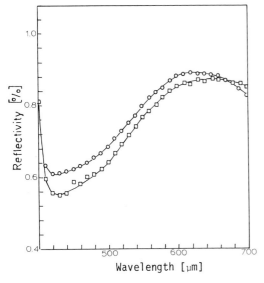

Fig.10.5. Spectral reflection of
the tapetum of the glasseye
Pseudopriacanthus altus (a bony
fish)

of the coelacanth reflects specularly; the spectrum spans the visible with
a broad maximum at about 475 nm [10.9,119]. In the Australian lungfish the
reflecting platelets are randomly oriented, making for a diffuse reflector
[10.35].

DARTNALL et al. [10.45] measured tapetal reflection of the bush baby and
found it to be maximal at long wavelengths (> 500 nm) (Fig.10.6).

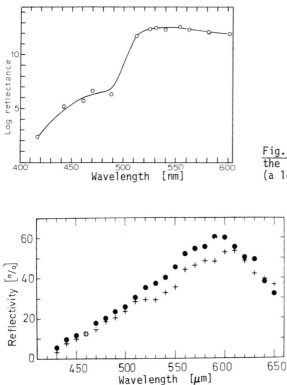

Fig.10.6. Spectral reflection of the tapetum of the bush baby (a lemur) [10.45]

Fig.10.7. Spectral reflection of the cat's tapetum. Upper curve, solid circles, initially; lower curve, crosses, 1 h later [10.120]

The spectra of cat tapeta vary somewhat from animal to animal, from green to yellow, indeed the color varies within one eye (Fig.10.7). Reflection spectra are broad, with maxima ranging from about 500 to 600 nm. Reflectance at the maximum is about 60% for "yellow" tapeta, and 49% for "green". The tapetum reflects diffusely, indeed acts like a perfect diffuser, and there is only a small decrease in the amount of light reflected at 35° relative to normal incidence [10.120]. Dehydration causes the tapetum to become blue, the change being reversible [10.120,121].

Diffusely reflecting tapeta of fishes have very diverse spectra (Fig. 10.8), blue in Welsh's cusk-eek (maximum 440 nm), light yellow to almost white in catfishes (reflection maximal >500 nm), orange in gars (reflection maximal >600 nm), red in *Malacosteus* [10.35]. Many are mat white, e.g., weakfishes (Fig.10.8). Values for maximal reflectance depend upon the method of measurement. They range from 50% in the sand trout to 66% in the gar-pike

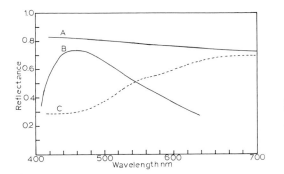

Fig.10.8. Reflectance of tapeta lucida of three fishes: A, sandtrout; B, crested cusk-eel [10.55]; C, shortnose gar [10.122]. A and B raised 0.3 unit

[10.53-55,122]. WEALE [10.123] has emphasized how tissue injury may change the reflection factor. The blue tapetum of Welsh's cusk-eel becomes white in hypotonic, and deeper blue in hypertonic media; these changes are reversible [10.55,56].

10.6 Reflection Mechanisms

As we have discovered, there is a variety of substances composing the reflecting particles of retinal tapeta lucida. Those examined have high refractive indices; besides guanine (n = 1.82) [10.124], they are uric acid 1.743, dihydroxanthopterin 1.73 [10.56], tridocosahexaenoin 1.50, catfish reflecting material 1.56 [10.54], gar reflecting material 1.59 [10.12], and cat tapetal material, about 1.58 [10.125].

The particles — tapetal spheres etc. — of diffusely reflecting tapeta generally have diameters of 300 to 400 nm (catfish, sandtrout, gar, poorwills, etc.); they are packed fairly densely, about 10 per μm^3 [10.54]. Particles having refractive indices greater than the surrounding medium and dimensions approaching a wavelength of visible light scatter light according to Mie theory [10.126]. Because of dense packing and thickness of the reflecting layer, a high degree of backscattering occurs, and the tapeta are efficient diffuse reflectors for all regions of the visible spectrum, (lipid tapeta), or of long wavelengths only (melanoid tapeta). In the latter, although the absorption of short wavelengths by each spherule is small, the cumulative effect of absorption by many spherules causes short wavelengths to be strongly absorbed.

No mathematical treatment of a model or analysis of a real tapetal system appears to have been published.

Chorioidal reflectors of fishes contain stacks of thin films, arranged
in an orderly manner, that reflect light by constructive interference [10.31].
The films are alternating layers of thin crystals and cytoplasmic lamellae,
thicknesses and refractive indices of which cause them to be quarter-wavelength plates. The reflectivities of such biological systems have been
treated by LAND [10.127], HUXLEY [10.128], DENTON [10.129], and DENTON and
LAND [10.130]. In a regular system of 23 films (the number in a shark reflecting cell), a reflectivity of 98% is achieved. Reflection occurs over a
wide spectrum, with a sharp decrease to zero and minor oscillations at short
and long wavelengths (Fig.10.9). The measured spectra of specular reflectors
are much broader than those of such model systems, and reflectivities are
high throughout the visible range (Figs.10.4 and 5). There is a fair amount
of variation in the orientation and spacing of films (selachian tapeta),
crystals vary in thickness (bigeye tapeta), and the tapetal surface is actually a mosaic of differently colored spots. Measurements of minute, singly
colored areas give spectral curves with restricted bands and expected side
oscillations. It appears, therefore, that the broad spectrum curves of these
tapeta, which are produced by constructive interference of light rays, are
caused by variable spacing between platelets, variable thicknesses of platelets, and a summation of disparate spectra from many differently colored
areas. The very large number of films allows high reflectivity to be achieved,
even when spacing is variable. Indeed, with many films and widely divergent
spacing, high reflectivity throughout the visible spectrum results.

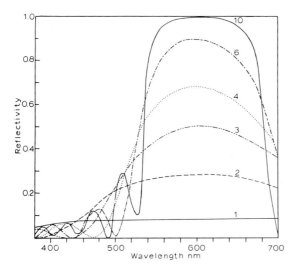

Fig.10.9. Computed reflectivities of stacks of 1 to
10 regularly arranged one-
quarter-wavelength plates
having physical characteristics and spacing to give
maximal reflection at 620 nm

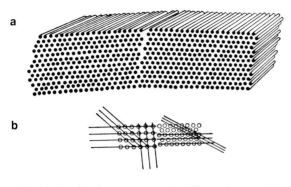

Fig.10.10a,b. Arrangment of rodlets in a rod bundle of the cat's tapetum.
a) A perspective representation of a bundle, divided into two domains whose
corresponding lattice planes are not quite parallel, b) idealized representation of a section perpendicular to the axes of the rodlets of parts of two
adjacent domains with different lattice forms in one bundle [10.125]

Earlier workers, noting the iridescence and spatially variable coloration
of the cat's tapetum, suggested that reflection might be due to interference
[10.19,130], a view which receives support from recent investigations. Arrays
of rodlets form the reflector; within a bundle and through a series of superposed bundles they are mutually parallel and very regularly disposed, and a
section through a bundle cutting their axes presents an array similar to a
crystal lattice (Fig.10.10).

PEDLER [10.42] has suggested that this structural array of the rodlets is
similar to that of a two-dimensional crystal and should exhibit Bragg reflection from the layers of rodlets forming the lattice planes, rather as X rays
are diffracted by atoms in a crystal. The wavelengths λ_n reflected most intensely are

$$\lambda_n = (2n_t d/p)\cos\theta \quad ,$$

where $n_t d$ is the optically effective distance of repeat of the structure,
p is an integer, and θ is the angle from the normal within the structure.
PEDLER estimated that the rodlets were spaced with their centers at 0.3 to
0.5 μm apart.

The overall color of the cat's tapetum is green to yellow. Observed microscopically, the bundles of rodlets near the surface reflect blue light, while
deeper cones reflect longer wavelengths (deep ones appear red). The angles
between lattice planes vary from one rod bundle to another, and reflections
can occur simultaneously from more than one set of lattice planes. Normal
reflection of wavelength λ occurs for optical layer spacings of $1/2\lambda, \lambda, 3/2\lambda,$

etc. In one case observed, there was simultaneous reflection from two surfaces mutually inclined at angles of 25°. The colors were blue and green, and second-order reflection from horizontal planes was deduced, i.e., spacing was λ, which is the spacing measured on electron micrographs [10.42]. It has been suggested [10.125] that the several colors seen through the stacked bundles could arise from changing of spacing through the thickness of the tapetum, although confirmation from electron micrographs is not available. COLES [10.125] has noted that more reflections according to Bragg's equation can be predicted than are observed, e.g., it is to be expected that there would be third-order reflection of blue light from deeper, red-reflecting bundles. He suggested that a rodlet may have a structure showing strong wavelength dependence, such that only those wavelengths close to the one appropriate for the lattice spacing are reflected.

10.7 Functional Considerations (Transmission Through Photoreceptors, Retinal Absorption, Reflection, and Visual Sensitivity)

Reflection from regularly organized chorioidal tapeta of fishes is very specular. The cone of light which falls upon the central retina of selachians (sharks, dogfishes) is about 22° (± 11°), which is the acceptance angle of the rods. Light is channelled or guided down the latter to the tapetum [10.30]. Generally, because of the changing orientation of tapetal plates towards the periphery, light falls upon them perpendicular to their surfaces (Fig.10.1). If the rods were aligned perpendicular to the retinal surface, as assumed by DENTON and NICOL [10.30], light would pass obliquely through the layer of rods and be reflected back by the tapetum in the original path. In sharks, however, the photoreceptors are directed towards the exit pupil, e.g., the smooth hound *Mustelus*, and recent work by A. Laties suggests that this arrangement may be a general one [10.29,132]. Towards the periphery of the eye, as well as centrally, light is channelled down the rods to the tapetal plates (see Chap.4).

In diffuse reflectors, such as that of the catfish, incident rays channelled down the rods must be backscattered over wide angles. Much reflected light must reenter the rods obliquely at angles greater than the critical angle, and pass across banks of rods. No published information seems to be available showing the angular distribution of reflected light. The number of rods relative to ganglion cells is very large in many animals having tapeta lucida, and each ganglion cell could have a large receptive field. Also, receptive fields with a center-surround arrangement must act to sharpen the image [10.133].

In the present state of the art little more of significance can be said concerning light-guide and waveguide effects on tapetal function and vice versa. Light is guided and/or funnelled by the inner segment into the outer segment, and nonabsorbed light is partially reradiated out of the receptor. It, together with light which passes between the photoreceptors, strikes the tapetum and is reflected in degree according to the reflective properties of the latter [10.123]. The relationship of the pigment epithelium to the cones in the cat is peculiar; apical processes of the former ensheath the cone outer segments. It has been suggested this sheath could direct light to a tapetal region in line with the longitudinal axis of the cone, and se-lect reflected rays having a narrow angle about the same axis [10.134]. In several fishes that have chorioidal tapeta (sturgeon, bigeye, etc.) the pig-ment cell processes, but not the cell bases, contain black pigment. An ar-rangement of this kind is one that can reduce scatter and lateral spread of light reflected from the tapetum.

The situation is peculiar in a fish such as *Scopelarchus* where the rods are grouped into bundles and are embedded in tapetal-lined pits. The reflect-ing platelets are parallel to the surface of the pit; in sum, they should perform like a reflecting hemisphere [10.13,64]. The rods in each bundle should function as a unit. Again, in the walleye pike the rods are arranged in groups optically isolated from each other by tapetal processes. The tape-tum reflects diffusely, and nonmotile cones are present; to reach the rods, light must pass through the latter whose ellipsoids operate as light funnels [10.51].

There is general agreement that spectral reflection from the tapetum of the cat does not much affect the shape of the scotopic sensitivity curve (Fig. 10.11) [10.120,135]. Reflection from other tapeta, however, may markedly af-fect the sensitivity curve or spectrum absorbance of the retina. In the bush baby the spectrum of total light absorbed in two transits through the retina has its maximum shifted to 518 nm, in contrast to the spectral absorbance of the rhodopsin, maximum 501 nm (Fig.10.12). The difference is due to tapetal reflectance [10.45]. A shift to long wavelengths may also be expected in the gar, which has an orange tapetum and a porphyropsin with maximal absorbance at 523 nm.

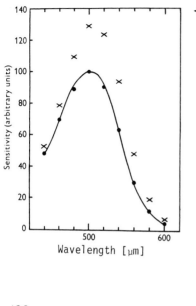

◄ Fig.10.11. The effect of a yellow tapetum on the absolute scotopic sensitivity in the cat. Circles and continuous curve, absorption curve of visual pigment 502. Crosses, absolute sensitivity in presence of 'yellow' tapetum [10.120]

Fig.10.12. Spectra of absorption in the visual pigment and of reflection from the tapetum of the bush baby (a lemur). A, absorption by visual pigment in the first retinal transit of incident light. B, reflection by tapetum back to retina. C, absorption by visual pigment in return transit of light. D, sum of curves A and C. E, as D, but with effective enhancement due to tapetal fluorescence [10.45]
▼

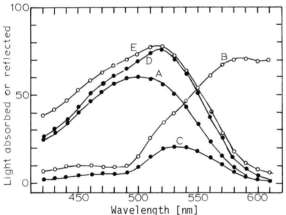

Some estimates are available concerning the effect of tapetal reflection on absorption by the retina and visual threshold. In selachians from coastal waters, retinal density (owing to visual pigment) is low, about 0.28. The tapetum, being specular, reflects light along its original path. Reflectivity approaches 90%, and the tapetum raises retinal absorption from 47% to 68%. In deep sea squaloids, where retinal density is about 0.50, the tapetum increases absorption from 70% to 88% [10.30]. Again, in the glasseye, retinal density is 0.21, the tapetum reflects maximally at long wavelengths. Owing to tapetal reflection, light absorption in the retina is increased from 40%

(one transit) to 60% (two transits) at the absorbance maximum for rhodopsin
[10.37].

The yellow tapetum of the cat, it has been estimated, increases scotopic
sensitivity (lowers absolute visual threshold) by about 30% (Fig.10.11). It
appears that the overall sensitivity of the cat's retina, as measured in behavioral experiments, is about the same as the scotopic sensitivity of the
periphery of the human eye [10.120,121,136].

Studying goldfish, WHEELER [10.136] has found that the dorsal retina of
the scotopic eye is more sensitive than the ventral; he points to the tapetum as the causative agent. The tapetum increases both the effective background energy L and the effective stimulus ΔL. Because of the Weber function,
$\Delta L/L = aL^b$, an increase of stimulus energy ΔL produces a greater response in
the dorsal retina, and tapetal reflection is more effective near scotopic
threshold.

The tapeta of some animals are fluorescent, e.g., selachians, cats, and
bush babies. The material in the cat's eye has been reported to be nekofla-
vine (having general properties of a flavine [10.138] and riboflavine [10.139].
It was suggested that the fluorescence may increase visual contrast in viewing dark objects against a night sky. Free riboflavin, as flat crystals, occurs in the tapetum of the bush baby. DARTNALL et al. [10.45] have calculated
the enhancement of sensitivity (rhodopsin absorbance) conferred by fluorescence. They found it to be significant, especially in the short-wavelength
region where it suffices to compensate for the lower spectral reflectance
of the tapetum (Fig.10.12). Short wavelengths absorbed by the tapetum in
this region generate fluorescence within the tapetal layer and, depending
on the quantum yield, some of the absorbed energy is reemitted as yellow
light. An alternative suggestion of direct energy transfer has been proposed
by WALKER and RADDA [10.140]. From model experiments they suggest that flavin, photoexcited by absorption of blue light, sensitizes a reaction of rhodopsin by a triplet-triplet transfer mechanism, causing cis-trans isomerization.

It has been found that some fishes can detect plane-polarized light [10.141,
142]. The arrangement of membranes in cones of the anchovy suggests that a
two-channel analyser system is present. Membranes in long cones are parallel
to the long axis of the cell and face the reflecting platelets of the tapetum
at an angle of about 70° [10.61,62]. Now, guanine platelets reflect polarized
light in one plane (e vector parallel to the surfaces of the crystals) more
strongly than light polarized at an angle perpendicular to it, and they polar-
ize more strongly at oblique angles of incidence [10.129]. If polarized light

with its e vector parallel to the membranes encounters the platelets parallel
to their surfaces, it would be reflected to a greater degree than light having its e vector perpendicular to these surfaces. The platelets, therefore,
could reinforce the polarized light sensitivity of the long cones.

10.8 Ecological Aspects

Much has been made of the apparent correlation (positive or negative) that
exists between the spectral composition of light in aquatic environments
and the wavelength of maximal absorbance of rod pigments of fishes (see,
for example, [10.143]). To achieve maximal effect it might be expected that
the most efficient tapetum would be one reflecting efficiently and equally
well across the visible spectrum, or reflecting maximally in the region of
visual pigment absorbance. The white lipid tapeta of fishes do perform in
the former manner. Also, there is an impressive correlation between the
spectrum reflectivity of the selachian tapetum, absorbance of the visual pigment, and spectrum transmittance of sea water in the regions inhabited by
the fish. But the disparities are just as notable: blue tapeta in some fishes
(cusk-eel), etc.), tapeta reflecting long wavelengths in catfishes, gars,
and priacanthids. It can be said that both catfishes and gars usually inhabit
turbid or stained waters where much of the subaquatic daylight lies at long
wavelengths. The red tapetum of *Malacosteus* appears to have some functional
relationship to the red-sensitive visual pigment and red bioluminescence of
the Malacosteidae [10.144].

Certainly many animals possessing tapeta are crepuscular or nocturnal in
habits, e.g., foxes, deer, lemurs, whip-poor-wills, cusk-eels, and glasseyes.
Others are active day and night, e.g., skipjacks and seatrout. Some fishes
with tapeta that inhabit waters having very low transmissivity, e.g., catfishes, gars, and weakfishes, encounter scotopic conditions at shallow depths
during the daytime. Deep-sea species having tapeta live in mesopelagic regions where ambient light levels are low [10.145], migrate upwards at night
(e.g., lanternfishes) or are bathybenthic, using luminescent light (e.g.,
deep-sea squaloid sharks and gadoids [10.13,144,146].

A diffusely reflecting tapetum, in theory, can cause degradation of the
image, but this factor may be of small or negligible significance at the low
light levels and under the environmental conditions that the animals experience. Furthermore, in the very turbid waters inhabited by many riverine and
coastal fishes, loss of image contrast due to forward scattering becomes so

great that the animals can distinguish objects at distances of only a few
meters or less [10.146,147]. Indeed, in some fishes lacking spectacles corneas may be cloudy, and resolution poor.

Acknowledgments. Some of the work reported in this Chapter was supported by
the National Eye Institute, National Institutes of Health (U.S. Public Health
Service). The author thanks Drs. J.M. Enoch and N.A. Locket for helpful information.

References

10.1   F. Leydig: Histologische Bemerkungen über den *Polypterus bichir.*
        Z. wiss. Zool. *5*, 40-74 (1854)
10.2   V. Franz: Zur Anatomie, Histologie und funktionellen Gestaltung des
        Selachierauges. Jena. Z. Naturwiss. *40*, 697-840 (1905)
10.3   V. Franz: "Höhere Sinnesorgane (Auge)", in *Handbuch der vergleichenden
        Anatomie der Wirbeltiere*, Vol.2, 2. Hälfte, ed. by L. Bolk, E. Göppert,
        E. Källius, W. Lubosch (Urban and Schwarzenberg, Berlin 1934) pp.989
        pp.989-1292
10.4   G.L. Walls: The vertebrate eye and its adaptive radiation. Bull. Cran-
        brook Inst. Sci. *19*, 785 (1942)
10.5   A. Pirie: "The Chemistry and Structure of the Tapetum Lucidum in
        Animals", in *Aspects of Comparative Ophthalmology*, ed. by O. Graham-
        Jones, Proc. Symposium, The British Small Animal Veterinary Association,
        1965 (Pergamon Press, Oxford 1966) pp.57-69
10.6   J.A.C. Nicol: The tapetum lucidum of the sturgeon. Contrib. Mar. Sci.
        14, 5-18 (1969)
10.7   R.D. Martin: Adaptive radiation and behaviour of the Malagasy lemurs.
        Philos. Trans. R. Soc. London B*264*, 295-352 (1972)
10.8   J.A.C. Nicol, E.S. Zyznar: The tapetum lucidum in the eye of the big-
        eye *Priacanthus arenatus* Cuvier. J. Fish Biol. *5*, 519-522 (1973)
10.9   N.A. Locket: The choroidal tapetum lucidum of *Latimeria chalumnae.*
        Proc. R. Soc. London B*186*, 281-290 (1974)
10.10   J.A.C. Nicol, H.J. Arnott: Tapeta lucida in the eyes of goatsuckers
        (Caprimulgidae). Proc. R. Soc. London B*187*, 349-352 (1974)
10.11   J.A.C. Nicol, H.J. Arnott: Studies on the eyes of gars (Lepisosteidae)
        with special reference to the tapetum lucidum. Can. J. Zool.*51* , 501-
        508 (1973)
10.12   J.A.C. Nicol, H.J. Arnott, A.C. Best: Tapeta lucida in bony fishes
        (Actinopterygii): a survey. Can. J. Zool. *51*, 69-81 (1973)
10.13   N.A. Locket:"Adaptations to the Deep-Sea Environment", in *The Visual
        System of Vertebrates*, ed. by F. Crescitelli, Handbook of Sensory
        Physiology, Vol.7/5 (Springer, Berlin, Heidelberg, New York 1977)
        pp.67-192
10.14   E. Murr: Die Tapetum-Bildungen in Wirbeltieraugen nach Bau und physio-
        logischer Bedeutung auf Grund eigener Präparate. Ms. 1922, in Händen
        des Verfasser, z.Z. Landwirtsch.-Zool. Inst., Berlin (quoted from
        [10.3]
10.15   D.M. Soemmerring: *A Comment on the Horizontal Section of Eyes in Man
        and Animals,* Thesis (1818), Lederle Laboratories, Copenhagen, ed. by
        S.R. Andersen, O. Munk, trans. H.D. Schepelern, Acta Opthalmol. Suppl.
        No. 110 (1971)

10.16 G.L. Johnson: Contributions to the comparative anatomy of the mammalian eye, chiefly based on ophthalmoscopic examination. Philos. Trans. R. Soc. London B*194*, 1-82 (1901)

10.17 G.L. Johnson: Ophthalmoscopic studies on the eyes of mammals. Philos. Trans. R. Soc. London B*254*, 207-220 (1968)

10.18 S. Delle Chiaje: Istituzioni di Anatomia comp. (Napoli 1836); in M.L. Verrier: Recherches sur les yeux et la vision des poissons. Bull. Sci. Fr. Belg. Suppl. XI (1928)

10.19 E. Brücke: Anatomische Untersuchungen über die sogenannten leuchtenden Augen bei den Wirbeltieren. Arch. Anat. Physiol. Wiss. Med. pp. 387-406 (1845)

10.20 H. Müller: Anatomisch-physiologische Untersuchungen über die Netzhaut bei Menschen und Wirbeltieren. Z. Wiss. Zool. *8*, Heft 1, 1-122 (1856)

10.21 W. Kühne, H. Sewall: Zur Physiologie des Sehepithels, insbesondere der Fische. Unters. Physiol. Inst. Univ. Heidelberg *3*, 221-277 (1880)

10.22 G. Abelsdorff: Über Sehpurpur und Augenhintergrund bei den Fischen. Arch. Anat. Physiol, Physiol. Abt. 1896, 345-347 (1896)

10.23 A. Brauer: Die Tiefseefische. II. Anatomischer Teil. Wiss. Ergebn. Dtsch. Tiefsee-Exped. 'Valdivia' *15*, 2 (1908)

10.24 A.J. Van Rossem: Eyeshine in birds, with notes on the feeding habits of some goatsuckers. Condor *29*, 25-28 (1927)

10.25 E.J. Slijper: *Whales, Trans. A.J. Pomerans* (Basic Books, New York 1962)

10.26 G. Abelsdorff: Physiologische Beobachtungen am Auge der Krokodile. Arch. Anat. Physiol. Physiol. Abt. 1898, 155-167 (1898)

10.27 S.R. Detwiler: *Vertebrate Photoreceptors* (Macmillan, New York 1943)

10.28 W. Wunder: Physiologisch-vergleichend-anatomische Untersuchungen an der Knochenfischnetzhaut. Z. Vgl. Physiol. *3*, 1-66 (1925)

10.29 J.M. Enoch: Retinal receptor orientation and the role of fiber optics in vision. Am. J. Optom. *49*, 455-471 (1972)

10.30 E.J. Denton, J.A.C. Nicol: The chorioidal tapeta of some cartilaginous fishes (Chondrichthyes). J. Mar. Biol. Assoc. U.K. *44*, 219-258 (1964)

10.31 A.C.G. Best, J.A.C. Nicol: Reflecting cells of the elasmobranch tapetum lucidum. Contrib. Mar. Sci. *12*, 172-201 (1967)

10.32 A. Rochon-Duvigneaud: *Les Yeux et la Vision des Vertébrés* (Masson, Paris 1943)

10.33 J.A.C. Nicol: The tapetum in *Scyliorhinus canicula*. J. Mar. Biol. Assoc. U.K. *41*, 271-277 (1961)

10.34 J.A.C. Nicol: Reflectivity of the chorioidal tapeta of selachians. J. Fish. Res. Board Can. *21*, 1089-1100 (1964)

10.35 N.A. Locket: Personal communication (1975)

10.36 H.J. Arnott, J.A.C. Nicol: Reflection of ratfish skin (*Hydrolagus colliei*). Can. J. Zool. *48*, 137-151 (1970)

10.37 R.T. Wang, J.A.C. Nicol, E.L. Thurston, J. McCants: Studies on the eyes of bigeyes (Teleostei Priacanthidae) with special reference to the tapetum lucidum. Proc. R. Soc. London B*112*, 499-512 (1979)

10.38 E. Murr: Über die Entwicklung und den feineren Bau des Tapetum lucidum der Feliden. Z. Zellforsch. Mikrosk. Anat. *6*, 315-336 (1927)

10.39 A. Bruni: Nuovi appunti sulla struttura del tappeto lucido cellulare. Monit. Zool. Ital. *39*, 223-232 (1929)

10.40 Y. Hosoya: Studien über das Tapetum lucidum choroideale. Tohoku J. Exp. Med. *12*, 119-145 (1929)

10.41 M.H. Bernstein, D.C. Pease: Electron microscopy of the tapetum lucidum of the cat. J. Biophys. Biochem. Cytol. *5*, 35-40 (1959)

10.42 C. Pedler: The fine structure of the tapetum cellulosum. Exp. Eye Res. *2*, 189-195 (1963)

10.43 R. Hebel: Entwicklung und Struktur der Retina und das Tapetum lucidum des Hundes. Ergeb. Anat. Entwicklungsgesch. *45*, 1-93 (1971)

10.44 L.R. Wolin, L.C. Massopust: "Morphology of the Primate Retina", in *The Primate Brain*, ed. by C.R. Noback, W. Montagna (Meredith Corp., New York 1970) pp.1-27

10.45 H.J.A. Dartnall, G.B. Arden, H. Ikeda, C.P. Luck, M.E. Rosenberg, C.M.H. Pedler, K. Tansley: Anatomical, electrophysiological and pigmentary aspects of vision in the bush baby: an interpretative study. Vision Res. *5*, 399-424 (1965)

10.46 F. Leydig: Lehrbuch der Histologie des Menschen und der Tiere (Meidinter Sohn & Co., Frankfurt am Main 1857)

10.47 A. Pirie: The biochemistry of the eye related to its optical properties. Endeavour *17*, 181-189 (1958)

10.48 S. Ito, E.L. Thurston, J.A.C. Nicol: Melanoid tapeta lucida in teleost fishes. Proc. R. Soc. London B*191*, 369-385 (1975)

10.49 G.A. Moore: The retinae of two North American teleosts, with special reference to their tapeta lucida. J. Comp. Neurol. *80*, 369-379 (1944)

10.50 V.B. Meyer-Rochow: The larval eye of the deep-sea fish Cataetyx memoribilis (Teleostei, Ophidiidae). Z. Morph. Ökol. Tiere *72*, 331-340 (1972)

10.51 E.S. Zyznar, M.A. Ali: An interpretative study of the visual cells and tapetum lucidum of *Stizostedion*. Can. J. Zool. *53*, 180-196 (1975)

10.52 A.C.G. Best, J.A.C. Nicol: Notes on the retina and tapetum lucidum of *Howella* (Teleostei:Cheilodipteridae). J. Mar. Biol. Assoc. U.K. *58*, 735-738 (1978)

10.53 H.J. Arnott, J.A.C. Nicol, C.W. Querfeld: Tapeta lucida in the eyes of the seatrout (Sciaenidae). Proc. R. Soc. London B*180*, 247-271 (1972)

10.54 H.J. Arnott, J.A.C. Nicol, A.C.G. Best, S. Ito: Studies on the eyes of catfishes with special reference to the tapetum lucidum. Proc. R. Soc. London B*186*, 13-36 (1974)

10.55 J.A.C. Nicol, H.J. Arnott, E.L. Thurston, R.T. Wang: The tapetum lucidum in the eyes of cusk-eels (Ophidiidae). Can. J. Zool. *53*, 1063-1079 (1975)

10.56 E.S. Zyznar: *A Study of the Tapeta Lucida of Eight Teleost Fishes*, Ph.D. Thesis, Univ. Texas at Austin (1973)

10.57 M.R. McEwan: A comparison of the retina of the mormyrids with that of various other teleosts. Acta Zool. (Stockholm) *19*, 427-465 (1938)

10.58 C.P. O'Connell: The structure of the eye of *Sardinops caerulea*, *Engraulis mordax*, and four other pelagic marine teleosts. J. Morphol. *113*, 287-330 (1963)

10.59 B.A. Fineran, J.A.C. Nicol: Novel cones in the retina of the anchovy (*Anchoa*). J. Ultrastruct. Res. *54*, 296-303 (1975)

10.60 B.A. Fineran, J.A.C. Nicol: Studies on the eyes of anchovies *Anchoa mitchilli* and *A. hepsetus* with particular reference to the pigment epithelium. Philos. Trans. R. Soc. London B*276*, 321-350 (1977)

10.61 B.A. Fineran, J.A.C. Nicol: Studies on the photoreceptors of *Anchoa mitchilli* and *A. hepsetus* (Engraulidae) with particular reference to the cones. Philos. Trans. R. Soc. London B*283*, 25-60 (1978)

10.62 N.A. Locket: Retinal anatomy in some scopelarchid deep-sea fishes Proc. R. Soc. London B*178*, 161-184 (1971)

10.63 R.D. Frederiksen: Retinal tapetum containing discrete reflectors and photoreceptors in the bathypelagic teleost *Omosudis lowei*. Vidensk. Medd. Dan. Naturhist. Foren. Khobenhavn *139*, 109-146 (1976)

10.64 O. Munk: The visual cells and retinal tapetum of the foveate deep-sea fish *Scopelosaurus lepidus* (Teleostei). Zoomorphologie *87*, 21-49 (1977)

10.65 V. Franz: Zur mikroskopischen Anatomie der Mormyriden. Dem Gedächtnis W. Stendell's gewidmet. Zool. Jahrb. Abt. Anat. Ontog. Tiere *42*, 91-148 (1920)

10.66 G.A. Moore, R.C. McDougal: Similarity in the retinae of *Amphiodon alosoides* and *Hiodon tergesus*. Copeia 1949, 298 (1949)

10.67 H.-J. Wagner, M.A. Ali: Retinal organization in goldeye and mooneye (Teleostei: Hiodontidae). Rev. Can. Biol. *37*, 65-83 (1978)

10.68 E.S. Zyznar, F.B. Cross, J.A.C. Nicol: Uric acid in the tapetum lucidum of mooneyes *Hiodon* (Hiodontidae Teleostei). Proc. R. Soc. London B*201*, 1-6 (1978)

10.69 A.C.G. Best, J.A.C. Nicol: On the eye of the goldeye *Hiodon alosoides* (Teleostei: Hiodontidae). J. Zool. *188*, 309-332 (1979)

10.70 W.T. O'Day, H.R. Fernandez: Vision in the lantenfish *Stenobrachius leucopsarus* (Myctophidae). Mar. Biol. *37*, 187-195 (1976)

10.71 S. Garten: "Die Veränderungen der Netzhaut durch Licht", in *Graefe-Saemisch-Handbuch der gesamten Augenheilkunde*, Part I, Vol. 3, ed. by T. Saemisch, 2.Ed.(Wilhelm Engelmann, Leipzig 1907) Chapt.12,pp.1-28

10.72 J.H. Chievitz: Untersuchungen über die Area centralis retinae. Arch. Anat. Physiol. Anat. Abt. 1889, Suppl., 139-194 (1889)

10.73 H. Laurens, S.R. Detwiler: Studies on the retina. The structure of the retina of Alligator mississippiensis and its photomechanical changes. J. Exp. Zool. *32*, 207-234 (1921)

10.74 G.L. Walls: Notes on the retinae of two opossum genera. J. Morphol. 64, 67-87 (1939)

10.75 C.R. Braekevelt: Fine structure of the retinal epithelium and tapetum lucidum of the opossum (*Didelphis virginiana*). J. Morphol. *150*, 213-226 (1976)

10.76 W. Kolmer: Zur Kenntnis des Auges der Macrochiropteren. Z. Wiss. Zool. *97*, 91-104 (1910)

10.77 W. Kolmer: Zur Frage nach der Anatomie des Makrochiropterenauges. Anat. Anz. *40*, 626-629 (1912)

10.78 G. Fritsch: Beiträge zur Histologie des Auges von *Pteropus*. Z. Wiss. Zool. *98*, 288-296 (1911)

10.79 P. Gérard, A. Rochon-Duvigneaud: L'oeil et la vision des Mégacheiroptères. Arch. Biol. *40*, 151-173 (1930)

10.80 C. Pedler, R. Tilley: The retina of a fruit bat (*Pteropus giganteus* Brünnich). Vision Res. *9*, 909-922 (1969)

10.81 R.A. Suthers: "Vision, Olfaction, Taste", in *Biology of Bats*, Vol.2 ed. by W.A. Wimsatt (Academic, New York, London 1970) pp.265-310

10.82 T.W. Engelmann: Über Bewegungen der Zapfen und Pigmentzellen der Netzhaut unter dem Einfluss des Lichtes und des Nervensystems. Pflügers Arch. Gesamte Physiol. Menschen Tiere *35*, 498-508 (1885)

10.83 S. Exner, H. Januschke: Das Verhalten des Guanintapetums von *Abramis brama* gegen Licht und Dunkelheit. Sitzungsber. Akad. Wiss. Wien Math. Naturwiss. Kl. Abt. 3 *114*, 693-714 (1905)

10.84 S. Exner, H. Januschke: Die Stäbchenwanderung im Auge von *Abramis brama* bei Lichtveränderungen. Sitzungsber. Akad. Wiss. Wien Math. Naturwiss. Kl. Abt. 3 *115*, 269-280 (1906)

10.85 S. Garten: "Die Veränderungen der Netzhaut durch Licht", in *Handbuch der gesamten Augenheilkunde*, Vol.3, ed. by T. Axenfeld, A. Elschnig (Springer, Berlin 1925) pp.1-250

10.86 L.B. Arey: The occurrence and the significance of photomechanical changes in the vertebrate retina — an historical survey. J. Comp. Neurol. *25*, 535-554 (1915)

10.87 L.B. Arey: The movements in the visual cells and retinal pigment of the lower vertebrates. J. Comp. Neurol. *26*, 121-201 (1916)

10.88 G.H. Parker: The movements of the retinal pigment. Ergeb. Biol. *9*, 239-291 (1932)

10.89 J.H. Welsh, C.M. Osborn: Diurnal changes in the retina of the catfish, *Ameiurus nebulosus*. J. Comp. Neurol. *66*, 349-359 (1937)

10.90 L.B. Arey, G.H. Mundt: A persistent diurnal rhythm in visual cones (Abstract) Anat. Rec. *79*, Suppl, 5 (1941)

10.91 M.A. Ali: Retinomotor responses in the enucleated eyes of the brown bullhead (*Ictalurus nebulosus*) and the goldfish (*Carassius auratus*). Rev. Can. Biol. *23*, 55-66 (1964)

10.92 I. Bäck, K.O. Donner, T. Reuter: The screening effect of the pigment epithelium on the retinal rods in the frog. Vision Res. *5*, 101-111 (1965)

10.93 W. Wunder: Die Bedeutung des Adaptationszustandes für das Verhalten der Sehelemente und des Pigmentes in der Netzhaut von Knochenfischen. Z. vgl. Phyiol. *3*, 595-614 (1926)

10.94 H. Kobayashi: Notes on retino-motor phenomena in some fishes under the various light conditions. J. Shimonoseki Coll. Fish. *7*, 169-177 (1957)

10.95 T. Tamura: "Fundamental Studies on the Visual Sense of Fish", in *Modern Fishing Gear of the World*, ed. by H. Kristjonsson (Fishing News, London (Books) 1959) pp.543-547

10.96 J.H.S. Blaxter: Effect of change of light intensity on fish. Int. Comm. Northwest Atl. Fish. Spec. Publ. *6*, 647-661 (1964)

10.97 M.A. Ali, R. Crouzy: Action spectrum and quantal thresholds of retino-motor responses in the brook trout *Salvelinus fontinalis* (Mitchill). Z. vgl. Physiol. *59*, 86-89 (1968)

10.98 J.A.C. Nicol: Migration of chorioidal tapetal pigment in the spur dog *Squalus acanthias*. J. Mar. Biol. Assoc. UK *45*, 405-427 (1965)

10.99 K.P. Kuchnow, P.W. Gilbert: "Preliminary in vivo Studies on Pupillary and Tapetal Pigment Responses in the Lemon Shark, *Negaprion brevirostris*", in *Sharks, Skates, and Rays*, ed. by P.W. Gilbert, R.F. Mathewson, D. P. Rall (Johns Hopkins Press, Baltimore, MD 1967) pp.465-477

10.100 K.P. Kuchnow: Threshold for the elasmobranch tapetal-pigment response. Vision Res. *9*, 187-194 (1969)

10.101 K.P. Kuchnow: Tapetal pigment response of elasmobranchs. Vision Res. *9*, 849-854 (1969)

10.102 A. Pirie, D.M. Simpson: Preparation of a fluorescent substance from the eye of the dogfish, *Squalus acanthias*. Biochem. J. *40*, 14-20 (1946)

10.103 J.A.C. Nicol, C. Van Baalen: Studies on the reflecting layers of fishes. Contrib. Mar. Sci. *13*, 65-88 (1968)

10.104 E.S. Zyznar, J.A.C. Nicol: Reflecting materials in the eyes of three teleosts, *Orthopristes chrysopterus*, *Dorosoma cepedianum* and *Anchoa mitchilli*. Proc. R. Soc. London B*184*, 15-27 (1973)

10.105 N.A. Locket, S. Ito: Work in progress (1975)

10.106 J.A.C. Nicol, H.J. Arnott, G.R. Mizuno, E.C. Ellison, J.R. Chipault: Occurrence of glceryl tridocosahexaenoate in the exe of the sand trout *Cynoscion arenarius*. Lipids *7*, 171-177 (1972)

10.107 A. Pirie: Cholesterol in the tapetum lucidum of the eye of the opposum, *Didelphis virginiana*. Nature (London) *191*, 708-709 (1961)

10.108 P.J. Herring: Personal communication (1975)

10.109 S. Ito, J.A.C. Nicol: 2,5-S-S-Dicysteinylodopa: a new amino acid in the eye of the gar and its enzymic synthesis. Tetrahedron Lett. No. 38, 3287-3290 (1975)

10.110 S. Ito, J.A.C. Nicol: A new amino acid, 3-(2,5-S-S-dicysteinyl-3, 4-dihydroxy-phenyl) alanine, from the tapetum lucidum of the gar (Lepisosteidae) and its enzymic synthesis. Biochem. J. *161*, 449-507 (1977)

10.111 S. Ito, J.A.C. Nicol: Isolation of oligomers of 5,6-dihydroxyindole-2-carboxylic acid from the eye of the catfish. Biochem. J. *143*, 207-217 (1974)

430

10.112 S. Ito, J.A.C. Nicol: Identification of decarboxylated S-adenosyl-methionine in the tapetum lucidum of the catfish. Proc. R. Soc. London B190, 33-43 (1975)

10.113 S. Ito, J.A.C. Nicol: Isolation of S-adenosyl-3-thiopropylamine from the eye of the sea catfish (Arius felis). Biochem. J. 153, 567-570 (1976)

10.114 G. Weitzel, E. Buddecke, A.-M. Fritzdorff, F.-J. Strecker, U. Roester: Struktur der Tapetum lucidum von Hund und Fuchs enthaltenen Zinkver-bindung. Z. Physiol. Chem. 299, 193-255 (1955)

10.115 G. Weitzel: Chemie und Physiologie biogener Zink-Verbindungen. Angew. Chem. 68, 566-573 (1956)

10.116 G. Weitzel, E. Buddecke: Zink im Augenhintergrund des Seehundes. Physiol. Chem. 304, 1-10 (1956)

10.117 L.R. Croft: Isolation of a zinc protein from the tapetum lucidum of the cat. Biochem. J. 130, 303-305 (1972)

10.118 A. Pirie: Crystals of riboflavin making up the tapetum lucidum in the eye of a lemur. Nature (London) 183, 985-986 (1959)

10.119 J. Lenoble, Y. Legrand: Le tapis de l'oeil du Coelacanthe (Latimeria anjouanae (Smith)). Bull. Mus. Nat. Hist. Nat. 26, 460-463 (1954)

10.120 R.A. Weale: The spectral reflectivity of the cat's tapetum measured in situ. J. Physiol. (London) 119, 30-42 (1953)

10.121 R. Gunter: The absolute threshold for vision in the cat. J. Physiol. London 114, 8-15 (1951)

10.122 R.T. Wang, J.A.C. Nicol: The tapetum lucidum of gars (Lepisosteidae) and its role as a reflector. Can. J. Zool. 52, 1523-1530 (1974)

10.123 R.A. Weale: "Natural History of Optics", in The Eye, Comparative Physiology, Vol.6, ed. by H. Davson (Academic, New York, London 1974) pp.1-110

10.124 L.M. Greenstein: Nacreous pigments and their properties. Proc. Sci. Sect. Toilet Goods Assoc. No. 45, 20-26 (1966)

10.125 J.A. Coles: Some reflective properties of the tapetum lucidum of the cat's eye. J. Physiol. London 212, 393-409 (1971)

10.126 M. Kerker: The Scattering of Light and Other Electromagnetic Radiation (Academic, New York 1959)

10.127 M.A. Land: A multilayer inference reflector in the eye of the scallop, Pecten maximus. J. Exp. Biol. 45, 433-447 (1966)

10.128 A.F. Huxley: A theoretical treatment of the reflexion of light by multilayer structures. J. Exp. Biol. 48, 227-245 (1968)

10.129 E.J. Denton: On the organization of reflecting surfaces in some marine animals. Philos. Trans. R. Soc. London B258, 285-313 (1970)

10.130 E.J. Denton, M.F. Land: Mechanism of reflexion in silvery layers of fish and cephalopods. Proc. R. Soc. London B178, 43-61 (1971)

10.131 M. Schultze: Über das Tapetum in der Chorioides des Auges der Raubthiere. Sitzungsber. Niederrhein. Ges. Nat. Heilk. 29, 215-216 (1872)

10.132 J.M. Enoch: Personal communication (1975)

10.133 N.W. Daw, R.D. Beauchamp: Unusual units in the golfish optic nerve. Vision Res. 12, 1849-1856 (1972)

10.134 R.H. Steinberg, I. Wood: Pigment epithelial cell ensheathment of cone outer segments in the retina of the domestic cat. Proc. R. Soc. London B187, 461-478 (1974)

10.135 R. Gunter, H.G.W. Harding, W.S. Stiles: Spectral reflexion factor of the cat's tapetum. Nature London 168, 293-294 (1951)

10.136 E. Murr: Über den Helligkeitssinn der Hauskatze und die Bedeutung des Tapetum lucidum. Zool. Anz. Suppl. 3, 254-265 (1928)

10.137 T.G. Wheeler: Goldfish retina: dorsal versus ventral areas. Vision Res. 18, 1329-1336 (1978)

10.138 K. Matsui: Nekoflavin, a new flavin compound in the choroid of cat's eye. J. Biochem. (Tokyo) *57*, 201-206 (1965)

10.139 J.H. Elliott, S. Futterman: Fluorescence in the tapetum of the cat's eye. Identification, assay, and localization of riboflavin in the tapetum and a proposed mechanism by which it may facilitate vision. Arch. Ophthalmol. *70*, 531-534 (1963)

10.140 A.G. Walker, G.K. Radda: Photoreactions of retinal and derivatives sensitized by flavins. Nature (London) *215*, 1483 (1967)

10.141 R.B. Forward, Jr., T.H. Waterman: Evidence for E vector and light intensity pattern discrimination by the teleost *Dermogenys*. J. Comp. Physiol. *87*, 189-202 (1973)

10.142 H. Kleerekoper, J.H. Matis, A.M. Timms, P. Gensler: Locomotor response of the goldfish to polarized light and its e vector. J. Comp. Physiol. *86*, 27-36 (1973)

10.143 F. Crescitelli: "The Visual Cells and Visual Pigments of the Vertebrate Eye", in *Photochemistry of Vision*, ed. by H.J.A. Dartnall, Handbook of Sensory Physiology, Vol.7/1 (Springer, Berlin, Heidelberg, New York 1972) pp.245-263

10.144 W.T. O'Day, H.R. Fernandez: *Aristostomias scintillans* (Malacosteidae): a deep-sea fish with visual pigments apparently adapted to its own bioluminescence. Vision Res. *14*, 545-550 (1974)

10.145 G.L. Clarke, E.J. Denton: "Light and Animal Life", in *The Sea*, Vol.1, Physical Oceanography, ed. by M.N. Hill (Wiley, New York 1962) pp.456-468

10.146 J.A.C. Nicol: "Bioluminescence and Vision", in *Bioluminescence in Action*, ed. by P.J. Herring (Academic, London, New York, San Francisco 1978) pp.367-398

10.147 C.C. Hemmings, J.N. Lythgoe: "The Visibility of Underwater Objects", in *Symposium of the Underwater Association for Malta 1965*, ed. by J.N. Lythgoe, J.D. Woods (1966) pp.23-30

10.148 B.K. Gazey: Visibility and resolution in turbid waters. Unterwat. Sci. Technol. J. *2*, 105-115 (1970)

11. A Comparison of Vertebrate and Invertebrate Photoreceptors

G. D. Bernard

With 7 Figures

This chapter is intended for readers who are unfamiliar with research on invertebrate photoreceptors. The theme of the following discussion is that compound eyes and vertebrate eyes have much in common and that knowledge of one enhances understanding of the other.

For certain kinds of experiments, invertebrates are easier to work with than vertebrates. They have visual systems that are less complex, though certainly not simple. Many kinds of experiments are feasible with invertebrates that are not with vertebrates. Therefore, basic studies of invertebrates can progress more rapidly than equivalent studies of vertebrates.

What is the importance of research on arthropods for one whose interest is restricted to vertebrates? It is the change of thinking that accompanies a comparative approach. Whether or not results of invertebrate investigations are exactly the same as for vertebrates is relatively unimportant. Understanding basic principles of invertebrate vision can prompt one interested in the vision of vertebrates to establish new lines of inquiry and develop new insights.

A superb example of this is the work of HARTLINE [11.1] on visual receptors and retinal interactions, summarized in his Nobel lecture. His earliest work on the electrophysiological responses of single optic-nerve fibers of *Limulus*, the horseshoe crab, prompted him to apply the same methods to the frog, with surprising and important results. This opened a line of research that is still being pursued in many laboratories. Despite the oft-spoken caveat that "one can't generalize from *Limulus*", studies of this primitive arthropod profoundly contribute to the understanding of basic visual mechanisms [11.2].

The following discussion introduces the reader to the invertebrate literature and its terminology, surveys the similarities and differences, and summarizes recent progress in several of the more active areas of research.

This is not a comprehensive review; a number of good publications already serve that purpose [11.3-12].

The references cited below are not intended to indicate priority of discovery. Rather, I tried to select the more recent publications that contain detailed discussions and appropriate references.

After describing anatomy, optics, waveguide effects, photochemistry, and physiology, this chapter summarizes some requirements for visual information processing that retinal photoreceptors must meet.

11.1 Anatomy

Many insects have two kinds of visual organs. The most important are the compound eyes, which are composed of thousands of similar units called *ommatidia* that occupy a rather thin layer at the surface of the head. The second kind of visual organ is the ocellus. Three of these tiny, simple eyes are usually grouped together on the vertex of the head between the two compound eyes. Relatively little is understood about the functional importance of ocelli and about how ocellar information is integrated with information from the compound eyes [11.13-15]. The discussion below is restricted to compound eyes.

Each *ommatidium*, tens of micrometers in diameter by hundreds of micrometers in length, is a little eye that contains optical elements consisting of a cornea and crystalline cone, a *retinula* comprising eight or more photoreceptor cells, and several types of pigmented cells. Figure 11.1 diagrams an ommatidium of the worker bee. The *cornea* is part of the invertebrate's cuticle, but is modified to be lenticular and transparent even in the near-ultraviolet. The so-called *crystalline cone* is not a photoreceptor, but a transparent bullet-shaped structure located immediately beneath the cornea.

The receptor cells of a retinula each contain a photoreceptor organelle called a *rhabdomere* that is composed of a closely packed stack of microvilli extending from the surface of the cell (Figs.11.2 and 5). A *microvillus* is a thin tube, roughly 500 Å in diameter, made of cell membrane that is loaded with molecules of visual pigment. The collection of rhabdomeres in a single retinula is called a *rhabdom*. The details of how the various rhabdomeres are packed into the rhabdom can be rather complicated and can be quite different in different regions of the same eye. Figures 11.1 and 2 show how some bee rhabdoms are both stratified and twisted [11.16]. The rhabdomeres of a retinula are packed together so that a rhabdom functions as a single optical waveguide, the *fused rhabdom*. An important exception to this generality is the

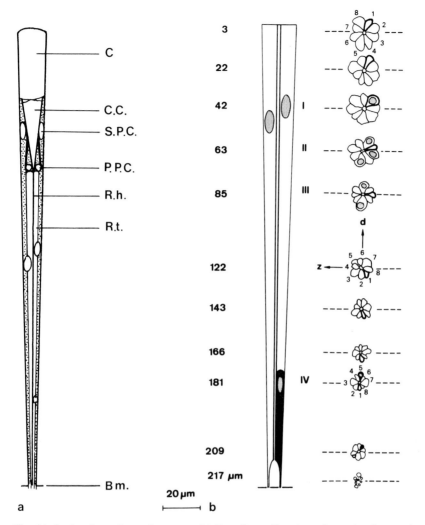

Fig. 11.1a,b. Geometry of an ommatidium from the dorsal part of a worker bee's eye. (a) A longitudinal section of one ommatidium drawn to scale. C. corneal lens; C.C. crystalline cone; S.P.C. secondary pigment cell; P.P.C. primary pigment cell; Rh. rhabdom; Rt. retinula; Bm. basement membrane. (b) A longitudinal section of a retinula and the series of cross sections are to the same scale. The series of numbers to the left are the levels at which the sections were taken, measured from the rhabdom's distal tip. Cell no. 1 is marked by heavy outline, and cell no. 9 is shown in black. I, II, III, and IV are nuclear layers. The dorsal direction is indicated by d [11.16]

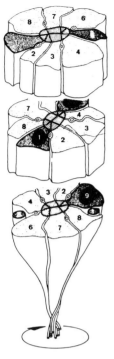

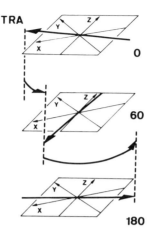

TRA

0

60

180

Fig.11.2. Three-dimensional model of the twisted
retinula of the worker bee. Cross sections are shown
for 0, 60, and 180 µm underneath the distal tip of
the retinula. The angular orientation of the transverse axis of the rhabdom (TRA) within the x-y-z
reference system of the eye is given for the dorsal
half of the eye [11.16]

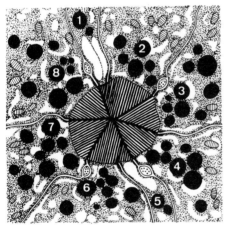

>430nm   **dark**   1µm

Fig.11.3. Cross sections through two retinulae of the desert ant, *Cataglyphis
bicolor*, after dark adaptation (right) and long-wavelength adaptation (left).
In the dark-adapted state of a cell, the vesicles of the endoplasmic reticulum form large vacuoles, the "palisade", around the rhabdom (cells no. 1
and 5 of the left retinula, and all cells of the right retinula). During
light adaptation (cells no. 2-4 and 6-8 of the left retinula), the large
vacuoles disappear and the screening pigment granules (black dots) move toward the rhabdom. Schematic drawings from EM micrographs [11.16]

open rhabdom of dipteran flies, hemipteran bugs, and some beetles. In this case the rhabdomeres are sufficiently separated so that a single rhabdomere is able to function in optical isolation of its neighbors.

Retinular cells contain other organelles that are optically important. The refractive index of the region immediately surrounding the rhabdom in the dark-adapted state can be lowered by the *palisade*, which is an aggregation of dilated cisternae of the endoplasmic reticulum (Fig.11.3). Its refractive index is close to that of cytoplasm because the volume occupied by the palisade is very watery, containing a relatively small fraction of membrane. The refractive index of the region surrounding the rhabdom can be high in the light-adapted state owing to aggregations of mitochondria or pigment granules.

Retinular cells in many species contain dense granules, of diameter 0.1 to 1 μm, that are pigmented with ommochromes or pteridines [11.17], but not melanin. These *pigment granules* mediate a mechanism for photomechanical adaptation, the so-called pupillary response, which is discussed in Sect. 11.4.1. It is important to understand that pigment granules do not contain rhodopsin and are not involved in phototransduction. Light absorbed by pigment granules does not cause the receptor potential. That function is performed by the visual pigment, which resides in the rhabdomeric membrane.

The proximal end (the end nearest the brain) of a retinular cell is a nerve fiber which usually terminates in the first visual neuropile, the *lamina ganglionaris*, although in some species a fraction of the retinular cell axons terminate in the second visual neuropile, the *medulla*. No one has reported a receptor cell axon terminating in the third visual neuropile, the *lobula*. The lamina, medulla, and lobula comprise the part of the insect brain that is primarily devoted to visual information processing.

Other cells of an ommatidium contain light-absorbing pigment granules, but do not have rhabdomeres nor do they possess axons. *Primary pigment cells* surround the crystalline cone and are located only in the distal part of the ommatidium (the part most distant from the brain). *Secondary pigment cells* surround the retinula and the primary pigment cells and run from the cornea all the way to the basement membrane which supports the proximal end of the retinula. The basal region of a retinula contains pigmented cells such as cone-cell processes, basal retinular cells, or tracheolar cells. The granules of all pigmented cells, called screening pigment or iris pigment, have the important optical function of controlling the spatial, spectral, and angular distributions of light within the volume of the eye.

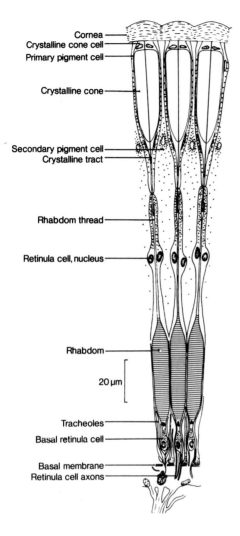

Cornea
Crystalline cone cell
Primary pigment cell

Crystalline cone

Secondary pigment cell
Crystalline tract

Rhabdom thread

Retinula cell, nucleus

Rhabdom

20 μm

Tracheoles
Basal retinula cell

Basal membrane
Retinula cell axons

Fig.11.4. Axial section through light-adapted ommatidia of *Ephestia*, a small moth that has a superposition eye. As the eye adapts to the dark, the pigment granules migrate distally, leaving the region between crystalline cones and rhabdoms relatively transparent. (From [11.19], modified by [11.20])

The terms *apposition eye* and *superposition eye*, commonly applied to compound eyes, originated in the work of EXNER [11.18], who pointed out the ecological correlation between diurnal/nocturnal habits and the structure of compound eyes. Species that are active during the day usually have ommatidia in which the distal tip of a rather long, slender rhabdom is very close to the proximal tip of the crystalline cone (Fig.11.1a). Species that are active at night usually have ommatidia in which the distal tip of a rather short, fat rhabdom is hundreds of micrometers from the cone tip (Fig.11.4). A thin, transparent column, the *crystalline tract*, runs from the cone tip to the rhabdom tip. The tract may be a process of the retinular cells, of the cone cells, or both.

Changes in the level of ambient illumination can trigger changes in both the position of pigment granules and the morphology of rhabdoms in some nocturnally specialized eyes, over time scales of tens or even hundreds of minutes. Diurnally-specialized apposition eyes exhibit much less dramatic photomechanical changes that are nevertheless physiologically important, having time constants as short as one or two seconds (see Sects.11.2.4 and 11.4.1).

11.2 Optics

The multiplicity of lenses of the compound eye is one of the most obvious differences when compared to the vertebrate eye. One of the earliest meaningful comparisons of these two kinds of visual organs was made in 1665 by Robert HOOKE [11.21], one of the first microscopists, who wrote:

".... onely the Picture of those parts of the external objects that lie in, or neer, the *Axis* of each *Hemisphere*, are discernably painted or made on the *Retina* of each *Hemisphere*, and that therefore each of them can distinctly sensate or see onely those parts which are very neer perpendicularly oppos'd to it, or lie in or neer its optick *Axis*. Now, though there may be by each of these eye-pearls (ommatidia), a representation to the Animal of a whole *Hemisphere* in the same manner as in a man's eye there is a picture or sensation in the *Retina* of all the objects lying almost in a *Hemisphere*; yet, as in a man's eye also, there are but some very few points which lying in, or neer, the optick *Axis* are distinctly discern'd: So there may be multitudes of Pictures made of an Object in the several Pearls, and yet but one, or some very few that are distinct; The representation of any object that is made in any other Pearl, but that which is directly, or very neer directly, oppos'd, being altogether confus'd and unable to produce a distinct vision.
So that we see, that though it has pleas'd the All-wise Creator, to indue this creature with such multitudes of eyes, yet he has not indued it with the faculty of seeing more than another creature;...."

This is a concise summary of the Mosaic Theory of vision which, for reasons I do not understand, has been attributed to MUELLER [11.22], circa 1826, rather than HOOKE [11.21], circa 1665.

Modern work has indeed established that an individual ommatidium samples only a restricted region of space. The extent of this region, the visual field, can vary considerably among the species, among the ommatidia of a given eye, and even with the state of adaptation. The functional properties of a photoreceptor cell depend on the optical constants and anatomical geometry of receptor cells and their associated structures.

11.2.1 Optical Constants

The optical constants of invertebrate photoreceptors are difficult to measure owing to their small diameters and complicated geometry. The fused rhabdom of most apposition eyes is only one or two micrometers in diameter by hundreds of micrometers in length and is composed of eight or nine rhabdomeres that may be stacked, tapered, interdigitated, and twisted. Fused rhabdoms of superposition eyes are ten times fatter and also have complicated geometry. With the exception of some crayfish and lobsters (Table 11.1), it has not been possible to measure optical constants of a fused rhabdom.

Table 11.1. Estimates of the optical constants of invertebrate photoreceptors based on direct optical measurements

Quantity	Estimate	Species	Source
Refractive index of rhabdom	1.365	fly	[11.23], corrected by [11.24]
	1.365	bee	[11.25], corrected by [11.26]
Refractive index of surround	1.339	many	[11.26]
Effect of anomalous dispersion on refractive index of rhabdom	10^{-3}	most	[11.27], corrected by [11.28]
Birefringence	-5×10^{-4}	fly	[11.29], corrected by [11.30]
	-1.2×10^{-3} to -6.7×10^{-3}	fly	[11.31]
Optical Density[a] (base 10)	$0.0033/\mu m$	fly	[11.32]
	$0.005/\mu m$	crayfish	[11.33]
Dichroic ratio	2	fly	[11.32]
	3	crayfish	[11.33]

[a]Monochromatic light, linearly polarized parallel to the microvilli, at wavelength for maximal density.

Eyes with open rhabdoms are more suited to measurements of optical constants because optical coupling among individual rhabdomeres is low. Most studies have used dipteran flies, which have long slender rhabdomeres of diameter between 1/2 - 2 μm. Because of such small diameters, accurate measurements require end-on illumination, in which case waveguide effects must be considered in calculating optical constants from raw data [11.26,34]. Table 11.1 contains estimates of optical constants of invertebrate photoreceptors that are based on direct optical measurements.

The macroscopic optical properties of the microvillar medium are those of a negatively-birefringent positively-dichroic uniaxial crystal in which the principal axis for both birefringence and dichroism is the microvillar direction (Fig.11.5). A singular exception is rhabdomere R7 of the fly, which is negatively dichroic [11.32,35].

The dichroic ratio of the microvillar medium may be as high as ten in some species, according to electrophysiological measurements of polarizational sensitivity of single retinular cells interpreted according to optical theory [11.33,36,37].

The vertebrate photoreceptors are also uniaxially anisotropic. However, they are not polarizationally sensitive because their principle axis is parallel to the long axis of the outer segment. This comparison is illustrated in Fig.11.5.

11.2.2 Visual Field of a Photoreceptor

The Mosaic Theory pertains to apposition compound eyes for which the internal focal plane of the ommatidial lens is near the distal tip of the rhabdom. Because the refractive index of a photoreceptor is greater than that of the surrounding medium (which is true also for vertebrates), the photoreceptors function as optical waveguides. Since the diameter of an ommatidial lens is rather small, usually < 50 μm, and since the focal length is two or three times the lens diameter, the diffraction-broadened image of a point object is roughly the same diameter as a rhabdomere, and the width of the visual field of a retinular cell is within a factor of two times the angular width of the Airy disk [11.8,10]. The angular location of the peak of a cell's visual field need not be parallel to the anatomical axis of the ommatidium [11.39], nor must it be the same for all retinular cells of a given ommatidium. In flies, which have open rhabdoms, the eight retinular cells of a single ommatidium sample seven distinct spatial directions. Furthermore, eight retinular cells in seven different ommatidia sample the same direction [11.32,40].

In eyes with fused rhabdoms, all retinular cells of a given ommatidium are thought to have identical visual fields, although in theory there can be significant differences if the rhabdom supports more than one type of waveguide mode [11.41]. The visual field of a retinular cell in most photopic

(A) OUTER SEGMENT OF VERTEBRATE ROD

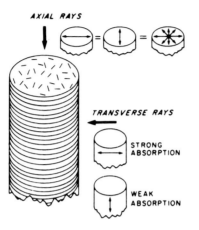

(B) RHABDOMERE OF DECAPOD CRUSTACEAN

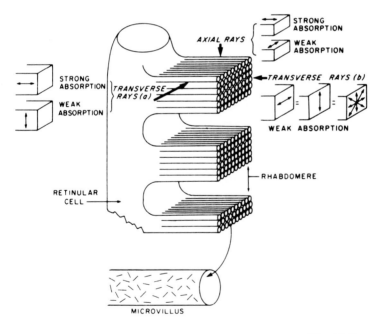

Fig.11.5.
Caption see opposite page

eyes has a half-width of at least a few degrees [11.8] , although the large visual predators, such as dragonflies, have some cells with much narrower visual fields [11.42].

The optics of nocturnally specialized compound eyes are more complicated. Most eyes of the superposition type undergo major morphological changes when adapting to large changes in the ambient level of illumination, with important optical consequences. When thoroughly dark-adapted, the region between the crystalline cones and the rhabdoms is free of pigment granules and so are the proximal tips of the crystalline cones. In this state of adaptation the optical stimulation of a given retinular cell is from light that has passed through the corneal lenses of many ommatidia, perhaps hundreds in a large eye. When illuminated by a distant point-source, a distribution of light at the tips of the rhabdoms is created, centered on the rhabdom that is pointed at the source. This situation is illustrated in Fig.11.6. The width of this so-called superposition image depends very much on the particular species; some have well-focused images and others do not. Consequently, the width of the visual field of a dark-adapted retinular cell depends on the species, varying from a few degrees to as much as forty degrees [11.8,9].

Light that enters the center of the patch of facets that is involved in forming the superposition image is more effective than light entering at the edge of the patch. Therefore, some invertebrate eyes may also exhibit a "ɔtiles-Crawford effect".

Fig.11.5. Orientation of visual pigment in the membranes of photoreceptors as revealed by measurements of dichroic absorption. (a) The chromophores of vertebrate rod and cone outer segments lie randomly oriented in the planes of the disk membranes. Consequently, absorption of axially incident rays is independent of the plane of polarization. When illuminated from the side, however, the outer segments are dichroic: Light polarized parallel to the plane of the disks (and thus to the axes of many of the chromophores) is strongly absorbed, whereas light polarized at right angles to the disks is only weakly absorbed. (b) The measurements of the dichroism of decapod crustacean rhabdomeres are consistent with the view that the chromophores are oriented more or less parallel to the microvilli. Shown here is a single retinular cell, of the type found in certain crustacea and insects, and several of its tongues of microvilli. In an intact rhabdom these tongues are interleaved with the rhabdomeres of other receptors. In some crustacea it is possible to irradiate a single band of microvilli of an isolated rhabdom, using a laterally incident microbeam. Absorption is isotropic when the beam is incident parallel to the microvillar axes, but is dichroic when the light is incident at right angles to the microvillar axes. This latter condition is equivalent to axial illumination, as occurs in the living eye. Absorption is strongest when the light is polarized parallel to the microvillar axes [11.38]

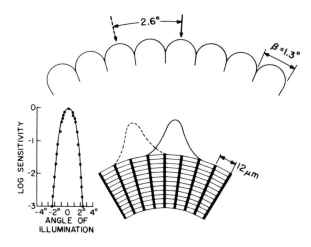

<u>Fig.11.6.</u> The graph of angular sensitivity, set to the left, is for a single retinular cell of the skipper *Epargyreus* [11.43]. Sensitivity at 1.3° (the interommatidial angle, β) is down by 1 log-unit, and at 2.6° (2β) is down by 3 log-units. The solid arrow near the cornea represents the axis on which a distant point source is located, oriented to stimulate maximally the center rhabdom. A rough approximation to the retinal-image distribution is shown by the solid curve. When the source is displaced by 2.6°, as indicated by the dashed arrow, the image distribution is shifted across the retina by about 24 μm (dashed curve). Since the center rhabdom is essentially unstimulated by a source 2.6° off axis, the image distribution can be no larger than three rhabdoms in diameter (about 36 μm). Therefore, the narrow angular sensitivity function is consistent with an image spot that illuminates one rhabdom well and its six neighbors weakly. This conclusion does not imply that the image is formed by the axial corneal facet and its six neighbors. Measuring the size of the corneal aperture involved in stimulating the rhabdom by a distant point source requires a different experiment, as described in [11.8], from which this figure is taken

When scotopic eyes are thoroughly light adapted, the region between cones and rhabdoms is loaded with pigment granules. Consequently, most of the light that stimulates a particular rhabdomere enters the eye through the corneal facet of the same ommatidium and propagates down the crystalline tract into the rhabdom. The sensitivity of a completely light-adapted retinular cell is low because the collecting aperture is small and because light is attenuated as it propagates down the pigment-cladded tract. The visual field of a retinular cell is usually much narrower when light adapted than when dark adapted [11.44].

11.2.3 Associated Structures

The optical properties of photoreceptors can be significantly influenced by tapetal reflectors, pigment granules, and corneal interference filters.

Just as the eyes of most nocturnal vertebrates contain a tapetum (see Chap.10), so do the eyes of nocturnal invertebrates. The tapetum of insects is usually a specialization of the tracheolar cells (the cells that distribute air throughout the animal), while the tapetum of marine invertebrates is usually made of highly reflecting pigment [11.45]. The most important optical effect of a tapetum is to return to the photoreceptors the light that would otherwise be spilled from their proximal ends, thereby enhancing sensitivity in conditions of dim illumination. Some invertebrate tapeta, as in crayfish, are specular reflectors which only minimally disturb the angular and polarizational properties of the incident light. Others, as in moths, are diffuse reflectors which scatter and depolarize the incident light.

Most butterfly eyes contain tapeta even though butterflies are strictly diurnal animals. Butterfly eyeshine is unusual because it is strongly colored and exhibits a heterogeneous mixture of hues over the eye. Each ommatidium contains a separate interference reflector composed of periodic platelets, which is optically isolated from the reflectors of neighboring ommatidia [11.46,47].

As described in Sect.11.1, there are several types of pigmented cells in compound eyes. Their pigment granules subserve several optical functions, the most important of which is optical isolation of neighboring ommatidia and reduction of scattered light. Granules of retinular cells can adjust the transmission of light into the photoreceptors, as discussed in Sect.11.4.1. Colored pigment granules can affect the spectral sensitivity of photoreceptors [11.48-50]. Highly reflecting granules can serve a dioptric function [11.51], and can increase the efficiency of light collection by the crystalline cones (see Chap.8).

Although the ocular media distal to the photoreceptors are spectrally neutral in most invertebrates, even in the near-ultraviolet, there are some exceptions. The anterior surface of the corneas of some dipteran flies contain periodically-layered interference filters that have narrow reflectance spectra [11.52]. Fish corneas also contain multilayered reflectors [11.53].

11.2.4 Waveguide Effects

Direct observations of waveguide modes in the fused rhabdom of the bee [11.25] and in the outer rhabdomeres of the fly's open rhabdom [11.40] have been reported. The waveguide properties of invertebrate photoreceptors can significantly influence the response of a photoreceptor cell to spectral, spatial, and angular parameters. Waveguide effects are particularly important in eyes having fused rhabdoms, but can be relatively unimportant in eyes having open rhabdoms.

Rhabdomeres in compound eyes that have open rhabdoms are similar to photoreceptors of many vertebrates in that they are long and slender and are optically isolated from neighboring photoreceptors. The waveguide effects that are important for open rhabdoms are a consequence of the fact that an optical fiber necessarily carries a fraction of its guided light in the region immediately external to the surface of the fiber. The magnitude of this fraction depends on the diameter, the refractive index of the fiber and the surrounding medium, and on the wavelength of the light being guided [11.8,34].

Pigment granules that are right next to the waveguide boundary can scatter and absorb light that would otherwise be guided, despite the fact that the granules are external to the waveguide. These effects have been directly observed in living insect eyes [11.39,40]. Granules must be closer to the boundary than 1/2 μm to have a significant effect, since the intensity of light decreases very rapidly with distance from the boundary. Therefore, radial movements of intracellular pigment granules over a distance of < 1 μm can mediate large changes in the transmission of light into the bulk of the photoreceptor. This would not be true if the photoreceptors were not waveguides.

The fraction of light carried in the surrounding medium is significantly greater at the long-wavelength end of the visual spectrum than it is at the short-wavelength end, if the diameter of the photoreceptor is < 1 μm. This property causes the spectral sensitivity of a retinular cell to differ somewhat from the absorption spectrum of its visual pigment because illumination of the rhodopsin molecules is more efficient at short wavelengths than at long wavelengths [11.54]. To gauge the importance of this effect, one must know the characteristic waveguide parameter V as a function of wavelength. This is difficult to determine because measurements of refractive index are insufficiently precise. KIRSCHFELD and SNYDER [11.55] have developed an approach for experimental determination of V that avoids this problem. Their approach involves measuring the effective birefringence and its variation with wavelength.

Because rhabdomeres are both dichroic and birefringent with an optic axis that is perpendicular to the waveguide axis, the response of a retinular cell varies with both the plane of polarization and the degree of polarization of the incident light. Most vertebrate photoreceptors are not sensitive to polarized light because their optic axis is parallel to the long axis of the outer segment (Fig.11.5). Polarizational sensitivity of a rhabdomere is independent of birefringence if the optic axis is the same throughout an optically isolated rhabdome. If the rhabdomere is twisted, the polarizational sensitivity is degraded unless the effective birefringence is sufficiently high [11.56,57].

Waveguide effects are much more important in eyes that contain fused rhabdoms because the photoreceptors within a single rhabdom are optically coupled. The response of a retinular cell depends on the properties of not only its rhabdomere but also those of the other rhabdomeres that share the same fused rhabdom [11.41,58]. Optical coupling influences how light is both propagated and absorbed within a photoreceptor. Photons that enter the tip of one photoreceptor can ultimately be absorbed in another receptor. Understanding of these kinds of effects may be best developed in terms of the following thought-experiment.

Consider two neighboring photoreceptor waveguides, one containing a rhodopsin of λ_{max} = 350 nm, and the other containing 550 nm rhodopsin. Assume that both receptors have an overall optical density of unity at their respective values of λ_{max}. Call the UV-absorbing photoreceptor "receptor U" and the other "receptor G". Illuminate the pair at 550 nm so that one unit of quantum flux enters each receptor. Compute the amount absorbed in each of two different circumstances. First, assume that there is no optical coupling between the modes of the two receptors, as in an open rhabdom. Second, assume that there is complete optical coupling, as in a fused rhabdom. Then repeat the calculations for 350 nm illumination. These four situations are depicted in Fig. 11.7.

In the open case, shown in Fig.11.7a, receptor U absorbs nothing because its rhodopsin is transparent at 550 nm. All 1.0 units of quantum flux is spilled from its proximal end. Receptor G absorbs 0.9 units while the remaining 0.1 units are spilled.

In the fused case, depicted in Fig.11.7b, the pair functions as a single waveguide in which the 2.0 units of quantum flux is carried by a single mode. The loss constants of photoreceptor waveguides are small enough so that absorption by visual pigments can be treated as a perturbation on the modes of the lossless waveguide [11.34,41]. In our case the overall density experienced

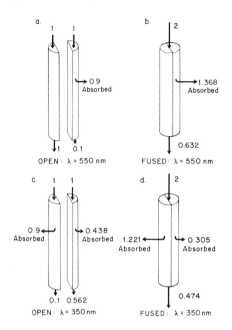

Fig.11.7. Calculations from a simple, theoretical model that demonstrate the effects of optical coupling on the spectral sensitivity and absolute sensitivity of photoreceptors. See Sect.11.2.4 for details

by the single mode is 0.5, which is the average of the overall densities of the two isolated rhabdomeres. Therefore, 1.37 units of flux are absorbed by receptor G, nothing is absorbed by receptor U, and 0.63 units are dumped from the proximal end of the pair. The effect of fusing the two receptors is to increase the absorbed flux (and the sensitivity) by 52%.

When the pair is illuminated with 350-nm quanta, both receptors absorb light. Suppose that the absorbance coefficient of rhodopsin in receptor G at 350 nm is one-fourth of that at 550 nm. Calculations for absorbed and spilled fractions of the 2.0 units of 350 nm quanta are shown in Fig.11.7c for the open case and in Fig.11.7d for the fused case. The effects of optical coupling are to increase the sensitivity of receptor U by 36% and to *decrease* that of receptor G by 30%.

Consider the effects of optical coupling on spectral sensitivity of receptor G by comparing ratios of absorbed quanta at the two wavelengths. The ratio Q(350)/Q(550) is 0.49 for the open case, which is substantially higher than 0.25, the absorbance ratio of the visual pigment. This is a manifestation of self-shielding. In the fused case, the sensitivity ratio is only 0.22, demonstrating that optical coupling within a fused rhabdom can eliminate the spectral broadening caused by self-shielding and still retain high absolute sensitivity.

Comparing the four situations of Fig.11.7 shows that the effects of fusing the photoreceptors are: a) to increase both the absorption of green light by receptor G and the absorption of UV light by receptor U; b) to decrease the absorption of UV light by receptor G; c) to decrease the amount of light spilled from the photoreceptors; and d) to reduce the effects of self-screening on the spectral sensitivity of receptor G.

That self-screening can be reduced by optical fusion of rhabdomeres was suggested by SHAW [11.36] and elaborated by SNYDER et al. [11.58]. The latter authors coined the term "lateral filtering" to apply to the redistribution of absorbed light that occurs among the rhabdomeres because of optical coupling.

Although most vertebrate photoreceptors are optically isolated from their neighboring receptors, there are situations in which optical coupling may be important, for example in fish retinas which contain closely apposed outer segments of twin cones, double cones, or even single cones. Optical coupling is usually thought to be undesirable because it degrades resolution. However, consideration of the invertebrate retina shows that optical coupling can be beneficial by eliminating undesirable effects of self-shielding.

The study of the physiological optics of invertebrate photoreceptors has recently been complicated by the discovery that rhabdoms can be twisted. The angular position of a given rhabdomere within the cross-section of a rhabdom can change with longitudinal position (Figs.11.1,2). Propagation of light in such a twisted, dichroic, birefringent waveguide is complicated for several reasons. Most important is that twisting causes optical coupling among the modes. If the effective birefringence is low, twisting of the rhabdom causes a reduction in polarization sensitivity, but only a small reduction if birefringence is high. Twisting of a rhabdom also has important consequences for the polarizational sensitivity of basal retinular cells [11.56,57].

The existence of higher-order modes in a rhabdom can have significant consequences for the spatial sensitivity of retinal cells; visual fields of different retinular cells in the same fused rhabdom need not be identical [11.41].

11.3 Photochemistry

Vertebrate and invertebrate rhodopsins are remarkably similar. Invertebrate pigments are not only based on Vitamin A, their choromophore is also retinaldehyde [11.59-61]. Detailed studies of cephlapods [11.62] and of the neuropteran *Ascalaphus* [11.63] have revealed that the chromophore is the 11-*cis*

isomer, while that of the stable photoproduct (metarhodopsin) is the *all-trans*
isomer. Although little is known about the opsin moiety of invertebrate pigments, there is evidence that the molecular weight of the protein is comparable to that of vertebrate rhodopsin [11.64].

There are some physiologically important differences in photochemistry,
the most important being that the isomerized chromophore does not separate
from the opsin. Invertebrate visual pigments are converted by light into
long-lived metarhodopsins that can be photoisomerized back to native rhodopsin [11.60,65]. Although it is a matter of some controversy, invertebrate
rhodopsins can also be regenerated metabolically in the dark, with a time
constant that depends on the species and on the preparatory procedures [11.66,
67]. An interesting invertebrate anomaly is that excessive conversion of
rhodopsin to metarhodopsin causes a very prolonged, depolarizing afterpotential, the so-called PDA (Sect.11.4.1).

Most invertebrates possess several spectral types of rhodopsins. One common pigment has a wavelength for maximal absorbance (λ_{max}) of about 525 nm.
Some eyes also contain a rhodopsin with λ_{max} in the range 430-450 nm. Very
few invertebrates have rhodopsins with λ_{max} greater than 550 nm, which ex-
plains why most invertebrates are practically blind in the orange and red
regions of the spectrum. Important exceptions are some species of butterflies
[11.67], which possess a rhodopsin of λ_{max} = 610 nm. As dehydroretinal is
not known from any invertebrate, the chromophore is most likely retinal,
which would make 610 nm the greatest λ_{max} yet reported for any retinal-based
visual pigment, vertebrate or invertebrate.

The visible spectrum for most invertebrates extends well into the near
ultraviolet because they possess a UV-absorbing rhodopsin, as in *Ascalaphus*,
with λ_{max} near 350 nm. Why this rhodopsin has λ_{max} *less* than for free retinal
is not understood. The relationship of chromophore to opsin must be very
special.

Although absorption spectra of invertebrate photopigments are similar in
shape to those of vertebrate pigments, some have a much wider spectrum with
two prominent peaks [11.59,68,69]. Dipteran flies are the most thoroughly
studied animals in this class. The basis for the two peaks is an active research topic [11.70-75]. There is agreement that a fly rhabdomere contains
only one spectral type of rhodopsin, which absorbs maximally between 440 nm
and 500 nm depending on the species, plus a photostable substance that absorbs ultraviolet light and somehow transfers energy to the blue-absorbing
rhodopsin. The identity of this so-called antenna pigment and the mechanism
for energy transfer have not yet been established. One important clue is that

both the antenna pigment and its effect on spectral sensitivity is abolished by Vitamin A deprivation [11.74]. Another unusual feature of fly rhodopsin is the huge bathochromic shift upon conversion to metarhodopsin, which can be as large as 90 nm.

A substantial difference between photoreceptors of invertebrates and vertebrates is in the fluidity of their membranes. Whereas rhodopsin molecules are able to both translate and rotate rather freely within rod and cone membranes, movement of pigment molecules within rhabdomeric membranes is quite restricted. Furthermore, the chromophores of rhodopsin molecules are preferentially aligned with respect to the microvillar axis [11.37].

Some invertebrates are particularly well suited to photochemical studies because it is possible to make measurements on visual pigments of the intact, living animal by exploiting the optical properties of the photoreceptors and their associated optical components. One very successful approach, developed by STAVENGA [11.76], is to illuminate the back of a fly's head, focus a microscope photometer on the deep pseudopupil, and measure transmittance spectra for light that has passed through the rhabdomeres. By comparing spectra taken after establishing photosteady states at various wavelengths it is possible to determine the absorption spectra of both rhodopsin and metarhodopsin [11.61,77].

The butterfly is also a good subject for in vivo photochemical studies owing to the presence of a retinal tapetum. However, this is a more difficult subject than the fly because the eye contains three or four spectral types of rhodopsin in its fused rhabdoms [11.67].

The noninvasive methods for studying invertebrate visual pigments have the important advantage that the retinular cells are completely intact, with all metabolic processes functioning, thereby avoiding the artifacts of the standard methods of microspectrophotometry caused by the preparatory procedures. Furthermore, one can make both physiological and photochemical measurements on the identical photoreceptor cells. The advantages over combined ERG and retinal densitometry of vertebrate eyes [11.78] are high signal-to-noise ratio and precise localization of measurements from very small groups of receptors.

Some interesting questions for comparative study are: a) What are the mechanisms for tuning λ_{max} of rhodopsin? Values for invertebrates range from 345 nm to 610 nm. b) Why does the isomerized chromophore remain attached to the opsin? c) How does UV-absorption by the antenna pigment of fly photoreceptors lead to isomerization of rhodopsin? d) What causes the PDA, and what can the PDA tell us about phototransduction?

11.4 Physiology

In general, the photoreceptor cells of invertebrates respond to steps of moderately bright light with a graded, depolarizing receptor potential due to an increase in membrane conductance to sodium ions. Receptor cells of vertebrates also respond with graded receptor potentials, but hyperpolarize due to a decrease in sodium conductance [11.79]. Based on intracellular recordings from many electrophysiological experiments on receptor cells from both kinds of animals, it is clear that there are strong similarities despite the difference in polarity of response [11.79-82]. The intensity-response functions are similar, as are the adaptational properties. One difference is that unitary events (quantum bumps), associated with absorption of individual photons, are observed routinely in intracellular recordings of responses from thoroughly-dark-adapted invertebrate photoreceptor cells to very dim illumination [11.83-85], but require special methods to observe them in recordings from vertebrate receptors [11.86].

There are some interesting exceptions to the above generalizations. Some primitive invertebrates have photoreceptors of ciliary origin that hyperpolarize in response to light. A few have receptors of microvillar origin that hyperpolarize. In some invertebrate receptors, changes in potassium conductance are important in generating the receptor potential [11.87].

11.4.1 Light Adaptation and Dark Adaptation

Eyes of most animals have a remarkable ability to function over a very wide range of ambient light intensity. Much of the control of visual sensitivity is peripheral, involving the receptor cells and neural interactions at their axon terminals. No single process is responsible for adaptation. Those processes involving the receptor cells may be neural, photochemical, or photomechanical in nature.

Several kinds of neural mechanisms contribute. The receptor potential created by turning on a moderately bright light consists of a large phasic peak followed by eventual return to a tonic plateau. The cell's inability to maintain the peak receptor potential is a manifestation of light adaptation. The processes at work during the first 100 milliseconds or so are largely neural, involving the membrane of the receptor cell. In the receptors of invertebrates, free intracellular calcium ions are involved in this process, but the relationship is not direct [11.88-90]. Studies of *Limulus* [11.91], squid [11.92], fly [11.93], and drone bee [11.81] show that this kind of neural adaptation is a local phenomenon.

Neural interactions between receptor cells and second-order cells are very important in rapidly adjusting sensitivity of invertebrate eyes as well as vertebrate eyes [11.82,94].

Sensitivity of photoreceptors is a monotonic function of visual pigment concentration for both vertebrates and invertebrates, but the functions are different. The relationship is linear for invertebrates, with a constant of proportionality that varies considerably among the species [11.65,95]. However, adaptational changes in the ventral photoreceptors of *Limulus* are relatively independent of pigment concentration [11.96].

The photoconversion of a large fraction of invertebrate rhodopsin to metarhodopsin somehow causes a prolonged depolarizing afterpotential (PDA) that can persist for several hours in the dark after the illumination has been removed. The photoreceptor cell can be repolarized at any time during the PDA by illuminating the cell with spectral light that photoconverts a large fraction of metarhodopsin back to native rhodopsin [11.97-101]. The mechanisms by which the PDA is established and abolished are matters of controversy. This is a fruitful area of comparative investigation for the understanding of phototransduction.

The eyes of many animals exhibit anatomical changes that influence the sensitivity of retinular cells. Examples are changes in shape and location of receptors, changes in distribution of intracellular organelles, migration of masses of pigment granules, and changes of pupillary aperture. These processes influence sensitivity by regulating the amount of light that reaches a photoreceptor. Time constants for mechanisms of photomechanical adaptation are much greater than for neural mechanisms, ranging from seconds to hours [11.8,102].

A well-known photomechanical mechanism for adaptation of the human eye is the pupillary response, which can adjust intensity by roughly one log-unit [11.103]. The terms "pupil" and "pupillary response" appear regularly in modern literature on compound eyes in several different contexts with different meanings [11.39].

When viewed under natural illumination, the compound eyes of many arthropods contain a central dark spot which moves across the eye as the observer changes her direction of observation. This spot has been called the pseudopupil since the 19th century. FRANCESCHINI [11.40] defines three types of pseudopupils (characterized by the adjectives deep, corneal, and reduced corneal) based on three distinct optical methods of viewing the eye through a microscope.

A second use of the word pupil originates with KUIPER [11.104] who applied the term 'longitudinal pupil' to mechanisms of photomechanical adaptation that involve accumulation of pigment granules at the boundary of a waveguide such as the crystalline tract or the rhabdom. The observable changes in optical properties caused by the migrating pigment granules is called the *pupillary response*. Both the decrease in transmittance through the rhabdomere and the increase in scattering by pupillary granules have been directly observed in many species [11.40,105,106].

A third use of the word pupil was introduced by KUNZE [11.20] in reference to the exit pupil of the dioptrics of an ommatidium of a scotopic eye. The diameter of this pupil can also change with state of adaptation.

The responses of longitudinal pupils of scotopic compound eyes are much slower than for photopic eyes, with time constants of tens or hundreds of minutes. Photomechanical mechanisms of vertebrate eyes have similar time constants.

The fly's eye exhibits rapid pupillary responses that have time constants of a few seconds and an intensity-response function that spans at least three log-units [11.40]. The pupillary response of an individual retinular cell depends primarily on the optical stimulation of its rhabdomere and not on stimulation of other cells in the same ommatidium [11.107].

Similar results have been obtained for Hymenoptera [11.105] and for butterflies [11.106]. However, Hymenoptera have somewhat slower pupils than flies and butterflies. Even so, the pupillary responses of photopic compound eyes are much faster than the pigment migrations found in eyes of vertebrates.

In flies, the decrease in transmittance caused by a large pupillary response is wavelength dependent, being effective in the spectral band where the rhodopsin absorbs well (420-520 nm) and ineffective where the metarhodopsin absorbs well (520-650 nm). Consequently, the metarhodopsin is selectively illuminated, thereby hastening recovery of sensitivity when the bright illumination is removed [11.39,108]. Another mechanism for keeping the titer of rhodopsin high is for the screening pigment granules to be transparent at long wavelength. This is why the house fly has a red eye [11.39].

The pupillary response of invertebrates is functionally similar to the change in optical stimulation caused by the vertebrate pupil, although the anatomy and physiology are quite different. The time constants for pupillary responses of active, diurnal insects are about the same as for the pupillary response of man.

Vertebrate retinas also exhibit photomechanical migration of pigment granules. However, the known mechanisms involve slow, longitudinal movements over large distances. Rapid, radial migrations over small distances, analogous to those of the insect retina, have not been described. Such subtle migrations are very difficult to demonstrate with ordinary histological methods. Mechanisms for rapid photomechanical adaptation may well be present in any vertebrate eye that has pigment granules in its photoreceptor cells, or that has pigment-containing cells close to outer segments, inner segments, myoids, or ellipsoids. In situations like these, pigment granules need only move 1 μm to cause a large change in transparency, leading to a significant change in sensitivity of photoreceptor cells.

11.4.2 Intracellular Optical Physiology

The optical effects that accompany the pupillary response can be used as tools to study the sensitivity of photoreceptor cells of intact, living invertebrates [11.40]. Few species are suitable for measurements of pupillary transmittance based on antidromic illumination of the eye, owing to dense pigmentation in the regions proximal to the basal ends of the rhabdomeres. However, many species exhibit observable pupillary scattering. Action spectra for increases of pupillary scatter from eyes of flies compare favorably to the spectral sensitivity functions that have been measured with intracellular electrophysiological methods [11.109]. By using appropriate adapting lights and short stimulating flashes it is possible to isolate the pupillary scatter from only one spectral type of retinular cell and to measure its action spectrum. Three spectral types of receptors have been demonstrated in the eyes of intact bees by this method [11.110,111].

Many butterfly species are particularly good subjects for noninvasive optical methods because their retinas contain tapetal reflectors. The eyeshine of butterflies can be used as a tool to study both the spectral sensitivities of retinular cells and the absorption properties of photopigments and photoproducts [11.112]. With such methods it has been shown that some butterfly eyes have four spectral types of receptors, one of which contains a rhodopsin absorbing maximally at 610 nm [11.67].

The advantages of the noninvasive optical methods compared to conventional methods are: a) the animal under study is completely intact and healthy (the 'alert-insect preparation') with undisturbed optics, morphology, and physiology; b) responses are stable and reproducible over very long periods of time — as long as five months in some butterfly species; c) the region of the eye that contributes to the response can be precisely localized and bounded; d) both physiological and photochemical measurements can be performed on identical photoreceptor cells.

11.4.3 Genetic Dissection

Many single-step mutants of the fly *Drosophila* are known to have specific defects of the photoreceptor cells. Some have normal anatomy and ultrastructure, but defective electrophysiological responses. Others have anatomical defects in only one of the three classes of photoreceptor cells.

For example, the receptor cells of most *norpA* alleles have normal ultrastructure, photochemistry, resting potential, and membrane resistance, but no receptor potential. The *norpA* lesion appears not to affect the machinery that generates quantum bumps, but interferes with a step after photoisomerization and before bump production. The photoreceptors of the *trp* mutant are normal except that the receptor potential is noisy and phasic, having no sustained component. The decay of the response to sustained illumination arises from a reduction in the quantum efficiency of bump production, individual bumps appearing normal [11.113,114].

Several mutations are known that eliminate specific classes of photoreceptor cells, leaving others both anatomically and physiologically intact [11.115, 116]. The mutation *ora* prevents formation of outer rhabdomers R1-R6, while *sev* prevents formation of the superior central cell, R7. The mutants *rdgA* and *rdgB* have normal retinal ultrastructure upon eclosion, but within one week, cells R1-R6 have completely degenerated leaving cells R7 and R8 intact. These two mutants differ in that the retinal degeneration of *rdgB* requires illumination while degeneration in *rdgA* does not. Electrophysiological and behavioral studies of these mutants have been valuable in understanding visually-mediated behavior [11.117-119].

11.5 Concluding Remarks

The most important task of the photoreceptor cells in any retina is the conversion of optical signals to neural signals. The illumination that impinges upon an eye is very complicated because the light is variable in direction, wavelength, polarization, and time. Only a fraction of the enormous amount of information carried by the light is "signal" of interest to an organism - the rest is "noise". The task of the photoreceptor cells, and their associated optics, is to both destroy the noise and convert the signal to neural information in the form of an ensemble of receptor potentials at the axon terminals.

Many comparable optical properties of vertebrate and invertebrate photoreceptors have evolved independently as adaptations to fulfill this task. The similarities are important as models or analogues. The many differences between invertebrate and vertebrate photoreceptors are attributable to constraints of size, distinct adaptive pressures, or developmental heritage. Although less obvious, the differences are important as well, for careful comparative studies can reveal fundamental principles for optical and neural processing of visual information.

Acknowledgement. This work is supported by grants EY01140 and EY00785 from the National Eye Institute, United States Public Health Service, by the Connecticut Lions Eye Research Foundation, Inc., and by Research to Prevent Blindness, Inc.

References

11.1   H.K. Hartline: Visual receptors and retinal interaction. Science *164*, 270-278 (1969)

11.2   F. Ratliff: *Studies on Excitation and Inhibition in the Retina* (Rockefeller, New York 1974)

11.3   C.G. Bernhard: *The Functional Organization of the Compound Eye* (Pergamon, London 1966)

11.4   W. Reichardt: *Processing of Optical Data by Organisms and by Machines* (Academic, New York 1969)

11.5   R.M. Eakin: "Structure of Invertebrate Photoreceptors", in *Photochemistry of Vision*, ed. by H.J.A. Dartnall, Handbook of Sensory Physiology, Vol.7/1 (Springer, Berlin, Heidelberg, New York 1972) pp.625-684

11.6   O. Trujillo-Cenóz: "The Structural Organization of the Compound Eye in Insects", in *Physiology of Photoreceptor Organs*, ed. by M.G.F. Fuortes, Handbook of Sensory Physiology, Vol.7/2 (Springer, Berlin, Heidelberg, New York 1972) pp.5-62

11.7   R. Wehner (ed.): *Information Processing in the Visual Systems of Arthropods*, Symposium held at the Department of Zoology, University of Zürich, 1972 (Springer, Berlin, Heidelberg, New York 1972)

11.8   T.H. Goldsmith, G.D. Bernard: "The Visual System of Insects", in *The Physiology of Insecta*, ed. by M. Rockstein, Vol.II, 2nd ed. (Academic, New York 1974) pp.165-272

11.9   G.A. Horridge: *The Compound Eye and Vision of Insects* (Oxford, London 1975)

11.10 A.W. Snyder, R. Menzel (eds.): *Photoreceptor Optics* (Springer, Berlin, Heidelberg, New York 1975)

11.11 F. Zettler, R. Weiler (eds.): *Neural Principles in Vision*, Proceedings in Life Sciences (Springer, Berlin, Heidelberg, New York 1976)

11.12 H. Autrum: "Vision in Invertebrates", in *Comparative Physiology and Evolution of Vision in Invertebrates*, ed. by H. Autrum, Vol.7/6A (Springer, Berlin, Heidelberg, New York 1979)

11.13 L.J. Goodman: The Neural Organization and Physiology of the Insect Dorsal Ocellus", in *The Compound Eye and Vision of Insects*, ed. by G.A. Horridge (Oxford, London 1975) pp.515-548

11.14 M. Wilson: The functional organization of locust ocelli. J. Comp. Physiol. *124*, 297-316 (1978a)

11.15 M. Wilson: Generation of graded potential signals in the second order cells of locust ocellus. J. Comp. Physiol. *124*, 317-331 (1978b)

11.16 R. Wehner: "Structure and Function of the Peripheral Pathway in Hymenopterans", in *Neural Principles in Vision*, ed. by F. Zettler, R. Weiler, Proceedings in Life Sciences (Springer, Berlin, Heidelberg, New York 1976) pp.280-333

11.17 H. Langer: "Properties and Functions of Screening Pigments in Insect Eyes", in *Photoreceptor Optics*, ed. by A.W. Snyder, R. Menzel (Springer, Berlin, Heidelberg, New York 1975) pp.429-455

11.18 S. Exner: *Die Physiologie der facettierten Augen von Krebsen und Insekten* (Franz Deuticke, Vienna 1891)

11.19 A. Fischer, G. Horstmann: Der Feinbau des Auges der Mehlmotte, *Ephestia kuehniella*. Z. Zellforsch. Mikrosk. Anat. *116*, 275-304 (1971)

11.20 P. Kunze: "Pigment Migration and the Pupil of the Dioptric Apparatus in Superposition Eyes", in *Information Processing in the Visual Systems of Arthropods*, ed. by R. Wehner (Springer, Berlin, Heidelberg, New York 1972) pp.89-92

11.21 R. Hooke: *Micrographia* (Royal Society, London 1665) (reprinted by Dover Publications, New York 1961)

11.22 J. Mueller: *Zur Vergleichenden Physiologie des Gesichtssinnes des Menschen und Tiere* (Cnobloch, Leipzig 1826)

11.23 G. Seitz: Der Strahlengang im Appositionsauge von *Calliphora erythrocephala* (Meig.). Z. Vgl. Physiol. *59*, 205-231 (1968)

11.24 D.G. Stavenga: Refractive index of fly rhabdomeres. J. Comp. Physiol. *91*, 417-426 (1974)

11.25 F.G. Varela, W. Wiitanen: The optics of the compound eye of the honeybee (*Apis mellifera*). J. Gen. Physiol. *55*, 336-358 (1970)

11.26 D.G. Stavenga: Waveguide modes and refractive index in photoreceptors of invertebrates. Vision Res. *15*, 323-330 (1975)

11.27 D.G. Stavenga, H.H. Van Barneveld: On dispersion in visual photoreceptors. Vision Res. *15*, 1091-1095 (1975)

11.28 W.S. Jagger, P.A. Liebman: Anomalous dispersion of rhodopsin in rod outer segments of the frog. J. Opt. Soc. Am. *66*, 56-59 (1976)

11.29 G. Seitz: Polarisationsoptische Untersuchungen am Auge von *Calliphora erythrocephala* (Meig.). Z. Zellforsch. Mikrosk. Anat. *93*, 525-529 (1969)

11.30 D.G. Stavenga: "Optical Qualities of the Fly Eye", in *Photoreceptor Optics*, ed. by A.W. Snyder, R. Menzel (Springer, Berlin, Heidelberg, New York 1975) pp.126-144

11.31 K. Kirschfeld, A.W. Snyder: "Waveguide Mode Effects, Birefringence and Dichroism in Fly Photoreceptors", in *Photoreceptor Optics*, ed. by A.W. Snyder, R. Menzel (Springer, Berlin, Heidelberg, New York 1975) pp.56-77

11.32 K. Kirschfeld: "Absorption Properties of Photopigments in Single Rods, Cones and Rhabdomeres", in *Processing of Optical Data by Organisms and by Machines*, ed. by W. Reichardt (Academic, New York 1969) pp.116-14

11.33 T.H. Goldsmith: "The Polarization Sensitivity - Dichroic Absorption Paradox in Arthropod Photoreceptors", in *Photoreceptor Optics*, ed. by A.W. Snyder, R. Menzel (Springer, Berlin, Heidelberg, New York 1975) pp.393-409

11.34   A.W. Snyder: "Photoreceptor Optics - Theoretical Principles", in
        Photoreceptor Optics, ed. by A.W. Snyder, R. Menzel (Springer, Berlin,
        Heidelberg, New York 1975) pp.38-55
11.35   K. Kirschfeld, R. Feiler, N. Franceschini: A photostable pigment within
        the rhabdomere of fly photoreceptors no.7. J. Comp. Physiol. *125*,
        275-284 (1978)
11.36   S.R. Shaw: Sense-cell structure and interspecies comparisons of
        polarized-light absorption in arthropod compound eyes. Vision Res. *9*,
        1031-1040 (1969)
11.37   T.H. Goldsmith, R. Wehner: Restrictions on rotational and translational
        diffusion of pigment in the membrane of a rhabdomeric photoreceptor.
        J. Gen. Physiol. *70*, 453-490 (1977)
11.38   T.H. Goldsmith: "Photoreception and Vision", in *Comparative Animal
        Physiology*, ed. by C.L. Prosser (Saunders, Philadelphia 1973) pp.577-632
11.39   D.G. Stavenga: "Pseudopupils of Compound Eyes", in *Comparative Physi-
        ology and Evolution of Vision in Invertebrates*, ed. by H. Autrum,
        Handbook of Sensory Physiology, Vol.7/6A (Springer, Berlin, Heidelberg,
        New York 1979) pp.357-439
11.40   N. Franceschini: "Sampling of the Visual Environment by the Compound
        Eye of the Fly", in *Photoreceptor Optics*, ed. by A.W. Snyder, R. Menzel
        (Springer, Berlin, Heidelberg, New York 1975) pp.98-125
11.41   G.D. Bernard: "Physiological Optics of the Fused Rhadom", in *Photore-
        ceptor Optics*, ed. by A.W. Snyder, R. Menzel (Springer, Berlin, Heidel-
        berg, New York 1975) pp.78-97
11.42   T.E. Sherk: Development of the compound eyes of dragonflies (Odonata).
        III. Adult compound eyes. J. Exp. Zool. *203*, 61-80 (1978)
11.43   K.B. Døving, W.H. Miller: Function of insect compound eyes containing
        crystalline tracts. J. Gen. Physiol. *54*, 250-267 (1969)
11.44   B. Walcott: Unit studies on light-adaptation in the retina of the
        crayfish, *Cherax destructor*. J. Comp. Physiol. *94*, 207-218 (1974)
11.45   E.S. Zyznar, J.A.C. Nicol: Ocular reflecting pigments of some malaco-
        straca. J. Exp. Mar. Biol. Ecol. *6*, 235-248 (1971)
11.46   G.D. Bernard, W.H. Miller: What does antenna engineering have to do
        with insect eyes? IEEE Student Journal *8*, 2-8 & color covers (1970)
11.47   W.H. Miller: "Ocular Optical Filtering", in *Comparative Physiology
        and Evolution of Vision in Invertebrates*, ed. by H. Autrum, Handbook
        of Sensory Physiology, Vol.7/6A (Springer, Berlin, Heidelberg, New
        York 1979) pp.69-143
11.48   K.L. Kong, T.H. Goldsmith: Photosensitivity of retinular cells in
        white-eyed crayfish (*Procambarus clarkii*). J. Comp. Physiol. *122*,
        273-288 (1977)
11.49   T.H. Goldsmith: The effects of screening pigments on the spectral sen-
        sitivity of some crustacea with scotopic (superposition) eyes. Vision
        Res. *18*, 475-482 (1978)
11.50   W.A. Ribi: A unique hymenopteran compound eye. The retina and fine
        structure of the digger wasp *Sphex cognatus* Smith (Hymenoptera,
        Sphecidae). Zool. Jahrb. Ab. Anat. Ontog. Tiere *100*, 299-342 (1978)
11.51   M.F. Land: Superposition images are formed by reflection in the
        eyes of some oceanic decapod crustacea. Nature (London) *263*, 764-765
        (1976)
11.52   G.D. Bernard: Evidence for visual function of corneal interference
        filters. J. Insect. Physiol. *17*, 2287-2300 (1971)
11.53   J.N. Lythgoe: "The Ecological Function and Phylogeny of Irridescent
        Multilayers in Fish Corneas", in *Light as an Ecological Factor: II*,
        ed. by G.C. Evans, R. Bainbridge, O. Rackham (Blackwell, London 1976)
        pp.211-247

11.54 A.W. Snyder, C. Pask: Spectral sensitivity of dipteran retinular cells. J. Comp. Physiol. *84*, 59-76 (1973)

11.55 K. Kirschfeld, A.W. Snyder: Measurements of a photoreceptor's characteristic waveguide parameter. Vision Res. *16*, 775-778 (1976)

11.56 R. Wehner, G.D. Bernard, E. Geiger: Twisted and non-twisted rhabdoms and their significance for polarization detection in the bee. J. Comp. Physiol. *104*, 225-245 (1975)

11.57 P. McIntyre, A.W. Snyder: Light propagation in twisted anisotropic media: Application to photoreceptors. J. Opt. Soc. Am. *68*, 149-157 (1978)

11.58 A.W. Snyder, R. Menzel, S.B. Laughlin: Structure and function of the fused rhabdom. J. Comp. Physiol. *87*, 99-135 (1973)

11.59 T.H. Goldsmith: "The Natural History of Invertebrate Visual Pigments", in *Photochemistry of Vision* , ed. by H.J.A. Dartnall, Handbook of Sensory Physiology, Vol.7/1 (Springer, Berlin, Heidelberg, New York 1972) pp.685-719

11.60 H. Langer (ed.): *Biochemistry and Physiology of Visual Pigments*, Symposium held at Institut für Tierphysiologie Ruhr-Universität, 1972 (Springer, Berlin, Heidelberg, New York 1973)

11.61 K. Hamdorf: "The Physiology of Invertebrate Visual Pigments", in *Comparative Physiology and Evolution of Vision in Invertebrates*, ed. by H. Autrum, Handbook of Sensory Physiology, Vol.7/6A (Springer, Berlin, Heidelberg, New York 1979) pp.145-224

11.62 R. Hubbard, R.C.C. St. George: The rhodopsin system of the squid. J. Gen. Physiol. *41*, 501-528 (1958)

11.63 R. Paulsen, J. Schwemer: Studies on the insect visual pigment sensitive to ultraviolet light: Retinal as the chromophoric group. Biochem. Biophys. Acta *283*, 520-529 (1972)

11.64 R. Paulsen, J. Schwemer: Proteins of invertebrate photoreceptor membranes. Eur. J. Biochem. *40*, 577-583 (1973)

11.65 K. Hamdorf, J. Schwemer: "Photoregeneration and the Adaptation Process in Insect Photoreceptor", in *Photoreceptor Optics*, ed. by A.W. Snyder, R. Menzel (Springer, Berlin, Heidelberg, New York 1975) pp.263-295

11.66 D.G. Stavenga: "Dark Regeneration of Invertebrate Visual Pigments", in *Photoreceptor Optics*, ed. by A.W. Snyder, R. Menzel (Springer, Berlin, Heidelberg, New York 1975) pp.290-295

11.67 G.D. Bernard: Red-absorbing visual pigment of butterflies. Science *203*, 1125-1127 (1979)

11.68 G.S. Wasserman: Invertebrate colour vision and the tuned receptor paradigm. Science *180*, 268-275 (1973)

11.69 R. Menzel: "Colour Receptors in Insects", in *The Compound Eye and Vision of Insects*, ed. by G.A. Horridge (Oxford, London 1975) pp.121-153 and 549-590

11.70 K. Kirschfeld, N. Franceschini: Photostable pigments within the membrane of photoreceptors and their possible role. Biophys. Struct. Mech. *3*, 191-194 (1977)

11.71 W.S. Stark: Sensitivity and adaptation in R7, an ultraviolet photoreceptor, in the *Drosophila* retina. J. Comp. Physiol. *115*, 47-59 (1977)

11.72 K. Kirschfeld, N. Franceschini, B. Minke: Evidence for a sensitizing pigment in fly photoreceptors. Nature (London) *269*, 386-390 (1977)

11.73 B. Minke, K. Kirschfeld: The contribution of a sensitizing pigment to the photosensitivity spectra of fly rhodopsin and metarhopsin. J. Gen. Physiol. *73*, 517-540 (1979)

11.74 W.S. Stark, K.L. Frayer, M.A. Johnson: Photopigment and receptor properties in *Drosophila* compound eye and ocellar receptors. Biophys. Struct. Mech. *5*, 197-209 (1979)

11.75 W.S. Stark, D.G. Stavenga, B. Kruizinga: Fly photoreceptor fluorescence is related to UV sensitivity. Nature (London) *280*, 581-583 (1979)

11.76 D.G. Stavenga: Fly visual pigments. Difference in visual pigments of blowfly and dronefly peripheral retinular cells. J. Comp. Physiol. *111*, 137-152 (1976)

11.77 D.G. Stavenga: Derivation of photochrome absorption spectra from absorbance difference measurements. Photochem. Photobiol. *21*, 105-110 (1975)

11.78 A. Knowles, H.J.A. Dartnall: "The Study of Visual Pigments in the Retina", in *The Eye*, ed. by H. Davson, Vol.2B, 2nd ed. (Academic, New York 1977) pp.557-579

11.79 T. Tomita: "Light-Induced Potential and Resistance Changes in Vertebrate Photoreceptors", in *Physiology of Photoreceptor Organs*, ed. by M.G.F. Fuortes, Handbook of Sensory Physiology, Vol.7/2 (Springer, Berlin, Heidelberg, New York 1972(pp.483-511

11.80 M.G.F. Fuortes, P.M. O'Brien: "Generator Potentials in Vertebrate Photoreceptors", in *Physiology of Photoreceptor Organs*,ed. by M.G.F. Fuortes, Handbook of Sensory Physiology, Vol.7/2 (Springer, Berlin, Heidelberg, New York 1972) pp.279-319

11.81 F. Baumann: "Electrophysiological Properties of the Honeybee Retina", in *The Compound Eye and Vision of Insects*, ed. by G.A. Horridge (Oxford, London 1975) pp.53-74 and 549-590

11.82 S.B. Laughlin: "Adaptations of the Dragonfly Retina and the Elucidation of Neural Principles in the Peripheral Visual System", in *Neural Principles in Vision*, ed. by F. Zettler, R. Weiler, Proceedings in Life Sciences (Springer, Berlin, Heidelberg, New York 1976) pp.175-193

11.83 M.G.F. Fuortes, P.M. O'Brien: "Responses to Single Photons", in *Physiology of Photoreceptor Organs*, ed. by M.G.F. Fuortes, Handbook of Sensory Physiology, Vol.7/2 (Springer, Berlin, Heidelberg, New York 1972) pp.321-338

11.84 E. Kaplan, R.B. Barlow: Energy, quanta, and *Limulus* vision. Vision Res. *16*, 745-751 (1976)

11.85 P.G. Lillywhite: Single photon signals and transduction in an insect eye. J. Comp. Physiol. *122*, 189-200 (1977)

11.86 K.-W. Yau, T.D. Lamb, D.A. Baylor: Light-induced fluctuations in membrane current of single toad rod outer segment. Nature (London) *269*, 78-80 (1977)

11.87 J.S. McReynolds: "Hyperpolarizing Photoreceptors in Invertebrates", in *Neural Principles in Vision*, ed. by F. Zettler, R. Weiler, Proceedings in Life Sciences (Springer, Berlin, Heidelberg, New York 1976) pp.394-409

11.88 J.E. Brown, J.R. Blinks: Changes in intracellular free calcium concentration during illumination of invertebrate photoreceptors: Detection with aequorin. J. Gen. Physiol. *64*, 643-665 (1974)

11.89 L.H. Pinto, J.E. Brown: Intracellular recordings from photoreceptors of the squid (*Loligo pealii*). J. Comp. Physiol. *122*, 241-250 (1977)

11.90 W.A. Hagins, S. Yoshikami: "Intracellular Transmission of Visual Excitation in Photoreceptors", in *Vertebrate Photoreception*, ed. by H.B. Barlow, P. Fatt (Academic, New York 1977) pp.97-139

11.91 A. Fein, J.S. Charlton: Local adaptation in the ventral photoreceptors of *Limulus*. J. Gen. Physiol. *66*, 823-836 (1975)

11.92 W.A. Hagins, H.V. Zonana, R.G. Adams: Local membrane current in the outer segments of squid photoreceptors. Nature (London) *194*, 844-868 (1962)

11.93 K. Hamdorf: Korrelation zwischen Sehfarbstoff und Empfindlichkeit bei Photorezeptoren. Verh. Dtsch. Zool. Ges. *64*, 148-158 (1970)

462

11.94 S.B. Laughlin, R.C. Hardie: Common strategies for light adaptation in the peripheral visual systems of fly and dragonfly. J. Comp. Physiol. *128*, 319-340 (1978)

11.95 S.N. Barnes, T.H. Goldsmith: Dark adaptation, sensitivity, and rhodopsin level in the eye of the lobster, *Homarus*. J. Comp. Physiol. *120*, 143-159 (1977)

11.96 A. Fein, R.D. DeVoe: Adaptation in the ventral eye of *Limulus* is functionally independent of the photochemical cycle, membrane potential, and membrane resistance. J. Gen. Physiol. *61*, 273-289 (1973)

11.97 J. Nolte, J.E. Brown: Ultraviolet-induced sensitivity to visible light in ultraviolet receptors of *Limulus*. J. Gen. Physiol. *59*, 186-200 (1972)

11.98 B. Minke, S. Hochstein, P. Hillman: Antagonistic processes as source of visible-light suppression of afterpotential in *Limulus* UV photoreceptors. J. Gen. Physiol. *62*, 787-791 (1973)

11.99 W.S. Stark, W.G. Zitzman: Isolation of adaptation mechanisms and photopigment spectra by Vitamin A deprivation in *Drosophila*. J. Comp. Physiol. *105*, 15-27 (1976)

11.100 R. Wright, D. Cosens: Blue-adaptation and orange-adaptation in white-eyed *Drosophila*: Evidence that the prolonged afterpotential is correlated with the amount of M580 in R1-6. J. Comp. Physiol. *113*, 105-128 (1977)

11.101 B. Minke, K. Kirschfeld: Microspectrophotometric evidence for two photointerconvertible states of visual pigment in the barnacle lateral eye. J. Gen. Physiol. *71*, 37-45 (1978)

11.102 B. Walcott: "Anatomical Changes During Light Adaptation in Insect Compound Eyes", In *The Compound Eye and Vision of Insects*, ed. by G.A. Horridge (Oxford, London 1975) pp.20-33 and 549-590

11.103 H.B. Barlow: "Dark and Light Adaptation: Psychophysics", in *Visual Psychophysics*, ed. by D. Jameson, L.M. Hurvich, Handbook of Sensory Physiology, Vol.7/4 (Springer, Berlin, Heidelberg, New York 1972) pp.1-28

11.104 J.W. Kuiper: The optics of the compound eye. Symp. Soc. Exp. Biol. *16*, 58-71 (1962)

11.105 D.G. Stavenga, J.W. Kuiper: Insect pupil mechanisms. I. On the pigment migration in the retinula cells of Hymenoptera (suborder Apocrita). J. Comp. Physiol. *113*, 55-72 (1977)

11.106 D.G. Stavenga, J.A.J. Numan, J. Tinbergen, J.W. Kuiper: Insect pupil mechanisms. II. Pigment migration in retinula cells of butterflies. J. Comp. Physiol. *113*, 73-93 (1977)

11.107 K. Kirschfeld, N. Franceschini: Ein Mechanismus zur Steuerung des Lichtflusses in den Rhabdomeren des Komplexauges von *Musca*. Kybernetik *6*, 13-21 (1969)

11.108 D.G. Stavenga, A. Zantema, J.W. Kuiper: "Rhodopsin Processes and the Function of the Pupil Mechanism in Flies", in *Biochemistry and Physiology of Visual Pigments*, ed. by H. Langer, Symposium held at Institute für Tierphysiologie Ruhr-Universität, (Springer, Berlin, Heidelberg, New York 1973) pp.175-180

11.109 G.D. Bernard, D.G. Stavenga: Spectral sensitivities of retinular cells measured in intact, living flies by an optical method. J. Comp. Physiol. *134*, 95-107 (1979)

11.110 G.D. Bernard, D.G. Stavenga: Spectral sensitivities of retinular cells measured in intact, living bumblebees by an optical method. Naturwissenschaften *65*, 442-443 (1978)

11.111 G.D. Bernard, R. Wehner: Intracellular optical physiology of the bee's eye: I. Spectral sensitivity. J. Comp. Physiol. *137*, 193-203 (1980)

11.112 D.G. Stavenga: Visual adaptation in butterflies. Nature (London) *254*, 435-437 (1975)

11.113 W.L. Pak: "Mutations Affecting the Vision of *Drosophila* melanogaster", in *Handbook of Genetics*, ed. by R.C. King (Plenum, New York 1975) pp.703-733

11.114 W.L. Pak, L.H. Pinto: Genetic approach to the study of the nervous system. Ann. Rev. Biophys. Bioeng. *5*, 397-448 (1976)

11.115 W.A. Harris, W.S. Stark, J.A. Walker: Genetic dissection of the photoreceptor system in compound eye of *Drosophila melanogaster*. J. Physiol. (London) *256*, 415,439 (1976)

11.116 W.A. Harris, W.S. Stark: Hereditary retinal degeneration in *Drosophila melanogaster*. A mutant defect associated with the phototransduction process. J. Gen. Physiol. *69*, 261-291 (1977)

11.117 R. Willmund, K.F. Fischbach: Light induced modification of phototactic behaviour of *Drosophila* melanogaster wildtype and some mutants in the visual system. J. Comp. Physiol. *118*, 261-271 (1977)

11.118 K.G. Hu, W.S. Stark: Specific receptor input into spectral preference in *Drosophila*. J. Comp. Physiol. *121*, 241-252 (1977)

11.119 M. Heisenberg: "Genetic Approach to a Visual System", in *Comparative Physiology and Evolution of Vision in Invertebrates*, ed. by H. Autrum, Handbook of Sensory Physiology, Vol.7/6A (Springer, Berlin, Heidelberg, New York 1979) pp.665-679

Additional References with Titles

Chapter 2

Alpern, M., Zwas, F.: The wavelength variation of the directional sensitivity of the Stiles π_r (μ). Vision Res. *19*, 1077-1087 (1979)

Anderson, D.H., Fisher, S.K.: The relationship of primate foveal cones to the pigment epithelium. J. Ultrastruct. Res. *67*, 23-32 (1979)

Anderson, D.H., Fisher, S.K., Erickson, P.A., Tabor, G.A.: Rod and cone disc shedding in the rhesus monkey retina: a quantitative study. Exp. Eye Res. *30*, 559-574 (1980)

Anderson, R.E., Maude, M.B., Kelleher, P.A.: Metabolism of phosphatidyl ethanolamine in the frog retina. Biochem. Biophys. Acta *620*, 227-235 (1980)

Applegate, R.A., Bonds, A.B.: Induced movement of receptor alignment toward a new pupillary aperture. Invest. Ophthal. and Vis. Sci. (submitted)

Balemans, M.G.M., Pevet, P., Legerstee, W.C., Nevo, E.: Preliminary investigations on melatonin and 5-methoxytryptophol synthesis in the pineal, retina and harderian gland of the mole rat and in the pineal of the mouse eyeless. J. Neurol. Transm. *49*, 247-256 (1980)

Bedell, H.E.: A functional test of foveal fixation based upon differential cone directional sensitivity. Vision Res. *20*, 557-560 (1980)

Bedell H.E.: Central and peripheral retinal photoreceptor orientation in amblyopic eyes as assessed by the psychophysical Stiles-Crawford function. Invest. Ophthal. and Visual Sci. *19*, 49-59 (1980)

Bedell, H.E., Enoch, J.M.: An apparent failure of the photoreceptor alignment mechanism in a human observer. Arch. Ophthal. *98*, 2023-2026 (1980)

Bedell, H.E., Enoch, J.M, Fitzgerald, C.R.: A graded disturbance of photoreceptor orientation bordering a region of choroidal atrophy. Arch. Ophthal. (in press)

Besharse, J.C., Phenninger, K.H.: Membrane assembly in retinal photoreceptors, I. freeze-fracture analysis of cytoplasmic vesicles in relationship to disc assembly. J. Cell Biol. *87*, 451-463 (1980)

Birch, D.G., Birch, E.E., Enoch, J.M.: Visual sensitivity resolution, and Rayleigh matches following monocular occlusion for one week. J. Opt. Soc. Amer. *70*, 954-958 (1980)

Bour, L.J., Verhoosel, J.C.M.: Directional sensitivity of photoreceptors for different degrees of coherence and directions of polarization of the incident light. Vision Res. *19*, 717-719 (1979)

Bunt, A.H., Klock, I.B.: Fine structure and radiautography of retinal cone outer segments in goldfish and carp. Invest. Ophthalmol. Vis. Sci. *19*, 707-719 (1980)

Bunt, A.H., Klock, I.B.: Comparative study of ^{3}H-Fucose incorporation into vertebrate photoreceptor outer segments. Vis. Res. *20*, 739-748 (1980)

Carroll, J.P.: Apodization model of the Stiles-Crawford effect. J. Opt. Soc. Amer. *70*, 1155-1156 (1980)

Enoch, J.M., Birch, D.G., Birch, E.E.: Monocular light exclusion for a period of days reduces directional sensitivity of the human retina. Science *206*, 705-707 (1979)

Enoch, J.M., Birch, D.G., Birch, E.E., Beneditto, M.D.: Alteration in directional sensitivity of the retina by monocular occlusion. Vision Res. *20*, 1185-1189 (1980)

Fisher, L.J., Easter, S.S., Jr.: Retinal synaptic arrays: continuing development in the adult goldfish. J. Comp. Neurol. *185*, 373-380 (1979)

Fitzgerald, C.R., Birch, D.G., Enoch, J.M.: Functional analysis of vision in patients after retinal detachment repair. Arch. Ophthal. *98*, 1237-1244 (1980)

Fitzgerald, C.R., Enoch, J.M., Birch, D.G., Beneditto, M.D., Temme, L.A., Dawson, W.W.: Anomalous pigment epithelial photoreceptor relationships and receptor orientation. Invest. Ophthal. and Vis. Sci. *19*, 956-966 (1980)

Fuld, K., Wooten, B.R., Katz, L.: The Stiles-Crawford hue shift following photopigment depletion. Nature *279*, 152-154 (1979)

Gold, G.H., Dowling, J.E.: Photoreceptor coupling in retina of the road, Bufo marinus I. anatomy. J. Neurophysiol. *42*, 292-310 (1979)

Goldman, A.I., O'Brien, P.J., Masterson, E., Israel, P., Teirstein, P., Chader, G.: A quantitative system for studying phagocytosis in pigment epithelium tissue culture. Exp. Eye Res. *28*, 455-467 (1979)

Goldman, A.I., Teirstein, P.S., O'Brien, P.J.: The role of ambient lighting in circadian disc shedding in the rod outer segment of the rat retina. Invest. Ophthalmol. Vis. Sci. *19*, 1257-1267 (1980)

Grün, G.: Developmental dynamic in synaptic ribbons of retinal receptor cells (Tilapia, Xenopus). Cell Tissue Res. *207*, 331-339 (1980)

Hayasaka, S., Lai, Y-L.: Effect of continous low-intensity light on the lysosomal enzymes in retina of albino rats. Exp. Eye Res. *29*, 123-129 (1979)

Hollyfield, J.E.: Membrane addition to photoceptor outer segments: progressive reduction in the stimulatory effect of light with increased temperature. Invest. Ophthalmol. Vis. Sci. *18*, 977-981 (1979)

Klyne, M.A., Ali, M.A.: Microtubules and 10 nm filaments in the Retinal Pigment Epithelium during the diurnal light-dark cycle. Cell Tissue Res. *214* (2), 397-406 (1981)

Lasansky, A.: Lateral contracts and interactions of horizontal cell dendrites in the retina of the larval Tiger Salamander. J. Physiol. *301*, 59-68 (1980)

Lo, W-K., Bernstein, M.H.: Daily patterns of the retinal pigment epithelium. Microperoxisomes and phagosomes. Exp. Eye Res. *32*, 1-10 (1981)

Nir, I., Hall, M.O.: Ultrastructural localization of lectin binding sites on the surface of retinal photoreceptors and pigment epithelium. Exp. Eye Res. *29*, 181-194 (1979)

O'Day, W.T., Young, R.W.: The effects of prolonged exposure to cold on visual cells of the goldfish. Exp. Eye Res. *28*, 167-187 (1979)

Pang, S.F., Tu, M.S., Suen, H.C., Brown, G.M.: Melatonin in the retina of rats — a diurnal rhythm. J. of Endocrinol. *87* (1), 89-94 (1980)

Petrosian, A.M., Gribakin, F.G.: New data on the heterogeneity of retinal rod photoreceptors membranes. Biofizika *25*, 342-343 (1980)

Pokorny, J., Smith, V.C., Ernest, J.T.: Monocular color vision defects: Specialized psychophysical testing in acquired and hereditary chorioretinal diseases. Int. Ophthal. Clinics *20*, 53-81 (1980)

Pourcho, R.G.: (3H) Taurine-accumulating neurons in the cat retina. Exp. Eye Res. *32*, 11-20 (1981)

Raynauld, J.P., Laviolette, J.R., Wagner, H.J.: Goldfish retina: a correlate between cone activity and morphology of the horizontal cell in cone pedicles. Science *204*, 1436-1488 (1979)

Remé, C.E., Knop, M.: Autophagy in frog visual cells in vitro. Invest. Ophthalmol. Vis. Sci. *19*, 439-456 (1980)

Schuschereba, S., Zwick, H.: The striated rootlet system of primate rods — a candidate for active photoreceptor alignment, OSA technical meeting, Sarasota, Fl., 1980

Sjöstrand, F.S., Kreman, M.: Freeze-fracture analysis of structure plasma membrane of photoreceptor cell outer segments. J. Ultrastruct. Res. *66*, 254-275 (1979)

Sjöstrand, F.S., Kreman, M., Crescitelle, F.: Freeze-fracture analysis of photoreceptor cell outer segment disks after minimal extraction of rhodopsin. J. Ultrastr. Res. *69*, 53-67 (1979)

Starr, S.J., Fitzke, F.W., Massof: The Stiles-Crawford effect in the central fovea. Invest. Ophthal. and Vision Sci. Suppl. *172*, 1979 (Abstract)

Steinberg, R.H., Fisher, S.K., Anderson, D.H.: Disc morphogenesis in vertebrate photoreceptors. J. Comp. Neurol. *190*, 501-518 (1980)

Teirstein, P.S., Goldman, A.I., O'Brien, P.J.: Evidence for both local and Ophthalmol. Vis. Sci. *19*, 1268-1273 (1980)

Ungar, F., Piscopo, I., Holtzman, E.: Calcium accumulation in intracellular compartments of frog retinal photoreceptors. Brain Res. *205*, 200-206 (1981)

Wagner, H.J.: Lightdependent plasticity of the morphology of horizontal cell terminals in cone pedicles of fish retinas. J. Neurocytol. *9* (5), 573-590 (1980)

Williams, D.R.: Visual consequences of the foveal pit. Invest. Ophthal. and Visual Sci. *19*, 653-667 (1980)

Wooten, B.R., Fuld, K., Moore, M., Katz, L.: "The Stiles-Crawford II Effect at High Bleaching Levels", in *Visual Psychophysics and Physiology*, ed. by Armington, J.C., Krauskopf, J., Wooten, B.R. (Academic, New York 1978) pp.245-256

Zwas, F.: Wavelength variation in directional sensitivity of the long- and mediumwave sensitive foveal cones of red-green dichromats. Vision Res. *19*, 1067-1076 (1979)

Zunz, Y.W.: Core mosaics in a teleost retina: Changes during light and dark adaptation. Experientia *36*, 1371-1374 (1980)

Chapter 6

DeGroot, P.J., Terpstra, R.E.: Millimeter-wave model of a foveal receptor. J. Opt. Soc. Am. *70*, 1436-1452

Nyquist, D.P., Johnson, D.R., Hsu, S.V.: Orthogonality and amplitude spectrum of radiation modes along open-boundary waveguides. J. Opt. Soc. Am. *71*, 49-54 (1981)

Snyder, A.W.: Weakly guiding optical fibers. J. Opt. Soc. Am. *70*, 405-411 (1980)

Snyder, A.W., Love, J.D.: *Optical Waveguide Theory* (Chapman and Hall, London 1981)

Chapter 9

Baylor, D.A., Fettiplace, R.: Light path and photon capture in turtle photoreceptors. J. Physiol. *248*, 433-464 (1975)

Bowmaker, J.K., Dartnall, H.J.A.: Visual pigments of rods and cones in a human retina. J. Physiol. *298*, 501-511 (1980)

Bowmaker, J.K., Dartnall, H.J.A., Mollon, J.D.: Microspectrophotometric determination of four classes of photoreceptors in an old world primate, Macaca fascicularis. J. Physiol. *298*, 131-143 (1980)

Burkhardt, D.A., Hassin, G., Levine, J.S., MacNichol, E.F., Jr.: Electrical responses and photopigments of twin cones in the retina of the walleye. J. Physiol. *309*, 215-228 (1980)

Fineran, B.A., Nicol, J.A.C.: Studies on the eyes of anchovies Anchoa mitchilli and A. hepsetus (Engraulidae) with particular reference to the pigment epithelium. Phil. Trans. R. Soc. London B*276*, 321-350 (1977)

Fineran, B.A., Nicol, J.A.C.: Studies on the photoreceptors of Anchoa mitchilli and A. hepsetus (Engraulidae) with particular reference to the cones. Phil. Trans. R. Soc. London B*283*, 25-60 (1978)

Shapley, R, Gordon, J.: The visual sensitivity of the retina of the conger eel. Proc. R. Soc. London B*209*, 317-330 (1980)

Webb, N.G.: Orientation of retinal rod photoreceptor membranes in the intact eye using X-ray diffraction. Vision Res. *17*, 625-631 (1977)

Chapter 10

Ali, M.A., Wagner, H.J.: Scanning electron microscopy of four teleostean retinas. Can. J. Zool. *35*, 199-210 (1976)

Best, A.C.G., Nicol, J.A.C.: Eyeshine in fishes. A review of ocular reflectors. Can. J. Zool. *58*, 945-956 (1980)

Eckelbarger, K.J., Scalan, R., Nicol, J.A.C.: The outer retina and tapetum lucidum of the snook (Centropomus undecimalis (Teleostei)). Can. J. Zool. *58*, 1042-1051 (1980)

Nicol, J.A.C.: Studies on the Eyes of Fishes: Structure and Ultrastructure in Vision in Fishes. *New Approaches in Research*, ed. by Ali, M.A. (Plenum Press, New York, London 1975) pp.579-607

Nicol, J.A.C.: Studies on the eyes of toad fishes Opsanus. Structure and reflectivity of the stratum argenteum. Can. J. Zool. *58*, 114-121 (1980)

Nicol, J.A.C.: Uric acid in the stratum argenteum fo toadfishes Opsanus. Can. J. Zool. *58*, 492-496 (1980)

Somiya, H.: Fishes with eyeshine: functional morphology of guanine type tapetum lucidum. Mar. Ecol. Prog. Ser. *2*, 9-26 (1980)

Zyznar, E.: "Tapeta Lucida and the Organization of Visual Cells in Teleosts" in *Vision in Fishes. New Approaches in Research*, ed. by Ali, M.A. (Plenum Press, New York, London 1975) pp.299-304

Zyznar, E.: "Theoretical Considerations about Tapeta Lucida", in *Vision in Fishes. New Approaches in Research*, ed. by Ali, M.A. (Plenum Press, New York, London, 1975) pp.305-312

Subject Index

Y. Le Grand, S. G. El Hage

Physiological Optics

1980. 118 figures, 9 tables. XVII, 338 pages
(Springer Series in Optical Sciences, Volume 13)
ISBN 3-540-09919-0

Photoreceptor Optics

Editors: A. W. Snyder, R. Menzel
1975. 259 figures. X, 523 pages
ISBN 3-540-07216-0

Frontiers in Visual Science

Proceedings of the University of Houston College of
Optometry Dedication Symposium, Houston, Texas,
USA, March, 1977

Editors: S. J. Cool, E. L. Smith, III

1978. 533 figures, 28 tables. XIV, 798 pages
(Springer Series in Optical Sciences, Volume 8)
ISBN 3-540-09185-8

Springer-Verlag
Berlin
Heidelberg
New York

Fiber Optics

A.B.Sharma, S.J.Halme, M.M.Butusov

Optical Fiber Systems and Their Components

An Introduction

1981. 125 figures. VIII, 246 pages
(Springer Series in Optical Sciences,
Volume 24)
ISBN 3-540-10437-2

Contents: Introduction. – Generation,
Modulation, and Detection. – Light Propagation in Waveguides. – Components for
Optical Fiber Systems. – Fiber Measurements. – Fiber Optical Systems and Their
Applications. – Selected Problems. – References.– Subject Index.

Semiconductor Devices for Optical Communication

Editor: H.Kressel

1980. 186 figures, 6 tables. XIV, 289 pages
(Topics in Applied Physics, Volume 39)
ISBN 3-540-09636-1

Contents: H.Kressel: Introduction. –
H.Kressel, M.Ettenberg, J.P.Wittke,
I.Ladany: Laser Diodes and LEDs for Fiber
Optical Communication. – D.P.Schinke,
R.G.Smith, A.R.Hartman: Photodetec-
tors. – R.G.Smith, S.D.Personick: Receiver
Design for Optical Fiber Communication
Systems. – P.W.Shumate, Jr.,
M.DiDomenico, Jr.: Lightwave Transmit-
ters. – M.K.Barnoski: Fiber Couplers. –
G.Arnold, P.Russer, K.Petermann: Modula-
tion of Laser Diodes. – J.K.Butler: The
Effect of Junction Heating on Laser Line-
arity and Harmonic Distortion. –
J.H.Mullins: An Illustrative Optical
Communication System.

Integrated Optics

Editor: T.Tamir

2nd corrected and updated edition. 1979.
99 figures, 11 tables. XV, 333 pages
(Topics in Applied Physics Volume 7)
ISBN 3-540-09673-6

Contents: T.Tamir: Introduction. –
H.Kogelnik: Theory of Dielectric Wave-
guides. – T.Tamir: Beam and Waveguide
Couples. – J.M.Hammer: Modulation and
Switching of Light in Dielectric Wave-
guides. – F.Zernike: Fabrication and
Measurement of Passive Components. –
E.Garmire: Semiconductor Components for
Monolithic Applications. – T.Tamir: Recent
Advances in Integrated Optics. – Additional
References with Titles. – Subject Index.

B.Saleh

Photoelectron Statistics

With Applications to Spectroscopy and
Optical Communication

1978. 85 figures, 8 tables. XV, 441 pages
(Springer Series in Optical Sciences,
Volume 6)
ISBN 3-540-08295-6

Contents: Tools from Mathematical Statis-
tics: Statistical Description of Random
Variables and Stochastic Processes. Point
Processes. – Theory: The Optical Field: A
Stochastic Vector Field or, Classical Theory
of Optical Coherence. Photoelectron
Events: A Doubly Stochastic Poisson
Process or Theory of Photoelectron Statis-
tics. – Applications: Applications to Optical
Communication. Applications to Spectros-
copy.

Springer-Verlag
Berlin
Heidelberg
New York